Sustainable Agriculture Reviews

Volume 52

Series Editor

Eric Lichtfouse
Aix-Marseille University, CNRS, IRD, INRAE, Coll France, CEREGE
Aix-en-Provence, France

Other Publications by Dr. Eric Lichtfouse

Books

Scientific Writing for Impact Factor Journals
https://www.novapublishers.com/catalog/product_info.php?products_id=42242

Environmental Chemistry
http://www.springer.com/978-3-540-22860-8

Sustainable Agriculture
Volume 1: http://www.springer.com/978-90-481-2665-1
Volume 2: http://www.springer.com/978-94-007-0393-3

Book series

Environmental Chemistry for a Sustainable World
http://www.springer.com/series/11480

Sustainable Agriculture Reviews
http://www.springer.com/series/8380

Journal

Environmental Chemistry Letters
http://www.springer.com/10311

Sustainable agriculture is a rapidly growing field aiming at producing food and energy in a sustainable way for humans and their children. Sustainable agriculture is a discipline that addresses current issues such as climate change, increasing food and fuel prices, poor-nation starvation, rich-nation obesity, water pollution, soil erosion, fertility loss, pest control, and biodiversity depletion.

Novel, environmentally-friendly solutions are proposed based on integrated knowledge from sciences as diverse as agronomy, soil science, molecular biology, chemistry, toxicology, ecology, economy, and social sciences. Indeed, sustainable agriculture decipher mechanisms of processes that occur from the molecular level to the farming system to the global level at time scales ranging from seconds to centuries. For that, scientists use the system approach that involves studying components and interactions of a whole system to address scientific, economic and social issues. In that respect, sustainable agriculture is not a classical, narrow science. Instead of solving problems using the classical painkiller approach that treats only negative impacts, sustainable agriculture treats problem sources.

Because most actual society issues are now intertwined, global, and fast-developing, sustainable agriculture will bring solutions to build a safer world. This book series gathers review articles that analyze current agricultural issues and knowledge, then propose alternative solutions. It will therefore help all scientists, decision-makers, professors, farmers and politicians who wish to build a safe agriculture, energy and food system for future generations.

More information about this series at http://www.springer.com/series/8380

Eric Lichtfouse
Editor

Sustainable Agriculture Reviews 52

Springer

Editor
Eric Lichtfouse
Aix-Marseille University
CNRS, IRD, INRAE, Coll France, CEREGE
Aix-en-Provence, France

ISSN 2210-4410 ISSN 2210-4429 (electronic)
Sustainable Agriculture Reviews
ISBN 978-3-030-73244-8 ISBN 978-3-030-73245-5 (eBook)
https://doi.org/10.1007/978-3-030-73245-5

This Springer imprint is published by the registered company Springer Nature Switzerland AG
The registered company address is: Gewerbestrasse 11, 6330 Cham, Switzerland

Contents

About the Editor

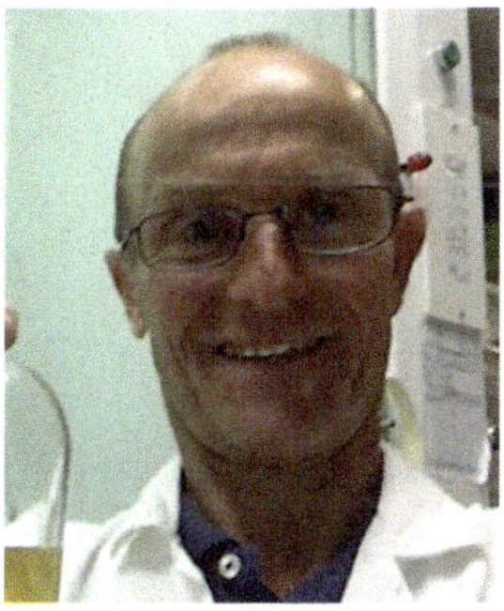

Dr. Eric Lichtfouse is professor of environmental chemistry and scientific writing at Aix Marseille University and Xi'an Jiaotong University, and chief editor and co-founder of the journal *Environmental Chemistry Letters*. He has invented carbon-13 dating following the discovery of various temporal pools of a single individual organic substance in soils. He has published the book *Scientific Writing for Impact Factors*, which includes a new tool – the Micro-Article – to identify the novelty of research results. Dr. Lichtfouse is a recipient of the analytical chemistry prize from the French Chemical Society, the grand prize of the Universities of Nancy and Metz, and a journal citation award by the Essential Indicators, and the Springer Nature award for editorial excellence. He is also top 2% world scientist in environmental sciences acording to the Stanford University ranking. He is world XTerra vice champion and ITU cross triathlon world bronze medal in age category.

Chapter 1
Organic Cultivation of Vegetables

Margit Olle and Ingrid H. Williams

Abstract Agriculture began organically. For many centuries, humans farmed without synthetic biocides or inorganic fertilizers, relying on fertilizers derived from plants and animals, and protecting crops from pests and diseases using naturally-occurring materials. From the second half of the nineteenth century, growers around the world successfully developed and refined farming systems that relied heavily on synthetic biocides and inorganic fertilizers. However, organic cultivation has emerged again during the last decades as result of consumers' demand for healthy food. Here we review the effects of organic cultivation on the production quantity and quality of vegetables. We found that organic cultivation affected the growth of vegetables positively in 43% of studies and negatively in 57% of studies. Organic cultivation affected the yield of vegetables 59% positively, 29% negatively and 12% did not show any significant influence. Organically grown vegetables have, in most studies (65%), better nutritional value than conventionally grown ones, 20% were not significantly different and only in 15% showed a reduction in nutritional value. Nitrate levels were lower in 86% of studies with organic cultivation and higher in only 14% of studies.

Organic cultivation of vegetables uses a variety of methods for disease and insect control, such as hot water, hot air and electron treatment, biological seed treatment with microorganisms, plant extracts and inducers of resistance, solarization for nematode control, biopesticides, and insect nets. Weed control is the most difficult part of vegetable production in organic cultivation. Good methods against weeds include tillage, mulching, flaming and hot water treatment. If the proper technology is used, the organic cultivation of vegetables is not so time- and money-consuming and produces vegetables of better quality and nutritional value without pesticide residues.

M. Olle (✉)
NPO Veggies Cultivation, Tartu, Tartumaa, Estonia

I. H. Williams
Institute of Agricultural and Environmental Sciences, Estonian University of Life Sciences, Tartu, Estonia

E. Lichtfouse (ed.), *Sustainable Agriculture Reviews 52*, Sustainable Agriculture Reviews 52, https://doi.org/10.1007/978-3-030-73245-5_1

Keywords Growth · Nutritional quality · Organic cultivation · Plant protection · Vegetables · Weeds control · Yield

1.1 Introduction

Agriculture began organically. For many centuries, humans farmed without synthetic biocides or inorganic fertilizers, relying on organic fertilizers derived from plants and animals, and protecting crops and animals from pests and diseases using naturally-occurring or minimally-processed materials (Kristiansen 2003). In 1842, Lawes started the first factory for the manufacture of artificial fertilisers in the UK, and, in 1843, he together with the chemist Gilbert, initiated the first of a series of long-term field experiments to measure the effect on crop yields of inorganic and organic fertilisers; some of these continue to this day at Rothamsted Research. Together Lawes and Gilbert laid the foundations of modern scientific agriculture and established the principles of crop nutrition. Throughout the following century, growers around the world successfully developed and refined farming systems that relied increasingly on synthetic biocides or inorganic fertilizers. During the past two or three decades the situation has begun to change. Increasing concerns about food quality, farm worker health, rural development, and the environmental impacts of farming systems, for example, have focused the attention of policy makers, consumers, researchers and farmers on alternative productions systems, including organics (Kristiansen 2003).

Organic farming is estimated to be growing at 30% a year worldwide in response to market forces (Ashley et al. 2007). The demand for certified organic produce, especially vegetables, currently exceeds supply and, in many cases, produce attracts premium prices (Ashley et al. 2007). Organic agricultural production is undergoing a rapid transformation as the demand for healthier food and more environmentally-sound production increases globally. Large producers are adopting organic practices to meet the growing demand (Smukler et al. 2008).

Organic fertilizers offer many benefits for horticulture. Firstly, they release nutrients slowly, which means that plants are fed over a long period. Secondly, this slow release avoids over-feeding thus avoiding providing too much nitrogen, phosphorous or potassium, all of which can be harmful in too high doses. Thirdly, they create a good environment for beneficial soil organisms, such as earthworms, which improve the soil structure by incorporating organic matter well down into the topsoil creating drainage and air tunnels while so doing (Stein 2009).

In organic farming, a number of methods are used to maintain soil fertility. These include: (1) crop rotation, which ensures that one crop does not deplete the soil of the nutrients that it uses most; (2) cover crops to protect against soil erosion; (3) the planting of special crops known as "green manures" that are ploughed back into the soil to enrich it; and (4) the addition of aged animal manures and plant wastes, also known as compost, to the soil. The distinguishing feature of these fertility management practices is the addition of organic matter to the soil, in the form of plant and animal wastes, to preserve soil structure and provide food for soil microorganisms. With these methods, plant nutrients are released slowly to the soil over time (Worthington 2001).

Well-managed organic vegetable production systems can provide food security and a healthy diet for humans, while being less harmful to the environment and more efficient in natural resource use. According to the online literature survey by Juroszek and Tsai (2008), tomato is the vegetable most commonly researched in organic farming, followed by lettuce, carrot and cucumber.

This review gives an overview of the effects of organic cultivation on the production quantity and quality characteristics of vegetables.

Germination and Growth

A germination survey of the first crop of radish showed that the germination ratio was lower in an organic plot than in a plot with reduced agricultural chemicals (Akira et al. 2003). The second germination survey compared mulch; organic mulch was not practical because it resulted in a low germination ratio (Akira et al. 2003).

Lettuce seedlings grown with organic fertiliser showed an increased growth rate compared to those grown with traditional fertiliser. By contrast, cabbage and cauliflower seedlings grown with organic fertiliser showed a reduced growth rate (Botrini et al. 2004). The optimum pH of growing media for lettuce in organic cultivation was close to 6 and the optimum electrical conductivity (EC) was lower than 2 dSm^{-1}. Increasing pH and especially EC reduced growth, and conductivity about 3 dSm^{-1} or higher had an inhibitory effect on growth when pH was about 7 or higher (Lončarić et al. 2009). The impact of rowcovers on cucumber plant growth was significant. Use of rowcovers increased vine length, flower count, leaf area and leaf count (Nair and Ngouajio 2010). Soil amendment treatments consisted of combinations of the following: poultry compost, poultry litter, dairy compost, dairy manure, blood meal, feather meal, and Fertrell™ 5-5-3. Poultry compost resulted in the greatest plant growth in all trials (Rauton 2007).

The application of aerated compost tea from organic compost based using MOVR (the mixture of rice straw compost, vermicompost, and Hinoki cypress bark compost) to the root zone increased the plant shoot and root growths and yield of the red leaf lettuce. Thus, compost tea could be used as an agent for promoting plant growth in organic cultivation of crops (Kim et al. 2015).

Improved growth effects were observed in vermicompost treated plants than the control plants. Fascinatingly, remarkable growth promoting effects were recorded in Abelmoschus esculentus and Solanum melongena. Consequently, the observed results influentially suggest that vermicompost and its beneficial microbes might be explored as efficient for the enhanced growth of vegetable crops in the near future (Illanjiam et al. 2019)..

Overall, organic cultivation positively affected vegetable growth in 43% of studies from the literature and negatively affected growth in 57% of studies.

1.2 Yield

Application of commercial organic fertilizer resulted in higher vegetable yields than application of commercial inorganic fertilizer when applied at the current recommended levels: inorganic at 151 kg·ha−1 and commercial organic at 660 kg·ha^{-1}. The Ca and Mg in the organic fertilizer, which are absent from the inorganic fertilizer, could have contributed to the higher yield (Alimi et al. 2007). Organic treatments resulted in higher carrot root production compared to conventional treatments (Bruno et al. 2007). Lettuce yield was significantly higher when organic fertilization was used, compared to a mineral nutrient supply (Porto et al. 2008). Cabbage and tomato yields produced in the organic system were higher than those in the conventional system (Ma et al. 2009). Applications of compost led to higher marketable yields of cucumber (Nair and Ngouajio 2010). All the associations of the lettuce and rocket as well as their sole crop had better productive performance under the organic fertilization and the regrowth of rocket increased the agronomic efficiency of the intercropping system (Oliveira et al. 2010). Soil amendment treatments consisted of combinations of the following: poultry compost, poultry litter, dairy compost, dairy manure, blood meal, feather meal, and Fertrell™ 5-5-3; poultry compost resulted in the greatest yield in all trials (Rauton 2007). Yield could be higher in organic cultivation of vegetables because organic management of soil fertility affects soil dynamics and plant metabolism, resulting in differences in plant composition and nutritional quality. Soil that has been managed organically has more microorganisms. These produce many compounds that help plants, including substances such as citrate and lactate that combine with soil minerals and make them more available to plant roots (Worthington 2001).

Mahmoud et al. (2009) obtained lower yields of cucumber from organic treatments than from conventional treatments (Table 1.1). Similarly organic onions gave lower yields than those grown conventionally (Ashley et al. 2007). Examples of yields of vegetables grown in integrated versus organic production systems are presented in Table 1.2 (Fjelkner-Modig et al. 2000). Boteva and Rankov (2007) found that plant mass was greatest in head lettuce varieties, when cultivated on alluvial-meadow soil after fertilizing with 10 t/da farmyard manure, and for celery-lettuce varieties after fertilization with 15 t/da (treatments were: 0, 1, 5, 10, 15 and 20 t/da

Table 1.1 Effect of compost and mineral fertilizer treatments on the average yield and nitrates in cucumbers harvested in 2007 and 2008

Treatment	Fruit yield kg per plot (20 m^2)	Nitrates in fruits (mg/kg)
100% mineral N	112.5	92.7
100% mixed compost	10.8	0.0
100% plant compost	37.5	0.0
100% animal compost	10.5	0.0
F test	**	**
L.S.D. 0.05	1.71	1.85

**Significant between treatments at $p < 0.05$ (Mahmoud et al. 2009)

Table 1.2 Annual yields for integrated and organically-grown vegetables on sandy loam soil (L) and sand (A) during 1994–1998

Crop	Yield (integrated) (t/ha)	Yield (organic) (t/ha)	(organic/integrated) yield ratio %
Cabbage (loam soil)	51.3 ± 10.5	36.1 ± 7.4	70
Carrot (sand)	47.2 ± 13.7	38.8 ± 12.8	82
Pea (loam soil)	4.5 ± 1.2	3.6 ± 1.2	80
Pea (sand)	2.8 ± 0.6	2.4 ± 0.4	85
Potato (sand)	31.4 ± 5.0	27.6 ± 7.1	88

Fjelkner-Modig et al. (2000)

Table 1.3 Yield, dry matter, and content of nitrates and nitrites of Monanta carrots from organic and intensive conventional farms in 1997 x ± SD (average value and standard deviation)

Analyzed parameter in relation to fresh mass	Organic 1, n 3 = 10	Conventional 2, n = 10
Yield in autumn 1997 in t/ha	37.5 ± 14.1 a 4	50.0 ± 23.1 a 4
Dry matter (g/100 g)	11.55 ± 0.84 a	11.10 ± 0.80 a
Nitrates (mg/kg)	52.2 ± 36.8 a	209.7 ± 175.0 b 5
Nitrites (mg/kg)	0.57 ± 0.42 a	1.06 ± 0.42 b

Rembialkowska (2003)
1 Torun and Plock provinces
2 Warszawa province
3 n number of samples; 1 sample = result for 1 farm
4 values in the same rows marked with the same letters do not differ significantly
5 values in the same rows marked with different letters are statistically significantly different

farmyard manure). Cabbage and cauliflower seedlings grown with organic fertiliser had lower fresh weight compared with traditional fertilization, particularly at the end of growth period (Botrini et al. 2004). The marketable yield of onions grown with chemical fertilizer was 64.5 ton ha^{-1}, while that of onions grown with organic fertilizer (applied 6 times in February and March) was 46.4 ton ha^{-1}, i.e. a 28% reduction (Lee et al. 2009). Conventional treatment also gave higher yields than organic in tomato and French bean (Thakur et al. 2010); this may be due to chemical fertilizers containing a few mineral substances, principally nitrogen, potassium and phosphorus and the addition sometimes of trace minerals. These fertilizers dissolve easily in the water present in soil (Worthington 2001).

The foliar application of seaweeds, a mixture of compost tea and argan by-product, and a mixture of seaweeds and Argan by-products had no effect on the quantity of bean, eggplant, tomato and melon crops produced (Bourgol 2005). Spent mushroom (*Pleurotus ostreatus*) cultivation substrate and mature compost amendments showed no significant effect on cucumber production (Ehaliotis et al. 2005). Rembialkowska (2003) found no significant differences in yield of carrots grown with organic vs. conventional cultivation (Table 1.3).

High application rates of mature compost, however, in a more fertile soil, resulted in yields comparable to those obtained with sheep manure (Ehaliotis et al. 2005);

application of sheep manure almost doubled cucumber production compared to the control (non-amended soil), whereas application of plain seaweed reduced it by 40% (Ehaliotis et al. 2005). Four organic transition strategies including tilled fallow, mixed-species hay, low intensity open-field vegetable production and intensive vegetable production under high tunnels, each with and without annual compost amendment, were analyzed for nematode communities and soil properties. In this study, tomato yield under high tunnels plots exceeded yield in other treatments potentially due to the season extension and higher N availability (Briar et al. 2011). The high tunnel growing strategy increased tomato yield. The results showed that the type of fertilizer applied significantly affected yield of lettuce (Masarirambi et al. 2010).

Organic mulches increase the production of cultivated crops. The reasons can be following (Ranjan et al. 2017):

- Mulching with organic materials increases the soil nutrients, maintains the optimum soil temperature, restrict the rate evaporation from the soil surface, restrict weed growth and prevent soil erosion. It improves the all properties of soil like physical, chemical and biological. The organic mulches are decomposing easily and increase the organic content like carbon etc., in the soil which helps to maintain the soil loose. These organic matters are beneficial for the growth of earthworms and soil microorganisms it is also food of these beneficial microorganisms. Organic mulches have also some advantageous properties like it has environment friendly and add beneficial nutrients in the soil. It could maintain the soil moisture, and increase water use efficiency. The organic system of lettuce production results in higher crop yields than the conventional production system (Kapoulas et al. 2017).

Plastic mulches can prove as a boon to enhance productivity in hills. A favorable soil, water-plant relation is created by placing mulch over the soil surface. The microclimate surrounding the plant and soil is significantly affected by mulch. Black plastic mulch (double coated 30 micron) could enhance soil moisture retention supers weed growth and enhanced crop yield. Therefore, mulching could be incorporated to enhanced yield in chilli (Narayan et al. 2017).

Yield losses due to pests and disease also affect the organic-conventional yield gap. The number of crop protection products approved for organic agriculture is very limited, and although they constitute an important input for reducing crop losses, especially in some horticultural crops, the lack of crop protection products or other effective crop protection measures limits organic yields. We should stress here that, although copper-based products are among the most widely used crop protection products in European organic farming and are important for controlling fungus attacks in, e.g. vines, fruit crops and potatoes, copper fungicides are prohibited in Scandinavian countries by national legislation (Röös et al. 2018).

Concerning horticultural crops, which are susceptible to many pests and pathogens, new crop protection strategies and development and increased use of a variety of biological control agents (e.g. bacteria, fungi and predatory arthropods) will be particularly important to reduce the yield gap. Increased use of resistant varieties is also crucial, but these varieties are however not fully resistant, implying that direct crop protection measures will be especially important to secure high yields and product quality in high value crops (Röös et al. 2018).

Strategies to increase yields in organic arable crops (applicable to field vegetables) that are applicable in Northern Europe (Röös et al. 2018):

1. Soil Fertility:
2. Crop rotation design and management including optimal management of legume
3. pre-crop effects and green manure crops
4. Increased crop diversity
5. Intercropping
6. New technologies for reduced tillage
7. Increased cooperation between livestock farms and stockless farms
8. Adding/promoting supportive microorganisms and fungi in soil
9. Plant nutrients:
10. Optimal use of legumes in rotations
11. Effective use of manures
12. Increased recycling and use of nutrients from society
13. Novel treatments of organic food wastes to produce high-quality composts
14. Technological solutions for safe sewage sludge treatments and recycling

Crop-weed competition

15. Crop rotation design and management
16. New physical weed control strategies and techniques including cover crop management
17. Use of the false seedbed technique
18. Precision farming and robots

Control of diseases:

19. Use of tolerant or resistant crop varieties
20. Crop rotation design and management
21. Preventative strategies like intercropping, deep ploughing, optimal planting date etc.
22. New techniques and products for preventing fungal infections, physical methods and biocontrol organisms
23. Replace copper that is currently used
24. Use of certified and dressed seeds

Control of pests

25. Crop rotation design
26. Habitat manipulation (hedgerows, wild flower strips etc.) to strengthen functional biodiversity (e.g. natural enemies)
27. Physical/biological methods like nets, traps and repellents
28. Selective pest control products with low negative side-effects

Organic cultivation affected the yield of vegetables 59% positively, 29% negatively and 12% did not have any significant influence.

1.3 Nutritional Quality

The nutritional and sanitary value of organic compared with conventional vegetables can be found in Table 1.4 (Lairon 2010). Table 1.5 gives results from studies of organic crops shown to have higher, lower or equal nutrient content compared with conventionally grown crops (Bordeleau et al. 2002). For leafy vegetables as well as for root vegetables and tubers, a trend for higher dry matter content in organic foodstuffs has been found while no significant difference has been obtained for fruit vegetables (Lairon 2010). Similarly, organically grown crops had higher dry matter content than integrated grown crops (Fjelkner-Modig et al. 2000). Organically grown cucumber contained higher dry matter (Ma et al. 2009). On the other hand, no statistical difference was observed in the percentage of dry matter among the following different treatments: control (no fertilization), chemical fertilization, chicken manure, cattle manure, worm manure, and organic compost, with the exception of the value observed at the organic compost treatment, which was the lowest (3.7%) (Abreu et al. 2010). Contrary to the findings of Lairon (2010), cabbage and cauliflower seedlings grown with organic fertilizer showed a reduction of dry weight compared with traditional fertilization, particularly at the end of growth period (Botrini et al. 2004).

Regarding vegetables (carrot, beetroot, lettuce, kale, leek, turnip, onion, celeriac and tomato) a trend has been observed for higher levels of iron and magnesium expressed on a nutritional quality and safety of organic food (Lairon 2010). Lettuce seedlings grown with organic fertiliser had a higher N and K uptake than those grown with traditional fertilisation. By contrast, cabbage and cauliflower seedlings grown with organic fertiliser showed a reduction of nutrient content, particularly at the end of growth period (Botrini et al. 2004).

Cabbage and cauliflower seedlings grown with organic fertiliser showed a reduction of chlorophyll content compared with traditional fertilization, particularly at the end of the growth period (Botrini et al. 2004). Regarding water-soluble vitamins, the most studied has been Vitamin C (ascorbic acid), a key vitamin for which higher daily intakes are recommended. Studies on tomato, celeriac and kale showed higher vitamin C levels in organically-grown products. In contrast, no difference was found during studies in leek, carrot or beetroot (Lairon 2010). The vitamin C

Table 1.4 Key items of nutritional and sanitary value of organic compared with conventional vegetables

Increased contents	Reduced contents	Comparable contents
Dry matter	Pesticide residues (mostly absent)	Most minerals
Some minerals (iron, magnesium)	Nitrates	Beta-carotene
Anti-oxidants: Vitamin C (potatoes), polyphenols in vegetables, salicylic acid in vegetables		

Lairon (2010)

Table 1.5 Numbers of studies of organic crops shown to have higher, lower or equal nutrient content compared with conventionally grown crops

Nutrient	Higher		Equal		Lower	
Reviewer	Woese	Heaton	Woese	Heaton	Woese	Heaton
Protein quality	3	–	0	–	0	–
Nitrate	5	0	10	2	25	14
Vitamin C	21	7	12	6	3	0
B-carotene	5	–	5	–	3	–
B-vitamins	2	–	12	–	2	–
Calcium	21	–	20	–	6	–
Magnesium	17	–	24	–	4	–
Iron	15	–	14	–	6	–
Zinc	4	–	9	–	3	–
Minerals	–	7	–	6	–	1
Dry matter	–	10	–	8	–	1

Bordeleau et al. (2002)

content of an organic fruit or vegetable is 27% more, on average, than a comparable conventionally grown fruit or vegetable (Worthington 2001). Similarly, in leafy vegetables, leaf concentrations of vitamin C were significantly higher in organic-fertilized than in chemically-fertilized vegetables (Xu et al. 2003). By contrast, the vitamin C content was not influenced by the growing system in the study of Fjelkner-Modig et al. (2000). These results may be due to the vitamin content of a plant depending on a number of factors such as climate, genetic properties, fertilizer and soil. Organically grown cucumber contained higher vitamin C (Ma et al. 2009). Nitrogen from any kind of fertilizer affects the amounts of vitamin C and nitrates as well as the quantity and quality of protein produced by plants. When a plant is presented with a lot of nitrogen, it increases protein production and reduces carbohydrate production. Because vitamin C is made from carbohydrates, the synthesis of vitamin C is reduced also. Because organically managed soils generally present plants with lower amounts of nitrogen than chemically fertilized soils, it would be expected that organic crops would have more vitamin C, less nitrates and less protein but of a higher quality than comparable conventional crops (Worthington 2001). Conventionally grown carrots showed higher content of nitrates than organic carrots (Rembialkowska 2003; Table 1.3). Mahmoud et al. (2009) found that nitrates were present in conventionally grown cucumber fruits but absent from those grown organically (Table 1.1).

Fat-soluble vitamin and carotenoid contents have been the subjects of some studies. A review of 27 related studies reported β-carotene levels in vegetables with no noticeable differences found overall between organic and conventional foodstuffs. By contrast, a positive relationship between N-fertilization and β-carotene levels has been reported in carrots, while a recent study on organic vs. conventional tomatoes showed higher contents of β-carotene in the former (Lairon 2010). Organically grown cucumber contained higher sugar (Ma et al. 2009). In leafy vegetables, leaf concentrations of sugars were significantly higher in organic-fertilized than in chemical-fertilized vegetables (Xu et al. 2003).

Depending on season, organic vegetables overall may contain at least 30–50% less nitrates than conventional ones (Worthington 2001; Porto et al. 2008; Lairon 2010). This was confirmed by a lower nitrate content of plants when the percentage of organic N increased (Mahmoud et al. 2009). The analysis of the scientific literature showed that, in most of the experiments, nitrate content clearly reduced by using organic procedures (Pimpini et al. 2005). In leafy vegetables, leaf concentrations of nitrate were lower in organic-fertilized than in chemical-fertilized vegetables (Xu et al. 2003). Similarly scientists found that conventional crops also have higher concentrations of nitrate; increased intakes of these compounds have been linked to negative health impacts (Barański et al. 2017).

Muramoto (1999) concluded the following from his research:

1. Nitrate levels exceed the maximum levels specified by European Commission Regulation much more often in conventional than in organic spinach.
2. Nitrate levels tend to be higher in organic spinach grown using guano and Chilean nitrate than in spinach grown using compost.
3. Nitrate levels in spinach are affected by the rate and type of nitrogen fertilizer applied, and also by soil nitrification activity, soil texture, and harvest time.
4. Organic growers may reduce nitrate concentration in spinach using methods such as pre-plant soil nitrate testing, compost based fertility management, afternoon to evening harvest, and petiole removal.
5. Nitrate levels in California-sampled Iceberg and Romaine lettuce are safe regardless of season and farming practice.

Organic crops contained significantly more iron, magnesium, and phosphorus than conventional crops (Worthington 2001). Differences in the nutritional content between organic and conventional vegetables can be seen in Table 1.6 (Worthington 2001). Plants produced by bounce back compost were higher in Ca, Fe and Zn contents on a fresh mass basis than plants produced by cattle manure, followed by those produced using inorganic fertilizers and lastly chicken manure (Masarirambi et al. 2010).

Potassium fertilizer can reduce the magnesium content and indirectly the phosphorus content of at least some plants. When potassium is added to soil, less magnesium is absorbed by plants and because phosphorus absorption depends on magnesium, less phosphorus is absorbed as well. Potassium is presented to plants differently by organic and conventional systems. Conventional potassium fertilizers dissolve readily in soil water presenting plants with large quantities of potassium while organically managed soils hold moderate quantities of both potassium and magnesium in the root zone of the plant. Given these plant responses, it would be expected that organic crops would contain larger amounts of magnesium and phosphorus than comparable conventional crops (Worthington 2001).

This study shows that alternative, organic fertilizers have similar or even better positive effects than farmyard manure and that they can contribute to the improvement of the nutritional value of the vegetables produced (Pavla and Pokluda 2008).

Table 1.6 Differences in nutrient content between organic and conventional vegetables: mean percent difference for four nutrients in five frequently studies vegetables

	Nutrient[a]			
Vegetable	Vitamin C	Iron	Magnesium	Phosphorus
Lettuce	+17	+17	+29	+14
Spinach	+52	+25	−13	+14
Carrot	−6	+12	+69	+13
Potato	+22	+21	+5	0
Cabbage	+43	+41	+40	+22

Worthington (2001)
[a]Plus and minus signs refer to conventional crops as the baseline for comparison. For example, vitamin C is 17.0% more abundant in organic lettuce (conventional 100%, organic 117%)

Conventional crops have higher levels of the toxic metal cadmium, and are four-times more likely to contain detectable pesticide residues; there are general recommendations to minimise the intake of pesticides and cadmium to avoid potential negative health impacts (Barański et al. 2017). Similarly Hadayat et al. (2018) found that for potato, lettuce, tomato, carrot and onion metal contents in conventional produce were slightly greater than organic produce, especially for Cd and Pb.

Organic crops have higher antioxidant activity and between 18 and 69% higher concentrations of a range of individual antioxidants; increased intakes of polyphenolics and antioxidants has been linked to a reduced risk of certain chronic diseases such as cardiovascular and neurodegenerative diseases and certain cancers (Barański et al. 2017). The nutrient composition differs only minimally between organic and conventional crops, with modestly higher contents of phenolic compounds in organic vegetables (Mie et al. 2017).

Organically grown vegetables have, in most cases (65%), better nutritional value than conventional ones. 15% of cases showed a reduction in nutritional value and 20% results were not significantly influenced. 86% of results showed a reduction and 14% of results showed an increase in nitrate levels in organic cultivation.

1.4 Plant Protection

Organic carrot and red beet experiments in Kurenurme, Estonia in 2011 produced very good quality crops with almost no disease and insect attack on plants (Fig. 1.1). Tomato spotted wilt was the most important disease in the organic tomato cropping system, resulting in smaller plant development, fewer flower clusters and lower yield. In the conventional tomato cropping system, the disease was kept under control, and the population of thrips, the virus vector, occurred at lower levels than in the organic tomato cropping system (Bettiol et al. 2004). Downy mildew incidence was similar between treatments with fertilizer and no fertilizer. Mineral fertilization did not increase downy mildew incidence compared to the organic treatments (Gonçalves et al. 2004). Overall, the physical seed treatments (hot water, hot air and

Fig. 1.1 Organic carrot and organic red beet experiments in Estonia in Kurenurme in August 2011. Experiments were carried through on the fields of self-employed entrepreneur Galina Rehkli. Photos are taken by Margit Olle

electron treatment) resulted in a moderate to good control of the respective diseases. Also, from all biological seed treatment groups (microorganisms, plant extracts and inducers of resistance) candidates with promising control properties could be identified. The influence of the biological methods on emergence was often more prominent under greenhouse than under field conditions (Schmitt et al. 2006).

The term of allelopathy was first used in 1937 by Molisch, defining the chemical interaction (both stimulation and inhibition) between all types of plants, including microorganisms. Allelopathic plants are basically used as catch crops or trap crops. They are used in plant protection of tropical regions against parasite weeds, reducing the parasite seed bank by 72%. Allelopathic compounds can act as insects repellent (Rodino et al. 2017).

Intercropping is the planned mixed growing of more than two species at one area of land in the same period of time. A prerequisite for the success of intercropping is the interdependence of the selected crops in growth and development due to their biological particularities. The method is based on complex interactions between companion species with good results in trap cropping, weed suppression, physical-spatial interactions. Intercropping effects are not only reducing pest populations, it proved to be efficient in controlling plant diseases. The grain intercrop reduced humidity in the canopy and reduced the raindrop splash effect, those two conditions being the favorable to fungal spores spread. Moreover, studies showed that intercropping system is more productive than single cultivation system due to complementing effects of the companion crops (Rodino et al. 2017).

The nematophagous fungus, *Pochonia chlamydosporia* (Goddard) Zare & Gams, has been investigated as a potential biological control agent for use in integrated pest management strategies for *Meloidogyne incognita* (Kof & White) Chitwood in vegetable crops. The fungus significantly reduced nematode infestations in soil following a tomato crop, in a strategy that combined the use of the fungus with crop rotation (Atkins et al. 2003). Solarization associated with organic fertilization has potential to be used in nematode control and to reduce the need for application of pesticides (Silva et al. 2006). A combined approach is needed for nematode management in organic farming systems. Measures such as appropriate crop rotation with less susceptible crop species and green manure crops, soil amendments with antagonistic crops, and consistent weed control are effective when used in together in concert. Identification of prevalent plant-parasitic nematode specie/s through nematode diagnostic services should be considered for choosing non-hosts before planning long-term crop rotation for organic farming. Additional research is needed most in the development of resistant cultivars suitable for organic farming, nematode-free vegetative-propagating materials, and further steps in commercialization of biological control products permitted in organic farming systems (Briar et al. 2016).

Aphids (Brevicoryne brassicae) are normally sufficiently controlled by naturally occurring antagonists, such as hoverflies. Flowering strips (e.g. buckwheat) and companion plants can further enhance their efficacy, whereas the use of broad spectrum insecticides (pyrethrins and Spinosad) negatively affect the antagonists. Mainly young plants, before and shortly after planting and especially during dry weather conditions, are susceptible and can be protected by nets. However, the use of nets might also exclude the natural enemies. Additional irrigation can mitigate aphid damage and promote quick plant growth. Vegetable oils and soaps are available for direct pest control and should be applied using dropleg technology in order to reach the insects on the lower leaf surfaces (Daniel et al. 2016).

The incidence of thrips on onions was similar between fertilized and not fertilized treatments. Treatments with mineral fertilization presented the same incidence of thrips on onions when compared to organic fertilization (Goncalves and Silva 2003). Results from one study suggest that a combination of using reflective mulch and host plant resistance can additively suppress whitefly infestations, which have particular importance in the fast-growing organic vegetable production industry (Simmons et al. 2010).

Poisonous plants like neem (Azadharacta indica), Aak (Calotropis procera), Arosa (Adhatoda vasica), chili (Capsicum annuum), garlic (Allium sativum) with insecticidal properties contribute well in the management of harmful pests. The biopesticides have remarkable effect against the leaf cutters along with repelling the sucking insects (Anjum et al. 2016).

There are control methods for disease and insect control in organic cultivation of vegetables (hot water, hot air and electron treatment, biological seed treatment groups like microorganisms, plant extracts and inducers of resistance, solarization for nematode control, biopesticides, insect net).

Weed management is a major constraint in organic production. It can be expensive and time-consuming and severe crop yield losses may be incurred when weeds are not adequately controlled. Research on organic weed management in herb and vegetable production is increasing internationally (Kristiansen 2003).

There are several ways to control weeds in organic vegetable production (Table 1.7). Nowadays, corn gluten meal is used in organic cultivation systems for weed control (Core 2006). Corn gluten meal is a non-selective pre-emergence or preplant-incorporated herbicide that inhibits root development, decreases shoot length, and reduces plant survival. The development of a mechanized application

Table 1.7 Summary of direct (or physical) and indirect (or cultural) weed control practices used in organic herb and vegetable production

Direct/physical methods	Indirect/cultural methods
Tillage: Mechanical cultivation of soil before and during the cropping phase **Hand weeding**: Manually hoeing or pulling weeds **Mulching**: Organic materials normally used, use of woven plastic 'weed mat' is restricted **Slashing**: Slashing or mowing using hand-operated or tractor-mounted implements **Grazing**: a wide range animals used, usually in rotation, rarely within cropping phase (e.g. poultry) **Biological control**: Classical; inundative, mycoherbicides methods available **Solarisation**: Requirements to be effective, limitations (e.g. selective control only) **Thermal methods**: Various flame, steam, hot water, infra red implements used, use of burning is restricted	**Rotation**: Varying crops, cover crops, fallows and grazing over time **Cover crops**: Green manure or other crop grown in fallow period to suppress weed growth by competition and allelopathy **Prevention**: Reducing weed seedling numbers prior to cropping phase, and avoiding weed seed production at all times **Timing**: Strategic timing of planting/sowing, tillage, fertilizing and irrigating, plant-back after cover crop **Planting density**: Increased usually, but some crops (e.g. cotton) use wider spacing to allow access for tillage implements **Intercropping**: Growing two or more crops in close proximity to improve resource capture **Crop and cultivar selection**: Sow vs. transplant, growth rate, canopy density & closure **Precision placement**: Irrigation and fertilizer applied close to crop (e.g. drippers) **Soil management**: Modify pH, fertility and specific nutrients

Kristiansen (2003)

system for the banded placement of corn gluten meal between crop rows (seed row not treated) has increased its potential use in organic vegetable production, especially in direct-seeded vegetables. This research determined that a corn gluten meal free planting strip (corn gluten meal applied between crop rows) provided increase crop safety for direct-seeded squash compared with broadcast applications. Furthermore, these results have implications for all direct-seeded organic vegetable crops once the optimum corn gluten meal application rates and corn gluten meal free strip width can be determined for specific vegetables to maximize crop safety, yields, and weed control efficacy (Webber et al. 2010). Webber and Shrefler (2006) added that the granulated formulation worked well at all application rates and application configurations. The powdered CGM did not flow easily, and its delivery was inconsistent and unreliable when used in the solid application configuration.

In addition more effective methods against weeds can be found: 1. Six methods of stale seedbed preparation were compared on certified and transitional organic land (Boyd et al. 2006). The flamer and the clove oil herbicide had the lowest number of weeds emerging with the crop following stale seedbed formation. 2. Field experiments were conducted from 1998 to 2000 to study the effect of summer cover crop and in-season management system on weed infestations in lettuce. The results indicate that prior summer cover crops can improve both conventional and organic vegetable production systems (Ngouajio et al. 2003). 3. Ammonium nonanoate provided consistent control of weeds across a large range of application volumes. The results indicate that ammonium nonanoate has excellent potential as an organic herbicide (Webber et al. 2011).

Soil Solarization, Mulching, Biodegradable Mulch, Natural Herbicides, Hot Water, and Agronomic Practices have been successfully adopted in many countries as safe methods for controlling weeds in the organic farming. In addition, there are some promising new and non-traditional measures such as Fresnel Lens, Electrical Weed Control, Lasers, etc. which could be employed for controlling the weeds in organic farming. Also the agronomic practices such as choice of competitive varieties, stale seedbeds had a significant impact on weeds. The growers in organic farming should keep these three points in mind: (1) start clean stay clean successful, (2) Prevention is always better than treatment and, (3) One year's seeds will lead to 7 year's weed infestation (Abouziena and Haggag 2016).

There has been done research, which confirms that the effects of different organic vegetable production practices on weed community structure and highlight the value of tilled fallow periods, cover crops, and prevention of weed seed rain for reducing weed populations (Jernigan et al. 2017).

Cover crop mixtures showed a strong potential for weed infestation reduction, given the high amount of biomass produced. As pointed out by previous research, a high quantity of cover crop biomass will ensure good weed suppression during subsequent cash crop cultivation (Ranaldo et al. 2016). Similarly high-residue clover and cereal rye cover crops provided substantial suppression of Palmer amaranth, large crabgrass, and yellow nutsedge in tomato production (Price et al. 2016).

Bioherbicide technology could be used as a component in integrated weed management strategies to help avoid herbicide resistance, reduce production costs, and increase crop yield in organic horticulture (Cai and Gu 2016).

The innovative machines for physical weed control can represent an important mean for organic vegetable growers in order to reduce costs and increase their income. Moreover, the innovative machines are low-tech equipment, which can be easily adjusted or modified depending on the different agronomic context and purchased according to their availability on the market at low cost (Fontanelli et al. 2015).

Weed control is the most difficult part of vegetable production in organic cultivation. Good methods against weeds can be: tillage, mulching, flaming and hot water treatment.

1.5 Conclusion

Findings from the literature suggest the following: 1. The growth of vegetables was affected positively in 43% and negatively in 57% of all studied cases from the literature on organic cultivation. 2. Organic cultivation affected the yield of vegetables 59% positively, 29% negatively and 12% did not have any significant influence. 3. Organically grown vegetables have, in most cases (65%), better nutritional value than conventional ones. 15% cases showed a reduction in nutritional value and 20%

of results were not significantly influenced. 86% of results showed reduction and 14% of results showed an increase in nitrate levels in organic cultivation. 4. Methods are available for disease and insect control in organic cultivation of vegetables (hot water, hot air and electron treatment, biological seed treatment groups like microorganisms, plant extracts and inducers of resistance, solarization for nematode control, biopesticides, insect net). 5. Weed control is the most difficult part of vegetable production in organic cultivation. 6. Good methods against weeds include tillage, mulching, flaming and hot water treatment. If the proper technology is used the organic cultivation of vegetables is not so time- and money-consuming and the trend is that, with organic cultivation, vegetables with better quality and better nutritional value with no pesticides residues can be produced.

Acknowledgements The present review was carried with financial support from Estonian Agricultural Registers and Information Board and with the help of the self-employed entrepreneur Galina Rehkli.

References

Abouziena HF, Haggag WM (2016) Weed control in clean agriculture: a review. Planta Daninha 34(2):377–392. https://doi.org/10.1590/S0100-83582016340200019

Abreu IMO, Junqueira AMR, Peixoto JR, Oliveira SA (2010) Microbiological quality and productivity of lettuce under chemical and organic fertilization. Ciênc Tecnol Aliment Campinas 30(1):108–118. https://doi.org/10.1590/S0101-20612010000500018

Akira M, Yasuhiro N, Masahiro T (2003) Development of standard techniques for organic cultivation of vegetables. (3). Fertilization management methods for organic cultivation. B. Effects of organic fertilization on growth and biological properties (radish), Annual report of the Gifu Prefectural Research Institute for Agricultural Sciences in hilly and mountainous areas, pp 261–264

Alimi T, Ajewole OC, Olubode-Awosola OO, Idowu EO (2007) Organic and inorganic fertilizer for vegetable production under tropical conditions. J Agric Rural Dev 1:120–136. https://hdl.handle.net/10568/40809

Anjum AS, Zada R, Tareen WH (2016) Organic farming: hope for the sustainable livelihoods of future generations in Pakistan. J Rural Dev Agric 1(1):20–29

Ashley R, Bishop A, Dennis J (2007) Intensive organic vegetable production integrated development. A report for the Rural Industries Research and Development Corporation. RIRDC Publication No 04/121 RIRDC Project No: DAT-37A, 46 pp

Atkins SD, Hidalgo-Diaz L, Kalisz H, Mauchline TH, Hirsch PR, Kerry BR (2003) Development of a new management strategy for the control of root-knot nematodes (*Meloidogyne* spp) in organic vegetable production. Pest Manag Sci 59:183–189. https://doi.org/10.1002/ps.603

Barański M, Rempelos L, Iversen PO, Leifert C (2017) Effects of organic food consumption on human health; the jury is still out! Food Nutr Res 61(1):1287333. https://doi.org/10.1080/16546628.2017.1287333

Bettiol W, Ghini R, Galvão JAH, Siloto RC (2004) Organic and conventional tomato cropping systems. Sci Agric (Piracicaba) 61(3):253–259. https://doi.org/10.1590/S0103-90162004000300002

Bordeleau G, Myers-Smith I, Midak M, Szeremeta A (2002) Food quality: a comparison of organic and conventional fruits and vegetables. Ecological Agriculture, Den Kongelige Veterinœr- og Landbohøjskole, Frederiksberg

Boteva K, Rankov V (2007) Effect of organic and organic-mineral fertilization on yield in some lettuce varieties. Plant Sci 44(6):539–541

Botrini L, Magnani G, Graifenberg A (2004) Effects of organic fertilization on growth and quality of lettuce (*Lactuca sativa* L.) and cabbage (*Brassica oleracea* L.) seedlings. Colture Protette 33(11):71–77

Bourgol A (2005) Effect of foliar fertilization on growth and yield of organic vegetables grown under plastic house in south of Morocco. Thèses et Masters, Centre International de Hautes Etudes Agronomiques Méditerranéennes Bari, 69 pp

Boyd NS, Brennan EB, Fennimore SA (2006) Stale seedbed techniques for organic vegetable production. Weed Technol 20:1052–1057. https://doi.org/10.1614/WT-05-109.1

Briar SS, Miller SA, Stinner D, Kleinhenz MD, Grewal PS (2011) Effects of organic transition strategies for peri-urban vegetable production on soil properties, nmatode community, and tomato yield. Appl Soil Ecol 47(2):84–91. https://doi.org/10.1016/j.apsoil.2010.12.001

Briar SS, Wichman D, Reddy GV (2016) Plant-parasitic nematode problems in organic agriculture. In: Organic farming for sustainable agriculture. Springer, Cham, pp 107–122

Bruno RLA, Viana JS, Silva VF, Bruno GB, Moura MF (2007) Production and quality of seeds and roots of carrot cultivated under organic and mineral fertilization. Hortic Bras 25:170–174

Cai X, Gu M (2016) Review – bioherbicides in organic horticulture. Horticulturae 2(2):3. https://doi.org/10.3390/horticulturae2020003

Core J (2006) Grower's choice: organic and conventional vegetable production explored. Agric Res 54(4):20

Daniel C, Hommes M, Koller M (2016) Plant protection in organic production of Brassica vegetables and oilseed rape. IOBC-WPRS Bull 118:110–115

Ehaliotis C, Zervakis GI, Karavitis P (2005) Residues and by-products of olive-oil mills for root-zone heating and plant nutrition in organic vegetable production. Sci Hortic 106(3):293–308. https://doi.org/10.1016/j.scienta.2005.04.006

Fjelkner-Modig S, Bengtsson H, Stegmark R, Nystrom S (2000) The influence of organic and integrated production on nutritional, sensory and agricultural aspects of vegetable raw materials for food production. Acta Agric Scand Sect B Soil Plant Sci 50:102–113. https://doi.org/10.1080/090647100750374250

Fontanelli, M, Frasconi C, Martelloni L, Pirchio M, Raffaelli M, Peruzzi A (2015) Innovative strategies and machines for physical weed control in organic and integrated vegetable crops, 2011–2016. http://hdl.handle.net/11568/749187

Goncalves PAS, Silva CRS (2003) Impact of the organic fertilization on onion thrips incidence. Hortic Bras 21(3):459–463

Gonçalves PAS, Silva CRS, Boff P (2004) Incidence of downy mildew in onion growing under mineral and organic fertilization. Hortic Bras 22(3):538–542

Hadayat N, De Oliveira LM, Da Silva E, Han L, Hussain M, Liu X, Ma LQ (2018) Assessment of trace metals in five most-consumed vegetables in the US: conventional vs. organic. Environ Pollut 243:292–300. https://doi.org/10.1016/j.envpol.2018.08.065

Illanjiam S, Sivakumar J, Sundaram CS (2019) Micobial diversity of vermicompost and its efficacy on organic vegetables. Res J Pharm, Biol Chem Sci 5:806–819. Life Science Informatics Publication. https://doi.org/10.26479/2019.0501.67

Jernigan AB, Caldwell BA, Cordeau S, DiTommaso A, Drinkwater LE, Mohler CL, Ryan MR (2017) Weed abundance and community composition following a long-term organic vegetable cropping systems experiment. Weed Sci 65(5):639–649. https://doi.org/10.1017/wsc.2017.33

Juroszek P, Tsai HH (2008) Research needs in organic vegetable production systems in tropical countries with a focus on Asia. 16th IFOAM Organic World Congress, Modena, Italy, June 16–20, 2008, 4 pp

Kapoulas N, Koukounaras A, Ilić ZS (2017) Nutritional quality of lettuce and onion as companion plants from organic and conventional production in North Greece. Sci Hortic 219:310–318. https://doi.org/10.1016/j.scienta.2017.03.027

Kim MJ, Shim CK, Kim YK, Hong SJ, Park JH, Han EJ, Kim JH, Kim SC (2015) Effect of aerated compost tea on the growth promotion of lettuce, soybean, and sweet corn in organic cultivation. Plant Pathol J 31(3):259. https://doi.org/10.5423/PPJ.OA.02.2015.0024

Kristiansen PE (2003) Sustainable weed management in organic herb and vegetable production. A thesis submitted for the degree of Doctor of Philosophy of the University of New England June 2003, 225 pp

Lairon D (2010) Nutritional quality and safety of organic food. A review. Agron Sustain Dev 30:33–41. https://doi.org/10.1051/agro/2009019

Lee JT, Ha IJ, Kim HD, Moon JS, Lee SD (2009) Times and frequencies of additional fertilization to improve nutrient efficiency of organic liquid fertilizer for onion organic cultivation. Korean J Hortic Sci Technol 27(1):30–36

Lončarić Z, Vukobratović M, Ragalyi P, Filep T, Popović B, Karalić K, Vukobratović Ž (2009) Computer model for organic fertilizer evaluation. Poljoprivreda (Osijek) 15(2):38–46

Ma CH, Chen JH, Yang RY, Palada MC, Ou SC, Lin YH, Chen LH (2009) Monitoring soil and vegetable quality under six fertilization strategies in organic and conventional farming systems. The 9th International Conference of the East and Southeast Asia Federation of Soil Science Societies, pp 373–374

Mahmoud E, El-Kader NA, Robin P, Akkal-Corfini N, El-Rahman LA (2009) Effects of different organic and inorganic fertilizers on cucumber yield and some soil properties. World J Agric Sci 5(4):408–414

Masarirambi MT, Hlawe MM, Oseni OT, Sibiya TE (2010) Effects of organic fertilizers on growth, yield, quality and sensory evaluation of red lettuce (Lactuca sativa L.) 'Veneza Roxa'. Agric Biol J N Am 1(6):1319–1324. https://doi.org/10.5251/abjna.2010.1.6.1319.1324

Mie A, Andersen HR, Gunnarsson S, Kahl J, Kesse-Guyot E, Rembiałkowska E, Quaglio G, Grandjean P (2017) Human health implications of organic food and organic agriculture: a comprehensive review. Environ Health 16(1):111

Muramoto J (1999) Comparison of nitrate content in leafy vegetables from organic and conventional farms in California. Center for Agroecology and Sustainable Food Systems University of California, Santa Cruz Santa Cruz, CA 95064 831/459-2506, 64 pp

Nair A, Ngouajio M (2010) Integrating rowcovers and soil amendments for organic cucumber production: implications on crop growth, yield, and microclimate. HortScience 45(4):566–574. https://doi.org/10.21273/HORTSCI.45.4.566

Narayan S, Makhdoomi MI, Malik A, Nabi A, Hussain K, Khan FA (2017) Influence of plastic and organic mulching on productivity, growth and weed density in chilli (Capsicum annuum L.). J Pharmacogn Phytochem 6(6):1733–1735

Ngouajio M, McGiffen MEJ, Hutchinson CM (2003) Effect of cover crop and management system on weed populations in lettuce. Crop Prot 22(1):57–64. https://doi.org/10.1016/S0261-2194(02)00111-4

Oliveira EQ, Souza RJ, Cruz MCM, Marques VB, Franca AC (2010) Productivity of lettuce and rocket in intercropping system under organic and mineral fertilization. Hortic Bras 28:36–40. https://doi.org/10.1590/S0102-05362010000100007

Pavla B, Pokluda R (2008) Influence of alternative organic fertilizers on the antioxidant capacity in head cabbage and cucumber. Not Bot Horti Agrobot Cluj 36(1):63–67. https://doi.org/10.15835/nbha36196

Pimpini F, Gianquinto G, Sambo P (2005) Organic vegetable production: evolution, base principles and quality of products. Italus Hortus 12(4):31–44

Porto ML, Alves JC, Souza AP, Araujo RC, Arruda JA (2008) Nitrate production and accumulation in lettuce as affected by mineral nitrogen supply and organic fertilization. Hortic Bras 26:227–230. https://doi.org/10.1590/S0102-05362008000200019

Price AJ, Duzy LM, Balkcom KS, Kelton JA, Kornecki TS, Sarunaite L (2016) Organic weed control and cover crop residue integration impacts on weed control, quality, yield and economics in conservation tillage tomato-a case study. In: Organic farming: a promising way of food production, p 69. https://doi.org/10.5772/61315

Ranaldo M, Carlesi S, Costanzo A, Bàrberi P (2016) Designing cover crop mixtures to enhance potential weed suppression in organic no-till vegetable systems. 14th ESA Congress 5–9 September 2016, Edinburgh, Scotland. 2 pp

Ranjan P, Patle GT, Prem M, Solanke KR (2017) Organic mulching-a water saving technique to increase the production of fruits and vegetables. Curr Agric Res J 5(3):371–380. https://doi.org/10.12944/CARJ.5.3.17

Rauton RSW (2007) Soil fertility management in organic vegetable production. In: Partial fulfillment of the requirements for the Degree Master of Science Plant and Environmental Sciences of the Graduate School of Clemson University, 67 pp

Rembialkowska E (2003) Organic farming as a system to provide better vegetable quality. Acta Hortic 604:473–479. https://doi.org/10.17660/ActaHortic.2003.604.52

Rodino S, Butu M, Fidler G, Marin A, Butu A (2017) Current strategies for the protection of organic crops in vegetable production. Sci Pap Ser B Hortic 61:303–306

Röös E, Mie A, Wivstad M, Salomon E, Johansson B, Gunnarsson S, Wallenbeck A, Hoffmann R, Nilsson U, Sundberg C, Watson CA (2018) Risks and opportunities of increasing yields in organic farming. A review. Agron Sustain Dev 38(2):14

Schmitt A, Jahn M, Kromphardt C, Krauthausen HJ, Roberts SJ, Wright SAI, Amein T, Forsberg G, Tinivella F, Gullino ML, Wikström M, Wolf J, Groot SPC, Werner S, Koch E (2006) STOVE: Seed treatments for organic vegetable production. Presented at the European Joint Organic Congress, Odense, Denmark, 2 pp

Silva MG, Sharma RD, Junqueira AMR, Oliveira CM (2006) Effect of solarization, chemical and organic fertilization on the control of nematode on greenhouse lettuce. Hortic Bras 24:489–494

Simmons AM, Kousik CS, Levi A (2010) Combining reflective mulch and host plant resistance for sweetpotato whitefly (Hemiptera: Aleyrodidae) management in watermelon. Crop Prot 29:898–902. https://doi.org/10.1016/j.cropro.2010.04.003

Smukler SM, Jackson LE, Murphree L, Yokota R, Koike ST, Smith RF (2008) Transition to large-scale organic vegetable production in the Salinas Valley, California. Agric Ecosyst Environ 126:168–188. https://doi.org/10.1016/j.agee.2008.01.028

Stein S (2009) A review of organic fertilizers. http://www.helium.com/items/1431343-a-review-of-organic-fertilizers

Thakur KS, Dharminder K, Amit V, Thakur AK, Mehta DK (2010) Effect of organic manures and biofertilizers on growth and yield of tomato and french bean under mid hills of Himachal Pradesh. J Hill Agric 1(2): pages not given

Webber CL, Shrefler JW (2006) Corn gluten meal applicator for weed control in organic vegetable production. J Veg Sci 12(4):19–26. https://doi.org/10.1300/J484v12n04_03

Webber CL, Shrefler JW, Taylor MJ (2010) Influence of corn gluten meal on squash plant survival and yields. HortTechnology 20(4):696–699. https://doi.org/10.21273/HORTTECH.20.4.696

Webber CL, Shrefler JW, Brandenberger LP, Taylor MJ, Carrier LK, Shannon DK (2011) Weed control efficacy with ammonium nonanoate for organic vegetable production. Int J Veg Sci 17(1):37–44. https://doi.org/10.1080/19315260.2010.501850

Worthington V (2001) Nutritional quality of organic versus conventional fruits, vegetables, and grains. J Altern Complement Med 7(2):161–173. https://doi.org/10.1089/107555301750164244

Xu HL, Wang R, Xu RY, Mridha MAU, Goyal S (2003) Yield and quality of leafy vegetables grown with organic fertilizations. Acta Hortic 627:25–33. https://doi.org/10.17660/ActaHortic.2003.627.2

Chapter 2
Reducing Arsenic Accumulation in Rice Using Physiology, Genetics and Breeding

Dibakar Das and Somnath Bhattacharyya

Abstract Arsenic accumulation in rice grains is a serious public health concern worldwide. Developing arsenic-depleted rice varieties is an effective solution to minimize arsenic exposure through food. Recently, progress has been made in understanding arsenic soil uptake, subsequent transport from root to shoot to grains, vacuolar sequestration and detoxification of arsenic in rice. Many genes controlling major physiological processes such as arsenic speciation within plants have been cloned and characterized. Quantitative trait loci responsible for arsenic accumulation have been identified from several crosses. Transcriptome studies have elucidated the roles of transcription factors, small and micro-RNAs, including long non-coding RNAs in genetic regulation of arsenic stress response in rice. Here we review arsenic uptake, speciation, vacuolar sequestration and transport within plants, including identification of genes controlling each step. We discuss the variation and genetic basis of arsenic accumulation including quantitative trait loci mapping, and the role of transcription factors and small and micro-RNAs in genetic regulation of arsenic stress response. Finally, we discuss the strategies for developing low grain-arsenic rice varieties emphasizing on marker assisted breeding. We also discuss the progress in transgenic approach including its limitations, and the possible applications of genome editing for reducing arsenic accumulation in rice grains.

Keywords Arsenic · Rice · Heavy metals · Transporters · Phytochelatins · Vacuolar sequestration · Nodulin 26- like intrinsic proteins · Arsenic reductase · Transgenics · Genome editing

D. Das (✉) · S. Bhattacharyya
Crop Research Unit, Genetics and Plant Breeding, Bidhan Chandra Krishi Viswavidyalaya, Mohanpur, West Bengal, India

E. Lichtfouse (ed.), *Sustainable Agriculture Reviews 52*, Sustainable Agriculture Reviews 52, https://doi.org/10.1007/978-3-030-73245-5_2

Abbreviations

ArsM	Arsenic methyltransferase
AtABCC1	*Arabidopsis thaliana* ATP binding cassette C type 1
AtABCC2	*Arabidopsis thaliana* ATP binding cassette C type 2
AtINT2	*Arabidopsis thaliana* Inositol transporter 2
AtINT4	*Arabidopsis thaliana* Inositol transporter 4
AtPHF1	*Arabidopsis thaliana* phosphate transporter traffic facilitator 1
cdPCS1	*Ceratophylium demersum* phytochetalin synthase 1
CRISPR/Cas	Clustered regularly interspersed short palindromic repeats/CRISPR associated protein
lncRNA	Long non-coding RNA
Lsi1	Low inorganic silicon 1
Lsi2	Low inorganic silicon 2
MATE	Multidrug and toxic compound extrusion proteins
OsABCC1	*Oryza sativa* ATP binding cassette C type 1
OsABCC7	*Oryza sativa* ATP binding cassette C type 7
OsARM1	*Oryza sativa* arsenite-responsive myeloblastosis 1
OsHAC1;1	*Oryza sativa* high arsenic content 1;1
OsHAC1;2	*Oryza sativa* high arsenic content 1;2
OsHAC4	*Oryza sativa* high arsenic content 4
OsNIP1;1	*Oryza sativa* nodulin 26- like intrinsic protein 1;1
OsNIP3;3	*Oryza sativa* nodulin 26- like intrinsic protein 3;3
OsNRAMP1	*Oryza sativa natural* resistance-associated macrophage protein 1
OsPHF1	*Oryza sativa* phosphate transporter traffic facilitator 1
OsPHR2	*Oryza sativa* phosphate starvation response 2
OsPHT8	*Oryza sativa* inorganic phosphate transporter 8
OsPIP2;4	*Oryza sativa* plasma membrane intrinsic protein 2;4
OsPIP2;6	*Oryza sativa* plasma membrane intrinsic protein 2;6
OsPIP2;7	*Oryza sativa* plasma membrane intrinsic protein 2;7
OsPT4	*Oryza sativa* phosphate transporter 4
OsPTR7	*Oryza sativa* putative peptide transporter 7
Pht1	Inorganic phosphate transporter 1
ppmv	Parts per million by volume
ScYCF1	*Sccharomyces cerevisiae* metal resistance protein 1
WaarsM	*Westerdyella aurantiaca* arsenic methyltransferase

2.1 Introduction

Arsenic is a toxic metalloid of ubiquitous origin in nature and a class I human carcinogen (Smith et al. 2000, 2002). Arsenic originates from both geogenic and anthropogenic sources (Garg and Singla 2011; Bibi et al. 2017). The wide occurrence of arsenic in ground-water and agricultural-soils directly and indirectly threatens the health of millions of people in many parts of the world. Human activities such as mining, use of arsenic based pesticides, herbicides and burning of coal also contribute towards arsenic release in the environment (Zhu et al. 2014). Due to non-threshold level of toxicity and wide occurrence in the human environment, arsenic is considered as one of the most hazardous elements in the environment.

The accumulation of arsenic through food and drinking water in the human body is a serious public health concern as arsenic contamination is a wide spread phenomenon, and accumulation results in a variety of diseases including cancer, cardiovascular diseases, skin lesions, and birth defects in newborn (Argos et al. 2011; Tseng 2007; Yu et al. 2006; Yuan et al. 2010). Arsenic accumulation in children impairs brain development and is associated with low IQ (Nahar et al. 2014; Signes-Pastor et al. 2019). A recent study found even low levels of inorganic arsenic exposure from rice is associated with mortality risks of cardiovascular diseases in England and Wales (Xu et al. 2020). The adverse health effects of arsenic on human health are well reported.

Rice, a major food grain of the world, is the most significant source of arsenic exposure next to contaminated drinking water. Rice is the primary source of arsenic exposure through food, especially in South East Asian countries where rice is a staple food. Rice contributes up to 50% of the total dietary arsenic consumption for West Bengal and Bangladesh and up to 60% for the Chinese population (Signes-Pastor et al. 2008; Li et al. 2011). High levels of arsenic in rice from Bangladesh and West Bengal are evident from several reports (Chowdhury et al. 2001; Roychowdhury et al. 2003; Meharg and Rahman 2003). Alarming levels of arsenic in rice has also been reported from many countries in the world. Therefore, there is urgent need to develop rice varieties that inherently accumulate less arsenic in grains to minimize arsenic exposure through food.

In recent years, significant advance in understanding the physiology and genetics of arsenic accumulation in rice has been achieved leading to improved understanding of arsenic uptake and accumulation in rice. Several quantitative trait loci and genes controlling important steps in arsenic uptake and accumulation have been identified. Progress is being made in understanding the genetic response to arsenic stress in rice by studying transcriptomes of plants exposed to arsenic. Although significant advance has been made, important gaps in knowledge still remain. Importantly, breeding for low grain arsenic accumulating rice has not moved in same pace as gene discovery, which demonstrate the complex physiological and genetic challenges associated with the goal.

In this review, we discuss the progress made in understanding arsenic uptake, speciation, vacuolar sequestration and transport within plant including identification of genes, which led to improved understanding of the overall physiology and genetics of arsenic accumulation in rice. We discuss the variation and genetic basis of arsenic accumulation including quantitative trait loci mapping, and the role of transcription factors and small and micro-RNAs in genetic regulation of arsenic stress response. Finally, we discuss the strategies for developing low grain-arsenic rice varieties emphasizing on marker assisted breeding while stressing that genes involved in vacuolar sequestration and phloem loading are important breeding targets for developing low grain-arsenic rice varieties. We also discuss the progress in transgenic approach including its limitations, and the probable applications of genome editing for reducing arsenic accumulation in rice grains.

2.2 Arsenic Uptake and Accumulation in Rice

2.2.1 Root Uptake

Rice accumulates both organic and inorganic species of arsenic efficiently than any other cereals (Ma et al. 2008). Inorganic species of arsenic include arsenite and arsenate; the two biologically relevant oxidation states. Organic species of arsenic include methylated species of arsenic; monomethylarsonic acid and dimethylarsinic acid. Inorganic species of arsenic is generally considered more toxic than organic species of arsenic. Arsenite predominates in rice grains grown anaerobicaly, i.e., under flooded condition, whilst arsenate is more abundant in rice grains grown aerobically, i.e., under non-flooded condition. Rice also accumulates little quantities of monomethylarsonic acid and dimethylarsinic acid in grains. As rice is traditionally cultivated under submergence, arsenite is the predominant species of arsenic mostly found in rice grains (Begum et al. 2016). A recent green house experiment indicates that as the concentration of pore water arsenite in the rhizosphere of rice increases under high temperature stress in future climate (38° C, 850 ppmv CO_2), arsenic concentrations in rice grains may increase by twofolds along with reduction in yield (Muehe et al. 2019). Thus, the problem of arsenic accumulation in rice may aggravate under high temperature stress in future climate.

Arsenic transport across biological membrane is a channel dependent process. In roots of rice and other arsenic accumulating species such as the model plant *Arabidopsis thaliana*, hereafter referred to as *A. thaliana*, arsenic enters through the silicic acid and phosphate transport pathways. Arsenite, the predominant species in anaerobic environment enters rice plant through the silicon transporters *Lsi1* and *Lsi2* (Ma et al. 2008). *Lsi1* and *Lsi2* belong to the nodulin 26- like intrinsic subfamily of aquaporins, which transport neutral molecules like water and urea. The similar structural and chemical properties of silicon and arsenic allow *Lsi* transporters to uptake arsenite as both arsenite and silicon is predominantly present as neutral molecules of arsenous acid [$As(OH)_3$] and silicic acid [$Si(OH)_4$] respectively below pH 8.

Both *Lsi1* and *Lsi2* are located to the plasma membrane in rice roots but their orientations differ (Yamaji and Ma 2010). *Lsi1* shows distal localization at polar ends of both exodermis and endodermis of the main and lateral roots. *Lsi2* is located on the proximal sides of exodermis and endodermis. *Lsi1* is permeable to arsenite, but not arsenate as expression studies in *xenopus* oocytes and yeast show markedly increased uptake of arsenite only (Ma et al. 2008). *Lsi1* shows both influx and efflux activities for silicon (Ma et al. 2006; Mitani et al. 2008), but only functions as an influx transporter in rice roots (Ma et al. 2006), whereas *Lsi2* functions as an efflux transporter (Fig. 2.1). Mutations in *Lsi1* and *Lsi2* result in reduced uptake of both silicon and arsenic. Mutation in *Lsi2* had a much greater impact; 66–75% reduction in grain arsenic accumulation than mutation in *Lsi1*. Arsenite concentration in xylem sap of *Lsi2* mutants was also significantly lower; 73–91% lower than that of their wild type plants (Ma et al. 2008). The difference in impacts of mutations in *Lsi1* and *Lsi2* may be due to the fact that *Lsi1* also works as an efflux transporter of arsenite when internal concentration of arsenite becomes greater than that of the

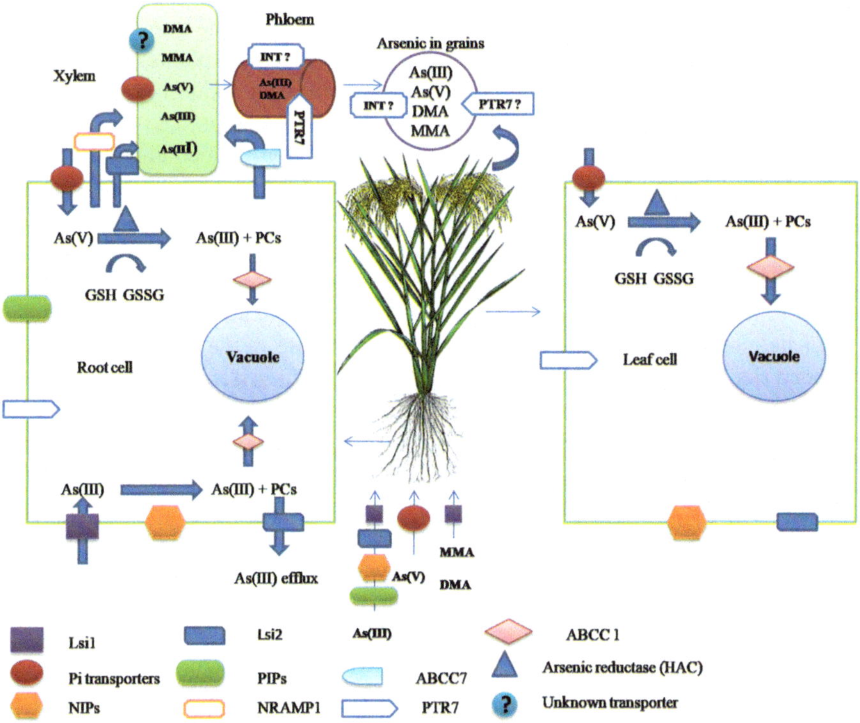

Fig. 2.1 Transporters involved in arsenic uptake and transport in rice. *Lsi1*, *Lsi2*, nodulin 26-like intrinsic proteins (NIPs) and plasma membrane intrinsic proteins (PIPs), and the phosphate transporters, *OsPHT1*, *OsPT4* and *OsPHT8* mediate the uptake of arsenite and arsenate respectively. *Lsi1* also mediate the transport of monomethylarsonic acid (MMA) and dimethylarsinic acid (DMA). *Lsi2* also functions as an effluxer of arsenite in rice roots. *Lsi2* and *NRAMP1* mediate xylem loading and root to shoot transport of arsenic. Inositol transporters (INTs) may be involved in phloem loading of arsenic as INTs are involved in phloem of loading of arsenic in *A. thaliana*. *OsPTR7* mediates the long distance transport of DMA from roots to grains. Arsenate is reduced to arsenite inside cells by three arsenate reductase (*HAC1;1*, *HAC1;2* and *HAC4*). Reduction of arsenate to arsenite is associated with conversion of glutathione (GSH) to glutathione disulfide (GSSG). *ABCC1*, a C type ATPase, mediates vacuolar sequestration and detoxification of arsenic by carrying arsenite-phytochelatin complexes into vacuoles. *ABCC1* is strongly expressed in the first internode of rice. Transporters mediating transport of arsenic in grains are yet to be identified. Question marks indicate current gaps in knowledge

external medium (Zhao et al. 2010). *Lsi1* mutants also show slower rate of monomethylarsonic acid and dimethylarsinic acid uptake.

Apart from the members of the nodulin 26-like intrinsic proteins subfamily of aquaporins, the members of the rice plasma membrane intrinsic subfamily, *OsPIP2;4* , *OsPIP2;6* and *OsPIP2;7* , are also involved in arsenite transport (Mosa et al. 2012). However, *Lsi1* and *Lsi2* are major routes through which arsenite enters the rice plant. The co-permeability of silicon transporters in rice to arsenic under the same environment explains why rice is especially good at absorbing arsenic than any other cereals. A recent experiment has demonstrated that plant nodulin 26-like intrinsic proteins subfamily of aquaporins have evolved from bacterial arsenic efflux channels that integrated into plant genomes by horizontal gene transfer and subsequently evolved to transport silicon and boron in higher seed plants (Pommerrenig et al. 2020). This may explain the co-permeability of plant nodulin 26-like intrinsic proteins to nutrients and arsenic from an evolutionary standpoint.

Arsenate, which is abundant in rice grown under non-flooded condition, enters rice plants through phosphate transporters. The inorganic phosphate transporter 1 (pht1) family, which function as Phosphate; H+ symporters with in the major facilitator family of proteins are permeable to arsenate as arsenate and phosphate are chemical analogue of each other. Under aerobic condition arsenite is rapidly oxidized to arsenate and the oxygen released by rice roots also converts arsenite to arsenate, contributing to the predominance of arsenate in the rhizosphere in non-flooded rice soils (Liu et al. 2006; Seyfferth et al. 2010). Arsenate is therefore more abundant in rice grains grown aerobically.

Physiological and genetic evidence have strongly established the role of phosphate transporters in arsenate uptake. Phosphate and arsenate compete to each other during uptake indicating the role of common transporters (Meharg and Hartley-Whitaker 2002). The isolation of arsenate resistant mutants of Arabidopsis defective in inorganic phosphate transporters have provided strong genetic evidence for the role of inorganic phosphate transporters in arsenate uptake (Catarecha et al. 2007; Shin et al. 2004; González et al. 2005). Mutation in *AtPHF1*, a regulator of Pht1s, results in retention of *Pht1;1* and other high affinity phosphate transporters in the endoplasmic reticulum (González et al. 2005). Mutation in *OsPHF1* results in reduced uptake of both inorganic phosphate and arsenate, and overexpression of *OsPHR2* or *OsPT8* leads to increased uptake and accumulation. *OsPHT1*, *OsPT4* and *OsPHT8* of the Pht1 family are involved in arsenate uptake in rice. (Kamiya et al. 2009; Ye et al. 2017; Wu et al. 2011; Wang et al. 2016a). Arsenate thus enters the rice plants through phosphate transporters only.

2.2.2 *Root to Shoot Translocation*

Arsenic is generally less mobile than silicon and phosphate in plants, except in hyperaccumulator species (Zhao et al. 2003). The mobility of arsenic also varies between and within species. In general, organic species of arsenic is more mobile than inorganic species of arsenic (Carey et al. 2010). This difference in mobility may be due to the complexion of inorganic arsenic with phytocheltins, which make them less mobile. Organic species of arsenic including silicon and phosphate, on the other hand, do not form complexes with phytochelatins.

Rice restricts most of the arsenic in roots leading to a low shoot to root ratio of 0.1–0.3 (Zhao et al. 2009). This is in contrast to hyperaccumulator species such as the fern *Pteris vittata,* which rapidly translocates accumulated arsenic from roots to the aerial parts. The ratio of arsenic concentration in xylem sap of rice to that of external medium is generally 0.3–0.6, which is many folds lower than that of silicon. Despite this, rice is an efficient accumulator of arsenic. This efficiency in accumulation is attributed to the specific localization of arsenic influx and efflux transporter in exodermis and endodermis cells in rice (Su et al. 2010). Transporter localization therefore appears to be an important determinant of metals/heavy metals accumulation in plants.

Arsenite is the predominant species of arsenic found in rice whether rice is exposed to arsenate or arsenite (Ma et al. 2008; Zhao et al. 2009; Su et al. 2010). However, when rice is exposed to arsenate, it is initially present in greater proportions in the xylem sap, after which the concentration of arsenite gradually increases suggesting a large amount of arsenate is reduced to arsenite in roots and shoots of rice. Recently, *OsHAC1;1* and *OsHAC1;2* have been identified as arsenic reductase in rice (Shi et al. 2016). *OsHAC1;1* and *OsHAC1;2* are predominantly expressed in roots. *OsHAC1;1* is abundant in the epidermis, root hairs, and pericycle cells, while *OsHAC1;2* is abundant in the epidermis, outer layer of cortex, and endodermis cells. The expression of these genes is induced by arsenite, and their loss results in decreased reduction of arsenate to arsenite , and loss of arsenic efflux to the external medium. Loss of arsenite efflux is associated with increased accumulation of arsenic in shoots. Overexpressing either of these genes also decreases arsenic accumulation in grains when grown aerobically. *OsHAC4,* which encodes a rhodnase-like protein, and expressed in root epidermis and exodermis, also functions as arsenate reductase in rice (Xu et al. 2017). These discoveries have disproved the widely held assumption that plant arsenate reductase are similar to bacterial arsenic reductase.

Recently, a C type ATP-binding cassette, *OsABCC7,* has been shown to transport arsenite from root to shoot of rice plants (Tang et al. 2019a). *OsABCC7* is strongly expressed in roots and its expression is strongly suppressed by arsenite exposure. The gene is predominantly expressed in the xylem parenchyma cells in the stele region of the primary and lateral roots and localized to the plasma membrane (Tang et al. 2019a). In *Xenopus* oocytes, *OsABCC7* showed an efflux activity for arsenite-phytochelatin and arsenite-glutathione complexes, but not for arsenite. Knock out of *OsABCC7* significantly reduces arsenite concentration in shoot-xylem with only little effect in root. *Lsi2* also plays a role in xylem loading of arsenic (Ma et al. 2008). Tiwari et al. 2014 suggested the role of *OsNRAMP1* in xylem loading of arsenic (Fig. 2.1).

Phloem loading of arsenic from xylem and further unloading in grains is a critical step in arsenic accumulation, although reports regarding their mechanisms are scanty in rice. However, some advances have been made in the model plant *A.*

thaliana. Two inositol transporters *AtINT2* and *AtINT4* in the plasma membrane of phloem companion cells mediate phloem loading of arsenic in *A. thaliana* (Duan et al. 2015). *AtINT2* and *AtINT4* are H^+- coupled symporters, which are strongly expressed in the vasculature, although little expression in root tissues is also detected (Schneider et al. 2006, 2007). Single mutations in *AtINT2* and *AtINT4* do not totally reduce arsenic translocation in grains, indicating phloem loading of arsenic is not mediated by *AtINT2* and *AtINT4* alone in plants (Duan et al. 2015). The functional analogues of *AtINT2* and *AtINT4* remain to be identified in rice. The identification of transporters responsible for phloem loading of arsenic would fill important gaps in knowledge of arsenic accumulation in rice.

2.2.3 Shoot to Grain Translocation

Grain unloading of arsenic from phloem represents the final step in arsenic accumulation in rice. Within the phloem, however, a large fraction of arsenic is detoxified and sequestered into the vacuole of phloem companion cells and thus restricted from translocation to grains (Song et al. 2014). The remaining fraction then diffuses via the sieve elements and finally reaches the grain. Within grains, arsenic accumulates both in husks and the endosperm with most arsenic being deposited in the aleurone layer of the endosperm. Like other nutrients, arsenic is unloaded from the ovular vascular trace to the nuceller tissue, and subsequently via the apoplast to the filial tissue; endosperm and the aleurone layer.

The arsenic load of rice grains is primarily contributed by phloem with 90% of arsenite and 55% of dimethylarsinic acid transport to the grain being accounted for by phloem transport alone (Carey et al. 2010). The contribution of xylem in grain arsenic unloading may not be significant as the xylem is primarily directed to vegetative parts rather than reproductive parts (Marschner 1995). Like in xylem, arsenite is the predominat species of arsenic along with little amounts of dimethylarsinic acid in the phloem. Feeding panicles of rice excised during grain feeling stage with arsenite, arsenate, glutathione- complexed arsenic or dimethylarsinic acid demonstrated that dimethylarsinic acid is translocated to grains with much greater efficiency than inorganic arsenic and is more mobile than inorganic arsenic both in phloem and xylem(Carey et al. 2010, 2011). Girdling the stem which limits phloem transport showed that phloem transport was primarily responsible for arsenite transport to the grain (Carey et al. 2010; Zhao et al. 2012). For dimethylarsinic acid, both phloem and xylem transport appear to be equal. There is also difference in the pattern in which arsenite and dimethylarsinic acid is unloaded in the grain; arsenite is mostly retained in the ovular vascular trace while dimethylarsinic acid is dispersed throughout the external grain and into the endosperm.

Stem translocation of inorganic arsenic may not solely rely on silicon transporters (Carey et al. 2011). Recently, *OsPTR7 (OsNPF8.1)* which encodes a putative peptide transporter has been shown to be involved in dimethylarsinic acid uptake in rice (Tang et al. 2017). *OsPTR7* shows significant expression during the grain feeling stage in roosts, leaves and the first node. The field grown mutants of *OsPTR7* had no detectable level of dimethylarsinic acid in grains compared with 35% of arsenic as dimethylarsinic acid in wild type plants. *OsPTR7* is considered as a long distance transporter of root to shoot translocation and grain transport of dimethylarsinic acid (Tang et al. 2017). As suggested earlier, inositol transporters along with other unidentified transporters may regulate grain unloading of arsenic in rice.

2.2.4 Arsenic Detoxification and Sequestration

After root uptake, a large fraction of arsenic is generally excluded back to the extracellular medium, and the remaining fraction is sequestered into root vacuoles or translocated via shoots to various organs (Ma et al. 2008). Inside the rice roots, most of arsenate is reduced to arsenite by three arsenate reductases *OsHAC1;1*, *OsHAC1;2*, and *OsHAC4* as discussed earlier. Reduction is associated with the oxidation of glutathione into glutathione disulfide. Transformed arsenite forms complex with thiol group (-SH) of phytochelatins, which are synthesized in plants by condensation of glutamate, cysteine and glycine residues through three sequential enzymatic reactions (Liu et al. 2010). The arsenite-phytochelatins complexes are then stored safely away from the cytosol to vacuoles. The acidic pH of vacuoles helps stabilize arsenite-phytochelatins complexes. The insoluble arsenite-phytochelatins complexes are less toxic than free anions of arsenic; arsenate and arsenite. The phytochelatin synthase *OsPCS1* and *OsPCS2* are involved in arsenic detoxification in rice (Das et al. 2017; Uraguchi et al. 2017). Phytochelatin synthases are the enzymes which synthesize phytochelatins.

Arsenite-phytochelatins complexes are immobilized in the cytosol which needs to be actively transported to the vacuole. A C type ATPase channel, *OsABCC1*, sequesters arsenite-phytochelatins complexes in roots, leaves and the first internode of rice (Song et al. 2014). *OsABCC1* is strongly expressed in the first internode; the node directly below the panicle and adjacent area of rice. *OsABCC1* is localized to the tonoplast of the phloem companion cells in the first internode and significantly reduces shoot to grain translocation of arsenic. In *A. thaliana ABCC1* functions redundantly with *ABCC2* (Song et al. 2010). Song et al. 2014 have clearly demonstrated that by limiting shoot to grain translocation of arsenic in the first internode, rice saves its next generation of progenies from toxicity.

In the rice genome, at least 45 Multidrug And Toxic Compound Extrusion (MATE) genes are present (Wang et al. 2016b). Seven of these genes show arsenic induced up-regulation in rice, while two are down regulated (Norton et al. 2008). Recently, a member of the rice MATE family protein, *OsMATE2*, has been shown to modulate arsenic accumulation in rice (Das et al. 2018). *OsMATE2* is transcriptionally upregulated in the developing seeds during arsenic stress and is positively correlated with arsenic content in mature grains. Endosperm specific silencing by RNAi of *OsMATE2* reduces arsenic accumulation in transgenic rice. However, physiological evidence of arsenic transport by *OsMATE2* is lacking and its cell-specific location is not known. The rest of the MATE proteins whose expression is affected by arsenic stress remain to be characterized. It is therefore tempting to suggest the role of MATE transporters in arsenic detoxification and sequestration given the widespread role of MATE proteins in metabolites/xenobiotic transport in all kingdoms of life (Upadhyay et al. 2019). However, there is no evidence of *OsMATE2* functioning as arsenic detoxification gene in rice.

Methylation of inorganic arsenic has long been suggested as possible way of detoxification as methylated arsenic is considered to be less toxic than inorganic arsenic. However, no arsenic methylase is known in plants (Ali et al. 2009). Recent experiments have shown that arsenic is methylated by bacteria in the rhizosphere of rice rather than within plants. A more recent study indicates that arsenite methylation constituted the core of microbial arsenic resistance systems before the rise of atmospheric oxygen (Chen et al. 2020). However, the physiological functions of arsenic methylation in anoxic Archean environment are still unclear. The higher toxicity of the trivalent methylated product methylarsenite has led Chen et al. 2020 to call into question the widely held assumption that methylation is a detoxification process. It has been hypothesized that the transient oxygenation of the Archean atmosphere and the existence of oxygen oases in local, shallow marine settings could have created niches where methylation could have operated as a detoxification pathway (Anbar et al. 2007; Crowe et al. 2013; Riding et al. 2014; Fakhraee et al. 2018). Alternatively, bacteria could have used methylation as a way to produce antibiotic in Archean environment with methylarsenite being a primitive antibiotic (Li et al. 2016; Chen et al. 2019). Further studies will clarify the physiological role of arsenic methylation in anoxic environments including the rhizosphere of rice.

2.3 Genetic Control of Arsenic Accumulation in Rice

2.3.1 Variation in Accumulation

Variation present in arsenic accumulation in rice grains is large (Norton et al. 2012). Arsenic content of rice grains vary with cultivar type, geographic regions and management practices. Arsenic in rice is dominated by inorganic species arsenite and arsenate and the organic species dimethylarsinic acid. Recent global survey of 1180 polished rice samples from 29 representative sites across 6 continents revealed the mean inorganic arsenic content was 66 microgram/kilogram (μg/kg), while that of dimethylarsinic acid was 21 μg/kg (Carey et al. 2020). Dimethylarsinic acid was more variable, ranging from less than 2–690 μg/kg. Inorganic arsenic ranged from less than 2–399 μg/kg. There is a strong linear relationship between inorganic arsenic and total arsenic in rice grown in Asia with a slope of 0.78, whereas rice grown in Europe and the United States shows a more variable, but generally hyperbolic relationship with dimethylarsinic acid being predominant in U.S. rice (Zhao et al. 2013). Overall, arsenite and dimethylarsinic acid are prominent species of arsenic in rice grains with different degrees of variation.

The influence of climate on arsenic speciation is considerable. Dimethylarsinic acid concentration in rice grains tends to be higher in temperate regions while inorganic arsenic concentrations remain relatively consistent across temperate, subtropical and northern hemisphere tropical regions (Carey et al. 2020). Low grain arsenic is found in East Africa ($^{<}$10 μg/kg) in southern hemisphere tropical regions, in the eastern hemisphere, and the Southern Indonesian Islands ($^{<}$20 μg/kg). Southern hemisphere South American rice tends to have universally high grain inorganic arsenic (Carey et al. 2020). Therefore, geographic origin of rice and rice products could provide some indication of their arsenic contents.

Rice grains from South East Asian countries have relatively high grain arsenic content. The higher arsenic content of rice from this region is attributed to high level of inorganic arsenic in ground water used for irrigation (Williams et al. 2006; Rahman and Hasegawa 2011). High levels of arsenic in rice from West Bengal in India and Bangladesh have been reported by various authors (Meharg and Rahman 2003; Meharg et al. 2009; Norton et al. 2012). Many high yielding varieties contain grain inorganic arsenic more than 0.2 mg/kg, which is the recommended safe limit as prescribed by the Codex Alimentarious Commission 2012. Total arsenic concentration in Chinese rice was found to be ranging from 0.11 to 0.44 mg/kg and from 0.17 to 0.42 mg/kg at two moderately contaminated sites (Duan et al. 2017). Rice grown in Hunan province in China accumulates more arsenic and poses great health risk to local population (Ma et al. 2016). Moderate to high arsenic content in rice varieties have been reported from many countries of the world (Table 2.1).

Genetic differences among rice varieties greatly influence grain arsenic accumulation although heritability of the trait is not very high. Norton et al. 2014 analyzed grain arsenic contents of rice varieties grown at four different sites and found cultivar differences could explain about 40–50% of phenotypic variation. In a field experiment at moderately contaminated sites in China, 38 and 31% of the total variation was explained by cultivar differences over two respective years (Duan et al. 2017). Zhang et al. 2008 reported relatively high estimates of heritability for grain arsenic content. It is likely that the homogeneous environment in the glasshouse could have led to the high estimate in the experiment of Zhang et al. 2008. The prevailing consensus is that the inheritance of grain arsenic accumulation in rice is moderate in nature.

Arsenic accumulation in rice grains also varies with subpopulations. Generally, *aus* types tend to accumulate highest grain arsenic, while either tropical or temperate *japonica* types accumulate lowest grain arsenic (Norton et al. 2014). Local aromatic rice types of Indian subcontinent also accumulate very less arsenic in grains compared with high yielding varieties. Local aromatic rice has much lower arsenic accumulation factor of 0.2–0.4% compared with 0.9–1% of high yielding varieties (Sandhi et al. 2017). There is also a significant negative correlation between heading date and grain arsenic content in rice (Duan et al. 2017). Local aromatic rice types, which are long duration in nature, have delayed heading date. The basis of this relationship and the reasons for low accumulation by local aromatic rice remain unexplored.

Table 2.1 Arsenic content of rice grains of different countries

Country	Total arsenic range(mg/kg)	Mean (mg/kg)	References
Bangladesh	0.058–1.835	0.496	Meharg and Rahman (2003)
	0.040–0.270	0.136	Das et al. (2004)
	0.108–0.331	0.183	Duxbury et al. (2003)
	0.072–0.170	0.117	
	0.030–0.300	0.130	Williams et al. (2005)
	0.002–0.557	0.143	Rahman et al. (2009)
	0.020–0.330	0.130	Meharg et al. (2009)
	0.058–1.835	0.496	Meharg and Rahman (2003)
	0.040–0.910	–	Williams et al. (2006)
	0.040–0.920	–	
	0.038–0.073[a]	0.056	Rahman et al. (2014)
India	0.120–0.663	0.358	Chowdhury et al. (2001)
	0.041–0.605	0.232	Roychowdhury et al. (2003)
	0.079–0.546	0.033	
	0.180–0.310	0.070	Meharg et al. (2009)
China	0.020–0.460	0.140	Meharg et al. (2009)
	0.052–0.253	0.129	Ma et al. (2016)
	0.033–0.437	0.143	Li et al. (2015)
	0.065–0.274	0.114	Liang et al. (2010)
Thailand	0.060–0.500	0.150	Adomako et al. (2011)
	0.010–0.390	0.140	Meharg et al. (2009)
Vietnam	0.032–0.465	0.208	Phuong et al. (1999)
Taiwan	0.010–0.630	0.100	Lin et al. (2004)
	0.010–0.140	0.050	
France	0.090–0.560	0.280	
	0.147-0.065	0.335	Frouin et al. (2019)
Spain	0.050–0.820	0.200	Lin et al. (2004)
			Torres-Escribano et al. (2008)
Italy	0.070–0.330	0.150	Lin et al. (2004)
Egypt	0.010–0.580	0.050	
USA	0.030–0.660	0.250	
	0.026–1.000	0.210	Heitkemper et al. (2009)
Australia	0.188–0.438	0.270	Rahman et al. (2014)
Peru	0.068–0.345	0.167	Mondal et al. (2020)

[a]Aromatic rice

A considerable effect of environment and genotype- environment interaction towards grain arsenic accumulation is also recognized (Duan et al. 2017). In a field experiment in china, environment explained 13 and 33% of total variation over two respective years, while Genotype X Environment interaction explained between 25 and 31% of total variation over the same years (Duan et al. 2017). Environmental factors that affect arsenic accumulation include soil pH, organic carbon content and microbial activities in the vicinity of rice roots (Nagajyoti et al. 2010; Majumder et al. 2013). Arsenic concentrations in rice grains vary significantly with growing seasons as winter rice accumulate much higher arsenic in grains than wet season rice due to the use of contaminated ground water for irrigation.

2.3.2 *Quantitative Trait Loci Associated with Arsenic Accumulation*

Quantitative trait loci associated with arsenic accumulation in grains has been identified from several crosses in rice. The first quantitative trait locus associated with arsenic accumulation in roots, AsTol on chromosome 6, was identified from a cross between *indica* and *japonica* rice varieties *Bala* and *Azucena* (Dasgupta et al. 2004). Subsequently, Zhang et al. 2008 also used *indica* X *japonica* crosses to identify quantitative trait loci associated with arsenic accumulation in roots, shoots and grains of rice both at seedling stage and maturity. One quantitative trait locus on chromosome 8 was co-located for arsenic concentrations in grains at maturity and shoot phosphorus content at seedling stage. Two quantitative trait loci on chromosome 6, and one quantitative trait locus on chromosome 8 associated with dimethylarsinic acid concentration in grains have also been identified (Kuramata et al. 2013). These three quantitative trait loci could explain approximately 73% variation in grain dimethylarsinic acid concentrations.

Recently, Murugaiyan et al. 2019 used backcross breeding populations to identify quantitative trait loci associated with arsenic accumulation in roots and shoots of rice at seedling stage. Although many quantitative trait loci were identified, only a very few quantitative trait loci could explain greater than or equal to 10% phenotypic variance. A large number of 74 quantitative trait loci including 6 quantitative trait loci stable across years and/water treatment have also been reported (Norton et al. 2019). Several quantitative trait loci associated with grain arsenic accumulation have also been reported by Descalsota-Empleo et al. 2019. Main effect quantitative trait loci associated with arsenic phytotoxicity tolerance from an F2 population derived from a cross between two *indica* rice varieties are also reported (Syed et al. 2016). Major quantitative trait loci responsible for arsenic accumulation in rice are presented in Table 2.2.

Although there are few reports of quantitative trait loci associated with arsenic accumulation in recent years, nonetheless reports of quantitative trait loci associated with grain arsenic accumulation are scanty in rice. Mapping studies especially those involving *indica* X *indica* crosses are very rare and as such the present studies have failed to capture the arsenic associated quantitative trait loci diversity in *indica* rice. Many of the identified quantitative trait loci have only small additive effects and are also co-localized with those previously reported.

Table 2.2 Major quantitative trait loci (QTL) responsible for arsenic accumulation in rice. Note that reports of QTLs responsible for grain arsenic accumulation are scanty.

QTL	Trait	Chromoseme	Marker interval	LOD	PVE (%)	Additive Effect	References
qAsR8.1	AsR	8	487SNP_8_6057678	4.652	10.5069	11.7883	Murugaiyan et al. (2019)
qAS5.2	AsS	5	296SNP_5_16808642	5.6649	12.6433	1.3646	
qAsS6	AsS	6	336SNP_6_1768006	5.323	11.929	1.4867	
qDMAs6.1	DMA-grain	6	AE06000003	11.06	36	0.014	Kuramata et al. (2013)
qDMAs6.2		6	AE06001721	6.90	25	0.007	
qDMAs8		8	AE08000162	3.89	12	0.0089	
qAs2.1	As-grain	2	2215213-id2011813	5.56	14.8	0.0196	Descalsota-Empleo et al. (2019)
qAs3.1		3	id3000913-2519506	4.60	12.1	0.0181	
qAs3.2		3	2696289-2713838	3.4285	10.3	0.0141	
qAs4.1		4	4333768-4349423	4.63	12.2	0.0187	
qAs8.1		8	8808054-id8005525	6.63	18.2	0.0271	
qAs8.2		8	8882427-8886338	3.2937	21.1	0.0155	
qAs9.1		9	9627787-rd9002686	5.78	15.2	0.0222	
AsS	AsS	2	RM318-RM450	5.7	24.4	1.38	Zhang et al. (2008)
AsR	AsR	3	RM1350RM3919	3.6	18.2	11.18	
AsSe1	As-grain	6	RM3-RM162	3.1	26.3	0.081	
AsSe2		8	RM1111-RM310	5.5	35.2	0.092	

AsR arsenic concentrations in roots, *AsS* arsenic concentrations in shoots, *DMA-grain* dimethylarsinic acid concentrations in grains, *As-grain* arsenic concentrations in grain, *PVE (%)* percentage phenotypic variation explained, *LOD* statistically significant values of logarithm of odds

2.3.3 Insights from Transcriptome Analysis

Recently, some studies have elucidated the roles of transcription factors, micro and small RNAs including long non-coding RNAs (lncRNAs) in regulation of arsenic stress response in rice by analyzing transcriptomes of plants exposed to arsenic stress. *OsARM1*, an R2R3 MYB transcription factor regulates arsenic-associated transporter genes and root to shoot translocation of arsenic (Wang et al. 2017). *OsARM1* binds to the conserved MYB binding sites in promoter or genomic regions of *Lsi1*, *Lsi2*, and *Lsi6*, a close homolog of *Lsi2* in the rice genome. *OsARM1* is expressed in the phloem of vascular bundles in upper and basal nodes of rice and its expression is found to be enhanced in response to arsenite treatment. *OsWRKY28*, a group IIa *WRKY* gene is also upregulated by arsenate exposure. *OsWRKY28* is localized in the nuclei and is strongly expressed in root tips, lateral roots and reproductive organs in response to arsenate and other oxidative stress (Wang et al. 2018). In hydroponic experiments, loss of function mutation in *OsWRKY28* results in lower accumulation of arsenate and phosphate accumulation in rice shoots. Loss of function mutations of *OsWRKY28* also result in poor root growth in rice.

Comprehensive analysis of sRNA-seq data sets from arsenic stressed rice indicates a large number of small RNAs are activated in roots and shoots of rice under arsenic stress(Tang et al. 2019b). A large number of small RNAs are also intensively repressed by arsenic treatment. Many of the small RNAs have their loci on tRNAs, rRNAs or long non-coding RNAs (lncRNAs). Some of these small RNAs are enriched in AGO1 and functionally involved in the development, stress response, reproduction or lipid metabolism. A number of lncRNAs are also targeted by few AGO1-enriched small RNAs, indicating the potential involvement of the lncRNAs in arsenic signaling.

Osa-mi156j, a specific micro-RNA is also downregulated in root tissues at seedling, tillering and flowering stages during 0–72 h of arsenite exposure as compared with other tissues in rice (Pandey et al. 2020). The potential targets of Osa-mi156j include DNA binding transcription factors and DNA binding proteins. *Osa-mi156j* modulates several metabolic activities in rice during arsenite stress. Transgenic plants overexpressing another micro-RNA *miR528* are also more sensitive to arsenite compared with wild type rice (Liu et al. 2015). Micro RNAs could play important roles in regulation arsenic stress response in rice and other plants.

The current literature discussed here underscores the complex nature of genetic responses involved in regulation of arsenic stress response in rice. Further studies will show how transcription factors, micro-RNAs including lncRNAs may specifically regulate arsenic stress response in rice. Experimental validation of miRNAs relating to transcriptional pathways of heavy metal tolerance is required (Dubey et al. 2018). Particularly, identification and characterization of novel transcription factors and regulatory RNAs that regulate important arsenic transporters would be significant advance in the field.

2.4 Breeding for Low Grain-Arsenic Rice

In the future climate, under high temperature stress (38° C, 850 ppmv CO_2), the concentration of pore water arsenite in rhizosphere of rice is likely to increase leading to augmented accumulation of arsenic in rice grains, and the coupled stress of climate and soil arsenic may reduce rice yields up to 42% (Muehe et al. 2019). This loss in productivity may not be compensated by the elevated levels of atmospheric carbon dioxide in future climate. Critically, as tested under 38° C, 850 ppmv CO_2, soil arsenic alone may account for twice as much as yield loss as combined temperature and carbon dioxide changes; indicating soil arsenic may be a stronger determinant of yield than climate. Under these projections, developing low arsenic accumulating variety with high yield potential should be the ideal breeding goal. Therefore, breeders should focus on combining low arsenic accumulating trait with high yielding ability. Selecting rice varieties with less than 0.2 mg/kg inorganic grain-arsenic may significantly limit arsenic exposure through food; although efforts should be directed to further reduce arsenic content in grains given arsenic is a non-threshold class I human carcinogen. As both yield and arsenic accumulation are influenced by several loci across the genome and subject to considerable environmental variations, developing stable high yielding low arsenic accumulating variety could be a challenging goal.

Breeding low arsenic accumulating rice represents complex physiological and genetic challenges. Although *Lsi1* and *Lsi2* were previously suggested for developing low arsenic accumulating rice varieties(Ma et al. 2008), the lack of identification of novel mutant alleles that accumulate less arsenic without reduced ability to uptake silicon has limited their use in breeding. Silicon is a highly beneficial element for rice and application of silicon improves rice yield, lodging resistance and also disease and pest resistance (Ma et al. 2006). Moreover, *Lsi1* and *Lsi2* largely control arsenic accumulation in the xylem (Ma et al. 2008), which is not connected to the reproductive parts. Therefore, the actual contribution of *Lsi* transporters to grain arsenic accumulation may not be significant.

Recently, the understanding of arsenic accumulation in rice has improved greatly as important genes controlling arsenic accumulation have been characterized (Song et al. 2014; Shi et al. 2016; Xu et al. 2017; Das et al. 2017, 2018). Although many authors have suggested overexpression/underexpression strategies to minimize arsenic accumulation in grains, these strategies may result in nutrient imbalance as transporters of arsenic may also be involved in transport of other nutrients. A safer strategy would be to identify novel alleles of these genes from available germplasm and introgress them in high yielding backgrounds. It is important to note here that despite the presence of large variation in arsenic accumulation in rice germplasms, the allelic status of genes controlling arsenic uptake and accumulation in rice is largely unreported. Natural variation in alleles of genes controlling arsenic accumulation can be used to develop low arsenic accumulating lines. Identification and development of molecular markers associated with superior alleles may be used to accelerate breeding (Fig. 2.2).

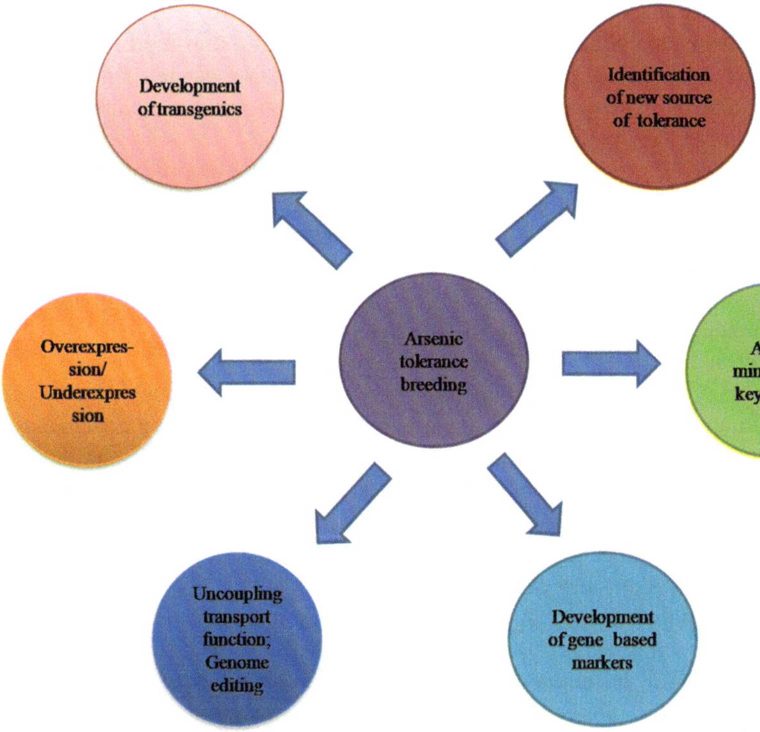

Fig. 2.2 Strategies for breeding low grain-arsenic accumulating rice varieties. Transgenics including overexpression/underexpression of genes involved in arsenic transport may help reduce arsenic accumulation in rice. However, some concern about the effect of overexpression/underexpression strategies on nutrient balance can not be ruled out. A safer alternative strategy would be identification of superior alleles of important genes and their introgression in elite genetic backgrounds. Identification of molecular markers such as gene based markers may greatly acclerate breeding without negative effect on nutrient balance and agronomic performance. Identification of new source of tolerance may facilitate identification of novel QTLs vis-à-vis novel genes. Uncoupling of transport function by genome editing such as CRISPR/Cas may allow engineering novel alleles that accumulate less arsenic without reduced uptake of essential nutrients

Identification of closely linked markers to quantitative trait loci associated with grain arsenic accumulation may also accelerate breeding. However, as discussed earlier, reports of quantitative trait loci associated with grain arsenic accumulation are scanty. Major quantitative trait loci with large additive variance that are stable across environments and years need to be identified. Identification of novel quantitative trait loci may facilitate the isolation of novel genes involved in arsenic accumulation. Validation studies in multiple populations and environments are important so that quantitative trait loci can be transferred to multiple diverse genetic backgrounds. Genome wide association studies and genomic selection should become a major advance in future.

Rice restricts arsenic accumulation in grains mainly by two ways, (a) exclusion of arsenite in extracellular space, either in root or shoot, and thus limiting arsenic in the vascular system and, (b) sequestration of arsenite in root and node vacuoles restricting arsenic upload in grains. Therefore, genes that determine shoot to grain translocation, especially those involved in detoxification and vacuolar sequestration such as *OsABCC1* could be valuable target in breeding experiments. *OsABCC1* reduces grain arsenic accumulation by more than 15 folds compared with its corresponding mutants (Song et al. 2014). New breakthroughs in breeding may also come from identification of genes responsible for phloem loading and grain unloading of arsenic as they are considered critical steps in arsenic accumulation in rice. Although many genes involved in arsenic accumulation are characterized and have been suggested for developing low arsenic accumulating rice, their effectiveness remain to be proven in breeding experiments. Therefore, studies which aim to validate quantitative trait loci and genes involved in arsenic accumulation in field experiments may help to prioritize breeding objectives by identifying important gene targets. Further, since several loci influence arsenic accumulation in rice, targeting more than one major gene that act in a synergistic manner may prove more effective than targeting a single gene in reducing arsenic accumulation.

Recently, few studies have demonstrated the potential of transgenics in reducing arsenic accumulation in rice. A novel fungal arsenic methyltransferase , *WaarsM* reduces grain arsenic accumulation in transgenic rice plants. WaarsM, from the soil fungus *Westerdyella aurantiaca* can convert toxic inorganic arsenicals to methylated arsenic species; thus reducing accumulation (Verma et al. 2018). Transgenic lines expressing *WaarsM* gwown in soil irrigated with arsenical containing water accumulated about 50% and 52% lower arsenic than non- transgenic plants in shoot and root respectively. Arsenic concentration in polished grains and husk was reduced by 52% (Verma et al. 2018). The improved arsenic resistance is provided via methylation and volatilization. However, Tang et al. 2016 transformed *A. thaliana* with *ArsM* from the eukaryotic alga *Chlamydomonas reinharditi* and found transgenic plants methylate most of inorganic arsenic to dimethylarsinic acid in shoots, resulting in greater phytotoxicity. Moreover, as discussed earlier the assumed less toxicity of methylated arsenic species is questionable (Chen et al. 2020). Therefore, the potential of reducing arsenic accumulation in transgenic plants expressing methylation related genes needs to be demonstrated by further studies.

Reduced arsenic accumulation in transgenic rice expressing two different arsenic vacuolar arsenic sequestration genes, *ScYCF1* and *OsABCC1* under the control of RCc3 promoter in the root cortical and internode phloem cells, along with a bacterial γ glutamylcysteine synthase driven by maize UBI promoter has also been observed (Deng et al. 2018). The transgenic plants exhibited reduced root to shoot and internode to grain translocation resulting in 70% reduction in arsenic accumulation in brown rice. The results also corroborate the findings of Song et al. 2014 indicating that vacuolar sequestration of arsenic is an important strategy to reduce arsenic accumulation in grains. Transgenic rice expressing *Ceratophylium demersum* phytochetalin synthase, *cdPCS1* also led to lower arsenic accumulation in grains (Shri et al. 2014). Transgenic plants expressing *cdPCS1* demonstrated increased phytochelatin synthase activity and enhanced synthesis of phytochelatins compared with non-transgenic plants. Transgenic plants expressing *cdPCS1* accumulated more arsenic in root and shoot but significantly lower arsenic in grains.

Recently, decreased root to shoot translocation of arsenite and shoot arsenic concentration was achieved by overexpressing two rice aquaporins *OsNIP1;1* and *OsNIP3;3* (Sun et al. 2018). The overexpressed *OsNIP1;1* and *OsNIP3;3* proteins were localized in the plasma membrane of root cells without polarity. When grown in arsenic contaminated soils, the overexpressed lines contained significantly lower arsenic in grains than wild type plants without adverse effect on plant growth and essential nutrient accumulation. Overexpression of *OsNIP1;1* and *OsNIP3;3* restricts arsenic loading into the xylem by allowing arsenic to leak out from the stele (Sun et al. 2018). However, knockout of either gene had little effect on arsenite uptake or translocation.

A potential issue with transgenic rice expressing arsenic related genes is achieving stable expression under field condition. Although the potential of transgenic technology in reducing arsenic accumulation in rice grain has been shown in controlled experiments, their effectiveness remains to be evaluated in field experiments. Regulatory issues may also significantly delay their introduction.

Since some arsenic transporters also transport other nutrients such as silicon and phosphate in rice, a radical way of engineering arsenic resistance could be by decoupling of transport function. Identification of important residues for arsenic transport which do not reduce silicon or phosphate transport could be used to engineer alleles that take up no or less arsenic from soil. Modern genome editing tools such as base editing by CRISPR/Cas may be successfully applied to engineer alleles with decoupled transport function and reduced arsenic accumulation (Zhu et al. 2020; Kersey et al. 2020). Although still in conceptual stage, the application of CRISPR/Cas in engineering heavy metal tolerance in plants may introduce new momentum in breeding.

2.5 Conclusion

Considerable advance in understating the physiology and genetics of arsenic uptake and accumulation has been made in rice over the recent years along with identification of genes such as *OsABCC1*, *OsABCC7*, *OsMATE2*, *OsPCS1*, and *OsPCS2* involved in arsenic uptake, transport, and vacuolar sequestration and detoxification. Identification of arsenate reductase such as *OsHAC1;1*, *OsHAC1;2* and *OsHAC4* has filled longstanding gaps in knowledge of arsenic speciation in plants. However, genes responsible for phloem loading and grain unloading, which are considered critical steps in arsenic accumulation in grains remain to be identified. Several other genes, in addition to those already known, may control arsenic uptake and accumulation, which also remain to be identified in near future.

Progress has also been made in elucidating the genetic basis of arsenic accumulation in rice with several quantitative trait loci underlying the trait being identified.

The roles of transcription factors, small and micro- RNAs including long non-coding RNAs (lncRNAs) have been elucidated in regulation of arsenic stress response in rice. Given the complex nature of genetic response to arsenic stress in rice, it is expected that future studies will shed more light on role of transcription factors, small RNAs and micro-RNAs including lncRNAs in regulation of arsenic stress response. Reports of large effect quantitative trait loci, especially quantitative trait loci with large additive effects, responsible for grain arsenic accumulation are still scanty. Identification of new source of tolerance may help identify novel quantitative trait loci vis-à-vis novel genes for arsenic accumulation.

Recent advances made in the understanding of physiology and genetics of arsenic accumulation are expected to accelerate the development of low grain arsenic accumulating varieties of rice. However, breeding low arsenic accumulating rice varieties represent complex physiological and genetic challenges. Identification of superior alleles of important genes and their molecular markers could lead to significant advance in breeding. Transgenic technology including overexpression/underexpression strategies is promising, although some concerns regarding their effect on nutrient balance and agronomic performance still remain. Achieving stable transgene expression under field condition could be an important issue, which will determine the success of this approach. Finally, modern tools of genome editing such base editing with CRISPR/Cas may be used to engineer novel alleles of genes to reduce arsenic accumulation in rice, although at present it remains in conceptual stage only.

Acknowledgements The authors are thankful to Indian Council of Agricultural Research - Niche Area of Excellence, Government of India for generous funding.

References

Adomako EE, Williams PN, Deacon C, Meharg AA (2011) Inorganic arsenic and trace elements in Ghanaian grain staples. Environ Pollut 159:2435–2442. https://doi.org/10.1016/j.envpol.2011.06.031

Ali W, Isayenkov SV, Zhao FJ, Maathuis FJ (2009) Arsenite (AsIII) transport in plants. Cell Mol Life Sci 66(14):2329–2239. https://doi.org/10.1007/s00018-009-0021-7

Anbar AD, Duan Y, Lyons TW, Arnold GL, Kendall B, Creaser RA, Kaufman AJ, Gordon GW, Scott C, Garvin J, Buick R (2007) A whiff of oxygen before the great oxidation event? Science 317(5846):1903–1906. https://doi.org/10.1126/science.1140325

Argos M, Kalra T, Pierce BL, Chen Y, Parvez F, Islam T, Ahmed A, Hasan R, Hasan K, Sarwar G, Levy D, Slavkovich V, Graziano JH, Rathouz PJ, Ahsan H (2011) A prospective study of arsenic exposure from drinking water and incidence of skin lesions in Bangladesh. Am J Epidemiol 174(2):185–194. https://doi.org/10.1093/aje/kwr062

Begum MC, Islam MS, Ismam M, Amim R, Parvez MS, Kabir AH (2016) Biochemical and molecular responses underlying differential arsenic tolerance in rice (Oryza sativa L.). Plant Physiol Biochem 104:266–277. https://doi.org/10.1016/j.plaphy.2016.03.034

Bibi S, Kamran MA, Sultana J, Farooqi A (2017) Occurrence and methods to remove arsenic and fluoride contamination in water. Environ Chem Lett 15:125–149. https://doi.org/10.1007/s10311-016-0590-2

Carey AM, Scheckel KG, Lombi E, Newville M, Choi Y, Norton GJ, Charnock JM, Feldmann J, Price AH, Meharg AA (2010) Grain unloading of arsenic species in rice. Plant Physiol 152(1):309–319. https://doi.org/10.1104/pp.109.146126

Carey AM, Norton GJ, Deacon C, Scheckel KG, Lombi E, Punshon T, Guerinot ML, Lanzirotti A, Newville M, Choi Y, Price AH, Meharg AA (2011) Phloem transport of arsenic species from flag leaf to grain during grain filling. New Phytol 192(1):87–98. https://doi.org/10.1111/j.1469-8137.2011.03789.x

Carey M, Meharg C, Williams P, Marwa E, Jiujin X, Gomes Farias J et al (2020) Global sourcing of low-inorganic arsenic rice grain. Expo Health 12:711. https://doi.org/10.1007/s12403-019-00330-y

Catarecha P, Segura MD, Franco-Zorrilla JM, García-Ponce B, Lanza M, Solano R, Paz-Ares J, Leyva A (2007) A mutant of the Arabidopsis phosphate transporter PHT1;1 displays enhanced arsenic accumulation. Plant Cell 19(3):1123–1133. https://doi.org/10.1105/tpc.106.041871

Chen J, Yoshinaga M, Rosen BP (2019) The antibiotic action of methylarsenite (AsIII) is an emergent property of microbial communities. Mol Microbiol 111(2):487–494. https://doi.org/10.1111/mmi.14169

Chen SC, Sun GX, Yan Y, Konstantinidis KT, Zhang SY, Deng Y, Li XM, Cui HL, Musat F, Popp D, Rosen BP, Zhu YG (2020) The great oxidation event expanded the genetic repertoire of arsenic metabolism and cycling. Proc Natl Acad Sci U S A 117(19):10414–10421. https://doi.org/10.1073/pnas.2001063117

Chowdhury UK, Rahman MM, Mandal B, Paul K, Lodh D, Biswas BK, Basu GK, Chanda CR, Saha KC, Mukherjee SC (2001) Groundwater arsenic contamination and human suffering in West Bengal, India and Bangladesh. Environ Sci 8:393–415. https://doi.org/10.1289/ehp.00108393

Codex Alimentarius Commission: proposed draft maximum levels for arsenic in rice (at step 3) (2012) http://usriceproducers.com/files/362_2012.01_WHO.Maximumlevelsforarsenic.pdf. Codex Alimentarius, Rome. Accessed April 2020

Crowe SA, Døssing LN, Beukes NJ, Bau M, Kruger SJ, Frei R, Canfield DE (2013) Atmospheric oxygenation three billion years ago. Nature 501(7468):535–538. https://doi.org/10.1038/nature12426

Das HK, Mitra AK, Sengupta PK, Hossain A, Islam F, Rabbani GH (2004) Arsenic concentrations in rice, vegetables, and fish in Bangladesh: a preliminary study. Environ Int 30(3):383–387. https://doi.org/10.1016/j.envint.2003.09.005

Das N, Bhattacharya S, Bhattacharyya S, Maiti MK (2017) Identification of alternatively spliced transcripts of rice phytochelatin synthase 2 gene OsPCS2 involved in mitigation of cadmium and arsenic stresses. Plant Mol Biol 94(1–2):167–183. https://doi.org/10.1007/s11103-017-0600-1

Das N, Bhattacharya S, Bhattacharyya S, Maiti MK (2018) Expression of rice MATE family transporter OsMATE2 modulates arsenic accumulation in tobacco and rice. Plant Mol Biol 98(1–2):101–120. https://doi.org/10.1007/s11103-018-0766-1

Dasgupta T, Hossain SA, Meharg AA, Price AH (2004) An arsenate (AsV) tolerance gene on chromosome 6 of rice. New Phytol 163(1):45–49. https://doi.org/10.1111/j.1469-8137.2004.01109.x

Deng F, Yamaji N, Ma JF, Lee SK, Jeon JS, Martinoia E, Lee Y, Song WY (2018) Engineering rice with lower grain arsenic. Plant Biotechnol J 16(10):1691–1699. https://doi.org/10.1111/pbi.12905

Descalsota-Empleo GI, Amparado A, Inabangan-Asilo MA, Tesoro F, Stangoulis J, Reinke R, Swamy BPM (2019) Genetic mapping of QTL for agronomic traits and grain mineral elements in rice. Crop J 7(4):560–572. https://doi.org/10.1016/j.cj.2019.03.002

Duan GL, Hu Y, Schneider S, McDermott J, Chen J, Sauer N, Rosen BP, Daus B, Liu Z, Zhu YG (2015) Inositol transporters AtINT2 and AtINT4 regulate arsenic accumulation in Arabidopsis seeds. Nat Plants 2(1):15202. https://doi.org/10.1038/nplants.2015.202

Duan G, Shao G, Tang Z, Chen H, Wang B, Tang Z, Yang Y, Liu Y, Zhao FJ (2017) Genotypic and environmental variations in grain cadmium and arsenic concentrations among a panel of high yielding rice cultivars. Rice 10(1):9. https://doi.org/10.1186/s12284-017-0149-2

Dubey S, Shri M, Gupta A, Rani V, Chakrabarty D (2018) Toxicity and detoxification of heavy metals during plant growth and metabolism. Environ Chem Lett 16:1169–1192. https://doi.org/10.1007/s10311-018-0741-8

Duxbury JM, Mayer AB, Lauren JG, Hassan N (2003) Food chain aspects of arsenic contamination in Bangladesh: effects on quality and productivity of rice. J Environ Sci Health A 38:61–69. https://doi.org/10.1081/ESE-120016881

Fakhraee M, Crowe SA, Katsev S (2018) Sedimentary sulfur isotopes and Neoarchean ocean oxygenation. Sci Adv 4(1):e1701835. https://doi.org/10.1126/sciadv.1701835

Frouin J, Labeyrie A, Boisnard A, Sacchi GA, Ahmadi N (2019) Genomic prediction offers the most effective marker assisted breeding approach for ability to prevent arsenic accumulation in rice grains. PLoS One 14(6):e0217516. https://doi.org/10.1371/journal.pone.0217516

Garg N, Singla P (2011) Arsenic toxicity in crop plants: physiological effects and tolerance mechanisms. Environ Chem Lett 9:303–321. https://doi.org/10.1007/s10311-011-0313-7

González E, Solano R, Rubio V, Leyva A, Paz-Ares J (2005) PHOSPHATE TRANSPORTER TRAFFIC FACILITATOR1 is a plant-specific SEC12- related protein that enables the endoplasmic reticulum exit of a high-affinity phosphate transporter in Arabidopsis. Plant Cell 17:3500–3512. https://doi.org/10.1105/tpc.105.036640

Heitkemper DT, Kubachka KM, Halpin PR, Allen MN, Shockey NV (2009) Survey of total arsenic and arsenic speciation in US-produced rice as a reference point for evaluating change and future trends. Food Addit Contam Part B Surveill 2(2):112–120. https://doi.org/10.1080/02652030903148298

Kamiya T, Tanaka M, Mitani N, Ma JF, Maeshima M, Fujiwara T (2009) NIP1;1, an aquaporin homolog, determines the arsenite (AsIII) sensitivity of *Arabidopsis thaliana*. J Biol Chem 284:2114–2120. https://doi.org/10.1074/jbc.M806881200

Kersey PJ, Collemare J, Cockel C, Das D, Dulloo EM, Kelly LJ, Lettice E, Malécot V, Maxted N, Metheringham C, Thormann I, Leitch IJ (2020) Selecting for useful properties of plants and fungi – novel approaches, oportunities, and challenges. Plants, People, Planet 2(5):409–420. https://doi.org/10.1002/ppp3.10136

Kuramata M, Abe T, Kawasaki A, Ebana K, Shibaya T, Yano M, Ishikawa S (2013) Genetic diversity of arsenic accumulation in rice and QTL analysis of methylated arsenic in rice grains. Rice 6(1):3. https://doi.org/10.1186/1939-8433-6-3

Li G, Sun GX, Williams PN, Nunes L, Zhu YG (2011) Inorganic arsenic in Chinese food and its cancer risk. Environ Int 37:1219–1225. https://doi.org/10.1016/j.envint.2011.05.007

Li X, Xie K, Yue B, Gong Y, Shao Y, Shang X, Wu Y (2015) Inorganic arsenic contamination of rice from Chinese major rice-producing areas and exposure assessment in Chinese population. Sci China Chem 58:1898–1905. https://doi.org/10.1007/s11426-015-5443-5

Li J, Pawitwar SS, Rosen BP (2016) The organoarsenical biocycle and the primordial antibiotic methylarsenite (AsIII). Metallomics 8(10):1047–1055. https://doi.org/10.1039/c6mt00168h

Liang F, Li Y, Zhang G, Tan M, Lin J, Liu W, Li Y, Lu W (2010) Total and speciated arsenic levels in rice from China. Food Addit Contam Part A Chem Anal Control Expo Risk Assess 27(6):810–816. https://doi.org/10.1080/19440041003636661

Lin H-T, Wong S-S, Li G-C (2004) Heavy metal content of rice and shellfish in Taiwan. J Food Drug Anal 12:167–174. https://doi.org/10.38212/2224-6614.2649

Liu WJ, Zhu YG, Hu Y, Williams PN, Gault AG, Meharg AA, Charnock JM, Smith FA (2006) Arsenic sequestration in iron plaque, its accumulation and speciation in mature rice plants (Oryza sativa L.). Environ Sci Technol 40(18):5730–5736. https://doi.org/10.1021/es060800v

Liu WJ, Wood BA, Raab A, McGrath SP, Zhao FJ, Feldmann J (2010) Complexation of arsenite (AsIII) with phytochelatins reduces arsenite (AsIII) efflux and translocation from roots to shoots in Arabidopsis. Plant Physiol 152:2211–2221. https://doi.org/10.1104/pp.109.150862

Liu Q, Hu H, Zhu L, Li R, Feng Y, Zhang L, Yang Y, Liu X, Zhang H (2015) Involvement of miR528 in the regulation of Arsenite (AsIII) tolerance in rice (*Oryza sativa* L.). J Agric Food Chem 40:8849–8861. https://doi.org/10.1021/acs.jafc.5b04191

Ma JF, Tamai K, Yamaji N, Mitani N, Konishi S, Katsuhara M, Ishiguro M, Murata Y, Yano M (2006) A silicon transporter in rice. Nature 440(7084):688–691. https://doi.org/10.1038/nature04590

Ma JF, Yamaji N, Mitani N, Xu XY, Su YH, McGrath SP, Zhao FJ (2008) Transporters of arsenite (AsIII) in rice and their role in arsenic accumulation in rice grain. Proc Natl Acad Sci U S A 105(29):9931–9935. https://doi.org/10.1073/pnas.0802361105

Ma L, Wang L, Jia Y, Yang Z (2016) Arsenic speciation in locally grown rice grains from Hunan Province, China: spatial distribution and potential health risk. Sci Total Environ 557–558:438–444. https://doi.org/10.1016/j.scitotenv.2016.03.051

Majumder A, Bhattacharyya K, Bhattacharyya S, Kole SC (2013) Arsenic-tolerant, arsenite (AsIII)-oxidising bacterial strains in the contaminated soils of West Bengal, India. Sci Total Environ 463–464:1006–1014. https://doi.org/10.1016/j.scitotenv.2013.06.068

Marschner H (1995) Mineral nutrition of higher plants, vol 2. Academic Press, London, pp 79–115

Meharg AA, Hartley-Whitaker J (2002) Arsenic uptake and metabolism in arsenic resistant and nonresistant plant species. New Phytol 154(1):29–43. https://doi.org/10.1046/j.1469-8137.2002.00363.x

Meharg AA, Rahman MM (2003) Arsenic contamination of Bangladesh paddy field soils: implications for rice contribution to arsenic consumption. Environ Sci Technol 37(2):229–234. https://doi.org/10.1021/es0259842

Meharg AA, Williams PN, Adomako E, Lawgali YY, Deacon C, Villada A, Cambell RC, Sun G, Zhu YG, Feldmann J, Raab A, Zhao FJ, Islam R, Hossain S, Yanai J (2009) Geographical variation in total and inorganic arsenic content of polished (white) rice. Environ Sci Technol 43(5):1612–1617. https://doi.org/10.1021/es802612a

Mitani N, Yamaji N, Ma JF (2008) Characterization of substrate specificity of a rice silicon transporter, Lsi1. Eur J Phys 456:679–686. https://doi.org/10.1007/s00424-007-0408-y

Mondal D, Periche R, Tineo B, Bermejo LA, Rahman MM, Siddique AB, Rahman MA, Solis JL, Cruz G (2020) Arsenic in Peruvian rice cultivated in the major rice growing region of Tumbes riverbasin. Chemosphere 241:125070. https://doi.org/10.1016/j.chemosphere.2019.125070

Mosa KA, Kumar K, Chhikara S, Mcdermott J, Liu Z, Musante C, White JC, Dhankher OP (2012) Members of rice plasma membrane intrinsic proteins subfamily are involved in arsenite (AsIII) permeability and tolerance in plants. Transgenic Res 21:1265–1277. https://doi.org/10.1007/s11248-012-9600-8

Muehe EM, Wang T, Kerl CF, Friedrich BF, Fendorf S (2019) Rice production threatened by coupled stresses of climate and soil arsenic. Nat Commun 10:4985. https://doi.org/10.1038/s41467-019-12946-4

Murugaiyan V, Ali J, Mahender A, Aslam UM, Jewel ZA, Pang Y, Marfori-Nazarea CM, Wu LB, Frei M, Li Z (2019) Mapping of genomic regions associated with arsenic toxicity stress in a backcross breeding populations of rice (Oryza sativa L.). Rice 12(1):61. https://doi.org/10.1186/s12284-019-0321-y

Nagajyoti PC, Lee KD, Sreekanth TVM (2010) Heavy metals, occurrence and toxicity for plants: a review. Environ Chem Lett 8:199–216. https://doi.org/10.1007/s10311-010-0297-8

Nahar MN, Inaoka T, Fujimura M (2014) A consecutive study on arsenic exposure and intelligence quotient (IQ) of children in Bangladesh. Environ Health Prev Med 19(3):194–199. https://doi.org/10.1007/s12199-013-0374-2

Norton GJ, Lou-Hing DE, Meharg AA, Price AH (2008) Rice–arsenate (AsV) interactions in hydroponics: whole genome transcriptional analysis. J Exp Bot 59(8):2267–2276. https://doi.org/10.1093/jxb/ern097

Norton GJ, Pinson SR, Alexander J, McKay S, Hansen H, Duan GL, Rafiqul Islam M, Islam S, Stroud JL, Zhao FJ, McGrath SP, Zhu YG, Lahner B, Yakubova E, Guerinot ML, Tarpley L, Eizenga GC, Salt DE, Meharg AA, Price AH (2012) Variation in grain arsenic assessed in a diverse panel of rice (Oryza sativa) grown in multiple sites. New Phytol 193(3):650–664. https://doi.org/10.1111/j.1469-8137.2011.03983.x

Norton GJ, Douglas A, Lahner B, Yakubova E, Guerinot ML, Pinson SR, Tarpley L, Eizenga GC, McGrath SP, Zhao FJ, Islam MR, Islam S, Duan G, Zhu Y, Salt DE, Meharg AA, Price AH (2014) Genome wide association mapping of grain arsenic, copper, molybdenum and zinc in rice (Oryza sativa L.) grown at four international field sites. PLoS One 9(2):e89685. https://doi.org/10.1371/journal.pone.0089685

Norton GJ, Travis AJ, Talukdar P, Hossain M, Islam MR, Douglas A, Price AH (2019) Genetic loci regulating arsenic content in rice grains when grown flooded or under alternative wetting and drying irrigation. Rice 12(1):54. https://doi.org/10.1186/s12284-019-0307-9

Pandey AK, Gedda MR, Verma AK (2020) Effect of arsenic stress on expression pattern of a rice specific *miR156j* at various developmental stages and their allied co-expression target networks. Front Plant Sci 11:752. https://doi.org/10.3389/fpls.2020.00752

Phuong TD, Chuong PV, Khiem DT, Kokot S (1999) Elemental content of Vietnamese rice. Part 1. Sampling, analysis and comparison with previous studies. Analyst 124(4):553–560. https://doi.org/10.1039/a808796b

Pommerrenig B, Diehn TA, Bernhardt N, Bienert M, Mitani-Ueno N, Fuje J, Bieber A, Spitzer C, Bräutigam A, Ma JF, Chaumont F, Bienert GP (2020) Functional evolution of nodulin 26-like intrinsic proteins: from bacterial arsenic detoxification to plant nutrient transport. New Phytol 225(3):1383–1396. https://doi.org/10.1111/nph.16217

Rahman MA, Hasegawa H (2011) High levels of inorganic arsenic in rice in areas where arsenic-contaminated water is used for irrigation and cooking. Sci Total Environ 409(22):4645–4655. https://doi.org/10.1016/j.scitotenv.2011.07.068

Rahman MM, Owens G, Naidu R (2009) Arsenic levels in rice grain and assessment of daily dietary intake of arsenic from rice in arsenic-contaminated regions of Bangladesh—implications to groundwater irrigation. Environ Geochem Health 31:179–187. https://doi.org/10.1007/s10653-008-9238-x

Rahman MA, Rahman MM, Reichman SM, Lim RP, Naidu R (2014) Arsenic speciation in Australian-grown and imported rice on sale in Australia: implications for human health risk. J Agric Food Chem 62(25):6016–6024. https://doi.org/10.1021/jf501077w

Riding R, Fralick P, Liang LY (2014) Identification of an Archean marine oxygen oasis. Precambrian Res 251:232–237. https://doi.org/10.1016/j.precamres.2014.06.017

Roychowdhury T, Tokunaga H, Ando M (2003) Survey of arsenic and other heavy metals in food composites and drinking water and estimation of dietary intake by the villagers from an arsenic-affected area of West Bengal, India. Sci Total Environ 308:15–35. https://doi.org/10.1016/S0048-9697(02)00612-5

Sandhi A, Greger M, Landberg T, Jacks G, Bhattacharya P (2017) Arsenic concentrations in local aromatic and high-yielding hybrid rice cultivars and the potential health risk: a study in an arsenic hotspot. Environ Monit Assess 189(4):184. https://doi.org/10.1007/s10661-017-5889-3

Schneider S, Schneidereit A, Konrad KR, Hajirezaei MR, Gramann M, Hedrich R, Sauer N (2006) Arabidopsis INOSITOL TRANSPORTER4 mediates high-affinity H+ symport of myoinositol across the plasma membrane. Plant Physiol 141(2):565–577. https://doi.org/10.1104/pp.106.077123

Schneider S, Schneidereit A, Udvardi P, Hammes U, Gramann M, Dietrich P, Sauer N (2007) Arabidopsis INOSITOL TRANSPORTER2 mediates H+ symport of different inositol epimers and derivatives across the plasma membrane. Plant Physiol 145(4):1395–1407. https://doi.org/10.1104/pp.107.109033

Seyfferth AL, Webb SM, Andrews JC, Fendorf S (2010) Arsenic localization, speciation, and co-occurrence with iron on rice (Oryza sativa L.) roots having variable Fe coatings. Environ Sci Technol 44:8108–8113. https://doi.org/10.1021/es101139z

Shi S, Wang T, Chen Z, Tang Z, Wu Z, Salt DE, Chao DY, Zhao FJ (2016) OsHAC1;1 and OsHAC1;2 function as arsenate (AsV) reductases and regulate arsenic accumulation. Plant Physiol 172(3):1708–1719. https://doi.org/10.1104/pp.16.01332

Shin H, Shin HS, Dewbre GR, Harrison MJ (2004) Phosphate transport in *Arabidopsis*: Pht1;1 and Pht1; 4 play a major role in phosphate acquisition from both low- and high-phosphate environments. Plant J39:629–642. https://doi.org/10.1111/j.1365-313X.2004.02161.x

Shri M, Dave R, Diwedi S, Shukla D, Kesari R, Tripathi RD, Trivedi PK, Chakrabarty D (2014) Heterologous expression of *Ceratophyllum demersum* phytochelatin synthase, *CdPCS1*, in rice leads to lower arsenic accumulation in grain. Sci Rep 4:5784. https://doi.org/10.1038/srep05784

Signes-Pastor AJ, Mitra K, Sarkhel S, Hobbes M, Burló F, de Groot WT, Carbonell-Barrachina AA (2008) Arsenic speciation in food and estimation of the dietary intake of inorganic arsenic in a rural village of West Bengal, India. J Agric Food Chem 56:9469–9474. https://doi.org/10.1021/jf801600j

Signes-Pastor AJ, Vioque J, Navarrete-Muñoz EM, Carey M, García-Villarino M, Fernández-Somoano A, Tardón A, Santa-Marina L, Irizar A, Casas M, Guxens M, Llop S, Soler-Blasco R, García-de-la-Hera M, Karagas MR, Meharg AA (2019) Inorganic arsenic exposure and neuropsychological development of children of 4–5 years of age living in Spain. Environ Res 174:135–142. https://doi.org/10.1016/j.envres.2019.04.028

Smith AH, Lingas EO, Rahman M (2000) Contamination of drinking-water by arsenic in Bangladesh: a public health emergency. Bull WHO 78:1093–1103

Smith AH, Lopipero PA, Bates MN, Steinmaus CM (2002) Public health: arsenic epidemiology and drinking water standards. Science 296:2145–2146. https://doi.org/10.1126/science.1072896

Song WY, Park J, Mendoza-Cózatl DG, Suter-Grotemeyer M, Shim D, Hörtensteiner S, Geisler M, Weder B, Rea PA, Rentsch D, Schroeder JI, Lee Y, Martinoia E (2010) Arsenic tolerance in Arabidopsis is mediated by two ABCC-type phytochelatin transporters. Proc Natl Acad Sci USA 107(49):21187–21192. https://doi.org/10.1073/pnas.1013964107

Song WY, Yamaki T, Yamaji N, Ko D, Jung KH, Fujii-Kashino M, An G, Martinoia E, Lee Y, Ma JF (2014) A rice ABC transporter, OsABCC1, reduces arsenic accumulation in the grain. Proc Natl Acad Sci U S A A111:15699–15704. https://doi.org/10.1073/pnas.1414968111

Su YH, McGrath SP, Zhao FJ (2010) Rice is more efficient in arsenite (AsIII) uptake and translocation than wheat and barley. Plant Soil 328:27–34. https://doi.org/10.1007/s11104-009-0074-2

Sun S-K, Chen Y, Che J, Konishi N, Tang Z, Miller AJ, Ma JF, Zhao F-J (2018) Decreasing arsenic accumulation in rice by overexpressing OsNIP1;1 and OsNIP3;3 through disrupting arsenite (AsIII) radial transport in roots. New Phytol 219:641–653. https://doi.org/10.1111/nph.15190

Syed MA, Iftekharuddaula KM, Mian MA, Khaleque RM, Golam R, Mustafizur GKM, Panaullah GM, Lauren G, Duxbury JM, Biswas PS (2016) Main effect QTLs associated with arsenic phyto-toxicity tolerance at seedling stage in rice (Oryza sativa L.). Euphytica 209:805–814. https://doi.org/10.1007/s10681-016-1683-5

Tang Z, Lv Y, Chen F, Zhang W, Rosen BP, Zhao FJ (2016) Arsenic Methylation in Arabidopsis thaliana expressing an Algal Arsenite (AsIII) methyltransferase gene increases arsenic phytotoxicity. J Agric Food Chem 64(13):2674–2681. https://doi.org/10.1021/acs.jafc.6b00462

Tang Z, Chen Y, Chen F, Ji YC, Zhao FJ (2017) OsPTR7 (OsNPF8.1), a putative peptide transporter in rice, is involved in dimethylarsenate (AsV) accumulation in rice grain. Plant Cell Physiol 58:904–913. https://doi.org/10.1093/pcp/pcx029

Tang Z, Chen Y, Miller AJ, Zhao F-J (2019a) The C-type ATP-binding cassette transporter OsABCC7 is involved in the root-to-shoot translocation of arsenic in rice. Plant Cell Physiol 60:1525–1535. https://doi.org/10.1093/pcp/pcz054

Tang Z, Xu M, Ito H, Cai J, Ma X, Qin J, Yu D, Meng Y (2019b) Deciphering the non-coding RNA-level response to arsenic stress in rice (*Oryza sativa*). Plant Signal Behav 14(9):1629268. https://doi.org/10.1080/15592324.2019.1629268

Tiwari M, Sharma D, Dwivedi S, Singh M, Tripathi RD, Trivedi PK (2014) Expression in Arabidopsis and cellular localization reveal involvement of rice NRAMP, OsNRAMP1, in arsenic transport and tolerance. Plant Cell Environ 37(1):140–152. https://doi.org/10.1111/pce.12138

Torres-Escribano S, Leal M, Vélez D, Montoro R (2008) Total and inorganic arsenic concentrations in rice sold in Spain, effect of cooking, and risk assessments. Environ Sci Technol 42(10):3867–3872. https://doi.org/10.1021/es071516m

Tseng CH (2007) Arsenic methylation, urinary arsenic metabolites and human diseases: current perspective. J Environ Sci Health C Environ Carcinog Ecotoxicol Rev 25(1):1–22. https://doi.org/10.1080/10590500701201695

Upadhyay N, Kar D, Deepak Mahajan B, Nanda S, Rahiman R, Panchakshari N, Bhagavatula L, Datta S (2019) The multitasking abilities of MATE transporters in plants. J Exp Bot 70(18):4643–4656. https://doi.org/10.1093/jxb/erz246

Uraguchi S, Tanaka N, Hofmann C, Abiko K, Ohkama-Ohtsu N, Weber M, Kamiya T, Sone Y, Nakamura R, Takanezawa Y, Kiyono M, Fujiwara T, Clemens S (2017) Phytochelatin synthase has contrasting effects on cadmium and arsenic accumulation in rice grains. Plant Cell Physiol 58(10):1730–1742. https://doi.org/10.1093/pcp/pcx114

Verma S, Verma PK, Meher AK, Bansiwal AK, Tripathi RD, Chakrabarty D (2018) A novel fungal arsenic methyltransferase, WaarsM reduces grain arsenic accumulation in transgenic rice (Oryza sativa L.). J Hazard Mater 344:626–634. https://doi.org/10.1016/j.jhazmat.2017.10.037

Wang P, Zhang W, Mao C, Xu G, Zhao FJ (2016a) The role of ospt8 in arsenate (AsV) uptake and varietal difference in arsenate (AsV) tolerance in rice. J Exp Bot 67(21):6051–6059. https://doi.org/10.1093/jxb/erw362

Wang L, Bei X, Gao J, Li Y, Yan Y, Hu Y (2016b) The similar and different evolutionary trends of MATE family occurred between rice and *Arabidopsis thaliana*. BMC Plant Biol 16:207. https://doi.org/10.1186/s12870-016-0895-0

Wang FZ, Chen MX, Yu LJ, Xie LJ, Yuan LB, Qi H, Xiao M, Guo W, Chen Z, Yi K, Zhang J, Qiu R, Shu W, Xiao S, Chen QF (2017) OsARM1, an R2R3 MYB transcription factor, is involved in regulation of the response to arsenic stress in rice. Front Plant Sci 8:1868. https://doi.org/10.3389/fpls.2017.01868

Wang P, Xu X, Tang Z, Zhang W, Huang XY, Zhao FJ (2018) OsWRKY28 regulates phosphate and arsenate (AsV) accumulation, root system architecture and fertility in rice. Front Plant Sci 9:1330. https://doi.org/10.3389/fpls.2018.01330

Williams PN, Price AH, Raab A, Hossain SA, Feldmann J, Meharg AA (2005) Variation in arsenic speciation and concentration in paddy rice related to dietary exposure. Environ Sci Technol 39:5531–5540. https://doi.org/10.1021/es0502324

Williams PN, Islam MR, Adomako EE, Raab A, Hossain SA, Zhu YG, Feldmann J, Meharg AA (2006) Increase in rice grain arsenic for regions of Bangladesh irrigating paddies with elevated arsenic in groundwaters. Environ Sci Technol 40(16):4903–4908. https://doi.org/10.1021/es060222i

Wu Z, Ren H, McGrath SP, Wu P, Zhao FJ (2011) Investigating the contribution of the phosphate transport pathway to arsenic accumulation in rice. Plant Physiol 157(1):498–508. https://doi.org/10.1104/pp.111.178921

Xu JM, Shi SL, Wang L, Tang Z, Lv TT, Zhu XL, Ding XM, Wang YF, Zhao FJ, Wu ZC (2017) OsHAC4 is critical for arsenate (AsV) tolerance and regulates arsenic accumulation in rice. New Phytol 215:1090–1101. https://doi.org/10.1111/nph.14572

Xu L, Polya DA, Li Q, Mondal D (2020) Association of low-level inorganic arsenic exposure from rice with age-standardized mortality risk of cardiovascular disease (CVD) in England and Wales. Sci Total Environ 743:140534. https://doi.org/10.1016/j.scitotenv.2020.140534

Yamaji N, Ma JF (2010) Further characterization of a rice silicon efflux transporter, Lsi2. Soil Sci Plant Nutr 57(2):259–264. https://doi.org/10.1080/00380768.2011.565480

Ye Y, Li P, Xu T, Zeng L, Cheng D, Yang M, Luo J, Lian X (2017) *OsPT4* contributes to arsenate (AsV) uptake and transport in rice. Front Plant Sci 8:2197. https://doi.org/10.3389/fpls.2017.02197

Yu HS, Liao WT, Chai CY (2006) Arsenic carcinogenesis in the skin. J Biomed Sci 13(5):657–666. https://doi.org/10.1007/s11373-006-9092-8

Yuan Y, Marshall G, Ferreccio C, Steinmaus C, Liaw J, Bates M, Smith AH (2010) Kidney cancer mortality: fifty-year latency patterns related to arsenic exposure. Epidemiology 21(1):103–108. https://doi.org/10.1097/EDE.0b013e3181c21e46

Zhang J, Zhu YG, Zeng DL, Cheng WD, Qian Q, Duan GL (2008) Mapping quantitative trait loci associated with arsenic accumulation in rice (Oryza sativa). New Phytol 177(2):350–355. https://doi.org/10.1111/j.1469-8137.2007.02267.x

Zhao FJ, Wang JR, Barker JHA, Schat H, Bleeker PM, McGrath SP (2003) The role of phytochelatins in arsenic tolerance in the hyperaccumulator *Pteris vittata*. New Phytol 159:403–410. https://doi.org/10.1046/j.1469-8137.2003.00784.x

Zhao FJ, Ma JF, Meharg AA, McGrath SP (2009) Arsenic uptake and metabolism in plants. New Phytol 181(4):777–794. https://doi.org/10.1111/j.1469-8137.2008.02716.x

Zhao FJ, Ago Y, Mitani N, Li RY, Su YH, Yamaji N, McGrath SP, Ma JF (2010) The role of the rice aquaporin Lsi1 in arsenite (AsIII) efflux from roots. New Phytol 186(2):392–399. https://doi.org/10.1111/j.1469-8137.2010.03192.x

Zhao FJ, Stroud JL, Khan MA, McGrath SP (2012) Arsenic translocation in rice investigated using radioactive 73As tracer. Plant Soil 350:413–420. https://doi.org/10.1007/s11104-011-0926-4

Zhao FJ, Zhu YG, Meharg AA (2013) Methylated arsenic species in rice: geographical variation, origin, and uptake mechanisms. Environ Sci Technol 47(9):3957–3966. https://doi.org/10.1021/es304295n

Zhu YG, Yoshinaga M, Zhao FJ, Rosen BP (2014) Earth abides arsenic biotransformations. Annu Rev Earth Planet Sci 42:443–467. https://doi.org/10.1146/annurev-earth-060313-054942

Zhu H, Li C, Gao C (2020) Applications of CRISPR–Cas in agriculture and plant biotechnology. Nat Rev Mol Cell Biol 21:661–677. https://doi.org/10.1038/s41580-020-00288-9

Chapter 3
Recent Advances in Speciation Analysis of Trace Antimony in Environmental and Biological Samples Based on Cloud Point Extraction and Spectrometric Methods

Ingrid Hagarová and Lucia Nemček

Abstract Antimony (Sb) is listed as a priority pollutant in many countries worldwide. Sb accumulates in the environment, and displays harmful effects. The toxical and biological properties of Sb vary depending on Sb chemical form and oxidation state. Here, Sb speciation analysis allows to distinguish the different Sb forms and oxidation states. Although Sb environmental science is increasing, Sb is still a poorly studied element in land and water ecosystems. Analytical methods for Sb include extraction, selective hydride generation, coprecipitation, chromatography, electrochemistry, kinetics and spectrometry. Cloud point extraction is the most widely used extraction method for reliable separation and preconcentration of one of the most toxic species of inorganic antimony, Sb(III). This extraction technique is based on phase separation when the temperature of the micellar solution of a nonionic surfactant is increased above a certain threshold value defined as the cloud point. There is little knowledge on cloud point extraction coupled to electrothermal atomic absorption spectrometry, hydride generation atomic absorption spectrometry, inductively coupled plasma optical emission spectrometry, inductively coupled plasma mass spectrometry and spectrophotometry.

Keywords Antimony · Biological samples · Cloud point extraction · Environmental samples · Extraction techniques · Preconcentration · Separation · Speciation analysis · Spectrometric methods · Surfactant

---

I. Hagarová (✉) · L. Nemček
Faculty of Natural Sciences, Comenius University in Bratislava, Bratislava, Slovak Republic
e-mail: ingrid.hagarova@uniba.sk; lucia.nemcek@uniba.sk

E. Lichtfouse (ed.), *Sustainable Agriculture Reviews 52*, Sustainable Agriculture Reviews 52, https://doi.org/10.1007/978-3-030-73245-5_3

Abbreviations

APDC	ammonium pyrrolidine dithiocarbamate
DDTC	diethyldithiocarbamate
DDTP	*O,O*-diethyl dithiophosphate
DLLME	dispersive liquid-liquid microextraction
DSPE	dispersive solid-phase extraction
ETAAS	electrothermal atomic absorption spectrometry
ETV-ICP-MS	electrothermal vaporization inductively coupled plasma mass spectrometry
ETV-ICP-OES	electrothermal vaporization inductively coupled plasma optical emission spectrometry
FAAS	flame atomic absorption spectrometry
GC-MS	gas chromatography mass spectrometry
HF-LPME	hollow fiber liquid phase microextraction
HG-AAS	hydride generation atomic absorption spectrometry
HG-AFS	hydride generation atomic fluorescence spectrometry
HS-SPME	headspace solid-phase microextraction
ICP-MS	inductively coupled plasma mass spectrometry
ICP-OES	inductively coupled plasma optical emission spectrometry
LPME	liquid-phase microextraction
SBME	solvent bar microextraction
SDME	single drop microextraction
SPME	solid-phase microextraction
TXRF	total reflection X-ray fluorescence
UA-IL-DLLME	ultrasound-assisted ionic liquid dispersive liquid-liquid microextraction

3.1 Introduction

Antimony (Sb) is a toxic metalloid with a wide range of industrial applications and annual world production of 18×10^4 tons (Jackson 2020). Consequently, considerable amounts of Sb are emitted to the environment every year and it is often present at elevated concentrations in each environmental compartment (Deng et al. 2020). There is a growing concern surrounding the possible adverse effects of Sb on natural ecosystems and human health, which leads to drafting of new analytical procedures for a reliable quantification of the element. The focus not only lies on quantification of the total concentration of Sb in different matrices, yet equally on reliable quantification of its individual species. Considering the speciation analysis of complex matrix samples, a combination of a suitable separation procedure with a reliable quantification method is required in order to deliver accurate results. A large variety of sample preparation procedures have been developed for separation and preconcentration of different Sb species, among which the extraction procedures

Fig. 3.1 Publications on Sb speciation in Scopus since January 2000. (Source: Scopus database; January 2020)

occupy an irreplaceable position in environmental and biological analysis (Gál et al. 2006). Primary techniques were based on liquid-liquid and solid-liquid extraction with different arrangements. Later on, procedures derived from cloud point extraction have been added to the repertoire of methods employed in a speciation analysis of Sb. For quantification of Sb, spectrometric methods, such as hydride generation atomic absorption spectrometry (HG-AAS), electrothermal atomic absorption spectrometry (ETAAS), inductively coupled plasma optical emission spectrometry (ICP-OES), inductively coupled plasma mass spectrometry (ICP-MS) and hydride generation atomic fluorescence spectrometry (HG-AFS) are amongst the most frequently used ones.

Since the year 2000, nearly 550 publications on speciation of Sb have been registered (Fig. 3.1). According to Scopus database information, on average 18 studies were published annually in the period of 2000–2009 and this number has doubled from 2010 to 2019. In comparison with arsenic, which it resembles both chemically and physically, antimony still remains the less studied element. A search on the Scopus database reveals that there are seven times more publications on speciation of arsenic than on speciation of antimony identified over the same period of interest. One reason might be that although belonging to the same periodic group, Sb appears to be much less toxic than its sister element compared on a molar basis (Maher 2009).

Whereas separation techniques based on liquid-liquid extraction and solid-phase extraction are well documented in the current literature (even for Sb speciation), the objective of this article is to provide a more detailed view on the recent cloud point extraction procedures combined with spectrometric methods that have been used in speciation analysis of Sb in environmental and biological samples.

3.2 Reviews on Antimony

The published literature contains different studies and comprehensive reviews on antimony-related topics. Some of these papers attempt to provide a clearer picture of natural biogeochemical cycles of Sb. Antimony is a global contaminant whose presence and continually increasing concentrations in each environmental compartment are of considerable interest. Nonetheless, many questions about the environmental chemistry of Sb still remain unanswered (Babula et al. 2008). It is important to note that the natural biogeochemical cycles of Sb need to be understood before an assessment of the impact of global contamination is made (Maher 2009). This is most definitely not possible without reliable quantification of total Sb and its species in all studied matrices. The literature reviewed in the course of this work can be further divided into the following sections: (1) antimony in the environment, (2) human exposure to antimony, (3) conventional and innovative techniques for antimony removal (4) different approaches to speciation of antimony and (5) antimony in bottled water. Papers dealing with particular fields are referred to in the appropriate sections below.

3.2.1 Antimony in the Environment

The key to understanding the behaviour of Sb in the environment lies in knowing its chemical form, concentration levels and speciation in various environmental compartments, its transformations, mobility, toxicity, possible interactions with other substances and many other aspects. Reviews on the presence of Sb in the environment are summarized in this section. Filella et al. (2002a, b, 2007) have focused their overviews mostly on Sb in natural waters. The authors discuss Sb occurrence (Filella et al. 2002a), its relevant solution chemistry (Filella et al. 2002b) and its relevant interactions with microbiota (Filella et al. 2007) in an aqueous environment. Their last paper on this topic (Filella et al. 2009) broadens the scope of the research on Sb behaviour in the different environmental compartments, its speciation, toxicity and cycling between compartments.

Smichowski (2008) aimed at providing a focused review on Sb present in volcanic ashes, fly ashes, road dust, airborne particulate matter, atmospheric aerosols and related matrices, and the methods for its reliable quantification in these types of samples. There is a lack of review papers on speciation analysis of Sb in atmospheric aerosols, most likely due to scarce information on this particular topic in the available literature. A possible reason might relate to the very low concentration of this element in atmospheric aerosols, which makes analytical determination of Sb in such matrices quite problematic.

The paper published by Oorts and Smolders (2009) reports a summary of the ecotoxicity data available for various Sb substances. Proceeding from the obtained data, they discuss the derivation of ecological threshold concentrations for Sb in freshwater and soils. However, ecotoxicity data are only accessible for a limited number of Sb substances. Furthermore, the information on the influence of abiotic factors on the bioavailability and toxic effects of Sb on either water or soil biota is also lacking.

Review by Tschan et al. (2009b) highlights that Sb is readily taken up by terrestrial plants, although little is known about the mechanism by which this occurs. The results of their original work (Tschan et al. 2009a) show that plant uptake of Sb increases linearly with Sb in solution or soluble Sb in soil over a wide range of concentrations until it is limited by toxicity. Antimony may thus be accumulated by plants on heavily contaminated soils at concentrations that may adversely affect human health. Later on, Feng et al. (2013) summarized studies in which possible mechanisms of plant uptake of Sb(V) and Sb(III) were described.

Natasha et al. (2019) bring a comprehensive look at a biogeochemical role of Sb in soil-plant system and the impact of Sb on human health related to consumption of crops and vegetables contaminated with this element. They aim to provide a conceivable link between exposure conditions, Sb chemical speciation, phytoavailability in soil, uptake and accumulation by plants, phytotoxicity, detoxification inside the plant and associated health risks.

Herath et al. (2017) present a critical overview of natural geochemical processes which trigger the mobilization of Sb from its host mineral constituents and related rocks to the surrounding areas. They underline the necessity of identification of regions and aquifer sediments most vulnerable to Sb-contamination in order to ensure the installation of safe drinking water wells. Recent advances in understanding the biogeochemical processes and ecological effects of Sb were summarized by He et al. (2019).

Article published by Vojteková et al. (2014) offers a quick overview on Sb levels in various environmental media, Sb chemistry and toxicity and interactions with organic and inorganic components. A short overview of chemical methods and procedures for speciation of Sb in different types of water samples, soils, sediments and fly ashes is also included. Recently, Bagherifam et al. (2019) published an interesting paper in which they evaluated and compared the central steps for derivation of soils, water and sediments regulatory guidelines with a focus on Sb. The authors point out that there are substantial variations in Sb toxicity guidelines (threshold values) in environmental systems that can be explained by the differences in geographical, ethnological, regulatory framework, scientific bases, site specific properties, selected software models and ecotoxicological criteria used for developing these guidelines. A number of factors suggest that there is a low to negligible likelihood for the potential global harmonization of all guideline values in the future.

3.2.2 Human Exposure to Antimony

One may be exposed to Sb by breathing air, drinking tap or bottled water, or eating food that contains Sb. Another possible route of exposure is skin contact with soil, water and other substances that contain Sb. Filella et al. (2009, 2012, 2013a, b) published several extensive reviews on human exposure to Sb. Their research largely focused on the Sb sources and its intake by humans (Belzile et al. 2011), Sb contents in some human tissues often used in biomonitoring (Filella et al. 2012), Sb contents in some human excreted biofluids (Filella et al. 2013a) and Sb concentration in human blood (Filella et al. 2013b). In their work, the authors present and critically discuss the literature information on Sb concentrations in water, air, food, hair, nails, teeth, urine, milk, saliva and human blood. Arain and Neitzel (2019) report on the risks associated with both formal and informal electronic waste (e-waste) recycling. Humans can be exposed to dangerous chemicals and (semi)metals, such as Sb, that are present in e-waste via multiple routes. Typically, exposure occurs by one of three exposure routes – ingestion, inhalation and dermal absorption (Nithya et al. 2020). The published data presented in reviewed studies showed that Sb concentrations in various types of biomarkers such as blood, hair and urine were consistently higher for occupationally exposed groups than for the reference and control groups.

3.2.3 Conventional and Innovative Techniques for Antimony Removal

Antimony is present in the environment as a result of natural processes and human activities both of which can lead to its increased concentration in the air, water, soils, sediments and other natural matrices (Escudero et al. 2019). Depending on the degree and nature of contamination, the effective removal techniques should be considered. Mubarak et al. (2015) provide a comprehensive overview of methods and procedures proposed for the removal of Sb from various contaminated matrices. Remediation techniques such as phytoremediation, bioremediation, coagulation-flocculation-sedimentation, adsorption, electrodialysis, ion exchange, chemical fixation, membrane separation and many others are included. Technologies stated above are also mentioned in a review published by Li et al. (2018). The authors show the weak points of particular removal procedures and conclude that new technologies are urgently needed to be developed and implemented to preserve sustainability of the environment. A closer look on As and Sb in wastewater and the options for their removal from this medium was taken by Ungureanu et al. (2015). The authors put a special emphasis on the adsorption and on the latest findings in the broad topic of alternative low-cost adsorbents. Deng et al. (2017) reviewed the application of iron-based materials for the removal of Sb from contaminated water. They discuss the interaction of Sb with the sorbents in relation to adsorption performance, influencing factors, mechanism, modelling of adsorption (isotherm, kinetic and thermodynamic models), advantages, drawbacks and the recent achievements in the field.

3.2.4 Different Approaches to Measure Speciation of Antimony

Reliable quantification of Sb species is an essential part of the studies aimed at elucidating Sb behaviour in various biogeochemical cycles. While speciation analysis of trace Sb in complex matrices is mostly based on combination of an efficient separation technique with a reliable quantification method, speciation analysis of inorganic antimony (iSb) in water samples relies on hydride generation process. In this case, optimization of the hydride generation process for selective quantification of Sb(III) is usually carried out and total iSb is quantified after an effective reduction of Sb(V) to Sb(III) (Ferreira et al. 2014). Strategies developed for speciation analysis of iSb using spectroanalytical techniques and hydride generation process were reviewed by Ferreira et al. (2014). Later on, Ferreira et al. (2019) published a summary of non-chromatographic and chromatographic procedures for speciation analysis of Sb in environmental matrices, where hydride generation atomic fluorescence spectrometry (HG-AFS) was used for Sb quantification.

The coupling of chromatographic techniques such as gas chromatography and high performance liquid chromatography with highly sensitive and selective quantification methods such as mass spectrometry, atomic fluorescence spectrometry, atomic absorption spectrometry or optical emission spectrometry has been widely exploited and accepted for the speciation analysis of trace Sb in complex matrices. Non-chromatographic techniques, mostly extractions, in combination with quantification methods listed above are also extensively used in speciation analysis of trace Sb in complicated matrices. The analytical methods for separation and/or quantification of Sb can be grouped in four main categories: chemical methods, chromatographic methods, electrochemical methods and kinetic methods (Smichowski et al. 1998).

Methodologies for the determination of total Sb in terrestrial environmental samples along with procedures for the speciation of Sb in aqueous solutions can be found in a review published by Nash et al. (2000). Krachler and Emons (2001a) focused on hyphenated instrumental techniques employed in speciation of both volatile and non-volatile Sb compounds. Related issues such as biomethylation, biological significance, occupational and environmental exposure were also discussed. Evaluation of selective hydride generation procedures, extraction techniques and chromatographic techniques in combination with atomic spectrometric methods for speciation of Sb in natural waters can be found in an article published by Hagarová and Kubová (2008). In the review on the speciation of As, Sb and Se published by Wu and Sun (2016), the performance of both chromatographic and non-chromatographic techniques was compared. Methods for quantification of total

Sb together with procedures suitable for its speciation, including electrochemical techniques, were reviewed by Chomisteková et al. (2016). Yu et al. (2019) aimed to update the recent research progress in the speciation analysis of trace As, Hg, Se and Sb in environmental and biological samples with an emphasis on the recent applications of high performance liquid chromatography hyphenated to atomic spectrometry or mass spectrometry.

3.2.5 *Antimony in Bottled Water*

Although some consider the bottled water industry a marketing trick of the century, there is no doubt that it has become a big business. According to The Guardian, the past two decades made clear that water in bottles has become the fastest growing drinks market in the world. Among the factors which might affect water quality, the type of material used for a bottle is very likely to alter the chemistry of water it holds. Regarding Sb contamination, elevated concentrations of Sb in water arise from the use of antimony trioxide (Sb_2O_3) as a catalyst in the manufacture of polyethylene terephthalate, the prevailing type of material used for water bottles (Shotyk and Krachler 2007). There are various catalysts that can be used in the synthesis, but Sb_2O_3 is probably the most suitable and widespread one. It has a high catalytic activity, does not colour the product and it is cost effective (Mihucz and Záray 2016). Bach et al. (2012) agree that Sb is the most relevant element being leached out of polyethylene terephthalate bottles, but according to the published data obtained from the studies reviewed the authors concluded that Sb concentration never exceeded specific migration limit established by European Union. The overview published by Diduch et al. (2011) shows the available information on levels of inorganic constituents (including Sb) and organic contaminants in bottled water samples in the context of sample preparation procedures and analytical techniques.

3.3 Extraction Procedures Used in Speciation Analysis of Antimony

According to IUPAC, "speciation analysis" is defined as analytical activities of identifying and measuring the quantities of one or more individual chemical species in a sample (Templeton et al. 2000). In the speciation analysis of (ultra)trace elements in complex matrices, a combination of a suitable separation technique with a reliable quantification method is usually necessary. Chromatographic techniques (such as gas chromatography or high performance liquid chromatography) are often coupled with highly sensitive and selective spectrometric methods. In this case, a major challenge is the remarkably high cost of instrumentation, which makes these methods almost inaccessible to a number of analytical laboratories. In an attempt to use cheaper alternatives that deliver reliable results, many analytical chemists focus on the development of new non-chromatographic separation techniques or modification of existing ones. Undoubtedly, extraction techniques remain the hot topic of research, particularly when sample containing only a few simple chemical forms is to be analysed (Pyrzynska 2020).

One of the latest trends in modification of traditional extraction procedures is towards miniaturization of analytical systems. Miniaturization and automation of the methodologies are the basic strategies for greening the analytical methods. In such arrangements, a strong reduction in energy and reagents consumption, waste generation and time taken is expected (Armenta et al. 2015). Traditional liquid-liquid extraction in its miniaturized form, so-called liquid-phase microextraction (LPME), was introduced by Rezaee et al. (2006). Depending on the separation mode, there are three main variants of LPME: single drop microextraction (SDME), dispersive liquid-liquid microextraction (DLLME) and membrane mediated liquid phase microextraction (such as hollow fiber liquid phase microextraction (HF-LPME) and solvent bar microextraction (SBME)) (Ibrahim et al. 2017; Nemček and Hagarová 2020). These techniques have many other modifications (Werner et al. 2018) and have been successfully applied to separation and preconcentration of various analytes, including (ultra)trace Sb and its species. The following provide some examples. A new sensitive SDME method for Sb(III) determination using N-benzoyl-N-phenylhydroxylamine and 1-butyl-3-methylimidazolium hexafluorophosphate has been presented by Huang et al. (2018). A simple and fast ultrasound-assisted ionic liquid DLLME (UA-IL-DLLME) method for preconcentration of trace Sb and Sn in beverage samples has been developed by Biata et al. (2017). Two modes of LPME, HF-LPME and DLLME have been investigated by Marguí et al. (2013) to preconcentrate Sb prior to its determination by total reflection X-ray fluorescence (TXRF) analysis.

The techniques based on liquid-phase microextraction are very diversified. Their modification often involves preconcentration based on the use of tailor-designed or carefully selected extractants, e.g. ionic liquids or low-melting-point organic solvents. However, some concerns were raised with regard to the usage of ionic liquids. This intensifies the search for new solvents such as deep eutectic solvents, which are commonly accepted in green separation processes and are now widely acknowledged as a new class of ionic liquids analogues because they share many characteristics and properties with ionic liquids (Smith et al. 2014). Regardless of the solvent used, the common advantages of all LPME techniques are lower volumes of solvents and less laborious operations in comparison to traditional liquid-liquid extraction, as well as lower waste generation.

Solid-phase extraction is a popular separation and preconcentration technique both in organic and inorganic analysis. The analytes are usually extracted on a solid sorbent by selective sorption. This technique allows sorbent to be packed inside the cartridges, syringe barrels, microcolumns or extraction disks. Owing to limitations of column arrangements, dispersive solid-phase extraction (DSPE) gains on popularity. There are various kinds of sorbents that can be used in DSPE methodology. In recent years, variety of nanomaterials have been evaluated as potential sorbents for the separation/preconcentration and speciation of various trace elements in a number of samples. The high surface area of nano-sized materials, their high chemical activity, high adsorption capacity, fast adsorption dynamics and good mechanical and chemical stability are all properties that are relevant for the development of effective DSPE procedures suitable for separation and preconcentration of various analytes (Hu et al. 2015). Considerable attention has been particularly directed to metallic nanoparticles, metal oxide nanoparticles, silica-based nanomaterials and carbon-based nanomaterials (Pyrzynska 2020). In regard to Sb speciation, Lopéz-García et al. (2017) described an interesting procedure that required magnetic particles covered with silver nanoparticles functionalized with the sodium salt of 2-mercaptoethanesulphonate. After separation of the solid by means of a magnetic field, the solid phase was directly introduced into ETAAS; or alternatively, the material was slurried and then injected into the atomizer. Speciation of Sb(III) and Sb(V) was achieved by means of two extractions carried out at different acidity.

Solid-phase microextraction (SPME), introduced by Arthur and Pawliszyn (1990), is a non-exhaustive technique based on the partition equilibrium of the analytes between the sample matrix and the extraction phase. In its most known configuration, the SPME device consists of an extraction phase coated onto a fused-silica rod; the extraction phase can be exposed directly to the sample media (direct immersion) or to its headspace (Souza-Silva et al. 2015). The application of headspace solid-phase microextraction (HS-SPME) for a direct analysis of volatile Sb compounds such as stibine (SbH_3), monomethylantimony, dimethylantimony and trimethylantimony was reported by Smith et al. (2002). The authors used SPME with polydimethylsiloxane fibres in combination with gas chromatography mass spectrometry (GC-MS). Another example of SPME involves the use of novel polystyrene oleic acid imidazole polymer in a micropipette tip of a syringe system and was described by Panhwar et al. (2018). Their simple, rapid and sensitive SPME in combination with ETAAS was used for speciation of iSb in mineral water, tap water and spring water samples and quantification of total Sb in acid digested soil and food samples.

In regard to SPME, various sorbents and configurations were described and used in different applications for a wide variety of sample matrices and analytes. This technique is considered a solvent-free because it is carried out without a solvent for isolation of the analytes. Another advantage is that it requires little or no solvent for the desorption step. Furthermore, most solid sorbents in SPME can be used repeatedly, thus greatly reducing waste generation and the cost of analysis. The possibility of an on-line set-up is an additional advantage that allows SPME to be performed at higher accuracy and precision, and lower labour consumption as a result of automation. Amongst obvious pros is also the miniaturized size and portability of SPME devices which greatly facilitates implementation of on-site sampling and optionally on-site analysis when coupled to portable analytical instruments. There is also a substantial reduction of errors associated with sample transportation and possible alterations during storage (Pérez-Fernández et al. 2017).

Cloud point extraction, first introduced by Watanabe (1974), is an excellent alternative to liquid-liquid extraction, where toxic organic solvents are replaced with nonionic surfactants. The phase separation is generally achieved by heating an aqueous solution containing nonionic surfactants, chelating reagents and target analytes over a critical temperature or using various additives to trigger the cloud point. Since the first work published, this extraction technique has gained popularity that is reflected in the thousands of papers and dozens of reviews published on this topic. Recent reviews on cloud point extraction are summarized by Gavazov et al. (2019).

Growing interest in Sb and its speciation in various matrices is illustrated by recent growth in the number of published studies concerning different analytical methods for the separation and quantification of Sb species. The extraction techniques are targeted in a fairly large group of studies as well. For this reason, it is impossible to mention all the papers on Sb speciation based on extraction procedures. Since procedures based on liquid-liquid extraction and solid-phase extraction (in their different miniaturized forms and modes) were discussed in the review recently published by Yu et al. (2019), the main objective of the present paper is to bring a comprehensive overview on recent cloud point extraction procedures used for separation and preconcentration of Sb in environmental and biological samples. Their relation to spectrometric methods is discussed further with several examples in the next section.

At this point, it could be useful to mention some Sb species which were identified and quantified in environmental and biological samples. In an aquatic environment, Sb predominantly exists in its inorganic forms as Sb(III) and Sb(V). In regard to methylantimony species, they have been found to be formed in both fresh and sea water. These organic species usually account for less than 10% of the total dissolved Sb (Krachler and Emons 2001a). Methylated Sb compounds were identified and quantified in biological samples (Krachler and Emons 2001b; Quiroz et al. 2011) as well as in soil and sediment samples (Krachler and Emons 2001b). The main Sb species identified in environmental and biological matrices can be seen in Table 3.1.

Table 3.1 Main antimony species commonly detected in environmental and biological samples

Species	Abbreviation	Chemical formula
Antimonite (Antimonous acid)	Sb(III)	$Sb(OH)_3$
Antimonate (Antimonic acid)	Sb(V)	$SbO(OH)_3$
Methylantimate (Methylantimonic acid)	MMSb	$CH_3SbO(OH)_2$
Dimethylantimate (Dimethylantimonic acid)	DMSb	$(CH_3)_2SbO(OH)$
Trimethylantimony dichloride	$TMSbCl_2$	$(CH_3)_3SbCl_2$
Trimethylantimony dihydroxide	$TMSb(OH)_2$	$(CH_3)_3Sb(OH)_2$
Trimethylantimony oxide	TMSbO	$(CH_3)_3SbO$

3.4 Speciation Procedures Based on Cloud Point Extraction and Spectrometric Quantification

Cloud point extraction is based on the following phenomenon: an aqueous solution of neutrally charged surfactant (nonionic or zwitterionic) at a concentration higher than critical micellar concentration becomes turbid and separates into two isotropic phases if an external condition is changed (e.g. temperature, pressure, pH or ionic strength). The surfactant solution becomes turbid because it attains the cloud point (i.e. incomplete solubilization) (Bezerra et al. 2005). At this point, the original surfactant solution separates into a surfactant phase of a small volume, which is rich in surfactant and contains an analyte trapped by micellar structures (so-called surfactant-rich phase) and a bulk diluted aqueous phase (so-called surfactant-poor phase or equilibrium solution).

The methodology for the separation and preconcentration of metallic ions can be briefly summarized as follows. In the first step, a chelating agent is usually added to a system to obtain the hydrophobic chelate which remains in the hydrophobic core of the micelles in surfactant-rich phase. This is followed by the addition of a surfactant and heating in a water bath to induce cloud point formation. In this moment, two isotropic phases are formed and their separation is accelerated by centrifugation. Afterwards, the system is usually cooled in an ice-bath in order to increase the viscosity of the surfactant-rich phase. The final step of cloud point extraction is decantation of an aqueous phase. The highly viscous surfactant-rich phase is obtained and conventionally diluted by methanol or ethanol solution of mineral acid (mainly HNO_3 or HCl). Such diluted sample is now ready for measurement. In order to increase the reliability of the procedure, some optimization of the experimental conditions needs to be done, including optimizing pH (which plays a crucial role in metal-chelate formation and subsequent extraction), concentration of a surfactant, concentration of a chelating agent, incubation temperature and time, and centrifugation time. In some cases, the addition of salts, alcohols, some other surfactants and some organic compounds may also be helpful (salting-out effect). Selection of a suitable diluting agent for decreasing viscosity of the surfactant-rich phase is another important step. In speciation studies, the experimental conditions must be optimized with respect to the selective separation and preconcentration of selected species. The particular steps of the speciation analysis of inorganic antimony (iSb) using cloud point extractions will be discussed further in this section.

Nearly all procedures described here are associated with the use of nonionic surfactant Triton X-114 (octyl phenoxy polyethoxy ethanol) as an extracting agent. In general, the majority (about 80%) of studies in which cloud point extraction procedures are employed for separation, preconcentration and speciation of inorganic analytes utilize this nonionic surfactant. It is due to its excellent physicochemical characteristics, such as low cloud point temperature (23–25 °C; which is particularly important for the extraction of thermally unstable metallic chelates) and high density of the surfactant-rich phase (1.052 g/ml; which facilitates phase separation

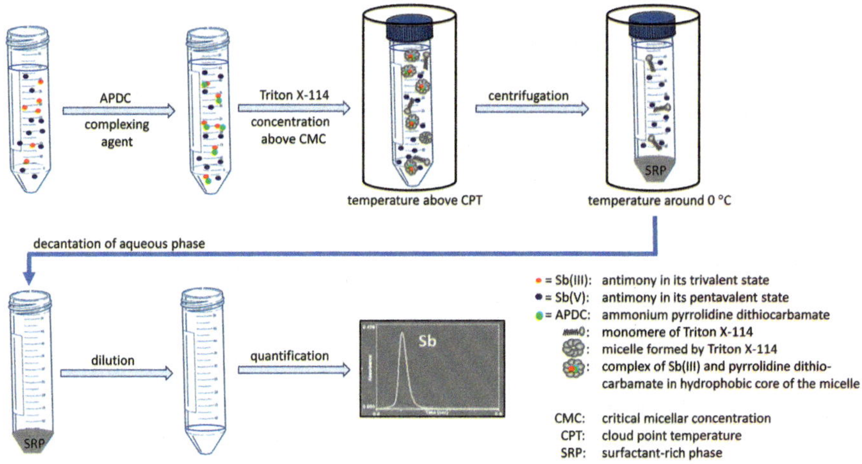

Fig. 3.2 Cloud point extraction developed for selective separation and preconcentration of Sb(III) in aqueous samples

by centrifugation). Triton X-114 is available in high purity at a relatively low price and shows low toxicity which favors it over other surfactants.

The majority of speciation studies dedicated to iSb using cloud point extraction requires optimization of the experimental conditions for selective separation and preconcentration of Sb(III). The total iSb is quantified by the same protocol after reduction of Sb(V) to Sb(III) using L-cysteine or iodide as reducing agents. Diagram showing the cloud point extraction adopted for selective separation and preconcentration of Sb(III) can be seen in Fig. 3.2. For quantification of Sb, spectrometric methods (such as flame atomic absorption spectrometry (FAAS), electrothermal atomic absorption spectrometry (ETAAS), hydride generation atomic absorption spectrometry (HG-AAS), inductively coupled plasma optical emission spectrometry (ICP-OES), inductively coupled plasma mass spectrometry (ICP-MS) or spectrophotometry) are usually used (Vojteková et al. 2014; Astolfi et al. 2006; Lebedev et al. 2003). Coupling of cloud point extraction with quantification methods listed above will be discussed in the following paragraphs.

3.4.1 Combination of Cloud Point Extraction and Flame Atomic Absorption Spectrometry

Since highly viscous surfactant-rich phase cannot be directly injected into the flame, a dilution of the surfactant-rich phase with a non-viscous solvent is required. The most commonly used eluents are dilute methanol or ethanol solutions of mineral acids. The effect of organic solvents on the flame in flame atomic absorption spectrometry (FAAS) has been a subject of several studies. It has been observed that low-surface-tension organic solvents may affect the nebulization process. There are reports indicating an appreciable enhancement in FAAS sensitivity, but at the same time, there are studies that found little or no sensitivity improvements in FAAS signal. Nevertheless, the application of low-surface-tension organic solvents is, in most cases, recommended (Stalikas 2002). Among these, methanolic or ethanolic solutions are proven to be efficient for lowering the viscosity of surfactant-rich phase.

One of the first researchers to apply cloud point extraction in the speciation analysis of inorganic antimony (iSb) in artificial seawater and wastewater was Fan (2005) and he did so in an attempt to extract and separate Sb(III) from an aqueous solution. The methodology was based on the formation of hydrophobic Sb(III)-N-benzoyl-N-phenylhydroxylamine complex trapped in a micellar phase of a nonionic surfactant. This complex was then extracted into the surfactant-rich phase at a temperature above the cloud point temperature, while Sb(V) remained in an aqueous phase. The concentration of Sb(III) in the surfactant-rich phase was then determined by FAAS. Total Sb was determined by the same procedure after reduction of Sb(V) to Sb(III) by L-cysteine; the amount of Sb(V) was calculated from the difference between values for total Sb and Sb(III).

Altunay and his co-workers published a series of papers where Victoria Pure Blue BO (Altunay and Gürkan 2015a), Pyronin B (Altunay and Gürkan 2015b), azomethine-H (Altunay and Gürkan 2016b), 2-(2-thiazolylazo)-p-cresol and 4-(2-thiazolylazo)resorcinol (Gürkan and Eses 2016) and morin (Altunay et al. 2016) were selected as complexing agents in cloud point extractions. The proposed procedures were tried out for separation and preconcentration of iSb species in various beverages, milk samples, fruit-flavored milk products and biological samples (such as blood serum and blood plasma). The authors highlight the benefits of these procedures which include minimum organic solvent consumption, experimental convenience and safety (for cloud point extraction), and fast and cheap analysis (for FAAS quantification). The combination of cloud point extraction and FAAS quantification was successfully utilized in their research work, showing good accuracy and precision, low quantification limits and high preconcentration factors in a wide linear range.

3.4.2 Combination of Cloud Point Extraction and Electrothermal Atomic Absorption Spectrometry

Since organic solvents and surfactants are both compatible with electrothermal atomic absorption spectrometry (ETAAS), no particular difficulties are anticipated in implementing ETAAS with cloud point extraction for element analysis. The contact angle of water with the surface of the graphite tube used in ETAAS is 85.7° while that of organic solvents is between 0 and 10°. Thus, the water-graphite system is expected to show shear behavior, whereas an organic solvent-graphite system has better interface compatibility. Both organic solvents and surfactants can efficiently decrease the contact angle between an aqueous solution and graphite material. These agents reduce the surface tension of liquid sample droplets allowing them to spread out evenly over the graphite surface before measurement. In order to use ETAAS for quantification of the target analyte following the cloud point extraction, the selection of an appropriate chemical modifier and an optimization of the furnace conditions need to be carried out.

To design a reliable cloud point extraction for selective quantification of Sb(III) in different types of natural waters, the experimental parameters were systematically investigated by Hagarová et al. (2008). In this study, ammonium pyrrolidine dithiocarbamate (APDC) was used as a complexing agent for Sb(III) separation and the ETAAS measurements were carried out in the presence of $Pd(NO_3)_2$ (used as a chemical modifier). In their later work, authors used the optimized procedure for systematic investigation of interferences caused by various coexisting ions and excess Sb(V) (Hagarová et al. 2012). While coexisting ions studied were found to have negligible effects on the extraction recovery rates, excess Sb(V) occurred to have a significant impact on recovery rates when Sb(V):Sb(III) ratio increased over 40:1; increasing the ratio led to increasing recovery rates, reaching as high as 160%. It is possible that certain amount of Sb(V) was reduced to Sb(III) which was then extracted.

The same complexing agent (APDC) was used in a procedure optimized by Jiang et al. (2010). Total Sb was determined after reduction of Sb(V) to Sb(III) using L-cysteine. The procedure was successfully applied for speciation of inorganic antimony (iSb) in leachates obtained from different food packaging materials. Andrade et al. (2017) optimized the cloud point extraction for the selective determination of Sb(III) in Brazilian mineral waters stored in polyethylene terephthalate bottles. Again, the preconcentration step was carried out by selective extraction of Sb(III) after its complexation with APDC, followed by separation of a surfactant-rich phase and subsequent quantification by ETAAS. The factors affecting separation and preconcentration were analyzed using univariate and multivariate approach. Optimization of an extraction procedure for selective separation and preconcentration of Sb(III) in water and blood serum samples using multivariate design was presented also by Souza and Tarley (2008). The cloud point extraction was based on the formation of Sb(III) complexes with *O,O*-diethyl dithiophosphate (DDTP) in the presence of Triton X-114. Total Sb was determined after reduction of Sb(V) to Sb(III) using L-cysteine. The latter two papers point out that using multivariate optimization reduces the number of experiments which results in lower reagent consumption, lower waste generation, and saving experimental time and cost.

3.4.3 Combination of Cloud Point Extraction and Hydride Generation Atomic Absorption Spectrometry

Hydride generation is probably the most popular technique for generation of volatile Sb compounds. In combination with atomic absorption spectrometry, HG-AAS offers fast measurement times and low detection limits for Sb in a wide variety of samples, including environmental and biological ones (Dědina 2007). It is based on the reaction of Sb compounds with tetrahydroborate in acid media to produce stibine (SbH_3). It is well-known that Sb(III) is reduced fast and easily in strongly acidic to near neutral media, while Sb(V) is only reduced at a low pH at much slower rates (Ferreira et al. 2014). This fact is of utmost importance in speciation studies of inorganic antimony (iSb) which require thorough control of experimental conditions for hydride generation. By setting the conditions (e.g. optimization of tetrahydroborate concentration, presence of acid) under which HG-AAS measurement is taken, it is possible to distinguish Sb(III) from total iSb. However, this sample pretreatment technique (which involves the use of conventional quartz tube atomizers) does not allow for sample preconcentration. An optimized cloud point extraction has emerged as a useful method not only for separation but also for the preconcentration of analytes. Complications associated with the procedure of cloud point extraction and hydride generation systems may arise from the presence of surfactant; it can promote foam formation during the hydride transport, thus making the release of hydride from solution more difficult (Bezerra et al. 2005).

In 2016, Altunay and Gürkan (2016a) proposed a simple and cost efficient method for simultaneous separation and preconcentration of iSb and Se in water, beverage and food samples by means of ultrasound-assisted cloud point extraction followed by their determination by HG-AAS analysis. The method involved the use of PONPE 7.5 as an extractant and neutral red as a complexing agent for Sb(III) and Se(IV). The authors investigated the main factors affecting extraction efficiency and the behavior of analytes during the hydride formation.

3.4.4 Combination of Cloud Point Extraction and Inductively Coupled Plasma Optical Emission Spectrometry

Introducing organic solvents into plasma-based techniques may give rise to a number of operating difficulties. Their adverse effects on various parameters including excitation conditions, plasma stability, nebulizer flows and torch dimensions have been well recognized and documented. As far as the presence of surfactants in the plasma was concerned, the reported effects on sample transport and analytical sensitivity were little or modest (Stalikas 2002). In case of plasma-based methods developed for an element quantification after cloud point extraction, the surfactant-rich phase is usually diluted directly by mineral acids.

Electrothermal vaporization is one of the sample introduction techniques employed in ICP-OES and ICP-MS that was reported to improve sample consumption, transport efficiency and absolute detection limit. An on-line cloud point extraction in combination with electrothermal vaporization inductively coupled plasma optical emission spectrometry (ETV-ICP-OES) for quantification of inorganic antimony (iSb) species in environmental and biological samples has been proposed by Li et al. (2006). Sb(III) forms a hydrophobic complex with ammonium pyrrolidine dithiocarbamate (APDC) and subsequently enters the surfactant-rich phase; surfactant-rich phase is then retained in a microcolumn packed with absorbent cotton. The surfactant-rich phase is then eluted with acetonitrile and determined by ETV-ICP-OES. In this particular case, APDC was employed not only as a chelating agent in cloud point extraction but also as a chemical modifier in ETV-ICP-OES. The method was applied to the speciation of iSb in water and urine samples. Utilizing the flow analysis has been shown to avoid the time-consuming process of batch procedures and to minimize waste. However, induction of the cloud point and retention of the surfactant-rich phase remain two of the most critical aspects of cloud point extraction coupled to flow injection analysis. Nevertheless, the search for solutions to technical difficulties arising from these two issues may lead to the invention and development of more time-efficient procedures.

The application of cloud point extraction and hydride generation for quantification of Sb and Se using inductively coupled plasma optical emission spectrometry (ICP-OES) was discussed by dos Santos Depoi and Pozebon (2012). The complexing agent used to separate both analytes was *O,O*-diethyl dithiophosphate (DDTP). Experimental conditions for selective separation and preconcentration of Sb(III) and Se(IV), reduction of Sb(V) and Se(VI), hydrides generation and interferences were investigated in detail to examine the possibilities of reliable and selective quantitative measurement of Sb(III) and Se(IV) in natural waters. Special attention was given to the performance of hydride generation and pneumatic nebulization, the two were compared with regard to their efficiency in introducing the surfactant-rich phase into the plasma. Samples tested for Sb(III) were taken from estuarine water, seawater near the Rio Grande do Sul's coast and mineral water stored in polyethylene terephthalate bottles. For the analysis of water samples and fruit juices, Dundar et al. (2018) used ICP-OES coupled to an ultrasonic nebulizer. Separation of Sb(III) from Sb(V) was achieved by complexation of Sb(III) with dithizone. The effects of coexisting ions as well as all experimental parameters were studied and optimized. Recently, Biata et al. (2019) presented an ultrasound-assisted cloud point extraction combined with ICP-OES as a tool for separation, preconcentration and quantification of Sb, Sn and Tl in food and water samples. The optimization of experimental conditions was carried out by means of fractional factorial design and response surface methodology. The complexing agent for the analytes was 1-2(-pyridylazo)-2-napthol and dissolution of the surfactant-rich phase was achieved by 10% HNO_3.

3.4.5 Combination of Cloud Point Extraction and Inductively Coupled Plasma Mass Spectrometry

From all the detection methods discussed above, inductively coupled plasma mass spectrometry (ICP-MS) is proven to be one of the most efficient and robust methods for element quantification. This technique is characterized by high sensitivity, large dynamic range, high capability of multi-element analysis and capability to perform isotopic measurements.

Coupling of flow injection analysis with ICP-ultrasonic nebulization represents a major breakthrough in the field of extraction technologies in terms of improvements in sample introduction into the plasma (without plasma fluctuation), transport and nebulization of the surfactant and organic phase (Silva et al. 2000). Other specific actions such as cooling of the nebulization spray chamber or use of a special plasma torch which can operate at a low argon flow rate can provide convenient remedies to possible issues arising from the presence of organic vapors in the plasma (Stalikas 2002). The combination of cloud point extraction and electrothermal vaporization inductively coupled plasma mass spectrometry (ETV-ICP-MS) is a powerful tool for simultaneous speciation of inorganic antimony (iSb) and Se in water samples (Li et al. 2008). Sb(III) and Se(IV) form complexes with diethyldithiocarbamate (DDTC) at pH 6.0 which can be quantitatively extracted into the surfactant-rich phase, whereas Se(VI) and Sb(V) remain in an aqueous solution. The proposed method is characterized with mild separation conditions (pH 6 is close to pH of natural water), simplicity, selectivity (good anti-disturbance ability), safety (no organic solvents) and low reagent cost.

3.4.6 Combination of Cloud Point Extraction and Spectrophotometry

In comparison with other spectrometric methods, spectrophotometry is still amongst most popular and attractive ones because of its simplicity, low-cost and compactness. Its main limitation is that it is not suitable for the measurement of (ultra)trace elements in complex matrices. If a reliable separation and preconcentration procedure is used, spectrophotometry can be successfully applied in (ultra)trace analysis.

Samadi-Maybodi and Rezaei (2012) reported on a new simple and versatile method for the cloud point extraction of ultra-trace amounts of Sb followed by spectrophotometric quantification of this element in seawater, human blood serum and Glucantime (anti-leishmania drug). The method was based on the color reaction of Sb(III) with iodide in acidic medium, in the presence of ascorbic acid as a reducing agent. An important detail about the experimental procedure is that researchers included a long thin neck container for extraction. Thanks to the special design of the vessel, no centrifugation and no cooling afterwards were required for separation of the extraction phase. As a result of different densities of the surfactant-rich phase and an aqueous phase, surfactant-rich phase separated as a top layer in a thin neck of the container. The surfactant-rich phase was then collected by a syringe-pipette and transferred to a quartz cuvette of the spectrophotometer. After adding methanol to the surfactant-rich phase in order to reduce viscosity, the absorbance of the solution was measured at the wavelength of 330 nm.

Cloud point extraction as a preconcentration step for Sb(III) detection prior to its spectrophotometric quantification was successfully tested also by El-Sharjawy and Amin (2016). The cloud point extraction was based on the reaction of Sb(III) with 3-dichloro-6-(3-carboxy-2-hydroxy-1-naphthylazo)quinoxaline in the presence of cetyltrimethylammonium bromide and KI at pH 4.5. The surfactant-rich phase was diluted with acetonitrile and the measurements were performed at 653 nm. The total amount of Sb in blood plasma, urine, water and biological samples was measured after Sb(V) to Sb(III) conversion using the same protocol.

A rapid, sensitive and more environmentally friendly analytical procedure utilizing cloud point extraction in a lab-in-syringe flow system aiming at the spectrophotometric quantification of Sb was developed by Frizzarin et al. (2016). The key feature of the method was the formation of an ion-pair between $[SbI_4]^-$ and H^+ which was subsequently extracted with Triton X-114. The multivariate optimization of some extraction parameters was carried out to identify interactions between the variables. The amount of total Sb in freshwater samples and antileishmanial drugs was determined after reduction of Sb(V) to Sb(III). The absorbance measurements were made at 345 nm. A lab-in-syringe flow system can be considered a revolutionary tool designed to make the procedure faster and easier; it allows for mechanization in all the steps involved in the analytical procedure, efficient mixing of multiple reagents solutions inside the syringe, and direct measurement in the surfactant-rich phase without the requirement for a filter or an eluent solution. It also reduces the risks of contamination. These characteristics make the proposed procedure more attractive compared to the existing batch or flow-based approaches.

3.5 Comparison of Analytical Characteristics of Procedures for Antimony Determination

The comparison of some experimental parameters of several reviewed extraction procedures can be seen in Table 3.2. For precision, relative standard deviations reported are in the range of 1.6–7.3% regardless of the detection method used and provide sufficient evidence of the reproducibility of the results. Traditional cloud point extractions generally involve a complex number of steps that must be undertaken to complete the task. After separation in the centrifuge, an aqueous phase has to be separated by decantation which is often considered the most problematic step. This step requires considerable attention so that reproducible results can be obtained. The reported relative standard deviations so far indicate that implementation of the optimized procedures would allow for precise and highly reproducible results.

Table 3.2 Comparison of some analytical characteristics of the methods proposed for Sb(III) separation, preconcentration and its quantitative determination

Sample	Detection system	Complexing reagent	Diluting agent	PF	LOD (μg/L)	RSD (%)	References
Waters	FAAS	BPHA	Met-OH	45	1.82	2.6	Fan (2005)
Beverages, blood plasma and serum	FAAS	VPB^{+}	THF	48	5.15	2.4	Altunay and Gürkan (2015a)
Beverages, milk samples	FAAS	Pyronin B	HNO_3	118	1.68	4.2	Altunay and Gürkan (2015b)
Beverages, dairy products	FAAS	Azomethine-H	Met-OH/ HNO_3	40	0.15	3.8	Altunay and Gürkan (2016b)
Beverages, milk, flavored milk products	FAAS	TAC	Met-OH/ HNO_3	60	0.28	3.5	Gürkan and Eser (2016)
Beverages, milk, flavored milk products	FAAS	TAR	Met-OH/ HNO_3	85	0.13	3.2	Gürkan and Eser (2016)
Beverages, food samples	FAAS	Morin	Et-OH/ HNO_3	115	0.03	5.2	Altunay et al. (2016)
Waters	ETAAS	APDC	Met-OH/ HNO_3	28	0.03	6.5	Hagarová et al. (2008)
Waters, blood serum	ETAAS	DDTP	Met-OH/ HNO_3	229	0.08	4.0	Souza and Tarley (2008)
Leachates from packaging materials	ETAAS	APDC	Met-OH/ HNO_3	NR	0.02	7.8	Jiang et al. (2010)
Waters	ETAAS	APDC	Met-OH/ HNO_3	108	0.02	5.2	Hagarová et al. (2012)
Waters	ETAAS	APDC	Met-OH/ HNO_3	12	0.06	5.3	Andrade et al. (2017)
Waters, urine samples	FI-ETV-ICP-OES	APDC	Acetonitrile	872	0.09	4.3	Li et al. (2006)
Waters, wine, nuts	HG-ICP-OES	DDTP	Met-OH/ HCl	5	0.08	NR	Depoi and Pozebon (2012)
Waters, fruit juices	ICP-OES	Dithizone	HNO_3	NR	0.04	2.6	Dundar et al. (2018)
Waters, food samples	ICP-OES	PAN	HNO_3	150	0.01	4.1	Biata et al. (2019)
Waters	ETV-ICP-MS	DDTC	Et-OH	50	0.03	4.2	Li et al. (2008)
Waters, blood serum, glucantime	UV/Vis	none	Met-OH	200	0.23	3.3	Samadi-Maybodi and Rezaei, (2012)

(continued)

Table 3.2 (continued)

Sample	Detection system	Complexing reagent	Diluting agent	PF	LOD (μg/L)	RSD (%)	References
Waters, blood plasma, urine samples	UV/Vis	DCHNAQ	Acetonitrile	50	0.05	1.8	El-Sharjawy and Amin (2016)
Waters, drugs	FI-UV/Vis	none	H_2SO_4	25	1.80	1.6	Frizzarin et al. (2016)
Waters, beverages, food samples	HG-AAS	NRH^+	HCl	120	0.0036	5.3	Altunay and Gürkan (2016a)

PF preconcentration factor, *LOD* limit of detection, *RSD* relative standard deviation, *NR* not reported, *BPHA* N-benzoyl-N-phenylhydroxylamine, *VPB*$^+$ Victoria Pure Blue BO, *TAC* 2-(2-thiazolylazo)-p-cresol, *TAR* 4-(2-thiazolylazo)resorcinol, *APDC* ammonium pyrrolidine dithiocarbamate, *DDTP O,O*-diethyldithiophosphate, *PAN* 1-2(-pyridylazo)-2-napthol, *DCHNAQ* 3-dichloro-6-(3-carboxy-2-hydroxy-1-naphthylazo)quinoxaline, *NRH*$^+$ neutral red, 3-amino-7-dimethylamino-2-methylphenazine hydrochloride, *Met-OH* methanol, *Et-OH* ethanol, *THF* tetrahydrofuran

The limit of detection is another important analytical parameter evaluated. It has been found to be highly dependent on the preconcentration factor (the higher the preconcentration factor the lower the limit of detection). For Sb(III) determined by cloud point extraction in combination with flame atomic absorption spectrometry, the obtained limits of detection were in the range of 0.03–5.15μg/l (Fan 2005; Altunay and Gürkan 2015a, b, 2016b; Gürkan and Eser 2016; Altunay et al. 2016). The lowest limit of detection reported for Sb(III) determined by such combination of techniques is comparable to limits of detection provided by combination of cloud point extraction and electrothermal atomic absorption spectrometry (range: 0.02–0.08μg/l) (Hagarová et al. 2008, 2012; Souza and Tarley 2008; Jiang et al. 2010; Andrade et al. 2017), cloud point extraction and inductively coupled plasma optical emission spectrometry or inductively coupled plasma mass spectrometry (range: 0.01–0.09μg/l) (Li et al. 2006, 2008; Depoi and Pozebon, 2012; Biata et al. 2019; Dundar et al. 2018). If the cost per analysis plays a critical role in deciding which detection method will be preferred for quantifying the analyte of interest, flame atomic absorption spectrometry seems to be the best choice.

A rapid, simple and low-cost detection method such as spectrophotometry has also been explored for its potential in speciation analysis of Sb after the selective separation and preconcentration of Sb(III) using optimized extraction procedures (Samadi-Maybodi and Rezaei 2012; El-Sharjawy and Amin 2016). From all methods listed in Table 3.2, the lowest limit of detection was obtained using cloud point extraction coupled to hydride generation atomic absorption spectrometry (Altunay and Gürkan 2016a). By means of hydride generation atomic absorption spectrometry, the hydride generation technique, the analyte can be separated from the liquid sample in the form of volatile hydride. The hydride generation efficiency can approach 100% under optimized experimental conditions and the limit of detection of the method is 20 times lower than the values reported for liquid sample nebulization when considering the efficiency of conventional nebulizers to be around 5% (Zurynková et al. 2018). The high preconcentration factor that can be obtained for Sb(III) using cloud point extraction leads to considerable improvement in a limit of detection, in particular when cloud point extraction is coupled to hydride generation atomic absorption spectrometry.

The remarkable outcomes were reported by Li et al. (2006) who applied the on-line flow-based cloud point extraction combined with electrothermal vaporization inductively coupled plasma optical emission spectrometry for Sb speciation in water and urine samples. In order to achieve the highest preconcentration factor attainable, 25, 50, 75, 100 and 150 ml volumes of sample were passed through the microcolumn at an optimum flow rate and different volumes of solutions were used for elution of the analyte. The preconcentration factor was 872 for analyte ions when the final sample volume was about 100 ml and the volume of acetonitrile was 100µl, which makes it probably the highest preconcentration factor ever obtained for inorganic Sb.

3.6 Conclusion

The majority of speciation studies on inorganic Sb that employ cloud point extraction methodology focus on the optimization of experimental conditions for selective separation and preconcentration of the analyte of interest. Of particular interest is the selective formation of a chelate-Sb(III) hydrophobic complex (at an optimal pH) which is subsequently retained in the hydrophobic cores of the micelles, while Sb(V) remains in an aqueous phase; thus, in the surfactant rich phase, only Sb(III) can be concentrated and separated. Total Sb is quantified by the same procedure after reduction of Sb(V) to Sb(III), and the amount of Sb(V) is calculated from the difference between the values for total Sb and Sb(III). In the case of relatively simple matrices such as natural waters, an error in the calculated concentrations of Sb(III) and Sb(V) seems to be negligible due to the fact that organic Sb species are identified very rarely in such types of samples.

Generally, only little attention is given to the stabilization of Sb species after water collection. Water samples are often analyzed directly with little or no pretreatment, samples for metal(loid) analysis are usually acidified with a drop of concentrated HNO_3. One of the few scientists who indicated that care and preservation of freshwater samples prior to analysis is a very important aspect of the sampling process were Frizarrin et al (2016). After the addition of ethylenediaminetetraacetic acid (EDTA) to stabilize the analyte, they stored the samples at 4 °C till further use. However, the majority of researchers agree that speciation analysis should be conducted as soon as possible after water collection for these species, with immediate analysis being preferred, in order to minimize the risk of oxidation of Sb(III) to Sb(V) before measurement.

Food and biological tissues must be decomposed prior to analysis. The most common way to solubilize biological sample is digestion with a concentrated acid.

The choice of acid is critical and so is reaction order, preferred acids being HNO_3 – $HClO_4$ mixture (Altunay and Gürkan 2015a), HNO_3 – HCl mixture (Altunay and Gürkan 2015b; Altunay et al. 2016), HNO_3 – H_2SO_4 – $HClO_4$ mixture (Altunay and Gürkan 2015a) and an external heat source. The mixture of HNO_3 and H_2O_2 (Altunay and Gürkan 2016a, 2016b; Gürkan and Eser 2016; Altunay et al. 2016; Depoi and Pozebon 2012; Biata et al. 2019) is commonly used in procedures carried out in closed vessels at elevated pressure, heated by microwave energy. After this step, Sb is present in its +5 oxidation state only and if optimized cloud point extractions were proposed for Sb(III), reduction of Sb(V) to Sb(III) must be done. For this purpose, a mixture of KI and ascorbic acid (Depoi and Pozebon 2012; El-Sharjawy and Amin 2016; Altunay and Gürkan 2016b; Gürkan and Eser 2016) or L-cysteine (Li et al. 2006, 2008; Souza and Tarley 2008; Altunay and Gürkan 2016a) are often used. By analysis of such decomposed samples only total Sb can be quantitatively determined and therefore the term "speciation analysis" should be avoided in this instance.

In traditional cloud point extractions, clouding is usually initiated by heating the mixture consisting of sample, complexing agent and surfactant in a water bath. The other type of energy which can be used to accelerate the reactions and clouding phenomena is ultrasound. The heat generated by the high-frequency sound waves enhances intensity and rate of interactions between the surfactant and an aqueous phase. The rise in popularity of ultrasound in chemistry can be attributed to its ease of use, low cost and green nature. Ultrasonic process plays an important role in an ultrasound-assisted cloud point extraction successfully applied in speciation analysis of Sb (Gürkan and Eser 2016; Altunay and Gürkan 2016b; Altunay et al. 2016; Biata et al. 2019). The faster the cloud point is induced, the more rapid extraction procedures can be envisioned for the future.

If duration of the procedure is an essential criterium for selecting the most appropriate pretreatment technique, then flow-based cloud point extraction seems to be a good choice (Li et al. 2006). A comprehensive review dedicated to cloud point extraction in flow-based systems was published by Melchert and Rocha (2016). This technique combines the advantages of flow analysis with that of cloud point extraction. The time-consuming steps such as centrifugation and cooling are omitted here. The use of cloud point extraction in a lab-in-syringe system can be considered a higher level of flow arrangement. The outstanding advantages of the method, such as mechanization of cloud point extraction, liquid-liquid extraction without organic solvents, mixing of reagent solutions with sample solution inside the syringe, reduction of the contamination risk and direct measurement of Sb in the surfactant-rich phase, make it rapid and environmentally friendly procedure. The pioneering exploitation of the lab-in-syringe system for cloud point extraction of Sb in water samples and antileishmanial drugs was discussed by Frizzarin et al. (2016).

Optimization of the cloud point extraction for reliable separation of the target analyte can take some time. A number of researchers use the univariate optimization strategy to find the best experimental conditions. This methodology is supposed to have an easier interpretation but requires more experiments which increase the consumption of reagents and time for analyses. Multivariate optimization is faster, more economical and effective, and allows for more than one variable to be optimized simultaneously. Several multivariate approaches such as fractional factorial design, Doehlert design and Box-Behnken design were exploited to identify interactions between the variables (Souza and Tarley 2008; Frizzarin et al. 2016; Biata et al. 2019).

There is no doubt that cloud point extraction represents a green analytical approach that limits or eliminates the use of toxic organic solvents. The optimization leads to development of new effective extraction procedures with high recovery rates and high preconcentration factors. As pointed out in the literature, cloud point extraction combined with spectrometric methods brings improvements in the limits of detection as well as in other analytical characteristics. The optimized cloud point extractions have a potential to be routinely used in many laboratories analysing environmental, biological, pharmaceutical and food samples for (ultra)trace organic and inorganic species. The potential of cloud point extraction for selective separation and preconcentration of Sb(III) has been documented by studies reviewed in this paper. Since the traditional cloud point extractions can be time- and labour-consuming and thus energy demanding, researchers are now investigating how to modify experimental conditions to ensure compliance with green chemistry principles. The processing time can be shortened by reducing the number of steps needed for a procedure (lab-in-syringe cloud point extraction); the consumption of chemicals can be reduced by miniaturization and/or using flow-based arrangement (flow-based cloud point extraction); an energy needed to accelerate the reaction and clouding phenomena can be produced in less environmentally harmful ways (ultrasound assisted cloud point extraction).

Acknowledgements The work was supported by the Scientific Grant Agency of the Ministry of Education, Science, Research and Sport of the Slovak Republic and the Slovak Academy of Sciences under the contract VEGA 1/0153/17.

References

Altunay N, Gürkan R (2015a) A new cloud point extraction procedure for determination of inorganic antimony species in beverages and biological samples by flame atomic absorption spectrometry. Food Chem 175:507–515. https://doi.org/10.1016/j.foodchem.2014.12.012

Altunay N, Gürkan R (2015b) A simple, inexpensive and convenient procedure for determination of inorganic Sb species in milk and beverage samples in PET containers by flame atomic absorption spectrometry. Anal Methods 7:9850–9860. https://doi.org/10.1039/C5AY01669J

Altunay N, Gürkan R (2016a) Separation/preconcentration of ultra-trace levels of inorganic Sb and Se from different sample matrices by charge transfer sensitized ion-pairing using ultrasonic-assisted cloud point extraction prior to their speciation and determination by hydride generation AAS. Talanta 159:344–355. https://doi.org/10.1016/j.talanta.2016.06.054

Altunay N, Gürkan R (2016b) Simultaneous determination of antimony and boron in beverage and dairy products by flame atomic absorption spectrometry after separation and pre-concentration by cloud point extraction. Food Addit Contam A 33:271–281. https://doi.org/10.1080/19440049.2015.1131335

Altunay N, Gürkan R, Yildirim E (2016) A new ultrasound assisted-cloud point extraction method for the determination of trace levels of tin and antimony in food and beverages by flame atomic absorption spectrometry. Food Anal Methods 9:2960–2971. https://doi.org/10.1007/s12161-016-0487-5

Andrade JK, de Andrade CK, Felsner ML, Quinia SP, dos Anjos VE (2017) Pre-concentration and speciation of inorganic antimony in bottled water and natural water by cloud point extraction with electrothermal atomic absorption spectrometry. Microchem J 133:222–230. https://doi.org/10.1016/j.microc.2017.03.043

Arain AL, Neitzel RL (2019) A review of biomarkers used for assessing human exposure to metals from e-waste. Int J Environ Res Public Health 16:1802. https://doi.org/10.3390/ijerph16101802

Armenta S, Garrigues S, de la Guardia M (2015) The role of green extraction techniques in Green Analytical Chemistry. TrAC Trends Anal Chem 71:2–8. https://doi.org/10.1016/j.trac.2014.12.011

Arthur CLP, Pawliszyn J (1990) Solid phase microextraction with thermal desorption using fused silica optical fibers. Anal Chem 62:2145–2148. https://doi.org/10.1021/ac00218a019

Astolfi ML, Canepari S, Catrambone M, Perrino C, Pietrodangelo A (2006) Improved characterisation of inorganic components in airborne particulate matter. Environ Chem Lett 3:186–191. https://doi.org/10.1007/s10311-005-0029-7

Babula P, Adam V, Opatrilova R, Zehnalek J, Havel L, Kizek R (2008) Uncommon heavy metals, metalloids and their plant toxicity: a review. Environ Chem Lett 6:189–213. https://doi.org/10.1007/s10311-008-0159-9

Bach C, Dauchy X, Chagnon MC, Etienne S (2012) Chemical compounds and toxicological assessments of drinking water stored in polyethylene terephthalate (PET) bottles: a source of controversy reviewed. Water Res 46:571–583. https://doi.org/10.1016/j.watres.2011.11.062

Bagherifam S, Brown TC, Fellows CM, Naidu R (2019) Derivation methods of soils, water and sediments toxicity guidelines: a brief review with a focus on antimony. J Geochem Explor 205:106348. https://doi.org/10.1016/j.gexplo.2019.106348

Belzile N, Chen YW, Filella M (2011) Human exposure to antimony. I. Sources and intake. Crit Rev Environ Sci Technol 41:1309–1137. https://doi.org/10.1080/10643381003608227

Bezerra AM, Arruda MAZ, Ferreira SLC (2005) Cloud point extraction as a procedure of separation and pre-concentration for metal determination using spectroanalytical techniques: a review. Appl Spectrosc Rev 40:269–299. https://doi.org/10.1080/05704920500230880

Biata NR, Nyaba L, Ramontja J, Mketo N, Nomngongo PN (2017) Determination of antimony and tin in beverages using inductively coupled plasma-optical emission spectrometry after ultrasound-assisted ionic liquid dispersive liquid-liquid phase microextraction. Food Chem 237:904–911. https://doi.org/10.1016/j.foodchem.2017.06.058

Biata NR, Mashile GP, Ramontja J, Mketo N, Nomngongo PN (2019) Application of ultrasound-assisted cloud point extraction for preconcentration of antimony, tin and thallium in food and water samples prior to ICP-OES determination. J Food Compos Anal 76:14–21. https://doi.org/10.1016/j.jfca.2018.11.004

Chomisteková Z, Culková E, Bellová R, Melicherčíková D, Durdiak J, Beinrohr E, Rievaj M, Tomčík P (2016) Methods and procedures for the determination of antimony as an environmentally important analyte. Chem List 110:671–677

Dědina J (2007) Atomization of volatile compounds for atomic absorption and atomic fluorescence spectrometry: on the way towards the ideal atomizer. Spectrochim Acta, Part B 62:846–872. https://doi.org/10.1016/j.sab.2007.05.002

Deng RJ, Jin CS, Ren BZ, Hou BL, Hursthouse AS (2017) The potential for the treatment of antimony–containing wastewater by iron-based adsorbents. Water 9:794. https://doi.org/10.3390/w9100794

Deng RJ, Tang ZE, Hou BL, Ren BZ, Wang ZH, Zhu CQ, Kelly S, Hursthouse A (2020) Microbial diversity in soils from antimony mining sites: geochemical control promotes enrichment. Environ Chem Lett 18:911–922. https://doi.org/10.1007/s10311-020-00975-1

Depoi SF, Pozebon D (2012) The use of cloud point extraction and hydride generation for improving the Sb and Se limits of detection in ICP OES. J Braz Chem Soc 23:2211–2221. https://doi.org/10.1590/S0103-50532012001200010
Diduch M, Polkowska Z, Namiesnik J (2011) Chemical quality of bottled waters: a review. J Food Sci 76:R178–R196. https://doi.org/10.1111/j.1750-3841.2011.02386.x
Dundar MS, Kaptan F, Caner C, Altundag H (2018) Speciation of antimony using dithizone ligand via cloud point extraction and determination by USN-ICP-OES. At Spectrosc 39:100–105. https://doi.org/10.46770/AS.2018.03.002
El-Sharjawy AAM, Amin AS (2016) Use of cloud point preconcentration for spectrophotometric determination of trace amounts of antimony in biological and environmental samples. Anal Biochem 492:1–7. https://doi.org/10.1016/j.ab.2015.08.008
Escudero LB, Quintas PY, Wuilloud RG, Dotto GL (2019) Recent advances on elemental biosorption. Environ Chem Lett 17:409–427. https://doi.org/10.1007/s10311-018-0816-6
Fan ZF (2005) Speciation analysis of antimony(III) and antimony(V) by flame atomic absorption spectrometry after separation/preconcentration with cloud point extraction. Microchim Acta 152:29–33. https://doi.org/10.1007/s00604-005-0426-4
Feng R, Wei C, Tu S, Ding Y, Wang R, Guo J (2013) The uptake and detoxification of antimony by plants: a review. Environ Exp Bot 96:28–34. https://doi.org/10.1016/j.envexpbot.2013.08.006
Ferreira SLC, dos Santos WNL, dos Santos IF, Junior MMS, Silva LOB, Barbosa UA, de Santana FA, Queiroz AFDS (2014) Strategies of sample preparation for speciation analysis of inorganic antimony using hydride generation atomic spectrometry. Microchem J 114:22–31. https://doi.org/10.1016/j.microc.2013.11.019
Ferreira SLC, dos Anjos JP, Assis Felix CS, da Silva Junior MM, Palacio E, Cerda V (2019) Speciation analysis of antimony in environmental samples employing atomic fluorescence spectrometry – review. TrAC Trends Anal Chem 110:335–343. https://doi.org/10.1016/j.trac.2018.11.017
Filella M, Belzile N, Chen YW (2002a) Antimony in the environment: a review focused on natural waters. I. Occurrence. Earth Sci Rev 57:125–176. https://doi.org/10.1016/S0012-8252(01)00070-8
Filella M, Belzile N, Chen YW (2002b) Antimony in the environment: a review focused on natural waters. II. Relevant solution chemistry. Earth Sci Rev 59:265–285. https://doi.org/10.1016/S0012-8252(02)00089-2
Filella M, Belzile N, Lett MC (2007) Antimony in the environment: a review focused on natural waters. III. Microbiota relevant interactions. Earth Sci Rev 80:195–217. https://doi.org/10.1016/j.earscirev.2006.09.003
Filella M, Williams PA, Belzile N (2009) Antimony in the environment: knows and unknows. Environ Chem 6:95–105. https://doi.org/10.1071/EN09007
Filella M, Belzile N, Chen YW (2012) Human exposure to antimony. II. Contents in some human tissues often used in biomonitoring (hair, nails, teeth). Crit Rev Environ Sci Technol 42:1058–1115. https://doi.org/10.1080/10643389.2011.556540
Filella M, Belzile N, Chen YW (2013a) Human exposure to antimony. III. Contents in some human excreted biofluids (urine, milk, saliva). Crit Rev Environ Sci Technol 43:162–214. https://doi.org/10.1080/10643389.2011.604257
Filella M, Belzile N, Chen YW (2013b) Human exposure to antimony. IV. Contents in human blood. Crit Rev Environ Sci Technol 43:2071–2105. https://doi.org/10.1080/10643389.2013.790741
Frizzarin RM, Portugal LA, Estela JM, Rocha FRP, Cerdà V (2016) On-line lab-in-syringe cloud point extraction for the spectrophotometric determination of antimony. Talanta 148:694–699. https://doi.org/10.1016/j.talanta.2015.04.076
Gál J, Hursthouse AS, Cuthbert SJ (2006) Chemical availability of arsenic and antimony in industrial soils. Environ Chem Lett 3:149–153. https://doi.org/10.1007/s10311-005-0022-1
Gavazov K, Hagarová I, Halko R, Andruch V (2019) Recent advances in the application of nanoparticles in cloud point extraction. J Mol Liq 281:93–99. https://doi.org/10.1016/j.molliq.2019.02.071

Gürkan R, Eser M (2016) Application of ultrasonic–assisted cloud point extraction/flame atomic absorption spectrometry (UA-CPE/FAAS) for preconcentration and determination of low levels of antimony in some beverage samples. J Iran Chem Soc 13:1579–1591. https://doi.org/10.1007/s13738-016-0874-2

Hagarová I, Kubová J (2008) Speciation of antimony in waters using separation coupled with atomic spectrometry. Chem List 102:782–790

Hagarová I, Kubová J, Matúš P, Bujdoš M (2008) Speciation of inorganic antimony in natural waters by electrothermal atomic absorption spectrometry after selective separation and preconcentration of antimony(III) with cloud point extraction. Acta Chim Slov 55:528–534

Hagarová I, Bujdoš M, Matúš P, Čanecká L (2012) The use of two extraction procedures in combination with electrothermal AAS for speciation of inorganic antimony in natural waters. Chem List 106:136–142

He M, Wang N, Long X, Zhang C, Ma C, Zhong Q, Wang A, Wang Y, Pervaiz A, Shan J (2019) Antimony speciation in the environment: recent advances in understanding the biogeochemical processes and ecological effects. J Environ Sci (China) 75:14–39. https://doi.org/10.1016/j.jes.2018.05.023

Herath I, Vithanage M, Bundschuh J (2017) Antimony as a global dilemma: geochemistry, mobility, fate and transport. Environ Pollut 223:545–559. https://doi.org/10.1016/j.envpol.2017.01.057

Hu B, He M, Chen B (2015) Nanometer-sized materials for solid-phase extraction of trace elements. Anal Bioanal Chem 407:2685–2710. https://doi.org/10.1007/s00216-014-8429-9

Huang XS, Guan MX, Lu ZLZ, Hang YP (2018) Determination of trace antimony(III) in water samples with single drop microextraction using BPHA-[C_4mim][PF_6] system followed by graphite furnace atomic absorption spectrometry. Int J Anal Chem 2018:8045324. https://doi.org/10.1155/2018/8045324

Ibrahim ASA, Al-Farawati R, Hawas U, Shaban Y (2017) Recent microextraction techniques for determination and chemical speciation of selenium. Open Chem 15:103–122. https://doi.org/10.1515/chem-2017-0013

Jackson T (2020) The periodic table: a visual guide to the elements. White Lion Publishing, London. ISBN 9781781319307

Jiang XM, Wen SP, Xiang GQ (2010) Cloud point extraction combined with electrothermal atomic absorption spectrometry for the speciation of antimony(III) and antimony(V) in food packaging materials. J Hazard Mater 175:146–150. https://doi.org/10.1016/j.jhazmat.2009.09.141

Krachler M, Emons H (2001a) Speciation of antimony for the 21st century: promises and pitfalls. TrAC Trends Anal Chem 20:79–90. https://doi.org/10.1016/S0165-9936(00)00065-0

Krachler M, Emons H (2001b) Urinary antimony speciation by HPLC-ICP-MS. J Anal At Spectrom 16:20–25. https://doi.org/10.1039/B007903K

Lebedev A, Sinikova N, Nikolaeva S, Poliakova O, Khrushcheva M, Pozdnyakov S (2003) Metals and organic pollutants in snow surrounding an iron factory. Environ Chem Lett 1:107–112. https://doi.org/10.1007/s10311-002-0004-5

Li YJ, Hu B, Jiang ZC (2006) On-line cloud point extraction combined with electrothermal vaporization inductively coupled plasma atomic emission spectrometry for the speciation of inorganic antimony in environmental and biological samples. Anal Chim Acta 576:207–214. https://doi.org/10.1016/j.aca.2006.06.018

Li YJ, Hu B, He M, Xiang GQ (2008) Simultaneous speciation of inorganic selenium and antimony in water samples by electrothermal vaporization inductively coupled plasma mass spectrometry following selective cloud point extraction. Water Res 42:1195–1203. https://doi.org/10.1016/j.watres.2007.09.002

Li J, Zheng B, He Y, Zhou Y, Chen X, Ruan S, Yang Y, Dai C, Tang L (2018) Antimony contamination, consequences and removal techniques: a review. Ecotoxicol Environ Saf 156:125–134. https://doi.org/10.1016/j.ecoenv.2018.03.024

Lopéz-García I, Rengevicova S, Muñoz-Sandoval MJ, Hernández-Córdoba M (2017) Speciation of very low amounts of antimony in waters using magnetic core-modified silver nanopar-

ticles and electrothermal atomic absorption spectrometry. Talanta 162:309–315. https://doi.org/10.1016/j.talanta.2016.10.044
Maher WA (2009) Antimony in the environment – the new global puzzle. Environ Chem 6:93–94. https://doi.org/10.1071/EN09036
Marguí E, Sagué M, Queralt I, Hidalgo M (2013) Liquid phase microextraction strategies combined with total reflection X–ray spectrometry for the determination of low amounts of inorganic antimony species in waters. Anal Chim Acta 786:8–15. https://doi.org/10.1016/j.aca.2013.05.006
Melchert WR, Rocha FRP (2016) Cloud point extraction in flow-based systems. Rev Anal Chem 35:41–52. https://doi.org/10.1515/revac-2015-0022
Mihucz VG, Záray G (2016) Occurrence of antimony and phthalate esters in polyethylene terephthalate bottled drinking water. Appl Spectrosc Rev 51:183–209. https://doi.org/10.1080/05704928.2015.1105243
Mubarak H, Chai LY, Mirza N, Yang ZH, Pervez A, Tariq M, Shaheen S, Mahmood Q (2015) Antimony (Sb) – pollution and removal techniques – critical assessment of technologies. Toxicol Environ Chem 97:1296–1318. https://doi.org/10.1080/02772248.2015.1095549
Nash MJ, Maskall JE, Hill SJ (2000) Methodologies for determination of antimony in terrestrial environmental samples. J Environ Monit 2:97–109. https://doi.org/10.1039/A907875D
Natasha SM, Khalid S, Dumat C, Pierart A, Niazi NK (2019) Biogeochemistry of antimony in soil–plant system: ecotoxicology and human health. Appl Geochem 106:45–59. https://doi.org/10.1016/j.apgeochem.2019.04.006
Nemček L, Hagarová I (2020) The recent strategies employed in chemical analysis of contaminated waters, sediments and soils as a part of the remediation process: extraction. In: Prasad R (ed) Environmental pollution and remediation. Springer, Singapore
Nithya R, Sivasankari C, Thirunavukkarasu A (2020) Electronic waste generation, regulation and metal recovery: a review. Environ Chem Lett. https://doi.org/10.1007/s10311-020-01111-9
Oorts K, Smolders E (2009) Ecological threshold concentrations for antimony in water and soil. Environ Chem 6:116–121. https://doi.org/10.1071/EN08109
Panhwar AH, Tuzen M, Hazer B, Kazi TG (2018) Solid phase microextraction method using a novel polystyrene oleic acid imidazole polymer in micropipette tip of syringe system for speciation and determination of antimony in environmental and food samples. Talanta 184:115–121. https://doi.org/10.1016/j.talanta.2018.03.004
Pérez-Fernández V, Rocca LM, Tomai P, Fanali S, Gentili A (2017) Recent advancements and future trends in environmental analysis: sample preparation, liquid chromatography and mass spectrometry. Anal Chim Acta 983:9–41. https://doi.org/10.1016/j.aca.2017.06.029
Pyrzynska K (2020) Nanomaterials in speciation analysis of metals and metalloids. Talanta 212:120784. https://doi.org/10.1016/j.talanta.2020.120784
Quiroz W, Arias H, Bravo M, Pinto M, Lobos MG, Cortés M (2011) Development of analytical method for determination of Sb(V), Sb(III) and TMSb(V) in occupationally exposed human urine samples by HPLC-HG-AFS. Microchem J 97:78–84. https://doi.org/10.1016/j.microc.2010.06.015
Rezaee M, Hosseini Y, Aghaee MMR, Ahmadi E, Sana BF (2006) Determination of organic compounds in water using dispersive liquid–liquid microextraction. J Chromatogr A 1116:1–9. https://doi.org/10.1016/j.chroma.2006.03.007
Samadi-Maybodi A, Rezaei V (2012) A cloud point extraction for spectrophotometric determination of ultra–trace antimony without chelating agent in environmental and biological samples. Microchim Acta 178:399–404. https://doi.org/10.1007/s00604-012-0852-z
Shotyk W, Krachler M (2007) Contamination of bottled waters with antimony leaching from polyethylene terephthalate (PET) increases upon storage. Environ Sci Technol 41:1560–1563. https://doi.org/10.1021/es061511+
Silva MAM, Frescura VLA, Curtius AJ (2000) Determination of trace elements in water samples by ultrasonic nebulization inductively coupled plasma mass spectrometry after cloud point extraction. Spectrochim Acta Part B 55:803–813. https://doi.org/10.1016/S0584-8547(00)00147-6

Smichowski P (2008) Antimony in the environment as a global pollutant: a review on analytical methodologies for its determination in atmospheric aerosols. Talanta 75:2–14. https://doi.org/10.1016/j.talanta.2007.11.005

Smichowski P, Madrid Y, Cámara C (1998) Analytical methods for antimony speciation in waters at trace and ultratrace levels. A review. Fresenius J Anal Chem 360:623–629. https://doi.org/10.1007/s002160050770

Smith LM, Maher WA, Craig PJ, Jenkins RO (2002) Speciation of volatile antimony compounds in culture headspace gases of *Cryptococcus humicolus* using solid phase microextraction and gas chromatography-mass spectrometry. Appl Organomet Chem 16:287–293. https://doi.org/10.1002/aoc.303

Smith EL, Abbott AP, Ryder KS (2014) Deep eutectic solvents (DESs) and their applications. Chem Rev 114:11060–11082. https://doi.org/10.1021/cr300162p

Souza OJM, Tarley CRT (2008) Preconcentration and speciation of Sb(III) and Sb(V) in water samples and blood serum after cloud point extraction using chemometric tools for optimization. Anal Lett 41:2465–2486. https://doi.org/10.1080/00032710802352522

Souza-Silva EA, Jiang R, Rodriguez-Lafuente A, Gionfriddok E, Pawliszyn J (2015) A critical review of the state of the art of solid-phase microextraction of complex matrices 1. Environmental analysis. TrAC Trends Anal Chem 71:224–235. https://doi.org/10.1016/j.trac.2015.04.016

Stalikas CD (2002) Micelle-mediated extraction as a tool for separation and preconcentration in metal analysis. TrAC Trends Anal Chem 21:343–355. https://doi.org/10.1016/S0165-9936(02)00502-2

Templeton DM, Ariese F, Cornelis R, Danielsson LG, Muntau H, Van Leeuwen HP, Łobiński R (2000) Guidelines for terms related to chemical speciation and fractionation of elements. Definitions, structural aspects, and methodological approaches (IUPAC Recommendations 2000). Pure Appl Chem 72:1453–1470. https://doi.org/10.1351/pac200072081453

Tschan M, Robinson BH, Nodari M, Schulin R (2009a) Antimony uptake by different plant species from nutrient solution, agar and soil. Environ Chem 6:144–152. https://doi.org/10.1071/EN08103

Tschan M, Robinson BH, Schulin R (2009b) Antimony in the soil – plant system – a review. Environ Chem 6:106–115. https://doi.org/10.1071/EN08111

Ungureanu G, Santos S, Boaventura R, Botelho C (2015) Arsenic and antimony in water and wastewater: overview of removal techniques with special reference to latest advances in adsorption. J Environ Manag 151:326–342. https://doi.org/10.1016/j.jenvman.2014.12.051

Vojteková V, Poperníková Z, Abusenaina AMM (2014) Antimony in various environment components. Chem List 108:135–140. https://doi.org/10.1007/s10311-010-0281-3

Watanabe H (1974) Spectrophotometric determination of cobalt with 1-(2-pyridylazo)-2-naphthol and surfactants. Talanta 21:295–302. https://doi.org/10.1016/0039-9140(74)80007-X

Werner J, Grzeskowiak T, Zgola-Grzeskowiak A, Stanisz E (2018) Recent trends in microextraction techniques used in determination of arsenic species. TrAC Trends Anal Chem 105:121–136. https://doi.org/10.1016/j.trac.2018.05.006

Wu D, Sun S (2016) Speciation analysis of As, Sb and Se. Trends Environ Anal Chem 11:9–22. https://doi.org/10.1016/j.teac.2016.05.001

Yu X, Liu C, Guo Y, Deng T (2019) Speciation analysis of trace arsenic, mercury, selenium and antimony in environmental and biological samples based on hyphenated techniques. Molecules 24:926. https://doi.org/10.3390/molecules24050926

Zurynková P, Dědina J, Kratzer J (2018) Trace determination of antimony by hydride generation atomic absorption spectrometry with analyte preconcentration/atomization in a dielectric barrier discharge atomizer. Anal Chim Acta 1010:11–19. https://doi.org/10.1016/j.aca.2018.01.033

Chapter 4
Groundwater Status and Challenges in Bangladesh

Md. Shajedul Islam and M. G. Mostafa

Abstract Bangladesh is a densely populated and developing country facing severe water pollution, crisis and security. Even though all population has access to freshwater, water is often contaminated as a result of over-exploitation, rapid urbanization, and pollution from industries, domestic and agricultural sources. Here, we review groundwater status and suitability for drinking and irrigation purposes. We present the status of groundwater aquifers, water quality, and sources of contamination. Groundwater is contaminated by elevated levels of dissolved Ca, Mg, Fe, Cr, and Pb in some areas. Arsenic is found in shallow aquifers at high levels throughout the country, and millions of consumers are suffering from severe and chronic poisoning from arsenic-enriched water. Moreover, groundwater is also contaminated by pathogens and agrochemicals. Additionally, in coastal areas, sea-level rise and increasing soil salinity make groundwater unsuitable for drinking and irrigation.

Keywords Bengal basin · Groundwater · Hand tube-well · Water quality parameters · Chemical contamination · Drinking water quality · Irrigation water quality · Arsenic · Agrochemicals · Water management

Abbreviations

ADB	Asian Development Bank
BADC	Bangladesh Agriculture Development Corporation
BBS	Bangladesh Burau of Statistics
BGS	British Geological Survey
BNDWQS	Bangladesh National Drinking Water Quality Standard
BWDB	Bangladesh Water Development Board
DDT	Dichlorodiphenyltrichloroethane
DPHE	Department of Public Health and Engineering

Md. S. Islam (✉) · M. G. Mostafa
Institute of Environmental Science, University of Rajshahi, Rajshahi, Bangladesh
e-mail: mgmostafa@ru.ac.bd

E. Lichtfouse (ed.), *Sustainable Agriculture Reviews 52*, Sustainable Agriculture Reviews 52, https://doi.org/10.1007/978-3-030-73245-5_4

DTW	Deep Tube-Well
ETP	Effluent Treatment Plant
FAO	Food and Agriculture Organization
GDP	Gross Domestic Product
GoB	Government of Bangladesh
HDL	High-Density Lipoprotein
IWM	Institute of Water Modeling
MICS-B	Multiple Indicator Cluster Survey-Bangladesh
NGO	Non-Government Organization
OXFUM	An International Organization
SDG	Sustainable Development Goal
STW	Shallow Tube-Well
UN	United Nations
UNDP	United Nations Development Program
UNICEF	United Nations International Children's Emergency Fund
US-EPA	United States-Environmental Protection Agencies
USGS	United States Geological Survey
WARPO	Water Resources Planning Organization
WB	World Bank
WHO	World Health Organization

4.1 Introduction

Water, especially groundwater, is the largest store of freshwater and vital resource for livings and the food safety of billions of people, and particularly in a thriving global agricultural economy (Freeze and Cherry 1979). About 97% of all unfrozen freshwater found beneath the earth surface as groundwater and globally, it offers about 50% of present drinkable water supplies, 40% of the industrial water demand, and 20% of the water used for irrigation purposes (UNESCO 2003; Zahid 2015). Though, some countries where fully depend on groundwater for drinking purposes. For instance, in Southeast Asian and Pacific nations, an average of 66% of households in municipal areas and 60% of households in rural areas rely on groundwater for drinking (Morris et al. 2003; Manasaki et al. 2018). Because of inadequate availability of surface water, which more worsen by contamination, industrialization, and urbanization, and by reason for the faith that groundwater is safer, most of the population in the Southeast Asian region depends on groundwater for drinking and household, irrigation, and industrial uses.

Groundwater present beneath the earth's surface in soil pore spaces and the fractures of rock formations. A unit of rock or an unconsolidated deposit is called an aquifer when it can yield a usable quantity of water. The depth at which soil pore spaces or fractures and voids in rock become completely saturated with water is called the water table. Groundwater recharged from the surface; it may discharge from the surface naturally at spring and seeps and can form oases or wetland.

Groundwater also often withdrawn for agriculture, municipal, and industrial use by constructing and operating extraction wells. This resource is often cheaper, more convenient, and less vulnerable to pollution than surface water. Therefore, it commonly used for public water supplies and groundwater provides the largest source of usable water storage in Bangladesh. But its quality, however, is failing at an alarming rate due to climate change and severe human activities, which poses substantial health risks to the public who drink it and take baths with it (Islam and Mostafa 2021a). Groundwater contamination occurs when undesirable harmful substances-often chemical or microbial-pollute aquifer water body, degrading water quality and rendering it toxic to humans or the environment. The contamination process of this resource is very difficult and sometimes impossible to clean up. The cent percent of peoples of Bangladesh use it for drinking purposes. So, it is very important to justify the quality of groundwater and find out the policy to remediate the challenges of groundwater management.

4.2 Groundwater Resources of Bangladesh

Bangladesh, a densely populated country of 2890 citizens/mile^2 with GDP of 274.03 b US dollars, is recently facing severe water pollution, crisis, and security (Fig. 4.1, BBS 2018). 76% of people do not have access to piped system and rely on hand tube wells, and 95% rural people fully depend on raw groundwater for domestic uses (ADB 2007b, c; Mojid et al. 2019). Even though 97% of the total population has access to freshwater, the purity of water is always disputed (WHO 2018).

This country is covered by alluvial sedimentary soil deposited by the Ganges, Brahmaputra, and Meghna rivers, making it one of the biggest deltas and widespread fertile cultivable land in the world (Ahmed et al. 2001). Bangladesh is a riverain country and carries 230 large and small rivers. Among those, 57 cross-boundary rivers, of which 54 comes from India and the remaining 3 shared with Myanmar (Chowdhury 2010). Bangladesh is situated at the bottom-most reaches of the Ganges, Brahmaputra, and Meghna river system which drains 1.72 million km^2 of land and though it contains only 8% of the watershed but has satisfactory surface water sources (Ahmad et al. 2001; Shamsudduha et al. 2009; Chowdhury 2010). Among the total water budget of Bangladesh, mostly, groundwater used for every purpose of the country needs such as irrigation, domestic and industrial, etc. In this country, about 32 km^3 of sub-surface water extracted annually, of which 90% uses in the agriculture sector, and 10% for household and industrial purposes collective that is equal to 4% of worldwide groundwater extraction (BADC 2013; Hanasaki et al. 2018; Shamsudduha et al. 2019).

The internal water resources projected to be 105,000 million cubic meters (Mm^3) in Bangladesh of which 84,000 Mm^3 of surface water formed within the country, and the rest of the volume is renewable groundwater resource (Rajmohan and Prathapar 2013). Bangladesh suffers from a hot monsoon climate with substantial variation in temperature and rainfall. The average rainfall varies from 1200 mm/

Fig. 4.1 Administrative map of Bangladesh

year in the west to over 4000 mm/year in the northeast part of the country (Chowdhury 2010). Approximately 80% of rainfall occurs during the monsoon (June to September) and in the winter period (December–February) and post-monsoon (October–November), only 20% of the rainfall offered. So, there is seasonal water-scarcity dependent on the period of the monsoon. In the winter-dry periods, over 70% of crop production are Boro rice, which can use up to 11,500 m^3 irrigated water per hectare, that comes from groundwater sources in the production process (Biswas and Mandal 1993).

In Bangladesh, throughout the dry time of the year, 58% of the country's available freshwater is to pay to irrigation, 41% used for aquaculture and navigation; while less than 1% used for domestic purposes (Chowdhury 2010). Besides, in the northern zone of the country, surface water deficiency and the easy access to groundwater through shallow tube-wells (STWs) has encouraged farmers to exploit groundwater for irrigation. Up until the 1970s, the government emphases the expansion of surface irrigation systems such as the Ganges-Kobadak (G-K) Project, Teesta Barrage (T-P) or Meghna-Dhonagoda (M-D) Projects. However, even with huge investment, achievement has been inadequate and only 7% of the total irrigable area covered by these schemes (Dey et al. 2017a). Currently, about 4.2 million hectares (M ha) of land are irrigated by groundwater however only 1.03 M ha by surface water (BADC 2018; CSISA-MI 2015).

Strictly for Bangladesh, groundwater is a life-sustaining reserve that plays a vital part in the irrigated agriculture sector and impacts the diversity of ecosystems. But the overexploitation of groundwater has been documented on both countrywide and regional scales. In current years, huge agricultural activities, over-exploitation, increasingly climate change, uneven rainfall, domestic and industrial discharge, and mismanagement of groundwater in that country have raised serious concern regarding groundwater pollution. The UN's report points out there are 80% human diseases caused by unsafe drinking water in a poor areas and globally every day about 25,000 people died from insecure drinking water (WB Group 2019). In addition, the report from WHO shown one-third people are living in the city of developing countries could not get harmless drinking water in the world (Wang 2006; Luo 2008). These reports are not easeful for Bangladesh. As the new epoch of the UN's SDGs has begun and Bangladesh committed to access to 'improved and safe drinking water supply' to address water quality (SDGs 6.3: *progress water quality by reducing pollution*) and water sustainability (SDGs 6.4: *ensure sustainable exploitation and supply of freshwater to address water scarcity*). Besides this, the government has clearly articulated its promise to '*ensure access to safe drinking water for all populations of Bangladesh'* in its 7th 5-year plane (2016–2020). For the achievement of these goals, an integrated multi-hazard groundwater risk model is crucially needed to guide water resources officials and policy-makers of the country (MICS-B 2018; WB Group 2019).

The quality of groundwater has become a serious concern in Bangladesh, by the last 2/3 decades (Raihan and Alam 2008). Almost all rural populations (31 million households) in Bangladesh intakes raw and untreated shallow-groundwater through privately-owned hand tube wells (depth 25–45 m) (BBS 2015). Potable water has been the main attention of regard since the early 1990s when extensive contamination of groundwater by Arsenic (As) was exposed in shallow aquifers (15–40 m) of Bangladesh (BGS-DPHE 2001b). At that time, WHO, UNICEF, and OXFUM were setting an extensive amount of shallow hand tube-well around the country. Now a day, arsenic contamination of groundwater is increasing at an alarming rate in the country. A comprehensive countrywide drinking water quality investigation has conducted by BBS which is titled 'Bangladesh National Drinking Water Quality Survey' (BNDWQS) of 2009 with the assistance of UNICEF. Throughout this time samples have collected from 3,00,000 household sources in 15,000 randomly selected blocks around the country. The results of this survey showed that 13.4% of the samples carried a higher concentration of arsenic (BBS 2009). Besides, Multiple Indicator Cluster Survey, Bangladesh (MICS-B 2018) was conducted in 2012–2013 in all the 64 district and Underground Drinking Water Thematic Report have exhibited, 41.7% of the households used drinking water sources that were faecal contaminated and arsenic concentration exceeded the Bangladesh standard of 0.05 ppm in 12.4% of samples. Nordstrom's (2002) investigation shown that, in the country, nearly 35 and 70 million people are at risk from arsenic, as contaminations in drinking water more than the native and WHO threshold limit respectively (Table 4.4). According to a WHO study, an assessed 35–77 million people have regularly been exposed to arsenic via drinking and that has pronounced as the *"largest mass poisoning in history"* (Flanagan et al. 2012). But then, high arsenic concentration is mainly found in groundwater from shallow aquifers with depths less than 40 meters. Not only that shallow water use in drinking purposes, it extensively used for irrigation in crop fields. According to groundwater arsenic statistics from 3205 locations and rice grain statistics from 595 sites, it observed that topographical variation in *arsenicosis* (arsenic toxicity) occurrence was better explained by arsenic consumption of drinking water than from rice (Panaullah et al. 2009; Ahmed et al. 2011b).

Aquifer's water is a significant concealed resource in the context of quality and quantity and that varies across hydrological and lithological environments (Islam and Mostafa 2021b). Generally, it is considered safe from infectious contamination, but with the presence of inorganic contaminants from the underlying rocks and minerals or any human activities, it is not easy to bring back the aquifers and this worsens the pollution impact (Yidana and Yidana 2010; Singh et al. 2013; Kumar and Singh 2015). Basically, hydro-geochemical processes that are precipitated, recharge-discharge, weathering of rocks, ion-exchange, redox, water mixing residence time, hydrolysis of minerals, etc. potentially influence the compositional status of groundwater (Belkhiri et al. 2010; Reghunath et al. 2002). As well, anthropogenic actives such as overexploitation, chemical spillages, industrial and transport wastes, leaching of sewage, landfill, chemical fertilizers, and pesticides, etc. influence groundwater quality. Practically, not only that arsenic contamination, there are huge possibilities to pollute the groundwater with respect to other heavy metals as well as some anions such as Pb, Cd, Co, Hg, F^-, NO_3^-, NO_2^-, PO_4^{3-} etc. by lithological or human activities. Previous numbers of research focused that Bangladesh's groundwater carries an excess of Ca, Mg, and Fe which those are not toxic metal, but excessive amounts are very harmful to the human body (Nerbrand et al. 2003; Heaney et al. 1982; Shajedul and Majumder 2020). In Bangladesh, there are 2.32 billion kg of chemical fertilizer, and 37,258 million tone pesticides are used in the agricultural field annually (Faruq 2018). Of these, the residual portions are leaching through topsoil and ultimately reach into groundwater aquifers and make water contamination. The chemical and microbial features of groundwater are very significant as they determine the suitability for domestic, industrial, or agricultural use (Mondal et al. 2010; Mallik et al. 2018).

4.3 Groundwater Aquifers in Bangladesh

Groundwater chemistry mostly depends on the local geological setting in aquifers. In this section, we draw a picture briefly of the aquifer lithology of Bangladesh. There are four major physiographic components such as (I) Pleistocene courtyard in the Barind and Madhupur zones (II) tertiary alluvial in the eastern and northern hills (III) the delta cover the rest of the country, and (IV) Holocene (recent) floodplains of the Ganges, Brahmaputra, and Meghna river exist at the surface in the country (BGS-DPHE 1999; Zahid 2015). The hydrogeological cross-sectional view of Bangladesh from north to the south showing the aquifer system and groundwater flow outline shown in Fig. 4.2. Historically, around 6000 years ago the sea level was much lower and the main rivers divided deep channels surrounding the Barind and Madhupur tract zones (Aggarwal et al. 2000). With some Pleistocene terraces, deltaic flood plains comprise the main portion of the basin and its sediments consist mostly of unconsolidated deltaic deposits and alluvial except for the mixed geology zone of pre-quaternary deposits that cover the north-eastern and southeastern mountainous areas of Bangladesh. Collected with the tertiary alluvial arrangements the maximum depth of the layout is mostly 20 km (Zahid and Hassan 2009). The country has promising hydrogeologic conditions and tropical monsoon weather indicates possible storing of groundwater in the aquifers. The uncombined nearby-surface Pleistocene to recent estuarine and fluvial sediments underlying most of the country usually form productive aquifers. Deep semi-contracted to the uncontracted fluvio-deltaic deposit of Miocene age to the recent forms many aquifers. But excluding the Dupi Tila sandstone construction of the plio-Pleistocene stage, others are too deep to consider for groundwater withdrawal except in the mountainous region (18% of total areas) (Aggarwal et al. 2000). Most of the groundwater extraction for irrigation or domestic purposes in the Barind and Madhupur upland areas are from the Dupi Tila aquifers. The floodplains of the active/inactive delta plain and the major rivers of the Ganges, Brahmaputra, and Meghna delta complex inhabit 82% of the country area (BGS-DPHE 2001a). The existing aquifer's geological studies indicate that the bulk of the best aquifers occurs from 30 to 130 m depth (BGS-DPHE 1999). These alluvial deposits are cyclic sediment of typically medium to fine sand, clay, silt, and siltstone. The separate layers cannot trace for long distances, vertically or horizontally (BGS-DPHE 2001b; Mojid et al. 2019). Hydrogeological cross-section of Bangladesh from north to south viewing aquifer system and groundwater flow outline described in Fig. 4.2. This figure was constructed from the literature of DPHE-BGS 2001 and Zahid 2015.

On an area basis, Bangladesh has three aquifers recognized by UNDP-BWDB (1982). Figure 4.2. shows that the composite aquifer or upper-shallow below the surface siltstone and clay unit, less than a few to several hundred meters thick and be made up of fine to very fine sand. The thickness of this zone, the main water zone, occurs at ranges from less than 5 meters in the northwest to more than 75 meters in the south and rest of the country (BGS 2001; Zahid et al. 2010). In the deep aquifer, water has no entry vertically downward and upward but flow very

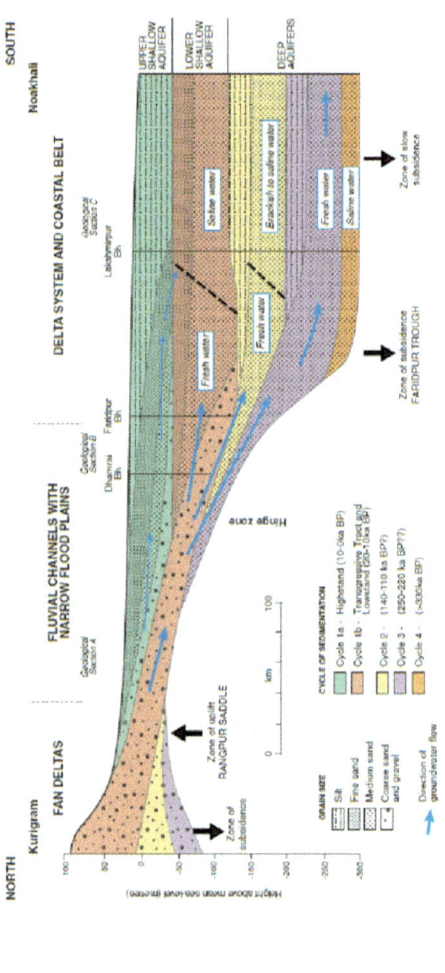

Fig. 4.2 Hydrogeological cross-section of the aquifer in Bangladesh

gradually along the slopes and dips of the aquifers (BWDB 2013). The depth of this aquifer in Bangladesh ranges from 190 to 960 m of the Dinajpur platform and 250–1500 meters in the basin of the Ganges, Brahmaputra, and Meghna including the sediment deposit of the Jaintia, Gondwana, Tipam, and Surma groups and a little portion of sandstone formation of the Dupi Tila (Khan 1991; Zahid and Hassan 2009).

Rain and floodwater are the principal sources of groundwater recharge in Bangladesh. Numbers of studies forecast that vast groundwater-fed irrigation would prolong net recharge in areas where soil properties and surface lithology are good permeable and thus favor recharge (UNDP-BWDB 1982; MPO 1987, 1991; WARPO 2000). The studies projected area wise annual usable and potential recharge between 183 to 1287 mm and 135 to 1910 mm respectively. The approximations of net recharge in Bangladesh has increased from 132 mm/year over a period from 1975 to 1980 to about 190 mm/year for the period 2002 to 2007 (WARPO 2000). Recent past studies confirmed that the ratio of groundwater recharge and abstraction in the country gradually downfall.

4.4 Groundwater Quality Parameters

The quality refers to the physical, chemical, bio-chemical, and biological features of water. It is a measure of the position of water relative to the necessities of one or more biotics, or to any human need (Johnson et al. 1997). Basically, water quality is a complicated or complex issue, in part, because water is an elaborate medium mostly tagged to the earth ecology. Hydrochemical and biological analysis and following water quality evaluation mostly make known quality of water that is suitable for domestic use, industrial, and agricultural purposes as well as helping in the management of the water resource (US-EPA 2016a, b). In this section of the text, the findings of countrywide various investigations about groundwater quality related parameters stated bellow.

4.4.1 Physico-Chemical Parameters

Physico-chemical parameters, such as temperature, pH, total dissolved solids (TDS); electrical conductivity (EC), salinity, alkalinity, total hardness (TH) of water are not water pollution indicator but those are very important and crucial properties and can affect the growth of biotic life in the water body and subsequently can cause an influence on the water (Soja and Wiejaczka 2014). Table 4.1 includes the result of different investigations on physico-chemical parameters in groundwater, which conducted in the varies zone of Bangladesh in a different period.

Even though pH has no straight impact on the consumer, but it is one of the most fundamental and essential functioning water quality properties (WHO 2011). The BDWS and WHO have recommended a maximum acceptable limit of pH for

Table 4.1 Physicochemical characteristics including pH, electrical conductivity (EC), total dissolved solids (TDS), salinity, alkalinity, and total hardness (mean value) of groundwater in Bangladesh

Sampling location	pH	Conductivity (μS/cm)	Dissolved solids (mg/L)	Salinity (mg/L)	Alkalinity (mg/L)	Total hardness (mg/L)	Reference
Dhaka Division							
Mymensingh Dist.	8.19	297.2	192.0	–	–	–	Shahidullah et al. (2000)
Faridpur District	7.53	788.8	748.6	–	–	–	Rahman et al. (2017a)
Dhaka District	7.08	471.0	309.2	–	180.0	179.3	Alam et al. (2019)
Faridpur District	6.91	788.0	–	–	–	–	Bodrud-Doza et al. (2016)
Gopalganj District							Islam et al. (2017a)
Wet Season	7.54	3207	1635	1690	–	–	
Dry Season	8.44	3218	1643	–	–	–	
Singair, M'ganj district							Mazeda et al. (2018)
Deep aquifer	6.37	349.0	–	–	–		
Shallow aquifer	6.97	614.0	–	–	–		
Chittagong Div.				–			
Chittagong city	7.41	452.0	–	2921	212.0	487.3	Mojumder and Shajedul (1995)
Noakhali District	7.29	3740	567.0	–	703.4	118.0	Miah et al. (2015)
Chittagong District	6.65	1063	406.9	–	175.6	132.0	Ahmed et al. (2010)
Comilla District	7.20	1078	676.0	–	132.1	337.1	Prodip et al. (2016)
Lakshimpue District	7.00	1135		–	–	–	Bhuiyan et al. (2010)
Cox's Bazar District	7.10	–	677.4	–	–	–	Seddique et al. (2013)
Brahmanbaria Dist.	7.23	327.5	233.0	–	–	–	Mahmud et al. (2007)
Rajshahi Division							
Rajshahi City, Shallow	6.91	445.4	297.4	–	–	302.3	Mostafa et al. (2017)

(continued)

Table 4.1 (continued)

Sampling location	pH	Conductivity (μS/cm)	Dissolved solids (mg/L)	Salinity (mg/L)	Alkalinity (mg/L)	Total hardness (mg/L)	Reference
Deep	7.00	455.0	241.0	–	–	304.2	
Rajshahi city	7.26	290.0	–	–	–	–	Rasul and Jahan (2010)
Rajshahi Dist. Deep	6.91	543.8	313.0	312.4	–	–	Rahman et al. (2017b)
Pabna District							Sarkar and Hassan (2006)
2002–03	6.94	790.0	–	–	–	225.5	
2003–04	7.97	790.0	–	–	–	195.5	
Bogra District	7.30	543.0	373.0	–	–		Islam and Shamsad (2009)
Dinajpur, Joypurhat, Thakurgaon, and Panchagarh District	7.6	220.0	118.3	–	–	–	Islam et al. (2016)
Thakurgaon District	7.53	203.5	166.0	–	–	–	Bhuiyan et al. (2015)
Khulna Division							
Khulna District	7.20	3018	1556	–	156.5	575.6	Dider-Ul et al. (2017)
Satkhira District							Tanvir et al. (2017)
North zone	7.75	2082	1098	4790	–	–	
Center zone	7.47	1594	905.4	3130	–	–	
South zone	7.51	614.0	250.9	650.0	–	–	
Satkhira District	7.61	1617	861.3	3380	–	214.8	Mirza et al. (2012)
Magura District	7.54	1344	–		–	435.4	Aminur et al. (2016)
Barisal city	7.58	4263	2190	3110	–	–	Sukhen et al. (2017)
Patuakhali District	7.42	897.3	4829	–	–	–	Islam et al. (2017a)

drinking from 7.5 to 8.5 (Table 4.4). Water with a pH higher than 8.5 or less than 6.5 can create staining etching or scaling. The overall countrywide pH result shows that the water source is within a suitable or desirable range from every perspective. Another physical parameter, total dissolved solids constitute inorganic salts or minerals and minor quantities of organic matter that dissolved in water. The water with a high range of total dissolved solids indicates that water is extremely mineralized (Meride and Ayenew 2016). The wanted value for total dissolved solids is 500 mg/L but the maximum limit is 1000 mg/L which is set for portable uses (Table 4.4). In coastal areas of Bangladesh, some investigations have found total dissolved solids and electrical conductivity levels exceed the desired boundary. A high value of total dissolved solids in drinking water may affect individuals who are suffering from constipation, kidney-functioning, and heart disease (Sasikaran et al. 2012).

The total dissolved solids (TDS) in water determines the electrical conductivity (EC), which measures the ionic loads in water that permits it to transmit current. No standard value proposed by WHO or BDWS but Zuane (1996) proposed that drinking water usually registers conductivity from 50 to 500μs/cm and with mineralized water recording values over 500. Various studies have shown that the groundwater in shallow-aquifers contains a high value of conductivity in the Southern part of Bangladesh (Table 4.1). Water with high conductivity may have an offensive taste, cause discoloration and precipitate scale in containers and pipe. Salinity in raised groundwater is common at shallow aquifers in the coastal part of southern Bangladesh and usually defined as dissolved solids or conductivity or chemical components such as sodium and chloride ions (Zahid and Hassan 2009; Zahid et al. 2015). Various studies showed that salinization is major environmental adversity affecting soil and water resources, agriculture and creating trouble in the natural ecosystem in the southern part of Bangladesh. But there is no national-scale regular monitoring of groundwater salinity in Bangladesh.

Increased groundwater salinity is also related to high concentrations of some parameters like sodium, boron, selenium, arsenic, fluoride, sulfate, and high radioactivity (IAEA 2004). Another physical parameter, alkalinity, is the power of water to defend against changes in pH that would make the water more acidic and it is usually a measure of dissolved bicarbonate and carbonate. As far, not enough data in this regard have found here. Among the physical parameters of groundwater, total hardness (TH) is very important and significant for drinking, irrigation, and industrial purposes. Total hardness is not caused by a single constituent but by a different type of dissolved multivalent metal ions, dominate Ca and Mg cations, while other cations such as Fe, Al, Mn, Ba, Sr, and Zn also contribute and associated anions are bicarbonate (temporary) and non-carbonate (permanent) (WHO 2011). Water containing $CaCO_3$ (equivalent value) at a concentration below 60 mg/L is usually considered as soft; 61–120 mg/L, moderately hard; 121–180 mg/L, hard; and more than 180 mg/L, very hard (McGowan 2000). The investigation has shown that the hardness value of groundwater samples of maximum areas of Bangladesh is higher than that of the acceptable range and the waters become hard to very hard.

4.4.2 *Chemical Parameters*

Major cations and anions in groundwater include Sodium (Na^+), Potassium (K^+), Calcium (Ca^{++}), Magnesium (Mg^{++}), Ammonium (NH_4^+), Chloride (Cl^-), Nitrite (NO_2^-), Nitrate (NO_3^-), Carbonate (CO_3^{2-}), Bicarbonate (HCO_3^-), Sulfate (SO_4^{2-}), Phosphate (PO_4^{3-}), etc. are not serious pollutants; also, these are essential for the human body functions in limited quantity. However, an additional amount of these ions can make the water unsafe for any living biotic (Hasan et al. 2019). Tables 4.2 and 4.3 represent the concise data of different ions in groundwater sources, which are taken out from frequent studies all over the country.

4.4.2.1 Cations

Sodium (Na^+) is one of the most common inorganic solutes in groundwater and is mostly found as chloride, bicarbonate, and sulfate salt (Whelton et al. 2007). It is an essential electrolyte that helps maintain the balance of fluid or water in the human body cell, proper muscle, and nerve faction, stable blood pressure level, maintains the osmotic pressure of the plasma, acid-base balance in the blood, etc. (Grollman 1961; Caldwell et al. 2010). Lacking sodium in the human body is also known as hyponatremia. However, the excess Na^+ increases blood pressure because it holds excess fluid in the body and increases the risk of stroke, heart failure, kidney disease, stomach cancer and osteoporosis (Kawasaki et al. 1978; Weinberger et al. 1986; Farquhar et al. 2015). According to World Health Organization (WHO) and BDWS, the standard concentration of sodium in potable water is 8.7 mEq/L or 200 mg/L; and for irrigational use (Food and Agriculture Organization, FAO standard) is up to 40 mEq/L (Table 4.4). Countrywide sodium levels in groundwater included in Table 4.3 The groundwater of southern districts such as Satkhira, Barisal, Bagerhat, Gopalganj, Khulna, Borguna, Chadpur, etc. is not suitable for drinking without proper desalinization.

In groundwater, potassium (K^+) is normally found as bicarbonate, sulfate and chloride and its concentration is found less than other key cations in water (Whelton et al. 2007). It is a vital element in the human diet, and along with sodium, it keeps the normal osmotic pressure in the cell. Besides, it is a co-factor for many enzymes and is essential for the ooze of insulin, creatinine phosphorylation, muscle contraction, carbohydrate digestion, nerve stimulation and protein dissolution (WHO 2009). Reducing sodium and increasing potassium in the human diet can help control hypertension and lower the risk of cardiac disease and death (Aaron and Sanders 2013). But the excess level in the human body, potassium may direct to muscle weakness, depression, heartbeat disorder, etc. (UKEVM 2003; He and MacGregor 2008). The blood plasma potassium level is usually 3.6 to 5.2 mmol/L and if blood holds 6.0 mmol/L can be dangerous, called hyperkalemia, and usually requires immediate treatment (Gosselin et al. 1984). Currently, there are no specific standard values for K^+ in drinking water. Countrywide K^+ levels in groundwater

Table 4.2 Major cationic mean concentrations in mg/L of groundwater in Bangladesh

Sampling location	Na^+	K^+	Ca^{++}	Mg^{++}	NH_4^+	References
13 selected zones of Bangladesh	83.97	7.78	56.39	19.28	0.57	Islam et al. (2015)
Dhaka Division						
Dhaka Metro	25.88	2.00	35.44	09.42	–	Nahar et al. (2014)
Dhamrai, Dhaka Dist.	25.20	3.00	78.10	24.30	–	Nahar et al. (2014)
Savar, Dhaka Dist.	7.50	2.00	06.40	01.54	–	Nahar et al. (2014)
Mymensingh District	11.27	2.34	41.20	10.92	–	Shahidullah et al. (2000)
Narayangonj District	22.74	3.58	67.99	16.00	1.05	Sarker and Zaman (2003)
Monshiganj District	64.75	10.1	87.50	44.30	–	Halim et al. (2009)
Mymensingh District	–	–	37.92	–	0.96	Ahmed et al. (2010)
Manikganj District	27.30	4.40	72.30	36.40	0.50	Halim et al. (2014)
Faridpur district	35.30	5.00	103.8	32.90	–	Bodrud-Doza et al. (2016)
Gopalganj Wet season	547.3	8.14	106.5	80.47	–	Atikul et al. (2018)
Dry season	639.0	8.53	90.51	76.56	–	
Manikganj, Deep	31.00	3.60	87.28	22.12	–	Mazeda et al. (2018)
Shallow	30.34	3.67	83.75	22.24	–	
Dhaka District	16.30	1.65	46.30	13.27	0.12	Alam et al. (2019)
Chittagong Division						
Chittagong city	395.6	2.89	106.5	51.91	3.71	Mojumder and Shajedul (1995)
Chadpur District	347.4	7.62	64.73	45.49	9.48	Zahid (2008)
Comilla District	123.5	9.52	24.27	33.70	9.29	Brömssen et al. (2014)
Homna, Comilla Dist.	05.00	0.003	02.79	04.01	–	Prodip et al. (2016)
Lakshimpue District	159.8	10.9	55.80	46.1	1.90	Bhuiyan et al. (2010)
Sylhet district	47.17	2.40	07.18	04.10	–	Islam et al. (2017c)
Brahmanbaria	6.67	3.12	10.40	06.36	–	Mahmud et al. (2007)
Khulna Division						
Khulna district	647.2	17.1	101.5	78.28	–	Dider-Ul et al. (2017)
Bagerhat District	437.5	6.15	16.77	18.19	–	IWM (2009)
Kushtia District	26.87	4.73	88.80	33.54	–	Hossain et al. (2013)
Eastern Zone of Bangladesh	90.41	9.14	36.75	23.88	–	Halim et al. (2010)
Sathkhira dist.	221.9	13.4	62.80	14.13	–	Rahman et al.(2011)
Satkhira, North Zone	344.6	18.5	47.98	11.40	–	Tanvir et al. (2017)
Central Zone	172.5	11.6	90.24	21.45	–	
South Zone	32.49	5.09	48.14	07.46	–	
Barisal city	642.7	9.46	31.80	74.10	–	Sukhen et al. (2017)
Satkhira District	221.9	13.40	62.80	11.13	1.13	Mirza et al. (2012)
Barguna District	863.0	16.16	136.0	155.0	–	Islam et al. (2017a)
Chuadanga District	117.9	9.50	294.0	70.90	–	Nahar et al. (2014)
Rajshahi Division						

(continued)

Table 4.2 (continued)

Sampling location	Na^+	K^+	Ca^{++}	Mg^{++}	NH_4^+	References
Rajshahi City, Shallow	54.90	13.5	166.1	86.80	–	Mostafa et al. (2017)
Pabna District, 2002–03	0.80	0.05	64.40	01.11	–	Sarkar and Hassan (2006)
2003–04	01.03	1.66	03.49	0.42	–	
Ishwardi, Dry season	26.13	7.05	65.03	41.53	–	Hossain et al. (2010)
Wet season	31.05	7.34	70.23	45.78	–	
Thakurgaon District	16.10	14.8	24.20	23.76	0.54	Bhuiyan et al. (2015)
Joypurhat District	12.17	0.30	54.70	04.04	–	Islam et al. (2017d)
Panchagarh, Joypurhat Thakurgaon Dinajpur	13.77	3.24	73–31	02.10	–	Islam et al. (2016)
Rajshahi District	31.42	2.22	87.52	13.47	–	Rahman et al. (2017b)

included in Table 4.2 and overall data indicate that, except Khulna, Thakurgaon. Satkhira and Rajshahi district, all district's groundwater contains K^+ normal in range. Maximum and minimum K^+ concentration found in Sathkhira (north zone, 18.52 mg/L) and Pabna district (0.05 mg/L) respectively.

Like Na and K, Mg is also abundant as a cation in groundwater and an essential nutrient for biotics, which regulates 300 biochemical reactions in the human body and, helps in the production of protein and energy (WHO 2011). It aids to maintain normal nerve and muscle system, keep the heartbeat steady and helps bone remain strong. Besides that, it also helps adjust blood glucose levels (Vallee et al. 1960; Soara et al. 2010; de Baaij et al. 2015; Gommers et al. 2016). Normal Mg levels in blood plasma are between 11.46 to 20.68 mg/L with levels less than 11.46 mg/L defining hypomagnesemia (Soara et al. 2010). According to WHO and BDWS, the standard values for magnesium in potable water should be 18 and 30 mg/L, respectively and for irrigation, this level is 60 mg/L (Table 4.4). Various studies (Table 4.2) showed that the maximum groundwater samples in Bangladesh contain an excess concentration of Mg (ranges from 32.9 to 155 mg/L); mainly in Faridpur, Chandpur, Monshiganj, Comilla, Pabna, Chuadanga, Manikganj, Lakshimpur, Barisal, Khulna, Gopalgang, Borguna, and Rajshahi district.

Calcium (Ca^{++}) is an abundant mineral in the human body and earth crust, and this metal ion plays an important role in the human cell functioning, hormones, cancer, heart disease, fluid balance in the body, muscle contraction, neurodegenerative disease, etc., as well as the descent of the testis (Tandogan and Ulusu 2005; Pravina et al. 2012). Even though calcium is good for bones and prevents osteoporosis, but bad for the brain and excessive consumption lids to hypercalciuria, kidney and arterial disease, urinary tract concretion and compression of bone restoration (Nerbrand et al. 2003; Heaney et al. 1982). Although the native and international standard values for a Ca in drinking water are 75 to 100 mg/L but the maximum groundwater sample holds excess of Ca. Calcium loaded along with Mg is mainly responsible for hardness, which is the main threat in the domestic and industrial water of Bangladesh. Table 4.2 shows that the groundwater of shallow aquifers in Rajshahi and Chuadanga contain 166.1 and 294 mg/L of Ca respectively.

Table 4.3 Anion mean concentrations in mg/L of groundwater in Bangladesh

Sampling location	Cl^-	F^-	NO_3^-	HCO_3^-	CO_3^{2-}	SO_4^{2-}	PO_4^{3-}	References
Selected 13 zones of Bangladesh	138.9	–	–	–	–	5.98	0.71	Islam et al. (2015)
Dhaka Division								
Dhaka Metro	35.15	–	7.42	132.0	–	12.8	BDL	Nahar et al. (2014)
Savar, Dhaka	02.40	–	9.01	27.05	–	0.10	–	Nahar et al. (2014)
Dhamrai, Dhaka	40.90	BDL	0.30	399.0	–	16.8	–	Nahar et al. (2014)
Manikganj District	38.12	–	1.40	179.0	–	17.7	2.60	Halim et al. (2014)
Narayangonj Dist.	35.46	–	0.04	272.6	–	1.54	3.55	Sarker and Zaman (2003)
Monshiganj District	69.05	–	3.90	478.2	–	5.20	1.80	Halim et al. (2009)
Mymensingh Dist.	0.27	–	0.06	02.60	0.40	0. 06	–	Shahidullah et al. (2000)
Faridpur District	23.91	0.10	–	542.30	–	5.30	–	Bodrud-doza et al. (2016)
Gopalganj District	847.0	0.40	9.96	263.0	12.9	–	1.11	Atikul et al. (2018)
Manikganj, Deep	11.60	–	1.69	302.8	–	2.41	–	Mazeda et al. (2018)
Shallow	14.70	–	1.56	315.8	–	8.70	–	
Dhaka District	38.05	–	0.55	–	–	13.2	0.30	Alam et al. (2019)
Chittagong Div.								
Chittagong District	169.3	–	1.00	–	–	20.0	1.90	Ahmed et al. (2010)
Comilla District	153.9	–	0.10	307.1	–	0.53	8.16	Brömssen et al. (2014)
Chadpur District	602.2	–	0.55	286.9	–	20.3	4.55	Zahid (2008)
Noakhali District	875.6	–	3.91	305.1	–	4.39	–	Ahmed et al. (2011a)
Brahmanbaria	56.80	–	0.25	103.0	–	0.20	–	Mahmud et al. (2007)
Sylhet District	12.3	–	3.20	147.0	–	2.95	–	Islam et al. (2017c)
Lakshimpur District	227.1	0.21	–	430.2	–	16.0	–	Bhuiyan et al. (2010)
Homna, Comilla	06.40	–	–	03.60	–	0.09	BDL	Prodip et al. (2016)

(continued)

Table 4.3 (continued)

Sampling location	Cl^-	F^-	NO_3^-	HCO_3^-	CO_3^{2-}	SO_4^{2-}	PO_4^{3-}	References
Khulna Division								
Khulna district	1776	–	2.61	510.0	1.23	4.97	–	Dider-Ul et al. (2017)
Barguna District	3513	–	4.12	254.0	–	19.5	–	Islam et al. (2017a)
Satkhira District	409.6	–	12.1	224.3	–	11.8	–	Mirza et al. (2012)
Bagerhat District	167.7	–	5.87	270.4	–	6.43	–	IWM (2009)
Eastern region of Bangladesh	135.0	–	9.10	166.8	–	11.4	3.20	Halim et al. (2010)
Sathkhira District	409.6	–	12.1	224.3	–	11.8	–	Rahman et al. (2011)
Satkhira Sadar								Tanvir et al. (2017)
North	544.8	BDL	14.8	278.7	–	15.9	–	
Center	436.5	–	12.8	169.9	–	10.1	–	
South	05.47	–	4.56	197.2	–	5.47	–	
Chuadanga Dist.	632.0	–	0.03	700.2	–	0.11	–	Nahar et al. (2014)
Kushtia District	64.04	–	–	388.0	–	3.19	5.03	Hossain et al. (2013)
Barisal city	1313	–	4.23	369.8	–	46.9	–	Sukhen et al. (2017)
Rajshahi Division								
Rajshahi City, Shallow	76.80	BDL	8.75	325.5	–	55.0	–	Mostafa et al. (2017)
Deep	44.90	–	1.67	188.4	–	36.6	–	
Thakurgaon District	36.56	–	6.66	82.66	–	14.9	1.58	Islam et al. (2013)
Pabna District								Sarkar and Hassan (2006)
2002	03.78	–	–	07.77	0.00	–	–	
2003	05.13	–	–	08.07	0.00	–	–	
Ishwardi, Pabna	74.66	–	1.42	386.9	–	2.95	5.86	Hossain et al. (2010)
Joypurhat District	–	0.05	–	136.9	–	1.50	–	Islam et al. (2017d)
Dinajpur, Thakurgaon, Panchagarh, Joypurhat	10.32	0.23	3.45	81.32	0.32	7.32	1.89	Islam et al. (2016)
Rajshahi Dist. shallow Deep	150.9 141.2	- -	2.80 2.08	428.5 246.8	8.40 34.0	1.17 1.03	- -	Rahman et al. (2017b)

BDL Below detection limit

Table 4.4 Drinking and irrigational water quality standards

	Drinking water quality standard*				Irrigational water quality standard*		
Parameters	*BDWS*[1]	*WHO*[2]	*US-EPA*[3]	*INDIA*[4]	*BIWS*[5]	*FAO*[6]	*US-EPA*[7]
pH	6.5–8.5	7.5–8.5	6.5–8.5	7–8.5	7.5–8.5	6.5–8.4	7.5–8.0
EC (μs/cm)	–	–	–	–	750	350–500	–
Salinity (mg/L)	Nil	–	–	–	–	–	–
Alkalinity (mg/L)	–	–	–	200	–	–	–
TDS (mg/L)	1000	600	500	500	–	450–2000	500–1000
T. hardness mg/L	200–500	500	–	300	–	–	–
Na (mEq/L)	8.7	8.6	1.3–2.6	8.0	–	0–40	–
K (mEq/L)	0.3	–	–	–	–	0–0.05	–
Ca (mEq/L)	3.75	5.0	–	3.75	–	0–20.0	–
Mg (mEq/L)	2.5	12.5	–	2.5	–	0–5.0	–
Cl^- (mEq/L)	4.2–17	7.0	7.0	7.0	17.0	0–30	–
NO_2^- (mg/L)	<1	0.2	1.0	–		–	–
NO_3^- (mg/L)	10	50(as N)	10(as N)	45	–	0–10 (as N)	–
SO_4^{--}(mg/L)	400	500	250	200	–	0–800	–
PO_4^{--}(mg/L)	6.0	–	–	–	0.2	0–2(as P)	–
NH_4^+ (mg/L)	0.2	–	–	–	3.0	0–5(as N)	
F^- (mg/L)	1.0	1.5	2.0	1.9	–	1.0	–
Al (mg/L)	0.2	–	–	0.03	1.0	–	5.0
B (mg/L)	1.0	–	–	0.01	<1.0	0–2.0	0.75
Mn (mg/L)	0.1	0.5	–	0.1	–	0.2	0.2
Fe (mg/L)	0.3–1.0	0.3	0.3	0.3	–	5.0	5.0
Co (mg/L)	0.05	–	–	–	–	0.05	0.05
Ni (mg/L)	0.1	0.02	–	–	0.5	0.2	0.2
Cu (mg/L)	1.0	2.0	1.3	0.05	0.2	0.2	0.2
Zn (mg/L)	5.0	3.0	–	–	–	2.0	2.0
Cr(6) (mg/L)	0.05	0.05	0.1	0.05	0.01	0.1	0.1
Cd (mg/L)	0.005	0.03	0.005	0.01	0.01	0.01	0.01
Pb (mg/L)	0.05	0.01	0.015	0.05	0.1	5.0	5.0
As (mg/L)	0.05	0.01	0.01	0.05	1.0	0.1	0.1
Hg (mg/L)	0.001	0.001	0.002	0.001	–	–	–

**BDWS* Bangladesh drinking water standard; *WHO* World health organization; *US-EPA* United state-environmental protection agencies; *BIWS* Bangladesh irrigation water standard; *FAO* Food and agriculture organization

[1]Department of Public Health and Engineering, Bangladesh (2019); [2]HWO-Drinking water standard, 4th ed. (2011); [3]US-EPA-Drinking water standard (2018); [4]Drinking water standard for India (IS10500, 2012); [5]Bangladesh irrigation water standard (2009); [6]FAO-Water quality for agriculture (1985); [7]US-EPA-Guidelines for water reuse (2004)

Ammonium (NH_4^+) is a vital nutrient in primary production, however high NH_4^+ loads can affect the metabolism by changing acid-base balance, troubling glucose tolerance, and dropping the tissue susceptibility to insulin (US-EPA 1989). Due to huge agricultural run-off as well as industrial effluents, the NH_4^+ loads in the groundwater of

Bangladesh are almost as high. A few studies investigate the NH_4^+ concentration in groundwater and among those the highest loads (9.48 mg/L) found in Chandpur district.

4.4.2.2 Anions

Chloride (Cl^-) is the most common and necessary anion in groundwater, which acts as a relatively minor contaminant in potable water. Every rock and mineral contains a small amount of Cl^- as NaCl, KCl and $CaCl_2$; and it is highly mobile in ground aquifers (WHO 1996a; Kelly et al. 2012). An adult human body holds approximately 81.7 gm of chloride ions and it maintains the cellular activities, fluid balance in the body, and metabolic actions in the human and plant body. Chloride has no toxicity in humans except in case of cardiovascular problems, but an excess amount of Cl^- as NaCl may cause human's hypertension and can corrode the metal pipes (Fadeeva 1971; WHO 2011). Although Cl^- is not a contaminant, because of the unpleasant taste and odor of water BDWS, WHO and US-EPA have set the secondary maximum level of 150 to 250 mg/L (Table 4.3). Various investigations showed that the shallow groundwater of southern coastal regions in Bangladesh carries an excessive amount of Cl^- which are very much far from standard levels. The highest concentration of Cl^- (3513 mg/L) was reported from the Borguna district. Also, the samples of Chittagong (463 mg/L), Satkhira (409.6 mg/L), Chandpur (602.2 mg/L), Chuadanga (632 mg/L), Noakhali (875.61 mg/L), and Khulna (1776 mg/L) carry an excess of Cl^- ion.

Fluoride (F^-) is a micro constituent in natural drinking water and its deficiency in water causes dental caries, particularly in children, and high concentration, on the water body, perhaps causes cavities and blackening of the teeth and weakness of the bones (Pizzo et al. 2007; Lamberg et al. 1997; Ali et al. 2016; Kumar et al. 2019). Every year, the US spends about $1.08 per person on fluoridated drinking water (US-CDC 2001). The standard level of F^- is from 0.5 to 1.5 mg/L in potable water, depending on the local environment, climate, and other sources of fluoride (WHO 2011). So far, there is no national-wide monitoring and evaluation for F^- levels in the drinking water of Bangladesh and has insufficient data for those investigations. A few studies showed (Table 4.3) that all water samples hold much less (0.05 to 0.4 mg/L) than the standard level.

Nitrate (NO_3^-) and nitrite (NO_2^-) is the general form of nitrogen in groundwater and with high concentrations in potable water may cause methemoglobinemia, also called 'blue baby syndrome', in which the capability of blood hemoglobin to transport oxygen reduced (Winton et al. 1971). The main sources of both NO_3^- and NO_2^- in water include inorganic nitrogen fertilizers, wastewater treatment, and septic tanks. The concentration levels of NO_3^- have been presented in Table 4.3. Except for Sathkhira Sadar (14.80 mg/L) the groundwater of all regions of Bangladesh is a safe nitrate level.

Bicarbonate (HCO_3^-) remains as Na, Mg, and Ca-bicarbonate and those are the most common and naturally occurring amplest minerals in the groundwater. It plays amazingly vital roles in various biological activities (Casey 2006). There are no native or international standard values for bicarbonate and it is associated with the total hardness of the water. Different countrywide studies found the HCO_3^- concentration in groundwater widely varied from 2.6 to 700.2 mg/L.

Sulfate (SO_4^{2-}) is another abundant and common anion of groundwater that comes from dissolution and weathering of Na, K, Mg and Ca-sulfate minerals and some types of surfactant (e.g. Sodium dodecyl sulfate) (de F. Araújo et al. 2018). There is no primary health-based standard value for SO_4^{2-} in domestic water proposed. But US-EPA set a secondary guideline value for SO_4^{2-} of 250 mg/L because water holding higher concentrations may generate an offensive taste that makes it inappropriate for household uses and it creates a high risk of dehydration from diarrhea (Heizer et al. 1997; US-EPA 2003). If the level of sulfate in drinking water over 600 mg/L the laxative effects reported (US-DHEW 1962; Chien et al. 1968). Table 4.3 shows that the levels of SO_4^{2-} in all the groundwater samples of Bangladesh are far below the recognized values (Table 4.4).

Phosphate (PO_4^{3-}) comes from phosphorus baring fertilizers and detergents into the water body. Like sulfate, PO_4^{3-} has no drinking water guideline concentration. Whoever, too much PO_4^{3-} can cause a health problem, such as kidney damage and osteoporosis and it can speed up eutrophication in the water body (De 2005; Zhang et al. 2011). The concentrations of phosphate in groundwater of Bangladesh have included in Table 4.3.

4.4.2.3 Trace Metals

Trace metals and metalloids, among an extensive limit of contaminations, are steady of a health concern due to their toxicity capacities at a very little concentration and can show an opposing effect on living existences, and tendency to bioaccumulate in lipids and tissues of biotics over time (Ikejimba and Sakpa 2014; Demir et al. 2015; Malik et al. 2019). These metals such as Cr, Pb, Hg, Cd, As, and Co have no useful effects in the body system, moreover, long time exposure may cause more acute interruptions in the normal operations of the human organ systems where the metals deposited (US-EPA 1986; Mominul et al. 2018). Though, some trace metals like Cu, Zn, Fe and Mn, as micronutrients, required by the body in limited amounts for metabolic actions, but the same elements, at higher amounts can cause opposing health effects (Valavanidis and Vlachogianni 2010; Wang et al. 2019). The key anthropogenic sources of trace metals in groundwater are natural matters leached into the soil or rocks, residue from agrochemicals, controlled release from the sewage treatment plant and industrial run-off, and unrestrained releases or escape from landfill spot and chemical accidents or calamities (Leeuwen et al. 2000; Chen et al. 2005). The groundwater contamination in Bangladesh with excessive trace metal, especially arsenic, become an alarming situation. The countrywide traces metal levels in groundwater included in Table 4.5.

Chromium (Cr^{3+} and Cr^{6+}) is a naturally occurring trace metal that is usually found in very trace concentrations in groundwater and not influenced by point-source contamination (WHO 2019; Zhou et al. 2019). The major sources of chromium discharge in Bangladesh are the tanning industry and landfills or other solid waste. Chromium (+6) easily enters cell membranes, and Cr (+3) does not (Slooff

Table 4.5 Trace metal mean concentrations in mg/L in groundwater of Bangladesh

Sampling location	Cr	Mn	Fe	Ni	Cu	Zn	Cd	Pb	B	References
Overall Bangladesh	–	0.5	3.01	–	–	–	–	–	0.501	BGS (2001)
24 selected Dist. of Bangladesh	–	1.69	1.43	–	BDL	–	–	0.011	–	Tahera et al. (2016)
Selected 13 zones of Bangladesh	–	1.70	5.68	–	–	–	0.02	–	–	Islam et al. (2015)
Dhaka Division										
Dhaka Metro		0.224	4.03	–	0.09	–	–	–	–	Nahar at. el. (2014)
Manikganj District	–	0.455	0.81	–	–	–	BDL	–	–	Rahman et al. (2016)
Gopalganj District	–	0.201	5.13	–	BDL	BDL	–	–	0.430	Atikul et al. (2018)
Narayangonj City	0.06	2.71	–	0.057	–	–	–	–	0.042	BGS (2000)
Narayangonj District	0.02	2.00	–	0.081	–	0.190	0.01	0.450	0.120	Seddique et al. (2004)
Faridpur District	–	BDL	5.93	0.003	–	0.011	–	0.006	–	Bodrud-Doza et al. (2016)
Rajarampur, N'ganj	–	–	–	BDL	.0002	0.030	–	0.0002	–	Islam et al. (2000)
Singair, M'ganj Dist.	0.093	2.08	7.11	0.043	.022	0.0643	–	0.019	–	Halim et al. (2014)
Dhaka District	0.011	–	–	0.037	0.056	0.034	–	–	–	Alam et al. (2019)
Chittagong Division										
Chittagong District	0.005	0.54	2.88	0.020	0.011	0.010	0.011	0.049	–	Ahmed et al. (2010)
Brahmanbaria Dist.	–	0.27	0.67	–	0.003	0.037	–	–	0.003	Mahmud et al. (2007)
Lakshipur District	–	0.65	3.23	0.002	–	0.216	–	0.004	BDL	Bhuiyan et al. (2016)
Moinamoti, Comilla	0.003	–	–	0.001	.0006	–	–	0.001	–	Islam et al.(2000)
Cox'sBazar sea beach	–	1.87	1.81	–	–	–	–	–	–	Seddique et al. (2016)
Lakshimpur District	–	0.65	3.24	0.0002	–	–	–	BDL	–	Bhuiyan et al. (2016)
Sylhet District	–	0.28	6.83	–	–	BDL	BDL	–	–	Islam et al. (2017c)

(continued)

Table 4.5 (continued)

Sampling location	Cr	Mn	Fe	Ni	Cu	Zn	Cd	Pb	B	References
Khulna Division										
Kushtia District	–	0.97	0.53	–	–	BDL	–	–	–	Hossain et al. (2013)
Andulia, Jhenaidah	–	–	–	BDL	0.078	–	–	0.0007	–	Islam et al. (2000)
Barishal metro	–	–	4.42	–	–	–	–	–	–	Sukhen et al. (2017)
Samta, Jessore Dist.	BDL	–	–	BDL	0.027	–	–	0.0005	0.001	Islam et al. (2000)
Chuadanga Dist.		0.29	28.9	–	--	–	–	–	–	Nahar et al. (2014)
Rajshahi Division										
Rajshahi City, shallow	–	1.47	3.12	–	0.081	0.190	0.014	1.170	–	Mostafa et al. (2017)
Deep	–	2.21	2.23		0.390	0.181	0.016	1.120	–	
Pabna Dist. 2002–03	-	-	0.74	-	-	-	-	-	-	Sarkar et al. (2006)
2003–04	-	-	0.74	-	-	-	-	-	-	
Rangpur District	–	0.68	7.73	BDL	–	0.033	–	–	–	Islam et al. (2017b)
C′ Nawabganj Dist.	0.001	1.44	–	–	BDL	–	0.010	0.096	BDL	Saha and Zaman (2011)

BDL Bellow detection limit

1989). In the human body, the maximum concentrations of Cr accumulated in lymph nodes, kidneys, liver, lungs, and spleen and continuing exposure can damage the liver and kidneys (Janus and Krajnc 1990; Shrivastava et al. 2002). WHO, US-EPS and BDWS recommended the permissible highest value of Cr in drinking water of 0.05 mg/L (Table 4.4). A few studies in different zones of Bangladesh focused on the Cr content in groundwater (Table 4.5). Results showed that the average concentration of Cr is lying within the threshold limits.

Mn is an element vital to the proper working of humans, animals, and plant metabolism, as it is obligatory for the operative of several cellular enzymes and can aid to activate hydrolases, kinases, transferases, decarboxylases (IPCS 2002). But excessive consumption (over 1.8 mg/L) of Mn-rich water, then showed neural symptoms that are alike to Parkinson's disease (Kondakis 1989). Memory damage, hallucinations, disorientating, and impulsive instability also concerning by manganese overdose (Gupta and Gupta 1987; ATSDR 2000; Dorman 2000). There is no US-EPA and WHO primary guideline for Mn in water supplies because they have no recognized serious health threats posed by it. However, a secondary extreme contaminant level of 0.5 mg/L for Mn because higher concentrations yield offensive taste, odor, color, staining and corrosion (WHO 2011). Table 4.5 shows that the maximum concentration of Mn is 2.71 mg/L in Narayonganj city and maximum samples are contained moderately high levels of Mn.

Iron (Fe) is the burning issue of rural drinking water in Bangladesh. It is the third most abundant metal in the earth's crust and it remains as ferrous and ferric oxides, hydroxides, sulfides, and carbonates (Elinder 1986; Knepper 1981). Although a low level of iron is essential in the human diet and plant metabolism and cannot do much harm, it encourages objectionable bacterial growth ('iron-bacteria') inside a waterworks and supply system, resulting in the deposition of a slushy coating on the piping (Can DNHW 1990). Besides, high iron content (over 0.3 mg/L) leads to an excess which can cause stomach problems, vomiting, diabetes, nausea, and hemochromatosis (Bothwell 1979; FAO-WHO 1988; Milman et al. 2001; Swaminathan et al. 2007; Toyokuni 2009). Several studies showed that the iron levels in the groundwater of Bangladesh are quite high (Table 4.5). The water samples of the maximum region hold over permissible concentration and reach to near about 6 mg/L of Fe. BNDWQS also informed that 40% of underground water conveys excess amounts of iron and the survey also showed the high average level of Fe present in the deep tube-well (1.37 mg/L) and shallow (2.65 mg/L) water all over the country.

Nickel (Ni) occurs mainly as the hydrated form [Ni $(H_2O)_6^{2+}$] in natural waters at pH 5–9 (IPCS 1991). The preliminary cause of nickel in drinking water is the dissolution of metals in contact with water, such as fittings and pipes (WHO 2005). When it accumulated in the human body through water, in small amounts, it is harmless and necessary in our nutrition. The serious harmful health effects from the consumption of nickel, such as lung function disorder, cancer of the lung and nasal sinus, and chronic bronchitis have occurred in individuals who have respired fine particles containing certain nickel species whiles employed in nickel melting plants (Sunderman Jr et al. 1989). According to WHO and US-EPA guidelines, the concentration of nickel should not exceed 0.02 mg/L in potable water. Studies reported the average concentration range between $<$0.00006 to 0.057 mg/L in groundwater, which is almost safe in ranges.

Copper (Cu) is an indispensable element in animals and plants which shows a significant role in metabolism (Bremmer and Beattie 1990). Widely it is used in plumbing pipes and fittings, and may dissolution from water pipes if the pH of acidic ranges. Copper-sulfate salt occasionally added to water-supply tanks to subdue unwanted algal growth, and copper compounds have used largely in agricultural insect killer sprays. Temporary exposure to Cu in potable water can lead to gastrointestinal suffering, long-time exposure can lead to copper toxicosis, which results in liver and kidney damage, anemia, hepatic cirrhosis, and deterioration of the basal ganglia (Harris and Gitlin 1996). An excess of copper in aquatic environments is seriously harmful to fish and other aquatic lives (Ali and Khan 2018). Countrywide various investigations showed that the Cu levels, with the ranges of 0.0002 to 0.39 mg/L, in groundwater (Table 4.5) are almost in safe positions.

Zinc (Zn) is a naturally occurring trace element and an essential nutrient for body metabolism and development, particularly for newborns and young children (Askary et al. 2011). However, drinking water containing high levels of Zn can lead to stomach cramps, neurological problems, vomiting; and chronic exposure to Zn is liable for depressed copper consumption, iron shortage, depressed levels of HDL cholesterol (Sandstead 1978; Hooper 1980). Water with Zn concentration of more than 3 mg/L tends to be opalescent, grows an oily film when boiled, and has an undesirable loathful taste (WHO 1996b). All the collected groundwater samples from various regions of Bangladesh showed that the levels of Zn (0.01 to o.941 mg/L) are much below the maximum acceptable ranges.

Cadmium (Cd) is a very toxic trace element with a very long half-life and it occurs naturally with zinc minerals (Jihen et al. 2008). This element can release to groundwater from buried wastes containing metal refinery byproducts and electronic components, and by coal-burning (Friberg et al. 1986). It can consume by eating vegetables grown in contaminated soil and fish or other seafood from contaminated or drinking water holds cadmium. Acute exposure can cause nausea, cancer, diarrhea, anemia, bone marrow disorders, muscle cramps, liver injury and kidney failure (Krajnc 1987). In Bangladesh, the mean concentrations of 0.011–0.053 mg/L Cd were found in groundwater samples, which are 5 to 10 times higher than the tolerable level (Table 4.5).

Lead (Pb) is another omnipresent toxic trace metal and substantial public health concern in the environment (WHO 2011; Flora et al. 2012). It can cause different biochemical effects when exposed to it for a relatively short time duration (Elom et al. 2014). These effects may comprise interfering with red-blood-cell chemistry, delays in usual physical and mental growth in an infant, hearing and learning capacities of children, scarcity in attention span, kidney disease, stroke, cancer, and rises in the blood pressure of adults (Moore 1988; Ong and Lee 1980). The highest permissible concentration of Pb in drinking water set by WHO and BDWS is 0.01 and 0.05 mg/L respectively. Narayanganj and Rajshahi city's groundwater samples contain 0.0971 and 1.17 mg/ L of Pb which are almost 10 to 1000 times higher than the WHO standard (Table 4.5).

Arsenic Contamination

Arsenic (As), a first-category carcinogenic substance (Driscoll et al. 2004; Edmunds et al. 2015), occurs naturally in groundwater supplies through all-out parts of Southeast Asia. In the Ganges plain of Bangladesh and northern India, severe contamination of groundwater by naturally occurring arsenic affects 25% of hand tube-well in the shallow of two regional aquifers (Ravenscroft 2007a, b). It is first recognized as a thoughtful problem in Bangladesh in the 1990s, and concentration levels in mostly overall places exceed the maximum level suggested by the WHO. Over 75 million people, from 59 out of 64 districts (Fig. 4.3), brought to be at a risk of drinking water contaminated by As in Bangladesh (Safiuddin 2001; Uttam et al. 2000). Every year, an assessed 43,000 people invaded by arsenic poisoning in the country (Jahan 2016). The authority has taken several steps and

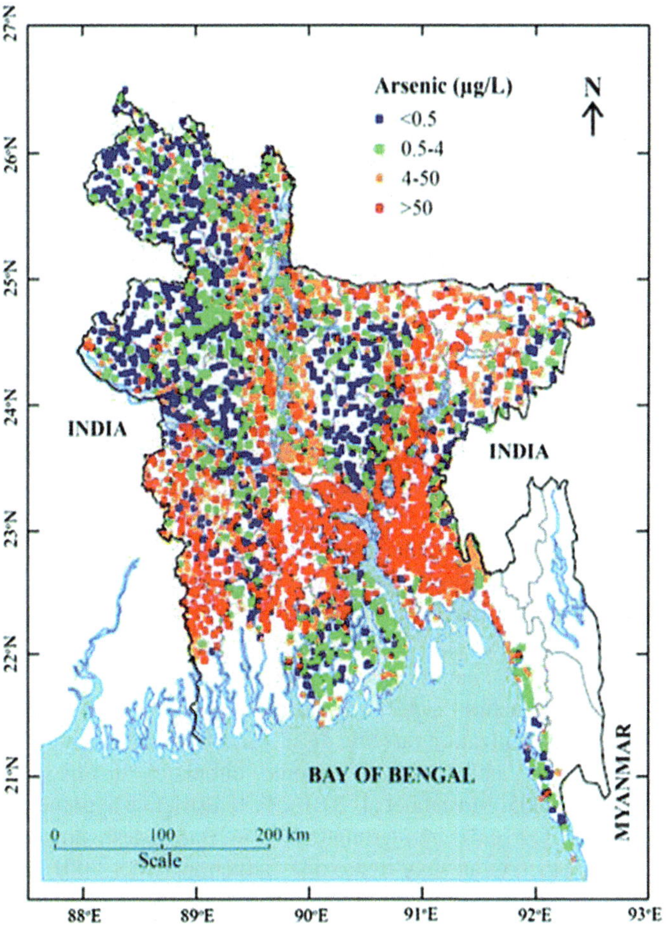

Fig. 4.3 Distribution of arsenic in groundwater of Bangladesh

made strategies to try to address the calamity. Nevertheless, of a countrywide movement and social mobilization activities by the government and NGO's consciousness and knowledge levels among the public remain far below the expectations. Several studies were conducted in Bangladesh, Southeast Asia, and other countries in the world as well, but the actual situation is still out of thorough control.

In the year 1992–93, the groundwater samples of some shallow tube-wells placed in the northern area, bordering to India, of Bangladesh were examined for As contamination and found to contain in the range of 59–388 ppb (part per billion) of arsenic, which was higher than the WHO and Bangladesh standard of 10 ppb (0.01 mg/L) and 50 ppb (0.05 mg/L) respectively in drinking water (Ahmed and Rahman 2000; Ahmed et al. 2001; DPHE 2009). The arsenic contamination has been detected in 59 out of 64 districts of Bangladesh since12 years of its first detection in the groundwater. Subsequent irregular finding and report of the contamination in groundwater, a countrywide investigation for arsenic concentration in tube-well water started in 1996. In 1998, the British Geological Survey (BGS) collected 2022 water samples from 41 arsenic prevalent districts (Smith et al. 2000). Lab-tests discovered that 35% of these samples were found to have arsenic levels above 50 ppb. The preliminary screening that guided in 1999 had verified 51,000 tube-wells, which quantified that arsenic toxicity was present in 211 out of 460 Upazillas (sub-districts), which are nearly one-third of the verified tube-wells. Then, in 2003, a countrywide comprehensive study directed by Arsenic Mitigation Water Supply Project (AMWSP), cover 57,482 villages in 271 Upazillas, where it showed 1.44 million tube-wells, out of total 4.95 million tube-wells, contaminated with arsenic (Table 4.6 & Fig. 4.3). The As levels in the water of the maximum affected tube-wells were found to be at the limit of 0.10–0.30 mg/L. The highest concentration of arsenic detected in the shallow tube-well water was 470 ppb and a previous study conducted by the experts of Bangladesh Council for Scientific and Industrial Research (BCSIR) have found also the highest-level of 1400 ppb in the same types of tube-well water of Pabna district (New Nation Report 1996). Another investigation has shown that water from 13,423 household tube-wells throughout the country revealed that arsenic concentration in 12.6% of tested samples crosses the Bangladesh standard (DPHE 2009; FAO, UNICEF, WHO, and WSP 2010; DGHS 2016). The complete scenario of the countrywide arsenic level of groundwater is summarized in Tables 4.6 and 4.7.

Arsenic contamination in aquifers water can rise both naturally and anthropogenically. This calamity arises in this region because of an unfortunate combination of three natural aspects: a source of arsenic (arsenic existing in the aquifer sediment), mobilization (arsenic dissolution from the arsenic-rich minerals to the groundwater) and transport (arsenic circulates in the groundwater) (Ravenscroft et al. 2005: Mostafa et al. 2010). Two restrictive hypotheses to narrate the mobilization of arsenic into groundwater in Bangladesh are one is pyrite oxidation and another is an Oxy-hydroxide reduction (BGS 2001; Zheng et al. 2004; Mostafa et al. 2011). The groundwater-table has declined gradually due to overexploitation of water for irrigation and domestic purposes, poor water management and insufficient recharge of the aquifers are the main anthropogenic

Table 4.6 Statistics of arsenic impact in Bangladesh

Basic Data	Value	Percentage (%)
Total area of Bangladesh	147,570 km^2	–
Total population in Bangladesh (June, 2019)	165.6 million	–
Total number of districts in Bangladesh (new)	64	–
WHO standard (arsenic in drinking water)	10 ppb	–
Bangladesh standard (arsenic in drinking water)	50 ppb	–
No. of districts surveyed for arsenic contamination	64	100
No. of districts having arsenic above 50 ppb in groundwater	59	92.2
Area of affected 59 districts	126,134 km^2	85.5
Population at risk	>80 million	>48
Potentially exposed population	>35 million	>21
No. of patients suffering from Arsenicosis	>38,000	–
No. of patients died	10	–
Estimated number of tube-wells in Bangladesh in 2002	**8.60 million**	**100**
Tube-wells tested for arsenic in 2003–2004	4.75 million	55
Tube-wells in safe (green marked)	3.30 million	39
Tube-well in unsafe (red marked)	1.40 million	16
Estimated total village in country	**87,319**	**100**
Village screened for arsenic	54.041	62
Village where <40% of the wells contaminated	70,610	81
Village where 40–80% of the wells contaminated	8331	10
Village where 80–99% of the wells contaminated	6062	07
Village where all wells contaminated	2316	03
Actions taken by people to avoid arsenic contamination		
Using arsenic free tube-wells	–	55
Using treated pond, canal, or river water	–	21
Using filtered water	–	05
Using rainwater tanks or sand filtered water	–	05
No action	–	32

Heavy contaminated districts: Chandpur (90%), Monshiganj (83%), Gopalganj (79%), Madaripur (69%), Noakhali (69%), Satkhira (67%), Comilla (65%), Faridpur (65%), Shariatpur (65%), Meherpur (60%), and Bagerhat (60%).
The least affected districts: Thakurgaon, Dinajpur, Panchagarh, Nilphamari, Natore, Lalmonirhat, Patuakhali, and Barguna.

References Karim (2000); Fazal et al. (2001a, b); Faruque and Alam (2002); Escobar et al. (2008); Chakraborti et al. (2010); ITNC (2008); BBS (2009); UNICEF (2010); Safiuddin (2011); and MICS-B (2018)

cause of arsenic contamination in Bangladesh (Karim et al. 1997; Mandal et al. 1997).

Arsenic poisoning is a serious threat and significant public health concern in Bangladesh. Long-term overexposure to arsenic in potable water has been linked to cancer of the lung, skin, bladder, nasal channels, kidneys, prostate, and liver. It also

Table 4.7 Arsenic concentration in groundwater of Bangladesh districts

Sample location (Old District)	Total no. of samples	Concentration (μgm/L) Distribution								% of Threat	
		<10	10 to 50	51 to 99	100 to 299	300 to 499	500 to 699	700 to >1000	Max. Conc.	BMACL[a]	WHO
Dhaka division	**13,597**	**8236**	**2024**	**983**	**1536**	**558**	**167**	**93**			
Dhaka District	4393	3850	169	90	214	61	09	00	**5533**	16.72	21.78
Faridpur District	3726	924	826	483	956	376	104	48	**1630**	41.86	65.63
Mymensingh District	2805	2186	411	110	61	16	08	10	**1172**	01.04	06.58
Narayanganj District	748	306	58	41	170	92	46	35	**1750**	76.70	86.89
Tangail District	1925	970	560	259	135	13	00	00	**365**	03.53	25.80
Rajshahi division	**18,581**	**12,532**	**3365**	**1181**	**1142**	**209**	**86**	**66**			
Bogra District	1165	995	135	20	16	01	00	01	**1040**	04.43	79.41
Pabna District	5395	1762	1886	815	695	124	52	41	**2108**	33.52	68.83
Rangpur District	2700	2049	524	59	37	19	10	02	**939**	14.01	38.58
Rajshahi District	5254	3735	732	280	393	65	24	22	**1600**	08.71	18.57
Dinujpur District	4087	3991	88	07	01	00	00	00	**130**	Σ0.00	01.10
Chittagong division	**6734**	**1514**	**601**	**613**	**2144**	**926**	**455**	**481**			
Chittagong District	551	504	26	13	08	00	00	00	**275**	05.74	12.84
Comilla District	1757	175	71	68	820	417	129	77	**1769**	74.50	79.27
Noakhali District	3691	367	324	471	1293	506	326	404	**4730**	95.14	99.42
Sylhet District	735	468	180	61	23	03	00	00	**302**	Σ0.00	15.35

Khulna division	**13,570**	**7766**	**2240**	**938**	**1664**	**570**	**230**	**162**			22.65
Jessor District	5465	4751	614	420	681	136	80	38	**1120**	17.38	
Khulna District	1300	940	378	145	453	239	24	24	**3143**	24.70	48.20
Kushtia Dist.	2065	1732	1051	322	274	80	51	36	**2190**	20.63	47.60
Barishal Dist.	803	295	187	51	256	115	75	64	**1770**	63.64	77.71
Patuakhali District	15	48	10	00	00	00	00	00	**15**	00	13.33
Total	**52,202**	**29,768**	**8230**	**3714**	**6484**	**2263**	**938**	**802**	–		
Percentage (%)	–	**57**	**15.8**	**7.1**	**12.4**	**4.3**	**1.8**	**1.6**	–		

[a]BMACL-Bangladesh maximum acceptable contamination levels

References DPHE/BGS/DFID (2000); Chakraborty et al. (2010); CSISA-MI (2015); MICS-B (2018); and World Bank Group (2019) relates to the injury of the cardiovascular, immunological, pulmonary, endocrine, and neurological system (Karim 2000; Smedley and Kinniburgh 2002). Collected data by the governmental authorities, NGO's and private bodies reveal that many peoples in Bangladesh are suffering from keratosis (68.3%) and melanosis (93.5%) which the more common appearances among the affected people. Besides, patients of hyper-keratosis (37.6%) and leucomelanosis (39.1%) have found in many cases. Some private bodies have analyzed 11,000 urine, skin, hair, and nail samples collected from affected villages in Bangladesh and the result shows that around 90% of people have arsenic in their urine, hair, and nail above the usual level (SOES and DCH 2000). The normal concentration of arsenic in hair, nail and urine is 0.08–0.25, 0.43–1.08 and 0.005–0.040 mg/kg respectively (Arnold et al. 1990; Dhar et al. 1997).

4.4.3 Microbial Contamination

There are numerous kinds of micro-organisms, including bacteria, algae, fungi, viruses, and protozoa (Azizullah et al. 2011). Like other third-world and less developed countries, Bangladesh faces serious waterborne diseases (Islam et al. 2001). Groundwater is well protected from the hazard of bacteriological contamination than surface water because of the topsoil which performs as a fence above it. Beyond that, microbial contamination of groundwater can occur when wells, offal holes, and septic tanks, poorly and roughly constructed (WHO 2011; Alwashali et al. 2014; Parvez 2016). In rural areas, communities or families using septic tanks are exclusively vulnerable to biological contamination of their drinking water. The most usual public health risk related to potable water is biological contamination from disease-causing microbes, such as bacteria, viruses, and protozoa. Intestinal bacteria from warm-blooded animals (E-coli) pose a serious threat to public health (Shiekh 2006; Ahmed 2013; Dey et al. 2017a, b). Drinking water guidelines state there should be no bacteria (E-coli) present, that is 0 coliform per 100 ml of water samples, in drinking water (WHO 2004).

Several investigations reveal overwhelming microbial infection of drinking water in Bangladesh which is included in Table 4.8. At Gakulnagar (village) under Dhaka district 60% of hand tube-well water is infected by the Coliform bacteria (Rahman 2009). Besides, another study found that 3 tube-wells are highly contaminated with total coliform in the Sardarkandi village under the same district (Ferguson et al. 2011). Likewise, among 207 underground water samples, collected from Sirajganj, Comilla, and Brahmanbaria districts, 41% samples were contaminated by total coliform, 13% samples by Escherichia coli and 29% samples by thermo-tolerate coliforms (Luby et al. 2008). Another investigation was conducted through 21 sallow tube-wells from a Shinduria village under the Savar sub-district of Dhaka and found total coliforms ranging from 1.5×10^1 to 4.98×10^4 count/100 ml and fecal coliform 0 to 3.49×10^3 count/100 ml (Rahman 2013). An examination carried out through collected 53 tube-well water samples from different

Table 4.8 Bacteriological contamination of groundwater in different areas of Bangladesh

Sampling location	No. of sample	Total coliforms counts/100 ml	Fecal coliforms counts/100 ml	Reference
Matlab, Chandpur District	05	05	03	Islam et al. (2001)
Dhaka Metro	19	1.0×10^1	0	Parveen et al. (2008)
Gakulnagar, Dhaka District	20	3	0	Rahman et al. (2009)
Dhaka District	19	1.0×10^1	0	Islam, et al. (2010)
Sinduria, Dhaka District	21	8.44×10^3	5.16×10^2	Rahman (2013)
Kushtia District	32	3.97	16.88	Rahaman and Hafezur (2018)
5 district of Bangladesh	248	3.7	—	Van Geen et al. (2011)
Chittagong District	18	1.3×10^1	31	Mojumder and Shajedul (1995)
Cox's Bazar Town	12	39.5	57	Mojumder and Shajedul (1995)
Comilla, Brahmanbaria and Sirajganj Districts	207	41% contaminant, n = 85	13% contaminant n = 27	Luby et al. (2008)
Chadpur and Narayonganj District	125	01–10: 30% sample 10–100: 9% sample >1000: 4% sample	54% contaminant, n = 67 20% contaminant, n = 25 26% contaminant, n = 8	Van Geen et al. (2011)
Different hydrological regions of Bangladesh, 10 old district.	26,229	21% contaminant, n = 550	58% contaminant, n = 15,215	Dey et al. (2017a)
Rural area of Bangladesh	896	–	4.8	Henry et al. (1990)
Mymensingh District	1726	41% contaminant 14% >10 cfu 3% >100 cfu	—	Ercumen et al. (2015)

37 districts of the country and found that 81.2% of them contaminated with antibiotic resistance coliform bacteria (Parvez 2016). In 2012–2013, a countrywide Multiple Indicator Cluster Survey (MICS) combinedly was conducted by UNICEF and Bangladesh Burau of Statistics (BBS). Underground water of Bangladesh covers the high level of coliform bacteria representing the poor quality of water and highlighting public awareness in this respect.

In this section, the total groundwater quality regarding the chemical aspect is discussed with reference. This study informed that the groundwater of Bangladesh is roughly safe for drinking purposes. But it is a fact that the arsenic concentrations in groundwater are in an alarming position. In addition, the levels of Hardness, Fe, Ca, and Pb in some samples are crosses the WHO and native acceptable value. So, we suggest that water should consume through proper treatment or filtration.

4.5 Drinking Water Quality Evaluation

From the point of views of the previous section in the text, the present concentration of some parameters in groundwater such as total hardness with carbonate and bicarbonate of calcium and magnesium through almost all over the zone, iron, and lead in some regions, salinity with sodium and chloride in coastal part, sulphate and phosphate in north-east zone, and arsenic in 85% area of the country are objectionable to drinking and other household purposes. Especially, without any treatment and proper filtration, the groundwater of 10 coastal districts in the country is completely unfit for potable purposes. Though groundwater is well protected from the hazard of microbial contamination, but various studies confirm that water of some region is contaminated with faecal coliform. With the help of several water quality indices, it can assess the water quality for drinking purposes.

Water quality index (WQI) is to abridge big quantities to water quality data into simple terms (e.g. good, fair, or poor) for reportage to the suitability of water use. It can assess based on various physical, chemical, and microbial parameters (Sahu and Sikdar 2008). Some water quality directories have been formulated all over the world which can easily judge-out the overall water quality inside the area efficiently and promptly (Tyagi et al. 2013; Landwehr and Deininger 1976). Groundwater quality index (GWQI), heavy metal evaluation index (HEI), heavy metal pollution index (HPI), the degree of contamination (C_d), etc. are the major evaluation indices for drinking water and some results of these indices of various groundwater samples in different places of the country included in Table 4.9. A water quality index delivers a single figure (like a grade) that expresses the overall water quality at a certain time and location based on several water quality parameters. The WQI range and type of water can classify as excellent (95–100), very good (89–94), good (80–88), fair (65–79), marginal (45–64), and poor (0–44); in which water quality index values placed within the bracket (Bhuiyan et al. 2010; Tirkey et al. 2013). An index aims to turn complex and complicated water quality data into information that is useable and comprehensible by the public. A single number of water quality index cannot state the complete story of water quality; many additional water quality parameters not comprised in the index. The index presented here is not aimed at public health or aquatic life guidelines. Nevertheless, a water quality index based on some very significant parameters can deliver a simple indicator of water quality (Bharti and Katyal 2011). Table 4.9 informed that concerning WQI, HPI, HEI, and $C_{d,}$ an average of 40 to 60% water not completely fit for potable or household uses. In Lakshimpur District, almost 50% of water samples are very poor quality, 15% are highly metal-polluted, and the degree of contamination is almost high.

Table 4.9 Water quality evaluation and health risk indices of groundwater in Bangladesh

Sample location	No. of samples	Depth (m)	Drinking Water Quality Evaluation Indices				Reference
			WQI	HPI	HEI	C_d	
Faridpur District	60	14–204	**100.70 (mean)** EW-28.33% GW-18.33% PW-48.33% RW-5.00%	**46.00(mean)** LP-66.67% MP-25% HP-8.33%	**8.55(mean)** LP-65% MP-31.67% HP-3.33%	**7.52(mean)** LP-71.7% MP-26.67% HP-1.67%	Bodrud-Doza et al. (2016)
24 selected sub-districts of Bangladesh	960	20–300	**11.8 to 371.5** EQ-16.67% GW-16.67% PW-41.66% RW-25%	—	—	—	Tahera et al. (2016)
Lakshmipur District	70	10–318	**114 (mean)** EW-18.57% GW-33.0% PW-40.0% RW-8.43%	**26.1(mean)** LP-67.14% MP-18.57% HP-14.29%	**7.44(mean)** LP-75.71% MP-22.86% HP-1.43%	**11.2(mean)** LP-60.0% MP- 24.29% HP-15.71%	Bhuiyan et al. (2010)
Singair, Manikganj Dist.	38	12–247	**23.76 to 257.18** EW-43% GW- 24% PW- 21% RW- 9%	—	—	—	Mazeda et al. (2018)
Moddopara, Dinajpur District	12	25–35	**19.2 to 44.1** EW-100% GW- 0% PW- 0% RW- 0%	**19.91(mean)** LH-81% MP-09.67% HP-09.33% *(calculated)*	**6.87(mean)** LH-78.34% MP-11.00% HP-10.66% *(calculated)*	**10.3(mean)** LH-73.09% MP-23.71% HP-02.39% *(calculated)*	Howladar et al. (2017)

(continued)

Table 4.9 (continued)

Sample location	No. of samples	Depth (m)	Drinking Water Quality Evaluation Indices				Reference
			WQI	HPI	HEI	C_d	
Gopalganj District	23	20–35	**11.3 to 90.5** EW-43.47% GW- 34.78% PW- 13.04% RW- 8.69%	**46.39(mean)** LH-59.98% MP-23.56% HP-16.46% *(calculated)*	**10.12(mean)** LH-78.37% MP-10.67% HP-11.96% *(calculated)*	**12.91(mean)** LH-67.00% MP-19.98% HP-13.02% *(calculated)*	Atikul et al. (2018)
Savar, Dhaka District	40	25–45	**46.2 to 85.5** EW-0% GW- 17.58% MW- 37.5% PW- 0%	—	—	—	Parvez and Pritul (2018)
Rajshahi city	190	45–90	EW-18.1% GW- 80.3% MW- 1.6% PW- 0%	**38.35(mean)** LP-66.67% MP-25% HP-8.33% *(calculated)*	**7.52(mean)** LP-67.80% MP-30.65% HP-1.55% *(calculated)*	**7.52(mean)** LP-62.92% MP-28.67% HP-08.32% *(calculated)*	Rahman et al. (2017b)

WQI: water quality index, HPI: heavy metal pollution index, HEI: heavy metal evaluation index

4.6 Assessing Irrigation Water Quality

In Bangladesh, groundwater is the main source of irrigation, which is around 30 to 40% of net tillable land (Shirazi et al. 2012; Huq and Naidu 2002). The groundwater contribution to the whole irrigated zone increased expressively from 41% in 1983 to 86% in 2002 (BADC 2002; Hasan et al. 2007; Islam and Shamsad 2009). Various studies assumed that the quality of agricultural water is deteriorating day by day due to heavy pumping and other anthropogenic activities. The water class used for irrigation is vital for the yield and number of crops, protection of the hydrological environment, and preservation of soil productivity. The physical and mechanical properties of the soil, soil permeability, and ex. soil structure (aggregational stability) are very sensorial to the type of transferrable chemical constituent present in irrigation water.

The pH, Na hazard, salinity hazard, some trace elements, bicarbonate, and carbonate in connection with the Mg and Ca content, toxic anion, free Cl, and nutrients are the most crucial factors to regulate the suitability of water use in irrigation (Wallace and Batchelor 1997). On the other hand, according to Michael (1978), Matthess (1982), Raghunath (1987), and Guy (2003), the agricultural water quality is decided by the 4 most admissible criteria such as total dissolved solids or electrical conductivity; the level of certain specific elements like Na, K, Mg, Ca, Cl or/and B contents; relative quantity of Na to other cation, denoted by Sodium Adsorption Ratio (SAR); and Residual Sodium Carbonate and Bicarbonate (RSC & RSBC). Apart from these, other measures like the presence of soil hardpan, depth of the water layer, $CaCO_3$ level in the soil, and K and NO_3 ions also influence the appropriateness of irrigation water indirectly (Michael 1978). Besides, Matthess (1982); and Ayers and Westcot (1985) described that poor-quality irrigation water creates 4 types of problems such as; specific ion toxicity (affects sensitive crops), salinity (effects crop water availability), water permeability (effects penetration rate of water into the soil), and miscellaneous (Atikul et al. 2018). The values of including previously discussed parameters, other parameters such as Soluble Sodium Percentage (SSP), Residual Sodium bicarbonate (RSC), Kelley's Ratio (KR), Permeability Index (PI), and Magnesium Adsorption Ratio (MAR) of countrywide irrigation groundwater samples have included in Table 4.11. Also, the limits of some important indices for rating water quality and its suitability in irrigational use presented in Table 4.10.

According to findings, high salinity of groundwater due to seawater intrusion in the coastal region of Bangladesh (15% of the total country area) makes the unfit, for irrigational uses (Tables 4.10 and 4.11) (Sarkar and Hassan 2006; Rahman and Majumder 2012; Rahman et al. 2017a; Brammer 2014; Atikul et al. 2018; Naus et al. 2019). Again, most shallow aquifer's water, including the coastal region, of 60 districts out of 64 contaminated with arsenic (BGS-DPHE 2001a; Chakraborty et al. 2010) and that used for irrigation. It has been estimated that yearly arsenic input with irrigation water is around 4.4 kg/ha in paddy field (Dittmar et al. 2010) and in rice grains arsenic found above 0.017 mg/gm (Meharg and Rahman 2003). EC and

Table 4.10 Limits of important parameter indices for rating groundwater quality and its suitability in irrigation use

Category (Water class)	Irrigation water quality indices*								Degree on restriction on use
	EC, μS/cm	*SAR (%)*	*SSP (me/L)*	*RSC (me/L)*	*PI (%)*	*MAR (%)*	*TDS (mg/L)*	*Salinity (mg/L)*	
I	<700	<10	<20	–	>75	–	–	–	***Excellent***
II	700–3000	10–18	20–40	<1.25	25–75	<50	>2000	<450	***Good***
III	>3000	18–26	40–80	1.25–2.5	<25	–	450–2000	450–2000	***Fair***
IV	–	>26	>80	>2.5	–	–	<450	>2000	***Poor or rejection***
Ref.	UCCC (1974)	Fipps (2003)	Wilcox (1955)	Gupta and Gupta (1987)	Hem (1970)	Gupta and Gupta (1987)	Bauder et al. (2011)	UCCC (1974)	

**EC* Electrical conductivity; *SAR* Sodium adsorption ratio; *SSP* Saturated sodium percentage; *PI* Permeability index; *MAR* Magnesium adsorption ratio; *TDS* Total dissolved solid

TDS are significant to consider in the findings of irrigation water quality, because various toxic solid materials may embed in the water, which may cause damage to the plants (Matthess 1982). Regarding, 'degree of restriction use,' TDS concentration < 450, 450–2000, and > 2000 mg/L represent the agricultural water as 'none', 'slight to moderate', and 'severe' respectively (UCCC 1974). According to this standard, in Bangladesh, the south zone is classified as 'slight to moderate' and 'severe' and both north and center areas as 'none' and 'slight to moderate' (Table 4.11).

Water permeability of soil is an important factor in the fertility rate of cultivable land. Permeability problem occurs when the usual percolation rate of soil significantly reduced and hampers moisture supply to crops, which is accountable for the two most water quality factors as the salinity of water and its Na content relative to Ca and Mg (Sarkar and Hassan 2006). High salinity increases permeability rate in water and low salinity or water with high Na to Ca ratio (Na:Ca) decreases infiltration (Ayers and Westcot 1985). Balance of some cations and anions or comparative proportions of other different cations well-defined by SAR, SSP, RSC, RSBC, MAR, KR, etc. are also the indicators of infiltration problem. A large proportion of Na in soil with chloride and carbonate or sulfate named saline or alkali soils, respectively (Todd 1980; Rahman and Majumder 2012). Percent of Na (SSP) in water reacts with soil to decrease soil permeability and its frequent use makes the soil impermeable, whereas high Na leads to grow of alkali soil (Raghunath 1987). High Na saturation also causes Ca deficiency. In contrast, the presence of Ca and/or Mg salt in irrigation water check the bad effects of Na by increasing the permeability of the soil (Asaduzzaman 1985). Besides, in the case of saltwater, it dissolves and leaches most of the soluble minerals, including Ca and Mg from the topsoil (Ayers and Westcot 1985). Sodium content comparative to Ca and Mg load,

Table 4.11 Irrigational groundwater quality of Bangladesh on different quality indices

Sampling Area	Irrigational water quality indices (mean value) and degree of water quality[a, b]												Reference
	SAR (%)	SSP me/L	RSBC me/L	RSC (%)	PI (%)	MAR mg/L	KR	H_T mg/L	TDS mg/L	EC µS/cm	Mg:Ca	Na:Ca	
Fulpur, M'singh Dist.	0.41	13.44	–	0.44	–	–	–	260	192	297	0.44:1	0.24:1	Shahidullah et al. (2000)
Pabna district 2002–03	0.53	16.0	4.37	3.26	68	24.60	0.18	225	553	790	0.32:1	0.23:1	Sarkar et al. (2006)
2002–03	0.74	41.0	4.58	4.16	78	10.74	0.26	195.5	553	790	0.12:1	0.29:1	
Satkhira Dist. North zone	11.62	82.2	2.17	–	94	26.97	4.62	166	1098	2082	0.44:1	6.63:1	Tanvir et al. (2017)
Center zone	4.22	52.6	1.72	–	68	24.24	1.28	313	905	1595	0.40:1	1.7:1	
South zone	1.14	32.95	0.83	–	78	21.96	0.48	150	251	614	0.29:1	0.64:1	
Comilla District	2.40	44.14	0.81	–	–	58.97	0.74	337	677	1079	1.45:1	1.79:1	Prodip et al. (2016)
Botiaghata, Khulna Dist.	12.64	71.37	3.29	–	–	54.78	6.82	576	1556	3019	0.77:1	6.38:1	Didar-Ul at el. (2017)
Gopalganj Dist. Wet-season	8.87	55.35	−1.0	−3.70	68.13	57.13	1.68	597	1635	3207	0.76:1	5.13:1	Atikul et al. (2018)
Dry-season	13.07	68.31	−0.15	−0.15	76.27	58.14	3.99	541	1644	3218	0.85:1	7.0:1	
Sunamganj District	15.29	38.01	0.7	–	92.71	58.0	1.74	3018	404	330	0.73:1	2.3:1	Raihan and Alam (2008)
Joypurhat District	0.43	14.91	−0.49	1.55	56.48	10.92	0.17	153	503	733	0.07:1	0.22:1	Islam et al. (2017d)
Patuakhali Dist. Pre-monsoon	91.27	75.38	−2.64	−15.3	72	53.26	16.97	977.4	4829	8973	1.14:1	6.35:1	Rahman et al. (2016)
Post-monsoon	70.63	26.34	−3.0	−14.7	76	51.81	9.38	922	4436	7887	1.07:1	6.2:1	
Brahmanbaria	0.41	21.50	1.17	0.6	–	49.5	0.002	52.2	233	327	1.0:1	0.56:1	Mahmud et al. (2007)

(continued)

Table 4.11 (continued)

Sampling Area	Irrigational water quality indices (mean value) and degree of water quality[a, b]												Reference
	SAR (%)	SSP me/L	RSBC me/L	RSC (%)	PI (%)	MAR mg/L	KR	H_T mg/L	TDS mg/L	EC µS/cm	Mg:Ca	Na:Ca	
Bogra District	0.23	24.42	–	2.26	89.8	–	0.27	–	335	549	–	–	Shammi et al. (2016)
Trisal, M'singh Dist.	1.94	47.80	0.75	–	79.28	–	–	103.6	–	–	–	–	Shirazi et al. (2012)
Khulna District	9.67	62.40	29.18	29.18	87.1	25.94	2.09	532	–	240.6	–	–	Sultan and Billah (2019)
Faridpur District	0.76	36.95	–	–0.62	–	–	0.49	147.65	268.78	460	1.81:1	1.40:1	Shaik et al. (2015)

[a]According to (chronologically): Ayers and Westcot (1985), Eaton (1950), Todd (1980), Clesceri (1981), Doneen (1962), Richards (1968), Gupta and Gupta (1987), Kelley (1963), Raghunath (1987), Freeze and Cherry (1979) and Wilcox (1955)

[b]*SAR* Sodium adsorption ratio; *SSP* Saturated sodium percentage; *RSBC* Residual sodium bi-carbonate; *RSC* Residual sodium carbonate; *PI* Permeability index; *MAR* Magnesium adsorption ratio; *KR* Kally's ratio; H_T Total hardness; *TDS* Total dissolved solid; *EC* Electrical conductivity

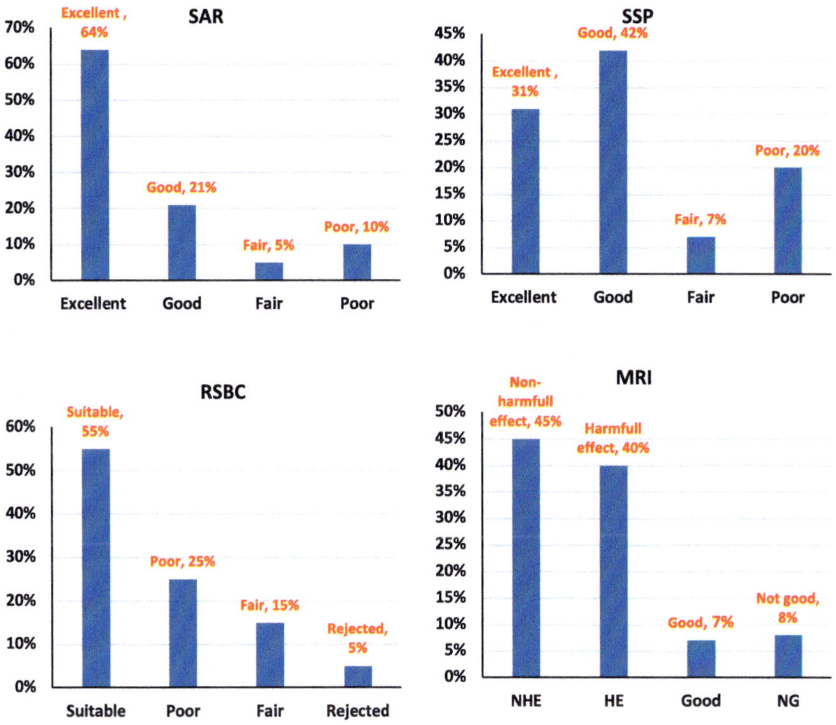

Fig. 4.4 The state of Sodium adsorption ratio (SAR), Saturated sodium percentage (SSP), Residual sodium bicarbonate (RSBC), and Magnesium ratio index (MRI) in groundwater samples of the various region of Bangladesh

usually known as sodium adsorption ratio (SAR) also impacts the penetration rate of water and so, low SAR is at all times desirable. In the maximum samples, except coastal areas, SAR values ranged below 1 and classified as excellent according to the conditions set by UCCC (1974) and Todd (1980) which showed in Table 4.10. Hence, rendering to sodium adsorption ratio, the groundwater tested was suitable for crop production. On the other hand, the PI value of water with 75% or more (>75%) and Kelley's ratio (KR) value with less than 1 (˂1) are suitable for irrigation. Table 4.11 shows that, except in the coastal areas, the Permeability index (PI) and Kelley's ratio (KR) value are good in the position. Concerning Mg:Ca and Na:Ca value, maximum water samples have good permeability characteristics and the information on Hazard Classes are almost acceptable for irrigation. Figure 4.4. shows that with respect to SAR, SSP, RSBC, and MRI 31–64% of water samples from different 11 districts of Bangladesh are very good in position, whereas 5–20% are very poor or rejected for the uses of irrigation purposes.

4.7 Sources of Groundwater Contamination in Bangladesh

Groundwater usually looks clean and clear because the water naturally filters out suspended particulate matter. But, natural and human-induced chemicals can find in groundwater (BGS and DPHE 2001a). Like other developing, densely populated, and agrarian countries; the main sources and mechanism of groundwater contamination in Bangladesh is geogenic, overexploitation, sewage, and sewage sludge, agrochemicals, commercial and industrial leaks, waste disposal, on-site sanitation systems, hydraulic fracturing, landfill leachate, etc. (Waller 1982; CSISA-MI 2015). Different physical factors influence the transport of pollutants, e.g. precipitation, diffusion, adsorption, decay in the groundwater. The contact of groundwater contamination with surface waters analyzed using hydrology transport models. Sources of groundwater contamination in Bangladesh are briefly described below.

4.7.1 Geogenic and Anthropogenic Sources

The geogenic process is the major cause of arsenic contamination of groundwater in Bangladesh, which is a serious threat to national public health (Mostafa et al. 2010; Fazal et al. 2001b). This process refers to naturally happens because of geological processes. Arsenic and fluoride pollution occurs because aquifer sediments hold organic matter that makes anaerobic conditions in the aquifers (WB Group 2019) and that conditions result in the bacterial dissolution of iron-oxides in the sediment and, hence, the release of the arsenic, usually tightly bonded to iron-oxides, into the water. So, arsenic-baring water is mostly Fe-rich, while secondary procedures often obscure the combination of dissolved both As and Fe (Ahmed et al. 2011a). A significantly high level of fluoride in groundwater is naturally caused by a deficiency of Ca in the aquifer systems (WHO 2006). Excessive concentrations of other parameters like salinity, iron, chromium, manganese, radon, and uranium in groundwater may also be of the geogenic source (Leeuwen 2000).

Bangladesh is the twelfth most densely populated and ninth most populous country in the world. Along with population growth, there is a cumulative problem of waste management all around the country. Solid waste generation in Bangladesh is around 22.4 million tons per year, 60,000 m^3/day effluents from 7000 big industries without any treatment just in and around Dhaka city and 30,000 m^3/day household liquid sewage generated by Dhaka Metropolitan (Memon 2002; Waste Atlas 2012; Abedin and Jahiruddin 2015). The tannery industries in Bangladesh produced large amounts of effluents containing toxic chemicals and discharged into nearby water bodies without any treatment that threatened the environment. Worldwide, it was estimated that discharged tannery waste effluent holds 300–400 million tons of heavy metals, toxic sludge, solvents, liquor, and other wastes materials, which dumped into water bodies every year (Wosnie and Wondie 2014;

Manjushree et al. 2015). Bangladesh has inadequate wastewater treatment infrastructure and no modern waste disposal management or there are organized failures of the on-site sewage dumping system (Enayetullah et al. 2005; Alamgir and Ahsan 2007). Along with nutrients and pathogens, untreated sewage can also have a significant load of heavy toxic metals and other inorganic and organic pollutants that may seep and leach into the groundwater system. The harmless effluent from Effluent Treatment Plants (ETP) may also reach into the aquifers by leaching if the effluent is discharged to local surface water bodies. Thus, those ingredients that are not removed in simple sewage treatment plants (STP) may reach the aquifer system as well (Philips et al. 2012). This is because, in local STP, micro-pollutants such as hospital wastes, medicinal residues, and other micro-contaminants contained in feces and urine only partly removed and the remaining dumping into surface watersheds, from it, may also accumulate in the groundwater. In agriculture, sewage sludge and/or spreading wastewater may also comprise causes of faecal pollution in groundwater.

In Bangladesh, 61.7% of the households used potable water were faecal contaminated (MICS-B 2018). Subsurface water contamination with pathogens can also happen from the liquids leaching and infiltrating into the groundwater layers from unhygienic on-site pit sanitation systems, which is the major cause of microbial contamination in Bangladesh (Wolf et al. 2015; MICS-B 2018). The outcome and transportation of microbes are very complex and the interaction among them not yet clear (WHO 2006). Bangladeshi soil is very coarse with sandy and easily liquids leach from the septic tank and pass the unsaturated soil zone and then finally enter the groundwater levels. Throughout the passageway in the soil, microbes can die off or adsorbs by soil particles, mostly depending on the travel time between the septic tanks and the well. Generally, pathogens destroy in 50 days of travel through the ground layer (DVGW 2006; Nick et al. 2012). The degree of pathogen elimination strictly varies with aquifer type, soil type, distance, and other environmental events, such as heavy rain or flooding (WHO 2006; Graham and Polizzotto 2013).

Landfill leachate, as another source of groundwater pollution in the country, from sanitary landfills can lead to groundwater contamination. Chemicals from leachate can spread into groundwater through runoff and precipitation. New landfills are required to line with clay or other synthetic substances, along with leachate to protect surrounding groundwater. Nevertheless, grown-up landfills do not have these measures and are often close to surface waters and in porous soils. Confined landfills can still pose a hazard to groundwater if they are not covered by waterproof material before closing to prevent the penetrate of hazardous substances (US-EPA 2019). Azimm et al. (2011) have published a laboratory test results in untreated grab leachate sample from Dhaka city with a high concentration of dissolved solids (734 mg/L), chemical oxygen demand (1631 mg/L), HCO_3^- (27,962 mg/L), NH_4^+ as N (1253 mg/L), Ni (1.05 mg/L), and Cr (0.74 mg/L) and have a very high potential for contaminating groundwater.

4.7.2 Industrial and Commercial Source

Meteorological processing and ore mining facilities are the primary responsibility for the existence of metals in groundwater of anthropogenic origin. The oil spills and gasoline leaked allied with underground pipelines, tanks and fuel stations can release benzenoid materials and other insoluble petroleum hydrocarbons that quickly infiltrate down into the groundwater layers (WHO 2006). Numbers of dangerous toxic chemicals, chlorinated hydrocarbon solvents, and especially wood preservatives used in chemical industries which may deposit on the top of less-permeability layers and then finally into ground layers through industrial seeping and leaking (WHO 2006; US-EPA 2018; Islam and Mostafa 2021c). Other related causes of groundwater pollution in Bangladesh is chemical spills from industrial operations, especially from leather, textile, and chemical factories; chemical spills occurring during transport, infiltration from mining operations or urban runoff, illegal waste dumping, deicing chemicals from airports, road salts; declining water-tables due to overexploitation (secondary cause), the burial of corpses and their subsequent degradation and even atmospheric pollution since groundwater is part of the hydrologic cycle (Shahid 2011; GWF 2015; SEPA 2005; Manjushree et al. 2015).

4.7.3 Agrochemicals and Pesticides

Globally, ever collective demand for food and agricultural resources has led to an increase of pesticide toxicity and environmental threat (Kaur and Goyal 2019; Kalyani et al. 2020). Agriculture is the determining factor of the Bangladesh economy. Nationally, around 300 kg of chemical fertilizers are consumed per hectare (FAO 2018), and annually 37,258 tons of pesticides used in farmland (Faruk 2018). Nitrate, phosphate, sulfate, and some semi-toxic metals can also pass in the groundwater via the overuse of fertilizers, as well as manure spreading. High application rates of P and N-containing fertilizers in the country combined with the rich water-solubility of phosphate and nitrate leads to increase runoff into surface water as well as leaching into groundwater, thus producing groundwater contamination (Rahman et al. 1995; Singh et al. 1995; Alam 2006; Jackson et al. 2008). Imperfect management practices in all types of fertilizer spreading can introduce both pathogens and nutrients (nitrate/phosphate) in the groundwater system (Suthar et al. 2009). The excess uses of animal manure or compost may also result in groundwater contamination with medicinal residues derived from veterinary drugs (UN-water 2015). The US-EPA, European Commission (EC) and Government of Bangladesh (GoB) also are dealing with the nitrate and phosphate hazard related to agricultural expansion, as a main water supply problem that requires suitable managing and governance (Custodio 2002; DoFE 2018). Runoff of pesticides, another serious hazard of Bangladesh, may leach into groundwater, causing public health problems from polluted water wells (WHO 2006). In general, more pesticide ingredients detected as groundwater quality monitoring programs have become more extensive; but, much less monitoring is conducted in developing countries, like Bangladesh, because of high investigation costs (WHO 2006).

Pesticides, are toxic and lethal substances used to kill animals, plants, or fungi that cause economic loss to crops and ornate plants, but are seriously harmful to the human and health of domestic animals also (Aktar et al. 2009; Chowdhury et al. 2013; Riah et al. 2014). In 1951, pesticides were first introduced in Bangladesh for controlling pests during cultivation and for improving production and then, pesticide use is quickly increasing day by day (Rahman 1997; Matin et al. 1998). Pesticide usage grew rapidly in the early 1970s, following the introduction of a modern and high productive rice variety Boro. Use dropped dramatically in 1973/'74 because the government has withdrawn the financial subsidy on pesticide by that time and which removed wholly in 1978. Pesticide use has increased again; sales of it doubled in the second half of the 1980s (Rahman 1997). Over the years, it has been seen that pesticide consumption in the agricultural sector has increased and this is projected to continue for the next periods due to technological and socio-economic development (Bempah 2011). In the early past, 37,258 million tones (MT) pesticide of 4500 varieties used in the agriculture of Bangladesh whereas this figure was 31,522 MT in 2006 (PPW and DAE 2018). Most of the farmers in Bangladesh use pesticides unreasonably, inappropriately, and purposelessly as they have inadequate knowledge and consciousness about the use (Miah 2014). Excess and unscientific use of pesticides can cause serious public health insecurity mainly as residues in food and water (Chowdhury et al. 2013).

Pesticide poisoning of groundwater is an issue of national consideration because groundwater is used for drinking purposes by about 95% of the nation's population, especially in rural areas (PPW-DAE 2018). Before the mid-1970s, it thought that soil acted as a natural protective filter that hindered pesticides from reaching unconfined groundwater aquifers (MICS-B 2018). Nowadays, studies have shown that this is not a fact. In Bangladesh, groundwater tables are high, and the soil is mainly sandy, coarse, and loose, so there is a big chance that groundwater can have contaminated by leaching of pesticide residues. Anwar and Bari (2004); Anwar and Yunus (2010); and Anwar and Saing (2010) confirmed that the Leaching Potential Index (LPI) of sediment and soil in the maximum zones of Bangladesh is sufficiently high to leaching pesticides into the subsurface environment. Pesticides, like most other chemical contaminants, can easily reach water-bearing layers from cultivation fields, seepage of polluted surface water, inappropriate disposal, unintentional or accidental spills and leaks, and even through injection waste substances into wells (Majewski and Capel 1995).

Accumulated pesticides can create a serious long-term risk to human body systems. In Bangladesh the pesticide exposure level (0.018 $mg.kg^{-1}$ day^{-1}) is much higher than that of the permissible level (0.005 $mg.kg^{-1}$ day^{-1}) (Alam et al. 2015). Thus, there is a substantial risk of exposure threats of pesticides to humans and other non-targeted other species in Bangladesh. Runoff, spillage, washing of pesticide containers in pond water non-contributor pesticide residues to surface and groundwater in Bangladesh. A study showed that 0.0448 mg/L concentration of

DDT arrived pond from the adjacent paddy field through water by rain wash (Hasanuzzaman et al. 2018). Some pesticides have denominated the maximum contaminant limit in potable water set by the WHO and US-EPA but many have not. Similarly, in drinking water, the consequence of combining more than one pesticide might be different than the things of each specific pesticide alone (US-EPA 2016). This is another situation where has no adequate systematic data to draw dependable conclusions.

Limited data be available concerning pesticide pollution of groundwater in Bangladesh. Inadequate financial grants and laboratory facilities are the main hindrances in the obtainability of data in the country. The results of separate studies detailing pesticide accumulation in groundwater of Bangladesh and the relevant world health organization (WHO) standard values are arranged in Table 4.12. Most of the farmers used mainly chlorinated hydrocarbons, organophosphates, chlorophenoxiacids, and carbamate pesticides in Bangladesh and a residual portion of those pesticides can accumulate in living beings through the food chain or drinking water (Dasgupta 2007). Organochlorinated pesticides like DDT, lindane, dieldrin, heptachlor was generally used in agricultural lands and control diseases like malaria in Bangladesh from the early '50s (Rahman 2000). In 1993, Bangladesh's government banned those pesticides, but there were reports that it used up illegally (Shoeb 2009, 2016). Organochlorine pesticide having high lipophilicity, lower polarity, less aqueous solubility, and a very stable half-life make it a serious threat to public health and the environment as it can bio-accumulate in the food chain (Nahar 2008; El-Mekkawi 2009; Shoeb 2009; Afful 2010; Valenzuela et al. 2020). On the other hand, organophosphorus pesticides are inexpensive and efficiency is very good, farmers are encouraged to use it. Nonetheless, these pesticides are very injurious to farmer's health, and those are also genotoxic and carcinogenic (Hayat 2010; Chowdhury 2012). As chlorpyrifos can cause attention to insufficient hyperactivity disease and development uncertainty, both in embryo and children (Rauh et al. 2006; Kaur and Goyal et al. 2019). On the other hand, carbamate especially carbofuran pesticides cause main problems in the reproductive system (Goad 2004) and it can cause vomiting, blurred vision, nausea, and breathing problems (Kamrin 1997).

After using the pesticides, rain washes the residues to the adjacent water bodies and they become polluted (Kreuger 1998). Matin et al., (1998) have collected 144 groundwater samples from around the country and found most of the samples carried 10 to 1000 times higher concentration of pesticide residue than WHO recommendation values (Table 4.12). In this investigation, all Dichlorodiphenyltrichloroethane (DDT) values were found ranging from 0.27 to 1.204 mg/L that was very much higher than recommended values. They also found heptachlor residues ranging from 0.025 to 0.789 mg/L. Another researcher (Islam 2007) collected 48 cultivated field water samples from various zones in the country and found water samples from 10 sites contaminated by DDT, lindane, and heptachlor. In this investigation, DDT found at Bogra and Rajbari districts; lindane found in Dhaka, Noakhali, Sylhet, and Shariatpur districts; and heptachlor found in Dhaka, Chittagong and Magura districts in which the highest level of DDT residue

Table 4.12 Pesticide contamination in mg/L of groundwater samples in Bangladesh

Sampling location	Water type and depth	Pesticide detected	Concentration	WHO (2011) Standard	Reference
Nayerhat, Dhaka District	Shallow 20–40 m	*p.p'-Dichlorodiphenyldichloroethane, DDE*	0	-	Rahman (1997)
		Dieldrin	Traces	0.00003	
Dhamrai upazila, Dhaka District	Shallow 25–40 m	*Malathion*	0.0042	0.0019	Hasanuzzaman et al. (2017)
		Diazinon	0.0003	0.0020	
Deferent region of Bangladesh	Shallow	*p.p'-Dichlorodiphenyldichloroethane, DDE*	0.010–0.084	0.001	Matin et al. (1998)
		p.p'-Dichlorodiphenyldichloroethane, DDD	0.014–0.365	0.001	
		p.p'-Dichlorodiphenyltrichloroethane, DDT	0.027–1.204	0.001	
		Heptachlor	0.025–0.789	0.00003	
Daudkandi, Comilla District	Shallow 07	*p.p'- Dichlorodiphenyldichloroethane, DDE*	BDL	0.001	Hasanuzzaman et al. (2019)
		p.p'- Dichlorodiphenyldichloroethane, DDD	BDL	0.001	
		p.p'- Dichlorodiphenyltrichloroethane, DDT	BDL	0.001	
		Malathion	BDL	-	
		Diazinon	BDL	-	
		Chlorpyrifos	BDL	0.03	
Ghior, Manikganj District	Shallow 21	*Cypermethrin*	BDL - 310	-	Shubhra at el. (2012)
		Chlorpyrifos	BDL - 34.6	0.03	
		Diazinon	BDL - 27	-	
Nagarpur, Tangail District	Shallow 42	*Malathion*	0.0258	-	Hasanuzzaman et al. (2018)
		Diazinon	0.0002	-	
		Chlorpyrifos	0.0037	0.03	
		p.p'- Dichlorodiphenyldichloroethane, DDE	BDL	0.001	
		p.p'- Dichlorodiphenyldichloroethane, DDD	BDL	0.001	
		p.p'- Dichlorodiphenyltrichloroethane, DDT	BDL	0.001	

BDL Bellow detection limit

estimated in Bogra, and it was 0.54 ppm. On the other hand, in Dhamrai and Savar Upazila, among the 27 well water samples, carbofuran and diazinon were found from Savar at 198.7 mg/L and 0.9 mg/L respectively. In Dhamrai Upazila malathion, carbaryl, and carbofuran were found at 105.2 mg/L, 14.1–18.1 mg/L, and 105.8 mg/L separately (Chowdhury 2012). Hossain et al. (2015) also conducted studies in the same places and find out the level of the same parameters. Chowdhury et al., (2012) collected water samples from paddy fields, lakes and well in Rangpur city and found carbofuran levels from 0.949 to 1.671 mg/L, chlorpyrifos ranging from 0.554 to 0.895 mg/L, and carbaryl was 0.195 mg/L in the lake water. Also, they found carbofuran in 7 samples ranging from 0.934 to 3.395 mg/L, chlorpyrifos in 7 samples ranging from 0.477 to 1.189 mg/L, and carbaryl in 2 samples at 0.055 and 0.163 mg/L in the paddy field water samples. Besides, Chowdhury et al., (2013) collected irrigated groundwater samples from 22 districts in Bangladesh and they detected DDT residue in Rajshahi, Feni, and Chapai Nawabganj districts. The highest DDT level was 8.29 mg/L. Also, they found both heptachlor and DDT residues in water samples from Chhatak at Sunamganj district and observed heptachlor residues in Madaripur, Sunamganj, and Natore districts with a maximum concentration of 5.24 mg/L. Sumon et al. (2018) quantify the residues of 10 most generally used organophosphorus pesticides, OPPs in water and sediment in north-west Bangladesh and to assess their ecological hazards for aquatic plants and animals.

Not only in Bangladesh, globally agrochemicals are the greatest threat to the aquatic environment. Several researchers agreed that, nanoscience may resolve those problems by using nano-agrochemicals. The N, P, B, K, Mn, Cu, Mo, Zn, Fe, Ni, and nano-carbon tubes show better efficiency as a nano-fertilizers. On the other hand, nano-pesticides as Zn, Mo, Ag, Cu, ZnO, SiO_2, and nano-designs show well broad-spectrum insect and pest defense efficacy in contrast with usual pesticides (Chhipa 2017). The main sources of groundwater pollution/contamination in Bangladesh are discussed in this section. Public unawareness, weak practice and policies, inadequate monitoring, and primitive acts and law are another cause of groundwater pollution. As a vast agrarian country, agrochemicals are one of the major sources of water contamination in Bangladesh.

4.8 Challenges in Groundwater Management

In the last three decades, the Government of Bangladesh tried to introduce strategies and policies to regulate and monitor groundwater resources, though sustainable and practical solutions have proven intangible. In Bangladesh, present laws and regulations concerning water resources do not cover adequately areas like the powers, rights, and duties of private users and the government; registering and leaving off right to water, and directorial structure to implement laws and regulations (ADB 2007a; Zahid 2015). However, in 1985 the Government of Bangladesh announced an ordinance, Ground Water Management Ordinance 1985, only for the supervision of agrarian groundwater resources. The ordinance is permitting introduced to control to the fitting of privately-owned tube wells in critical areas, where the groundwater was dwindling at quick rates, and where groundwater quality was worsening (Shah et al. 2009; Qureshi et al. 2010). Following regulations such as Irrigation Act, 1876; Irrigation Water Rate Ordinance, 1983; The National Environmental Policy, 1992; The Water Resources Planning Act, 1992; National Policy for Safe Water and Sanitation, 1998; National Water Policy, 1999; and National Policy of Arsenic Mitigation in Bangladesh, 2004 insist the need for the conservation of groundwater capitals. Lastly, the Bangladesh Water Act, 2013 have approved by the Government of Bangladesh for management, integration, development, use, distribution, extraction, preservation, and protection of water resources. In this act, any individual must get a permit or license for large-scale extraction of groundwater by the individuals or non-government organizations (NGOs) beyond domestic and agricultural use. Many of these acts and policies would necessitate up-to-date and extension.

A landmark in planning and formulating policy for the water division was the adoption of the National Water Policy (NWP) in 1999. This is done in consultation with NGOs, civil society, stakeholders, social activists, and major donors. Important water resource management and regularity issues addressed by the NWP are, water rights and provisions; river basin managing; water supply and sanitation; public and private venture; water and agriculture; stakeholder participation; water for the environment, regulative framework, etc. The major objective of the NWP is to progress water resource supervision and conservation of the environment and to give direction to all delegations working in the water sector for the accomplishment of itemized objectives. In the NWP, the necessity of collective use of surface and groundwater highlighted also and this act has provided comprehensive principles of development of water capitals and their rational exploitation under diverse constraints. This act has considered environmental restoration, protection, and improvement measures consistent with the National Water Management Plan (NWMP) and National Environmental Management Action Plan, NEMAP (Zahid 2015).

In Bangladesh, several acts, ordinances, or/and laws regarding sustainable groundwater management and its use, protection, and management are recognized. But, installing permit and licensing systems, enforcing laws, and the establishment of tradable asset rights has so far recognized largely unsuccessful and ineffective, perhaps owing to complications in accountability and enforcement, which are typical especially in rural areas (CSISA-MI 2015). Most acts, including the Groundwater Management Ordinance (1985), speak of dealing with groundwater at the rural Upazilla (Sub-district) level. Though, the supervision of subsurface beyond administrative restrictions has contrariwise ignored. Significant experience of what does not work in progressing groundwater legislation learn from the regulation knowledge of neighboring countries, such as China, Pakistan, and India, where licensing or permit systems also unsuccessful to yield the required results (Qureshi et al. 2010). Furthermore, to the reduced spread of the state into remote rural areas and weak responsibility and implementation, the sheer number of tube-well operators is also a key reason for the failure of permitting strategies in these countries (Shah et al. 2003; CSISA-MI 2015). In these countries, the population of the shallow pump and socio-economic situations bear great similarities with Bangladesh. Traditional groundwater regulation through the initiative of licensing systems and groundwater usage rights, direct and imperceptible pricing, and offering water on a volumetric basis is thus not alike to attain in Bangladesh also (Shah et al. 2009).

Now a day, Bangladesh is facing a crisis of fresh and arsenic-free groundwater due to overexploitation and overpressure on shallow aquifer water as well as inadequate knowledge on water management and lack of implication to groundwater-related acts. So, it needs to work on consciousness growing through educational and training programs for all stakeholders, improvement of alternative crop production, and programs to grow strong production markets for these alternative crops. For the improvement of groundwater management, it should be privatized on significant investments in modern agricultural technologies and water e.g. charging water on a volumetric base or crop definite pricing, simplify markets for non-rice crops, fixing allocations for groundwater withdrawal for different users, up-gradation of alternative cropping patterns, and extra backing through a financial subsidy for cultivator making the shift to less water demanding crops.

4.9 Recommendations

Resembling long-term extraction of groundwater to recharge and quality monitoring is the major objective of sustainable planning of groundwater resources. Keeping the water steadiness of withdrawals and recharge is essential for managing the anthropological impact on ecological and water resources. Quality of groundwater regarding drinking, irrigation, and industrial uses; projecting the next development potentials and socio-economic along with the environmental impact assessment, and groundwater resources management can achieve covering following features;

- Obsessive extraction of groundwater for domestic, irrigation, and industrial uses need to control;
- Should be frequently monitored the chemical and bacteriological characteristics of countrywide groundwater;
- Because of excessive demand of water and to reduce dependence on confined groundwater resources, consumption of obtainable surface water and cohesive use should be stressed, according to NWP and other strategies of the government; that will reduce the seasonal variation rate of the water table and minimize stress on groundwater stock;
- Assessment of the most valuable exploitation of groundwater resources by developing priority for long time use considering extensive droughts, shifting inhabitants and agricultural growth to minimize the increasing pressure on groundwater stock in an area. Evaluate groundwater contamination and substitute measures of keeping the resource in the future and protect the public health;

- Regional geologic modeling of the groundwater resources must develop for effective water management to plan agricultural and domestic water supplies and to forecast the groundwater circumstances in advance for dry seasons;
- Improved maintenance and operation of shallow tube-wells (STW), crop diversification, build-up the management efficiency, ensure electrification of deep tube-well (DTW), install new DTWs in the potential zones may increase crop production.

4.10 Conclusion

In a densely populated and developing country like Bangladesh, groundwater is the most valuable unlimited natural resources for livelihood, food security; and plays a vital role in the development process. The review observed that the hardness and iron level of underground drinking water is a key problem in Bangladesh. Besides, heavy metal, microbial, and pesticide contamination of groundwater is the main threat for the people of Bangladesh. Among the trace heavy metals, As found in groundwater at an alarming concentration in the country and now that is a great threat to public health. In some cases, a high level of Ca, Mg, Fe, Cr, Pb, and salinity, especially in the coastal zone, is also worrying. Other metals like Cu, Zn, Co, Cd, and Ni found below the standards limit in most of the samples. The review illustrated that groundwater of some areas found contaminated with heavy metals and unsafe for domestic and drinking purposes. Furthermore, except for bicarbonate and chloride in the coastal zone, the anions found in water sources are below the permissible level. As an agrarian country every year huge amounts of chemical pesticides used for crop production and few reports showed the groundwater in some areas contaminated with excessive amounts of residual pesticides. In coastal areas, the climate change and rising sea levels resulting in the saline water over intrusions into shallow aquifers and in those areas groundwater is unfit for irrigation uses. But except in some cases, overall irrigation water quality in another part of the country is in good condition. The study found that the lack of modern water treatment, and supply facilities, quality monitoring, as well as the reluctance of law implementation are the main challenges for safe water supply in the country. Moreover, public awareness and publicity programs towards groundwater contamination are seriously needed to ensure safe water for all. Further, advanced research and survey-based works are too enhanced to get better scenarios of safe water status and water-related problems in Bangladesh.

References

Aaron KJ, Sanders PW (2013) Role of dietary salt and potassium intake in cardiovascular health and disease: a review of the evidence. Mayo Clin Proc 88(9):987–995. https://doi.org/10.1016/j.mayocp.2013.06.005

Abedin MA, Jahiruddin M (2015) Waste generation and management in Bangladesh: an overview. Asian J Med Biol Res 1(1):114–120. www.ebupress.com/journal/ajmbr

ADB (2007a) Recent advances in water resources development and management in developing countries in Asia, Asian water development outlook 2007 discussion paper. Asian Development Bank (ADB), Manila. https://www.adb.org/sites/default/files/publication/27971/awdo.pdf

ADB (2007b) Asian water development outlook 2007 country paper Cambodia. Asian Development Bank (ADB), Manila. https://www.adb.org/sites/default/files/publication/27971/awdo.pdf

ADB (2007c) Asian water development outlook 2007 country paper Bangladesh. Asian Development Bank (ADB), Manila. https://www.adb.org/sites/default/files/publication/27971/awdo.pdf

Afful SA-A (2010) Spectrum of organochlorine pesticide residues in fish samples from the Densu Basin. Res J Environ Earth Sci 2(3):133–138. https://pdfs.semanticscholar.org/

Aggarwal PK, Basu AR, Kulkarni KM, Froehlich K, Tarafdar SA, Ali M, Ahmed N, Hussain A, Rahman M, Ahmed SR (2000) A report on isotope hydrology of groundwater in Bangladesh: implications for characterization and mitigation of arsenic in groundwater. Work performed within an IAEA-TC Project (No. BGD/8/016), 61. https://inis.iaea.org/search/search.aspx?orig_q=RN:34065511

Ahmad QK, Biswas AK, Rangachari R, Sainju MM (2001) Ganges–Brahmaputra–Meghna region: a framework for sustainable development. The University Press, Dhaka. https://link.springer.com/chapter/10.1007/978-3-540-74928-8_6

Ahmed WY (2013) Fecal indicators and bacterial pathogens in bottled water from Dhaka, Bangladesh. Braz J Microbiol 44(1):97–103. https://doi.org/10.1590/S1517-83822013005000026

Ahmed MF, Rahman MM (2000) Water Supply & Sanitation – rural and low-income urban communities. International Training Network Center-Bangladesh, Dhaka. p 409–428. https://www.scirp.org/

Ahmed KM, Imam MB, Akhter SH, Hasan MA, Khan AA (2001) Sedimentology and mineralogy of arsenic contaminated aquifers in the Bengal Delta of Bangladesh. Groundwater arsenic contamination in the Bengal Delta plain of Bangladesh. In: Jacks G, Bhattacharya P, Khan AA (ed) Proceedings of KTH-Dhaka University Seminar, KTH Special Publication, TRITA-AMI Report 3084, p 97–108

Ahmed MJ, Haque MR, Ahsan A, Siraj S, Bhuiyan MHR, Bhattacharjee SC, Islam S (2010) Physicochemical assessment of surface and groundwater quality of the greater Chittagong region of Bangladesh. Pak J Anal Environ Chem 11(2):1–11. https://www.researchgate.net/publication/266355309

Ahmed ZU, Panaullah GM, Gauch H Jr, McCouch MR, Tyagi W, Kabir MS, Duxbury JM (2011a) Genotype and environment effects on rice (Oryza sativa L.) grain arsenic concentration in Bangladesh. Plant Soil 338:367–382. https://doi.org/10.1007/s11104-010-0551-7

Ahmed ZU, Panaullah GM, DeGloria SD, Duxbury JM (2011b) Factors affecting paddy soil arsenic concentration in Bangladesh: prediction and uncertainty of geo-statistical risk mapping. Sci Total Env 412–413:324–335. https://doi.org/10.1016/j.scitotenv.2011.10.008

Aktar MW, Sengupta D, Choudhury A (2009) Impact of pesticides use in agriculture: their benefits and hazards. Interdisc Toxicol 2(1):1–12. https://doi.org/10.2478/v10102-009-0001-7

Alam AM (2006) Training needs of farmers on rice production in Dhamrai upazila under Dhaka district. M.S. thesis, Department of Agricultural Extension and Information System, Sher-e-Bangla Agricultural University, Dhaka

Alam M, Chowdhury MMS, Mijanur MR, Rahman MA, Gan SH (2015) Detection of residual levels and associated health risk of seven pesticides in fresh eggplant and tomato samples from Narayanganj district. Bangladesh J Chem:243574. https://doi.org/10.1155/2015/243574

Alam MS, Han B, Al-Mizan PJ (2019) Assessment of soil and groundwater contamination at a former tannery district in Dhaka, Bangladesh. Environ Geochem Health. https://doi.org/10.1007/s10653-019-00457-6

Alamgir M, Ahsan A (2007) Municipal solid waste and recovery potential: Bangladesh perspective. Iran J Environ Health Sci Eng 4(2):67–76. https://www.researchgate.net/publication/43562547

Ali H, Khan E (2018) Bioaccumulation of non-essential hazardous heavy metals and metalloids in freshwater fish. Risk to human health. Environ Chem Lett 16:903–917. https://doi.org/10.1007/s10311-018-0734-7

Ali S, Thakur SK, Sarkar A et al (2016) Worldwide contamination of water by fluoride. Environ Chem Lett 14:291–315. https://doi.org/10.1007/s10311-016-0563-5

Alwashali E, Jghalef B, Fadli M, Ashraf C, Abdelhak G (2014) Assessment of microbial contamination of groundwater in Oualidia area, Morocco. Eur Sci J 10(14) https://eujournal.org/index.php/esj/article/view/3393

Aminur MR, Nahar K, Al-Reza SM (2016) Chemical analysis of drinking water samples of some primary schools from Magura District, Bangladesh. J Water Res Ocean Sci 5(5):73–77. https://doi.org/10.11648/j.wros.20160505.12

Anwar AHMF, Bari MF (2004) Prioritising areas for groundwater monitoring based on pesticide leaching. J Appl Hydro 17(1):43–52

Anwar AH, Saing LA (2010) Aquifer vulnerability to pesticide leaching and its impact on groundwater. J Environ Res Dev 5(1):12–22. https://www.jerad.org/ppapers/dnload.php?vl=5&is=1&st=12

Anwar AHMF, Yunus A (2010) Pesticide leaching potential in a shallow unconfined aquifer. J Water Environ Technol 8(1) https://www.jstage.jst.go.jp/article/jwet/8/1/8_1_1/_pdf

Araújo KC de F, Barreto JP de P, Cardozo JC et al (2018) Sulfate pollution: evidence for electrochemical production of persulfate by oxidizing sulfate released by the surfactant sodium dodecyl sulfate. Environ Chem Lett 16:647–652. https://doi.org/10.1007/s10311-017-0703-6

Arnold HL, Odam RH, James WD (1990) Disease of the skin. Inc: clinical demography. Saunders, Philadelphia, pp 121–122

Asaduzzaman M (1985) Handbook of groundwater and Wells, 1st edn. BRAC Prokashana, Dhaka, pp 196–207

Askary VR, Jahan NA, Sabbagh A, Jahani FS, Dourandish N, Kamachali AR (2011) A potential medicinal importance of zinc in human health and chronic diseases. Clin Biochem 44(13), Supp-S. https://scholar.google.com/

Atikul MI, Rahman MR, Bodrud-Doza M, Muhib MI (2018) A study of groundwater irrigation water quality in South-Central Bangladesh: a geo-statistical model approach using GIS and multivariate statistics. Acta Geochimica, Springer https://doi.org/10.1007/s11631-017-0201-3

ATSDR (2000) Toxicological profile for manganese. United States Department of Health and Human Services, Public Health Service, Atlanta, GA; Agency for Toxic Substances and Disease Registry. https://www.atsdr.cdc.gov/ToxProfiles/tp.asp?id=102&tid=23

Ayers RS, Westcot DW (1985) Water quality for agriculture, irrigation and drainage. Paper No. 29. Food and Agriculture Organization (FAO) of the United Nations, Rome, Italy, p 1–117. https://www.fao.org/3/T0234E/T0234E00.htm

Azimm M, Rahman MM, Khan RH, Kamal ATMM (2011) Characteristics leachate generated at landfill sites and probable risks of surface and groundwater pollution in the surrounding areas: a case study of Matuail landfill site, Dhaka. J Bangladesh Acad Sci 35(2):153–160. https://citeseerx.ist.psu.edu

Azizullah A, Khattak MN, Richter P, Häder DP (2011) Water pollution in Pakistan and its impact on public health – a review. Environ Int 37:479–497. https://doi.org/10.1016/j.envint.2010.10.007

BADC (2002) Survey report on irrigation equipment and irrigated area in Boro/2001 season. Bangladesh Agricultural Development Corporation (BADC), Dhaka. https://badc.portal.gov.bd/

BADC (2013) Minor irrigation survey report 2012–2013. Bangladesh Agricultural Development Corporation, Ministry of Agriculture, Government of Bangladesh. https://badc.portal.gov.bd/

BADC (2018) Summary of irrigation equipment used, area irrigated and benefited farmers, Shech Bhavan, Dhaka. https://www.hindawi.com/journals/jse/2018/8471715/ref/

Bauder TA, Waskom RM, Sutherland PL, Davis JG (2011) Irrigation water quality criteria. Colorado State University extension, Fact Sheet No. 0.506. https://extension.colostate.edu/docs/pubs/crops/00506.pdf

BBS (2009) Bangladesh Bureau of Statistics. Agricultural statistical yearbook. Ministry of Planning, Dhaka. www.bbs.gov.bd

BBS (Bangladesh Bureau of Statistics) (2015) Agricultural statistical yearbook. Dhaka. www.bbs.gov.bd

BBS (Bangladesh Bureau of Statistics) (2018) Annual report. Ministry of Planning, Dhaka. www.bbs.gov.bd

Belkhiri L, Boudoukha A, Mouni L, Baouz T (2010) Application of multivariate statistical methods and inverse geological modeling for characterization of groundwater – a case study. Ain Azel plain (Algeria). Geoderma 159:390–398. https://doi.org/10.1016/j.geoderma.2010.08.016

Bempah CK-F (2011) A preliminary assessment of consumer's exposure to organochlorine pesticides in fruits and vegetables and the potential health risk in Accra Metropolis, Ghana. Food Chem 128(4):1058–1065. https://doi.org/10.1016/j.foodchem.2011.04.013

BGS (2000) Annual report. Bangladesh Geological Survey

BGS (2001) Arsenic contamination of groundwater in Bangladesh. British Geological Survey (BGS) Technical Report, WC/00/19. https://www.bgs.ac.uk/arsenic/bangladesh/

BGS-DPHE (2001a) Arsenic contamination of groundwater in Bangladesh. British Geological Survey (BGS) and Department of Public Health Engineering (DPHE), Bangladesh. Rapid Investigation Phase, Vol. 1, Final Report. https://www.bgs.ac.uk/downloads/start.cfm?id=2221

BGS-DPHE (2001b) Arsenic contamination of groundwater in Bangladesh. British Geological Survey and Department of Public Health Engineering, vol. 2, Final Report. BGS Technical Report WC/00/19. https://www.bgs.ac.uk/downloads/start.cfm?id=2251

Bharti N, Katyal D (2011) Water quality indices used for surface water vulnerability assessment. Int J Environ Sci 2(1):154–173. https://ssrn.com/abstract=2160726

Bhuiyan MAH, Bodrud-Doza M, Islam ARMT, Rakib MA, Rahman MS, Ramanathan AL (2010) Assessment of groundwater quality of Lakshimpur district of Bangladesh using water quality indices, geostatistical methods, and multivariate analysis. Environ Earth Sci 75. https://doi.org/10.1007/s12665-016-5823-y

Bhuiyan MAH, Ganyaglo S, Suzuki S (2015) Reconnaissance on the suitability of the available water resources for irrigation in Thakurgaon District of northwestern Bangladesh. Appl Water Sci 5:229–239. https://doi.org/10.1007/s13201-014-0184-8

Bhuiyan MAH, Bodrud-Doza M, Islam ARMT et al (2016) Assessment of groundwater quality of Lakshimpur district of Bangladesh using water quality indices, geostatistical methods, and multivariate analysis. Environ Earth Sci 75:1020. https://doi.org/10.1007/s12665-016-5823-y

Biswas MR, Mandal MAS (1993) Irrigation management for crop diversification in Bangladesh. University Press Ltd, Dhaka

Bodrud-Doza M, Islam ARMT, Ahmed F, Das S, Saha N, Rahman MS (2016) Characterization of groundwater quality using water evaluation indices, multivariate statistics and geostatistics in Central Bangladesh. Water Sci 30:19–40. https://doi.org/10.1016/j.wsj.2016.05.001

Bothwell TH (1979) Iron metabolism in man. Blackwell, Oxford

Brammer H (2014) Bangladesh's dynamic coastal regions and sea-level rise. Clim Risk Manag Elsevier 1:51–61. https://doi.org/10.1016/j.crm.2013.10.001

Brömssen MV, Markussen L, Bhattacharya P, Ahmed KM, Hossain M, Jacks G, Sracek O, Thunvik R, Hasan MA, Islam MM, Rahman MM (2014) Hydrogeological investigation for assessment of the sustainability of low-arsenic aquifers as a safe drinking water source in regions with high-arsenic groundwater in Matlab, southeastern Bangladesh. J Hydrol https://doi.org/10.1016/j.jhydrol.2014.05.055

BWDB (2013) Hydrogeological study and mathematical modelling to identify sites for installation of observation well nests, selection on model boundary, supervision of pumping test, slug test, assessment of different hydrogeological parameters collection and conduct chemical analysis of surface water and ground water, Final Report (Main Volume, Package 3), Bangladesh Water Development Board (BWDB) and Institute of Water Modelling (IWM). https://books.google.com.bd/books

Caldwell DJ, Mastrocco F, Nowak E et al (2010) An assessment of potential exposure and risk from estrogens in drinking water. Environ Health Perspect 118(3):343. http://doi:10.1289/ehp.0900654

Can DNHW (1990) Nutrition recommendations. The report of the Scientific Review Committee, Department of National Health and Welfare (Canada), DSS Cat. no. H49-42/1990E, Ottawa. https://catalogue.nla.gov.au/Record/1785247

Casey JR (2006) Why bicarbonate? Biochem Cell Biol 84(6):930–939. https://doi.org/10.1139/o06-184

Chakraborty D, Rahman MM, Das B, Murrill M, Dey S, Mukherjee SC (2010) Status of groundwater arsenic contamination in Bangladesh: a 14-year study report. Water Res 44(19):5789–5802. https://doi.org/10.1016/j.watres.2010.06.051

Chen T-B, Zheng Y-M, Lei M, Huang Z-C, Wu H-T, Chen H, Fan K-K, Yu K, Wu X, Tian Q-Z (2005) Assessment of heavy metal pollution in surface soils of urban parks in Beijing, China. Chemosphere 60:542–551. https://www.igsnrr.cas.cn/xwzx/jxlwtj/200607/P020090715579333559064.pdf

Chhipa H (2017) Nanofertilizers and nanopesticides for agriculture. Environ Chem Lett 15:15–22. https://doi.org/10.1007/s10311-016-0600-4

Chien L, Robertson H, Gerrard JW (1968) Infantile gastroenteritis due to water with high sulfate content. Can Med Assoc J 99:102–104. https://www.ncbi.nlm.nih.gov/pmc/articles/PMC1924382/

Chowdhury NT (2010) Water management in Bangladesh: an analytical review. Wat Policy 12:32–51. https://doi.org/10.2166/wp.2009.112

Chowdhury MA (2012) Organophosphorus and carbamate pesticide residues detected in water samples collected from paddy and vegetable fields of the Savar and Dhamrai Upazilas in Bangladesh. Int J Environ Res Public Health 9(9):3318–3329. https://doi.org/10.3390/ijerph9093318

Chowdhury AZ, Islam MN, Moniruzzaman M, Gan SH, Alam MK (2013) Organochlorine insecticide residues are found in surface, irrigated water samples from several districts in Bangladesh. Bull Environ Contam Toxicol 90(2):149–154. https://doi.org/10.1007/s00128-012-0897-z

Clesceri LS, Greenberg AE, Trussell RR (1981) Standard methods for the examination of water and wasp water, 17th edn. American Public Health Association, Washington DC, pp 1–175

CSISA-MI (2015) Groundwater management in Bangladesh: an analysis of problems and opportunities. Central System Initiative for South Asia – mechanization and Irrigation, USAID, Research Report No. 2. Dhaka. https://doi.org/10.13140/2.1.2381.9040

Custodio E (2002) Aquifer overexploitation: what does it mean? Hydrogeol J 10:254–277. https://doi.org/10.1007/s10040-002-0188-6

Dasgupta SM (2007) A pinch or a pint? Evidence of pesticide overuse in Bangladesh. J Agric Econ 58(1):91–11. https://doi.org/10.1111/j.1477-9552.2007.00083.x

De AK (2005) Environmental chemistry, 5th edn. New Age, New Delhi

de Baaij JH, Hoenderop JG, Bindels RJ (2015) Magnesium in man: implications for health and disease. Physiol Rev 95(1):1–46. https://doi.org/10.1152/physrev.00012.2014

Demir V, Dere T, Erginc S, Cakıra Y, Celika F (2015) Determination and health risk assessment of heavy metals in drinking water of Tunceli, Turkey. Water Resour 42:508–516. https://doi.org/10.1134/S0097807815040041

Dey N, Saha R, Parvez M, Bala S, Islam AKM, Paul J, Hossain M (2017a) Sustainability of groundwater use for irrigation of dry-season crops in Northwest Bangladesh. Groundw Sustain Dev 4. https://doi.org/10.1016/j.gsd.2017.02.001

Dey NC, Parvez M, Dey D, Saha R, Ghose L, Barua MK, Islam A, Chowdhury MR (2017b) Microbial contamination of drinking water from risky tubewells situated in different hydrological regions of Bangladesh. Int J Hyg Health 220(3):621–636. https://doi.org/10.1016/j.ijheh.2016.12.007

DGHS (2016) Health Bulletin-2016. Directorate General of Health Services, Dhaka. https://dghs.gov.bd/index.php/en/home/4821-health-bulletin-2016

Dhar RK, Biswas BK, Samanta G (1997) Groundwater arsenic calamity in Bangladesh. Curr Si 73:48–59. http://jstor.org/stable/24098145

Dider-Ul SMI, Bhuiyan MAH, Rume T (2017) Hydrogeochemical investigation of groundwater in shallow coastal aquifer of Khulna District, Bangladesh. Appl Water Sci 7:4219–4236. https://doi.org/10.1007/s13201-017-0533-5

Dittmar J, Voegelin A, Maurer F, Roberts LC, Hug SJ, Saha GC (2010) Arsenic accumulation in a paddy field in Bangladesh: seasonal dynamics and trends over a three-year monitoring period. Environ Sci Technol 44:2925–2931. https://doi.org/10.1021/es903117r

DoFE (2018) Department of Forest and Environment. Ministry of Environment, Bangladesh. Annual Report, 2018

Doneen LD (1962) Notes on water quality in agriculture. Published as a water science and engineering, Paper 4001, Department of Water Sciences and Engineering, University of California

Dorman DC (2000) Neurotoxicity of manganese chloride in neonatal and adult CD rats following sub chronic (21- day) high-dose oral exposure. J Appl Toxicol 20:179–187. https://doi.org/10.1002/(sici)1099-1263(200005/06)20:3%3C179::aid-jat631%3E3.0.co;2-c

DPHE (2009) Bangladesh national drinking water quality survey. Department of Public Health and Engineering, Government of Bangladesh. https://washdata.org/

DPHE/BGS/DFID (Department for International Development), UK (2000) Groundwater studies of arsenic contamination in Bangladesh. Final Report. Dhaka. www.bgs.ac.uk/arsenic/bangladesh

Driscoll T, Steenland K, Prüss-Üstün A, Nelson DI, Leigh J (2004) In: Annette Prüss-Üstün DC-L (ed) Occupational carcinogens: assessing the environmental burden of disease at national and local levels, vol 6. World Health Organization, Geneva. https://www.who.int/quantifying_ehimpacts/publications/9241591471/en/

DVGW (2006) Guidelines on drinking water protection areas – part 1. Bonn, Deutsche Vereinigung des Gas-und Wasserfaches e.V. Technical rule number W101. http://www.beuth.de/en/technical-rule/dvgw-w-101/92562980

Eaton FM (1950) Significance of carbonate in irrigation water. Soil Sci 62(2):123–133

Edmunds WM, Ahmed KM, Whitehead PG (2015) A review of arsenic and its impacts in groundwater of the Ganges–Brahmaputra–Meghna delta, Bangladesh. Environ Sci Processes Impacts 17:1032–1046. https://doi.org/10.1039/C4EM00673A

Elinder C-G (1986) Iron. In: Friberg L, Nordberg GF, Vouk VB (eds) Handbook on the toxicology of metals, vol II. Elsevier, Amsterdam, pp 276–297. https://www.elsevier.com/books/

El-Mekkawi HD (2009) Determination of chlorinated organic pesticide residues in water, sediments, and fish from private fish farms at Abbassa and Sahl Al-Husainia, Sharkia Governorate. Aust J Basic Appl Sci 3(4):4376–4383. https://www.researchgate.net/publication/267720994

Elom NI, Entwistle J, Dean JR (2014) Human health risk from Pb in urban street dust in northern UK cities. Environ Chem Lett 12:209–218. https://doi.org/10.1007/s10311-013-0436-0

Enayetullah I, Khan SSA, Sinha M (2005) Urban solid waste management. Senecio of Bangladesh: Problem and prospects. Waste concern technical documentation. https://www.semanticscholar.org/paper/

Ercumen A, Naser AM, Unicomb L, Arnold BF, Colford JM Jr, Luby SP (2015) Effects of source- versus household contamination of tubewell water on child Diarrhea in rural Bangladesh: a randomized controlled trial. PLoS One 10(3):e0121907. https://doi.org/10.1371/journal.pone.0121907

Escobar MEO, Hue NV, Cutler WG (2008) Recent developments on arsenic: contamination and remediation. Researchgate 4:1–32. https://www.researchgate.net/publication/247161898

Fadeeva VK (1971) Effect of drinking water with different content of chlorides on the organism of experimental animals. Gig Sanit 36(6):11–15. https://www.researchgate.net/publication/18051949

FAO (2018) Fertilizer consumption in Bangladesh, A report. https://www.fao.org/3/a-i5890e.pdf

FAO, UNICEF, WHO, and WSP (2010) Towards an arsenic safe environment in Bangladesh. Executive summary of the WORLD WATER DAY-2010. https://www.unicef.org/infobycountry/files/

FAO-WHO (1988) Requirements of vitamin A, iron, folate, and vitamin B_{12}. Report of a Joint FAO/WHO Expert Consultation, Rome, (FAO Food and Nutrition Series, No. 23)., and Food and Agriculture Organization of the United Nations. https://trove.nla.gov.au/version/20830415

Farquhar WB, Edwards DG, Jurkovitz CT, Weintraub WS (2015) Dietary sodium and health: more than just blood pressure. J Am Coll Cardiol 65(10):1042–1050. https://doi.org/10.1016/j.jacc.2014.12.039

Faruq AN (2018) Agriculture and pesticide consumption in Bangladesh (conference paper). Conference on "Effluent Control and Waste Disposal in Pesticide Industry – with Especial Agenda on – Integrated Pest Management with Special Emphasis on Bio-pesticide Component," at Bogor, Indonesia. https://www.researchgate.net/publication/327592585

Faruque HSM Alam S (2002) State of arsenic contamination in Bangladesh. In: Proceedings of the international water conference held at Hanoi, Vietnam. pp 8

Fazal MA, Kawachi T, Ichion E (2001a) Validity of the latest research findings on causes of groundwater arsenic contamination in Bangladesh. Water Int 26(2):380–389. https://doi.org/10.1080/02508060108686930

Fazal MA, Kawachi T, Ichion E (2001b) Extent and severity of groundwater arsenic. Water Int 26(3):370–379. https://doi.org/10.1080/02508060108686929

Ferguson AS, Mailloux BJ, Ahmed KM, van Geen A, McKay LD, Culligan PJ (2011) Hand-pumps as reservoirs for microbial contamination of well water. J Water Health 9(4):708–717. https://doi.org/10.2166/Fwh.2011.106

Fipps G (2003) Irrigation Water Quality Standards and Salinity Management Strategies. Available electronically from: http://hdl.handle.net/1969.1/87829

Flanagan SV, Johnstan RB, Zheng Y (2012) Arsenic in tube well water in Bangladesh: health and economic impacts and implementations for arsenic mitigation. Bull World Health Organ 90:839–846. https://doi.org/10.2471/BLT.11.101253

Flora G, Gupta D, Tiwari A (2012) Toxicity of lead: a review with recent updates. Interdiscip Toxicol 5(2):47–58. https://doi.org/10.2478/Fv10102-012-0009-2

Freeze RA, Cherry JA (1979) Groundwater. Prentice Hall Inc., Englewood Cliffs. https://www.un-igrac.org/sites/default/files/resources/files/Groundwater book-English.pdf

Friberg L, Nordberg GF, Vouk VB (1986) Handbook of the toxicology of metals, vol II. Elsevier, Amsterdam, pp 130–184. eBook ISBN: 9780123973399. https://www.elsevier.com/books/

Goad RT (2004) Carbofuran-induced endocrine disruption in adult male rats. Toxicol Mech Methods 14(4):233–239. https://doi.org/10.1080/15376520490434476

Gommers LM, Hoenderop JG, Bindels RJ, de Baaij JH (2016) Hypomagnesemia in type 2 diabetes: a vicious circle. Diadates 65(1):3–13. https://doi.org/10.2337/db15-1028

Gosselin RE, Smith RP, Hodge HC (1984) Clinical toxicology of commercial products, 5th edn. Williams & Wilkins, Baltimore. https://doi.org/10.1002/jps.2600741037

Goyal R, Kim S, Manohar N, Waters B, Wu DJ (2019) Watermarking public-key cryptographic primitives. In: Boldyreva A, Micciancio D. (eds) Advances in cryptology–CRYPTO 2019. CRYPTO 2019. Lecture Notes in Computer Science, vol 11694. Springer, Cham. https://doi.org/10.1007/978-3-030-26954-8_12

Graham JP, Polizzotto ML (2013) Pit latrines and their impacts on groundwater quality: a systematic review. Environ Health Perspect 121(5):521–530. https://www.ncbi.nlm.nih.gov/pmc/articles/PMC3673197

Grollman A (1961) The role of salt in health and disease am. J Cardiol 8(4):593–601

Gupta UC, Gupta SC (1987) Trace element toxicity relationships to crop production and livestock and human health: implications for management. Commun Soil Sci Plant Anal 29(11–14):1491–1522. https://doi.org/10.1080/00103629809370045

Guy F (2003) Irrigation water quality standards and salinity management strategies. Texas A&M University, Texas AgriLife Extension Service: College Station, Texas. Publication B-1667

GWF (2015) Global Water Forum. UNESCO, Autralian National University

Halim MA, Majumder RK, Nessa SA, Oda K, Hiroshiro Y, Saha BB, Hassain SM, Latif SA, Islam MA, Jinno K (2009) Groundwater contamination with arsenic in Sherajdikhan, Bangladesh: geochemical and hydrological implications. Environ Geol 58(1):73–84. https://doi.org/10.1007/s00254-008-1493-8

Halim M, Majumder R, Nessa S et al (2010) Arsenic in shallow aquifer in the eastern region of Bangladesh: insights from principal component analysis of groundwater compositions. Environ Monit Assess 161:453–472. https://doi.org/10.1007/s10661-009-0760-9.23

Halim MA, Majumder RK, Rasul G, Hirosiro Y, Sasaki K, Shimada J, Jinno K (2014) Geochemical evaluation of arsenic and manganese in shallow groundwater and core sediment in Singair Upazila, Central Bangladesh. Arab J Sci Eng 39(7):5585–5601. https://doi.org/10.1007/s13369-014-1111-x

Hanasaki N, Pokhrel Y, Kanae S, Yoshikawa S (2018) A quantitative investigation of the thresholds for two conventional water scarcity indicators using a state-of-the-art global hydrological model with human activities. Water Resour Res 54(10):8279–8294. https://doi.org/10.1029/2018WR022931

Harris ZL, Gitlin JD (1996) Genetic and molecular basis for copper toxicity. Am J Clin Nutr 63:836–841. https://doi.org/10.1093/ajcn/63.5.836

Hasan MA, Ahmed KM, Sracek O, Bhattacharya P, Bromssen MV, Broms S (2007) Arsenic in shallow groundwater of Bangladesh: investigations from three different physiographic settings. Hydrol J 15:1507–1522. https://doi.org/10.1007/s10040-007-0203-z

Hasan MK, Shabbir A, Jim KU (2019) Water pollution in Bangladesh and its impact on public health. Heliyon 5:1–23. https://doi.org/10.1016/j.heliyon.2019.e02145

Hasanuzzaman M, Rahman MA, Salam MA (2017) Identification and quantification of pesticide residues in water samples of Dhamrai Upazila, Bangladesh. Appl Water Sci 7:2681–2688. https://doi.org/10.1007/s13201-016-0485-1

Hasanuzzaman M, Rahman MA, Islam MS, Salam MA (2018) Pesticide residues analysis in water samples of Nagarpur and Saturia Upazila, Bangladesh. App Water Sci 8. https://doi.org/10.1007/s13201-018-0655-4

Hasanuzzaman M, Tania S, Wahida H, Uddin M, Rahman M (2019) Study on Organochlorine and organophosphorus pesticide residues in surface and underground water samples of Daudkandi Upazila in Bangladesh, p 271–278. https://doi.org/10.5829/idosi.aejaes.2019.271.278

Hayat KA (2010) Determination of pesticide residues in blood samples of villagers involved in pesticide application at district Vehari (Punjab). Pak Afr J Environ Sci Technol 4(10):666–684. https://www.academicjournals.org/AJEST

He FJ, MacGregor GA (2008) Beneficial effects of potassium on human health. Physiol Plant 133(4):725–735. https://doi.org/10.1111/j.1399-3054.2007.01033.x

Heaney RP, Gallagher JC, Johnston CC, Neer R, Parfitt AM, Whedon GD (1982) Calcium nutrition and bone health in the elderly. Am J Clin Nutr 36:986–1013. https://doi.org/10.1093/ajcn/36.5.986

Heizer WD, Sandler RS, Seal E, Murray SC, Busby MG, Schliebe BG, Pusek SN (1997) Intestinal effects of sulfate in drinking water on normal human subjects. Dig Dis Sci 42(5):1055–1061. https://doi.org/10.1023/a:1018801522760

Hem JD (1970) Study and Interpretation of the chemical characteristics of natural water, 2nd ed., U.S. Geo. Survey. Water Supply, Paper 1473

Henry FJ, Patwary Y, Huttly SR, Aziz KM (1990) Bacterial contamination of weaning foods and drinking water in rural Bangladesh. Epidemiol Infect 104(1):79–85. https://doi.org/10.1017/Fs0950268800054558

Hooper PL (1980) Zinc lowers high-density lipoprotein-cholesterol levels. J Am Med Assoc 244(17):1960. https://doi.org/10.1001/jama.1980.03310170058030

Hossain MG, Selim RAHM, Lutfun-Nessa M, Ahmed S (2013) Factor and cluster analysis of water quality data of the groundwater wells of Kushtia, Bangladesh: implication for arsenic enrichment and mobilization. J Geol Soc India 81:377–384. https://doi.org/10.1007/s12594-013-0048-0

Hossain MS, Chowdhury MA, Pramanik MK, Rahman MA, Fakhruddin AN, Alam MK (2015) Determination of selected pesticides in water samples adjacent to agricultural fields and removal of organophosphorus insecticide chlorpyrifos using soil bacterial isolates. Appl Water Sci 5(2):171–179. https://doi.org/10.1007/s13201-014-0178-6

Hossain MZ, Kabir GMA, Quraishi SB (2010) Contamination of Dhanmondi and Gulshan lake waters with some metals and inorganic pollutants. Bang J Academy of Science 34(1):89–93

Howladar MF, Al-Numan MA, Faruque MO (2017) An application of water quality index (WQI) and multivariate statistics to evaluate the water quality around Maddhapara granite mining industrial area Dinajpur, Bangladesh. Environ Syst Res 6:13. https://doi.org/10.1186/s40068-017-0090-9

Huq SMI, Naidu R (2002) Arsenic in groundwater of Bangladesh: contamination in the food chain. Arsenic Contamination in Bangladesh, ITN Bangladesh, Dhaka, p 69–72

IAEA (2004) Annual Report. Article VI.J of the International Atomic Energy Agency's (IAEA's) Statute requires the Board of Governors to submit "An Annual Report to the General Conference concerning the affairs of the Agency and any projects approved by the Agency." https://www.iaea.org/sites/default/files/anrep2004_full.pdf

Ikejimba CC, Sakpa S (2014) Comparative study of some heavy metals' concentrations in water and Tympanotonus fuscatus var radula samples of Egbokodo River, Warri, Nigeria. Int J Mod Biol Res 2:7–15. https://www.bluepenjournals.org/ijmbr/pdf/2014/January/Ikejimba and Sakpa.pdf

IPCS (1991) Nickel, nickel carbonyl and some nickel compounds. International programme on chemical safety health and safety guide no. 62. WHO, Geneva. https://www.inchem.org/documents/hsg/hsg/hsg062.htm

IPCS (2002) Principles and methods for the assessment of risk from essential trace elements. World Health Organization, Geneva, international programme on chemical safety (environmental health criteria 228). https://www.inchem.org/documents/ehc/ehc/ehc228.htm

Islam MN (2007) Monitoring of organochlorine pesticide residues in surface water samples of the agricultural field of Bangladesh. J Subtrop Agric Res Dev 5(5):357–360. https://www.researchgate.net/

Islam MS, Mostafa MG (2021a) Groundwater quality and risk assessment of heavy metal pollution in middle-west part of Bangladesh. Journal of Earth and Environmental Science Research 3(2):1–15. https://doi.org/10.47363/JEESR/2021(3)143

Islam MS, Mostafa, MG (2021b) Hydro-geochemical evaluation of groundwater for irrigation in the Ganges river basin areas of Bangladesh. Research square https://doi.org/10.21203/rs.3.rs-161359/v1

Islam MS, Mostafa MG (2021c) Environmental hazards of petroleum refinery in Bangladesh: a review. Petro Chem Indus Intern 4(1):15–21. https://doi.org/10.33140/PCII.03.01

Islam M, Shamsad S (2009) Assessment of irrigation water quality of Bogra district in Bangladesh. Bangladesh J Agric Res 34(4):507–608. www.researchgate.net/publication/312820337

Islam MR, Salminen R, Lahermo PW (2000) Arsenic and other toxic elemental contamination of groundwater, surface water, and soil in Bangladesh and its possible effects on human health. Environ Geochem Health 22:33–53. https://doi.org/10.1023/A:1006787405626

Islam MS, Siddika A, Khan MN, Goldar MM, Sadique MA, Kabir AN (2001) Microbiological analysis of tube-well water in a rural area of Bangladesh. Appl Environ Microbiol 67(7):3328–3330. https://doi.org/10.1128/FAEM.67.7.3328-3330.2001

Islam S, Susskind LE et al (2013) Water diplomacy: a negotiated approach to managing complex water networks. RFF Press Routledge, 711 Third Avenue, New York, NY. 10017

Islam ARMT, Rakib MA, Islam MS, Jahan K, Patwary MA (2015) Assessment of health Hazard of metal concentration in groundwater of Bangladesh. Am Chem Sci J 5(1):41–49. https://doi.org/10.9734/ACSj/2015/13175

Islam K, Rahman MS, Ali MH, Hossain AFMA (2016) Evolution of the aquifers system and groundwater quality of the North-West District of Bangladesh for development potential. BRAG Univ J XI(2):85–100

Islam SM-U, Majumder RK, Uddin MJ, Khalil MI, Alam MF (2017a) Hydrochemical characteristics and quality assessment of groundwater in Patuakhali district, southern coastal region of Bangladesh. Water Qual Expo Health 9(1):43–60

Islam ARM, Shuanghe S, Bodrud-Doza M, Atiqur MR, Samiran D (2017b) Assessment of trace elements of groundwater and their spatial distribution in Rangpur district, Bangladesh. Arab J Geosci 10(4). https://doi.org/10.1007/s12517-017-2886-3

Islam ARM, Nasir A, Bodrud-Doza M, Ronghao C (2017c) Characterizing groundwater quality ranks for drinking purposes in Sylhet district, Bangladesh, using entropy method, spatial autocorrelation index, and geostatistics. Environ Sci Pollut Res 24. https://doi.org/10.1007/s11356-017-0254-1

Islam ARM, Shuanghe S, Muhammed H, Bodrud-Doza M, Khin M, Habib M (2017d) Assessing groundwater quality and its sustainability in Joypurhat district of Bangladesh using GIS and multivariate statistical approaches. Environ Dev Sustain 19:1–25. https://doi.org/10.1007/s10668-017-9971-3

ITNC (International Training Network Centre) (2008) Position paper on Bangladesh response to arsenic contamination of groundwater. Bangladesh University of Engineering and Technology (BUET), Dhaka, Bangladesh, 75

IWM (2009) News Letter. Institute of Water Modeling, No. 38

Jackson LE, Burger M, Cavagnaro TR (2008) Roots, nitrogen transformations, and ecosystem services. Ann Rev Plant Biol 59(1):341–363. https://doi.org/10.1146/annurev.arplant.59.032607.092932. PMID18444903

Jahan H (2016) Arsenic in Bangladesh: How to Protect 20 Million from the World's Largest Poisoning. https://www.theguardian.com/global-development-professionals-network/2016/oct/18/

Jahan CS, Mazumder QH, Akter N, Adham MI, Zaman MA (2010) Hydrogeological environment and groundwater occurrences in the plio-pleistocene aquifer in Barind area, Northwest Bangladesh. Bangladesh GeoSci J 16:23–37. https://scholar.google.com/

Janus JA, Krajnc EI (1990) Integrated criteria document chromium: effects. Appendix to report no. 758701001, Bilthoven, Netherlands, National Institute of Public Health and Environmental Protection, 21

Jihen EH, Imed M, Fatima H, Abdelhamid K (2008) Protective effects of selenium (Se) and zinc (Zn) on cadmium (Cd) toxicity in the liver and kidney of the rat: histology and Cd accumulation. Food Chem Toxicol 46(11):352. https://doi.org/10.1016/j.fct.2008.08.037

Johnson DL, Ambrose SH, Bassett TJ, Bowen ML, Crummey DE, Isaacson JS, Johnson DN, Lamb P, Saul M, Winter-Nelson AE (1997) Meanings of environmental terms. J Environ Qual 26:581–589. https://doi.org/10.2134/jeq1997.00472425002600030002x

Kalyani N, Goel S, Jaiswal S (2020) On-site sensing of pesticides using point-of-care biosensors: a review. Environ Chem Lett. https://doi.org/10.1007/s10311-020-01070-1

Kamrin MA (1997) Pesticide profiles: toxicity, environmental impact, and fate. CRC Press 14(4):233–239. https://www.crcpress.com/

Karim MM (2000) Arsenic in groundwater and health problem in Bangladesh. J Water Res 34(1):304–310. https://www.researchgate.net/

Karim MA, Komori Y, Alam M (1997) Subsurface arsenic occurrence and depth of contamination in Bangladesh. J Environ Chem 7(4):783–792. https://doi.org/10.5985/jec.7.783

Kaur R, Goyal D (2019) Toxicity and degradation of the insecticide monocrotophos. Environ Chem Lett 17:1299–1324. https://doi.org/10.1007/s10311-019-00884-y

Kawasaki T, Delea CS, Bartter FC, Smith H (1978) The effect of high-sodium and low-sodium intakes on blood pressure and other related variables in human subjects with idiopathic hypertension. Am J Med 64(2):193–198. https://doi.org/10.1016/0002-9343(78)90045-1

Kelley WP (1963) Use of saline irrigation water. Soil Sci 95:355–391. https://journals.lww.com/soilsci/Citation/1963/06000/

Kelly WR, Panno SV, Hackley K (2012) The sources, distribution, and trends of chloride in the waters of Illinois. Illinois State Water Survey, Prairie Research Institute, University of Illinois. https://www.isws.illinois.edu/pubdoc/B/ISWSB-74.pdf

Khan FH (1991) Geology of Bangladesh. The University Press Limited, Dhaka

Knepper WA (1981) Iron. In: Kirk-Othmer encyclopedia of chemical technology, vol 13. Wiley Interscience, New York, pp 735–753

Kondakis XG (1989) Possible health effects of high manganese concentration in drinking water. Arch Environ Health 44:175–178. https://doi.org/10.1080/00039896.1989.9935883

Krajnc EI (1987) Cadmium – effects. Integrated criteria document, National Institute of public health and environmental protection, Bilthoven, report no. 758476004. https://www.worldcat.org/

Kreuger J (1998) Pesticides in stream water within an agricultural catchment in southern Sweden. Sci Total Environ 216(3):227–251. https://doi.org/10.1016/s0048-9697(98)00155-7

Kumar A, Singh CK (2015) Characterization of hydrogeological processes and fluoride enrichment in groundwater of South-Western Punjab. Water Qual Expo Health:1–15. https://doi.org/10.1007/s12403-015-0157-7

Kumar PS, Suganya S, Srinivas S et al (2019) Treatment of fluoride-contaminated water. A review. Environ Chem Lett 17:1707–1726. https://doi.org/10.1007/s10311-019-00906-9

Lamberg M, Hausen H, Vartiainen T (1997) Symptoms experienced during periods of actual and supposed water fluoridation. Community Dent Oral Epidemiol 25(4):291–295. https://doi.org/10.1111/j.1600-0528.1997.tb00942.x. PMID 9332806

Landwehr JM, Deininger RA (1976) A comparison of several water quality indexes. J Water Pollut Control Fed 48(5):954. https://www.jstor.org/stable/25039260

Leeuwen FX (2000) Safe drinking water: the toxicologist's approach. Food Chem Toxicol 38:51–58. https://doi.org/10.1016/s0278-6915(99)00140-4

Leeuwen vC, Trégoat O, Choné X et al. (2010) Vine water status is a key factor in grape ripening and vintage quality for red Bordeaux wine. How can it be assessed for vineyard management purposes?. Journal International des Sciences de la Vigne et du Vin 43. http://doi.10.20870/oeno-one.2009.43.3.798

Luby SP, Gupta SK, Sheikh MA, Johnston RB, Ram PK, Islam MS (2008) Tubewell water quality and predictors of contamination in three flood-prone areas in Bangladesh. J Appl Microbiol 105:1002–1008. https://doi.org/10.1111/j.1365-2672.2008.03826.x

Luo M (2008) Analysis of the relationship between drinking water and health. Guide Chin Med 6(14):160–161

Mahmud R, Inoue N, Sen R (2007) Assessment of irrigation water quality by using principal component analysis in arsenic affected area of Bangladesh. J Soil Nat 1(2):08–17. https://www.researchgate.net/publication/267795382

Majewski MS, Capel PD (1995) Pesticides in the atmosphere; distribution, trends, and governing factors. Open-File Report 94–506, USGS Numbered Series. https://doi.org/10.3133/ofr94506

Malik LA, Bashir A, Qureashi A et al (2019) Detection and removal of heavy metal ions: a review. Environ Chem Lett 17:1495–1521. https://doi.org/10.1007/s10311-019-00891-z

Mallik J, Sigh CK, AlMesfer MK, Kumar A, Khan RA, Islam S, Rahman A (2018) Hydrogeochemical assessment of groundwater quality in Aseer region, Saudi Arabia. Water 10:1847. https://doi.org/10.3390/w10121847

Manasaki N, Yoshikawa S, Pokhrel Y, Kanae S (2018) A global hydrological simulation to specify the sources of water used by humans. Hydrol Earth Syst Sci 22:789–817. https://doi.org/10.5194/hess-22-789-2018

Mandal BK, Chowdhury TR, Samanta G (1997) Chronic arsenic toxicity in West Bengal. Curr Sci 72:114–117. https://ci.nii.ac.jp/naid/10029235983

Manjushree C, Mostafa MG, Biswas TK, Mandal A, Saha A, K. (2015) Characterization of the effluents from leather processing industries. Env Processes 2(1):173–187. https://doi.org/10.1007/s40710-015-0065-7

Matin MA, Malek MA, Amin MR, Rahman S, Khatoon J, Rahman M, Aminuddin M, Mian A (1998) Organochlorine insecticide residues in surface and underground water from different regions of Bangladesh. Agri Ecosyst Environ 69(1):11–15. https://doi.org/10.1016/S0167-8809(98)00094-2

Matthess G (1982) The properties of ground water. Wiley, New York, p 397

Mazeda I, Bashar K, Ahmed N, Rasul R (2018) Hydrogeologic characteristics and groundwater potentiality of lower aquifer of Singair Upazila, Manikganj District, Bangladesh. J Bangladesh Acad Sci 42(1):25–40. https://doi.org/10.3329/jbas.v42i1.37830

McGowan W (2000) Water processing: residential, commercial, light-industrial, 3rd edn. Water Quality Association, Lisle

Meharg AA, Rahman MM (2003) Arsenic contamination of Bangladesh paddy field soils: implications for rice contribution to arsenic consumption. Environ Sci Technol 37:229–234. https://doi.org/10.1021/es0259842

Memon MA (2002) Solid waste management in Dhaka, Bangladesh. Innovation in community driven composting. https://kitakyushu.iges.or.jp/docs/demo/dhaka_bangladesh/spdhaka.pdf

Meride Y, Ayenew B (2016) Drinking water quality assessment and its effects on residents health in Wondo genet campus, Ethiopia. Environ Syst Res 5(1). https://doi.org/10.1186/s40068-016-0053-6

Miah SJ (2014) Unsafe use of pesticide and its impact on health of farmers: a case study in Burichong Upazila, Bangladesh. IOSR J Environ Sci Toxicol Food Technol 8(1):57–67. www.iosrjournals.org

Miah MY, Robel FN, Bhowmik S, Bottacharjee S (2015) Assessment of the coastal area water quality in Noakhali, Bangladesh. Int J Sci Eng Res 6(2):1116

Michael AM (1978) Irrigation theory and practices. Vikash Publishing House Pvt. Ltd, New Delhi, pp 686–740

MICS-B (2018) Water quality thematic report: 2012–2013, Multiple Indicator cluster survey, UNICEF and BBS, Bangladesh. https://www.unicef.org/bangladesh/en/reports/multiple-indicator-cluster-survey-2012-2013

Milman N, Pedersen P, Steig Tá, Byg K-E, Graudal N, Fenger K (2001) Clinically overt hereditary hemochromatosis in Denmark. 1948–1985: epidemiology, factors of significance for long-term survival, and causes of death in 179 patients. Ann Hematol 80(12):737–744. https://doi.org/10.1007/s002770100371

Mirza ATM, Rahman T, Rahman SH, Majumder RK (2012) Groundwater quality for irrigation of deep aquifer in southwestern zone of Bangladesh. Songklanakarin J Sci Technol 34(3):345–352. http://www.sjst.psu.ac.th

Mojid MA, Parvez MF, Mainuddin M, Hodgson G (2019) Water table trend—a sustainability status of groundwater development in north-West Bangladesh. Water 11:1182. https://doi.org/10.3390/w11061182

Mojumder SMMH, Shajedul MI (1995) A study on water quality parameters of surface and groundwater of Chittagong and Cox's Bazar town, Bangladesh. In: Proceedings of an international workshop on coastal aquaculture and environmental management, Cox's Bazar, Bangladesh

Mominul MI, Karim MR, Zheng X, Li X (2018) Heavy metal and metalloid pollution of soil, water and foods in Bangladesh: a critical review. Int J Environ Res Public Health 15:2825. https://doi.org/10.3390/ijerph15122825

Mondal NC, Singh VP, Singh VS, Saxena VK (2010) Determining the interaction between groundwater and saline water through groundwater major ions chemistry. J Hydrol 388:100–111. https://doi.org/10.1016/j.jhydrol.2010.04.032

Moore MR (1988) Hematological effects of lead. Sci Total Environ 71:419–431. https://doi.org/10.1016/0048-9697(88)90214-8

Morris BL, Lawrence ARL, Chilton PJC, Adam B, Calow RC, Klinck BA (2003) Groundwater and its susceptibility to degradation: a global assessment of the problem and options for management. Early Warning and Assessment Report Series, RS 03-3. United Nations Environment Programme, Nairobi, Kenya. https://www.unep.org/dewa/Portals/67/pdf/Groundwater_INC_cover.pdf

Mostafa MG, Chen Y-H, Jean J-S, Liu C-C, Teng H (2010) Adsorption and desorption properties of arsenate onto nano-sized iron oxide-coated quartz. Water Sci Technol 62(2):378–386. https://doi.org/10.2166/wst.2010.288

Mostafa MG, Chen Y-H, Jean J-S, Liu C-C, Lee Y-C (2011) Kinetics and mechanism of arsenate removal by nanosized iron oxide-coated perlite. J Hazard Mater 187:89–95. https://doi.org/10.1016/j.jhazmat.2010.12.117

Mostafa MG, Helal SMU, Haque ABMH (2017) Assessment of hydro-geochemistry and groundwater quality of Rajshahi City in Bangladesh. Appl Water Sci 7:4663–4671. https://doi.org/10.1007/s13201-017-0629y

MPO (1991) National Water Plan Project-Phase II. Master Plan Organization (MPO), Final Report, Dhaka. Hazra, USA; Sir M MacDonald, UK; Meta, USA; EPC, Bangladesh

MPO (Master Plane Organization) (1987) The groundwater resource and its availability for development. Technical Report No. 5, Ministry of Irrigation, Water Development and Flood Control, Dhaka

Nahar NM (2008) Analysis of pesticide residues in some local fish and vegetable. Dhaka Uni J Sci 56(2):1–4

Nahar MS, Zhang J, Ueda A, Yoshihisa F (2014) Investigation of severe water problem in urban areas of a developing country: the case of Dhaka, Bangladesh. Environ Geochem Health 36:1079–1094. https://doi.org/10.1007/s10653-014-9616-5

Narottam S, Zaman MR (2011) Concentration of selected toxic metals in groundwater and some cereals grown in Shibganj area of Chapai Nawabganj, Rajshahi, Bangladesh. Curr Sci 101(3):427–431

Naus FL, Schot P, Groen K, Ahmed KM, Griffioen J (2019) Groundwater salinity variation in Upazila Assasuni (southwestern Bangladesh), as steered by surface clay layer thickness, relative elevation and present- day land use. Hydrol Earth Syst Sci 23:1431–1451. https://doi.org/10.5194/hess-23-1431-2019

Nerbrand C, Agréus L, Lenner RA, Nyberg P, Svärdsudd K (2003) The influence of calcium and magnesium in drinking water and diet on cardiovascular risk factors in individuals living in hard and soft water areas with differences in cardiovascular mortality. BMC Public Health 3(3) https://www.biomedcentral.com/1471-2458/3/21

New Nation Report (1996) Immediate Government Steps Needed, Millions Affected by Arsenic Contamination. The New Nation, A daily newspaper of Bangladesh. https://thedailynewnation.com/

Nick A, Foppen JW, Kulabako R, Lo D, Samwel M, Wagner F, Wolf L (2012) Sustainable sanitation and groundwater protection – factsheet of working group 11. Sustainable Sanitation Alliance (SuSanA) https://www.susana.org/en/resources/library/details/98

Nordstrom DK (2002) Worldwide occurrences of arsenic in ground water. Science 296:2143–2145. https://doi.org/10.1126/science.1072375

Ong CN, Lee WR (1980) High affinity of lead for foetal hemoglobin. Br J Ind Med 37:292–298. https://doi.org/10.1136/oem.37.3.292

Panaullah GM, Alam T, Hossain MB, Loeppert RH, Lauren JG, Meisner CA, Ahmed ZU, Duxbury JM (2009) Arsenic toxicity to rice (Oryza sativa L.) in Bangladesh. Plant Soil 317:31–39. https://doi.org/10.1007/s11104-008-9786-y

Parveen S, Ahmed MSU, Nasreen T (2008) Microbial contamination of water in around Dhaka City. Bangladesh J Sci Ind Res 43(2):273–276. https://doi.org/10.3329/bjsir.v43i2.972
Parvez AK (2016) Bacteriological quality of drinking water samples across Bangladesh. Arch Clin Microbiol 7(1):9. https://www.acmicrob.com/microbiology/
Parvez N, Pritul AS (2018) Analysis of ground water quality: a case study in the Savar (Akrain) area. Journal of Water Resources and Pollution Studies 3(3):67–75. https://doi.org/10.5281/zenodo.1626349
Parvez N, Pritul A Hasan, Md (2018) Analysis of Maximum Temperature and Rainfall over Dhaka City as a Climatic Factor. https://doi.org/10.5281/zenodo.2009335
Philips PJ, Chalmers AT, Gray JL, Kolphin DW, Foreman WT, Wall GR (2012) Combined sewer overflows: an environmental source of hormones and wastewater micropollutants. Environ Sci Technol 46(10):5336–5343. https://doi.org/10.1021/Fes3001294
Pizzo G, Piscopo MR, Pizzo I, Giuliana G (2007) Community water fluoridation and caries prevention: a critical review. Clin Oral Investig 11(3):189–193. https://doi.org/10.1007/s00784-007-0111-6. PMID 17333303
PPW-DAE (2018) Annual report-2018. Pesticide Protection Wing and Directorate of Agriculture Extension, Bangladesh
Pravina P, Sayaji D, Avinash M (2012) Calcium and its role in human body. Int J Res Pharm Biomed Sci 58:–144. www.ijrpbsonline.com
Prodip KR, Roy B, Roy BC (2016) Assessment of groundwater quality for drinking and irrigation purposes in Comilla District of Bangladesh. Int J Sci Res Publ 6(7):52–59. https://ijsrp.org/
Qureshi AS, McCornick PG, Sarwar A, Sharma BR (2010) Challenges and prospects for sustainable groundwater management in the Indus Basin, Pakistan. Water Res Manag 24:1551–1569. https://doi.org/10.1007/s11269-009-9513-3
Raghunath HM (1987) Groundwater. Wiley Eastern Ltd., New Delhi, pp 344–369
Rahman MA, Hasegawa H (2011) High levels of inorganic arsenic in rice in areas where arsenic-contaminated water is used for irrigation and cooking. Sci Total Environ 409(22):4645–4655. https://doi.org/10.1016/j.scitotenv.2011.07.068
Rahaman A, Hafezur R (2018) Contamination of arsenic, manganese and coliform bacteria in groundwater at Kushtia District, Bangladesh: human health vulnerabilities. J Water Health 16:782–795. https://doi.org/10.2166/wh.2018.057
Rahman MH (1997) Risk assessment of pesticides used in Bangladesh. J Civ Eng 25(1):97–106. https://www.scopus.com/inward/record.url?eid=2-s2.0-84861696133&partnerID=10&rel=R3.0.0
Rahman MM (2000) Pesticides: their uses and problems in context of Bangladesh. National Workshop on Conventional and Nuclear Technique for Pesticide Residues Studies in Food and Environment at IFRB, Savar, Bangladesh, pp 1–25
Rahman SH (2009) Investigation of shallow tube-well water quality considering the influence of nearby latrines in a Rural Village of Bangladesh. IAHS Publ 329:299–304. https://www.semanticscholar.org/
Rahman SH (2013) Water quality of shallow tube wells as affected by sanitary latrines and groundwater flow. J Bangladesh Acad Sci 37(2):231–243. https://doi.org/10.3329/jbas.v37i2.17565
Rahman SH, Majumder RK (2012) Groundwater quality of irrigation of deep aquifer in southwestern zone of Bangladesh. Songklanakarin J Sc Technol 34(3):345–352. http://www.sjst.psu.ac.th
Rahman M, Malek M, Matin M (1995) Trend of pesticide usage in Bangladesh. Sci Total Environ 159(1):33–39. https://doi.org/10.1016/0048-9697(94)04206-3
Rahman SH, Ahmed S, Zakariya M (2009) Investigation of shallow tube-well water quality considering the influence of nearby latrines in a rural village of Bangladesh. Trends and Sustainability of Groundwater in Highly Stressed Aquifers (Proc. of Symposium JS.2 at the Joint IAHS & IAH Convention, Hyderabad, India, September 2009). *IAHS Publ.* 329. http://hydrologie.org/redbooks/a329/IAHS Publ. 329/37 299-304 329

Rahman MM, Sultana R, Shammi M, Bikash J, Ahmed T, Maruo M, Kurasaki M, Uddin MK (2016) Assessment of the status of groundwater arsenic at Singair Upazilla, Manikganj, Bangladesh; exploring the correlation with other metals and ions. Expo health 8:217–225. https://doi.org/10.1007/s12403-016-0196-8

Rahman MS, Saha N, Islam ARMT, Shen S, Bodrud-Doza M (2017a) Evaluation of water quality for sustainable agriculture in Bangladesh. Water Air Soil Pollut 228(10):385. https://doi.org/10.1007/s11270-017-3543-x

Rahman MATMT, Saadat AHM, Islam MS, Al-Mansur MA, Ahmed S (2017b) Groundwater characterization and selection of suitable water type for irrigation in the western region of Bangladesh. Appl Water Sci 7:233–243. https://doi.org/10.1007/s13201-014-0239-x

Raihan F, Alam JB (2008) Assessment of groundwater quality in Sunamganj of Bangladesh. Iran J Environ Health Sci Eng 5:155–166. https://www.researchgate.net/publication/

Rajmohan N, Prathapar SA (2013) Hydrogeology of the Eastern Ganges Basin: an overview. Colombo, Sri Lanka: International Water Management Institute (IWMI). 42p. (IWMI Working Paper 157). https://doi.org/10.5337/2013.216

Rasul T, Jahan MS (2010) Quality of ground and surface water of Rajshahi City area for sustainable drinking water source. J Sci Res 2(3):577–584. https://doi.org/10.3329/jsr.v2i3.4093

Rauh VA, Garfinkel R, Perera FP, Andrews HF, Hoepner L, Barr DB (2006) Impact of prenatal chlorpyrifos exposure on neurodevelopment in the first 3 years of life among inner-city children. Pediatrics 118(6):1845–1859. https://doi.org/10.1542/peds.2006-0338

Ravenscroft P (2007a) Predicting the global distribution of natural arsenic contamination of groundwater. Symposium on Arsenic: the geography of a global problem, Royal Geographical Society, London. https://www.geog.cam.ac.uk/research/projects/arsenic/symposium/S1.2_P_Ravenscroft.pdf

Ravenscroft P (2007b) Predicting the global extent of arsenic pollution of groundwater and its potential impact on human health. Report for UNICEF, New York. https://www.researchgate.net/publication/313628997

Ravenscroft P, Burgess WG, Ahmed KM, Burren M, Perrin J (2005) Arsenic in groundwater of the Bengal Basin, Bangladesh: distribution, field relations and hydrogeological setting. Hydrogeol J 13:727–751. https://doi.org/10.1007/s10040-003-0314-0

Reghunath R, Murthy TRS, Reghavan BR (2002) The utility of multivariate statistical techniques in hydrogeochemical studies: an example from Karnataka, India. Water Res 36:2437–2442. https://doi.org/10.1016/s0043-1354(01)00490-0

Riah W, Laval K, Laroche-Ajzenberg E et al (2014) Effects of pesticides on soil enzymes: a review. Environ Chem Lett 12:257–273. https://doi.org/10.1007/s10311-014-0458-2

Richards LA (1968) Soil and water conservation research branch. Agricultural Research Service. Agriculture Handbook No. 60

Safiuddin M (2011) Arsenic contamination of groundwater in Bangladesh: a review. Int J Phys Sci 6(30). https://doi.org/10.5897/IJPS11.1300

Saha N, Zaman M (2011) Concentration of selected toxic metals in groundwater and some cereals grown in Shibganj area of Chapai Nawabganj, Rajshahi, Bangladesh. Curr Sci 101(3):427–431. http://www.jstor.org/stable/24078523

Sahu P, Sikdar PK (2008) Hadrochemical framework of the aquifer in and around East Kolkata wetlands, West Bengal, India. Environ Geol 55:823–835. https://doi.org/10.1007/s00254-007-1034-x

Sandstead HH (1978) Zinc interference with copper metabolism. J Am Med Assoc 240(20):2188. https://doi.org/10.1001/jama.1978.03290200066029

Sarkar AA, Hassan AA (2006) Water quality assessment of a groundwater basin in Bangladesh for irrigation use. Pak J Biol Sci 9:1677–1684. https://doi.org/10.3923/pjbs.2006.1677.1684

Sarker B, Zaman MW (2003) Extent of pollution assessment in drinking water supplies of Narayanganj District in Bangladesh. J Biol Sci. https://doi.org/10.3923/jbs.2003.854.863

Sarkar RK, Reddy JN, Sharma SG, Ismail AM (2006) Physiological basis of submergence tolerance in rice and implications for crop improvement. Curr Sci 91:899–906

Sasikaran S, Sritharan K, Balakumar S, Arasaratnam V (2012) Physical-chemical and microbial analysis of bottled drinking water. Ceylon Med J 57(3):111–116. https://doi.org/10.4038/cmj.v57i3.4149. PMID: 23086026

Seddique AA, Ahmed KM, Shamsudduha M, Aziz Z, Hoque MA (2004) Heavy metal pollution in groundwater in and around Narayanganj town, Bangladesh. Bangladesh J Geol 23:1–12. https://www.researchgate.net/publication/234038920

Seddique S, Ashraf A, Hossain M et al (2013) Land use changes over 21 years and impact on groundwater aquifer in and around Cox's bazar town. Bangladesh 3:20–27

Seddique AA, Masuda H, Hoque A (2016) Radionuclide and heavy metal contamination in the paleo beach groundwater, Cox's bazar, Bangladesh: potential impact on the environment. Arab J Geosci 9(7):488

SEPA (2005) State Environmental Protection Administration of the People's Republic of China. Notice of Strengthening the Management of Environmental Impact Assessments to Prevent Environmental Risk. Beijing, SEPA Order No. 152. (in Chinese).

Shah T, Roy AD, Qureshi AS, Wang J (2003) Sustaining Asia's groundwater boom: an overview of issues and evidence. Nat Res Forum 27:130–141. https://doi.org/10.1111/1477-8947.00048

Shah T, Bruke J, Villholth K, Angelica M, Custodio E, Daibes F, Hoogesteger J, Giordano M, Girman J, van der Gun J et al (2009) Groundwater: a global assessment of scale and significance. Annu Rev Environ Resour 34:7.1–7.26. https://doi.org/10.1146/annurev.environ.030308.100251

Shahid S (2010) Impact of climate change on irrigation water demand of dry season Boro rice in Northwest Bangladesh. Clim Chang 105:433–453. https://doi.org/10.1007/s10584-010-9895-5

Shahid S (2011) Impact of climate change on irrigation water demand of dry season Boro rice in Northwest Bangladesh. Clim Chang 105:433–453. https://doi.org/10.1007/s10584-010-9895-5

Shahidullah SM, Hakim MA, Alam MS, Shamsuddoha ATM (2000) Assessment of groundwater quality in a selected area of Bangladesh. Pak J Biol Sci 3(2):246–249. https://doi.org/10.3923/pjbs.2000.246.249

Shaik MB, TIPU MMH, Raunak J, Mokhlesur RM (2015) Groundwater quality and its suitability assessment for irrigation and drinking in different aquifers of Faridpur, Bangladesh. J Water Poll Purif Res 2:5–14

Shajedul MI, Majumder SMMH (2020) Alkalinity and hardness of natural waters in Chittagong City of Bangladesh. Int J Sci Bus 4(1):137–150. https://doi.org/10.5281/zenodo.3606945

Shammi M, Rahman R, Rahman MM, Moniruzzaman M, Bodrud-Doza M, Karmakar B, Uddin MK (2016) Assessment of salinity hazard in existing water resources for irrigation and potentiality of conjunction uses: a case report from Gopalganj District, Bangladesh. Sustain Water Res Mang 2(4):369–378. https://doi.org/10.1007/s40899-016-0064-5

Shamsudduha M, Chandler RE, Taylor R, Ahmed KM (2009) Recent trends in groundwater levels in a highly seasonal hydrological system: the Ganges-Brahmaputra-Meghna Delta. Hydrol Earth Syst Sci 13:2373–2385. https://doi.org/10.5194/hess-13-2373-2009

Shamsudduha M, Joseph G, Haque SS, Khan MR, Zahid A, Ahmed KMU (2019) Multi-hazard groundwater risks to the drinking water supply in Bangladesh: challenges to achieving the sustainable development goals. World Bank Group 8922. https://doi.org/10.1007/s12403-019-00325-9

Shiekh MA (2006) High degree of fecal contamination in river, lake and pond waters in/and around Dhaka city of Bangladesh. Pak J Biol Sci 9(1):141–144. https://doi.org/10.3923/pjbs.2006.141.144

Shirazi SM, Akib S, Salman FA, Alengaram UJ, Jamneel M (2012) Agroecological aspect of groundwater utilization: a case study. Sci Res Essay 5:2786–2795. http://www.academicjournals.org/SRE

Shoeb MN (2009) Nitrofuran antibiotics and organochlorine pesticides in fish and poultry fish, feed ingredients, and water samples collected from shrimp cultivation area. J Bangladesh Chem Soc 22(1):1–8. http://lib.ewubd.edu/article_index/

Shoeb MM (2016) Organochlorine pesticide residues in poultry meats of Bangladesh. Croat J Food Sci Technol 8(1):30–33. https://doi.org/10.17508/CJFST.2016.8.1.04

Shrivastava R, Upreti R, Seth P, Chaturvedi U (2002) Effects of chromium on the immune system. Immunol Med Microbiol 34(1):1–7. https://doi.org/10.1111/j.1574-695X.2002.tb00596.x

Singh B, Singh Y, Sekhon GB (1995) Fertilizer-N use efficiency and nitrate pollution of groundwater in developing country. J Cont Hydrol 20(3–4):167–184. https://doi.org/10.1016/0169-7722(95)00067-4
Singh CK, Kumari R, Singh RP, Mukherjee S (2013) Geochemical characterization and heavy metal contamination of groundwater in Satluj River basin. Environ Earth Sci 71:201–216. https://doi.org/10.1007/s12665-013-2424-x
Slooff W (1989) Integrated criteria document chromium. Bilthoven, Netherlands, National Institute of Public Health and Environmental Protection, Report no. 758701002
Smedley PL, Kinniburgh DG (2002) A review of the source, behavior, and distribution of arsenic in natural waters. Appl Geochem 17:517–568. https://doi.org/10.1016/S0883-2927(02)00018-5
Smith AH, Lingas EO, Rahman M (2000) Contamination of drinking-water by arsenic in Bangladesh: a public health emergency. Bull World Health Organ 78(9):1093–1110. https://www.who.int/bulletin/archives/78%289%291093.pdf
Soara A, Perkinsb GD, Abbasc G, Alfonzod A, Barelli AE, Bierens OJLMF, Bruggerg H, Deakinh CD, Dunning JI, Georgiouj M, Handleyk NJ, Lockey DJI (2010) European Resuscitation Council Guidelines for Resuscitation 2010 section 8. Cardiac arrest in special circumstances: Electrolyte abnormalities, oisoning, drowning, accidental hypothermia, hyperthermia, asthma, naphylaxis, cardiac surgery, trauma, pregnancy, electrocution. Medicine. http://doi:10.1016/j.resuscutation.2010.08.015
SOES-DCH (2000) Groundwater arsenic contamination in Bangladesh. A) A summary of 239 days field survey from August 1995 to February 2000. B) Twenty-seven days detailed field survey information from April 1999 to February 2000, 70. www.soesju.org
Soja R, Wiejaczka Ł (2014) The impact of a reservoir on the physicochemical properties of water in a mountain river. Water Environ J 28(4):473–482. https://doi.org/10.1111/wej.12059
Sukhen G, Rahman A, Fuad M, Dhiman R, Rahim M (2017) Groundwater quality assessment of Barishal city corporation in Bangladesh. Barisal Uni J 1(4):339–350. www.researchgate.net/publication/332463988
Sultan MT, Billah SM (2019) Irrigation water quality of surface and ground water used for Boro rice cultivation in Khulna district in Bangladesh. Bangladesh J Sci Ind Res 54:263–274. https://doi.org/10.3329/bjsir.v54i3.42679
Sumon KA, Rashid H, Edwin TH, Peeters M, Bosma R (2018) Environmental monitoring and risk assessment of organophosphate pesticides in aquatic ecosystems of north-West Bangladesh. Chemosphere 206. https://doi.org/10.1016/j.chemosphere.2018.04.167
Sunderman FW Jr, Hopfer SM, Sweeney KR, Marcus AH, Most BM, Creason J (1989) Nickel absorption and kinetics in human volunteers. Proc Soc Exp Biol Med 191(1):5–11. https://doi.org/10.3181/00379727-191-42881
Suthar S, Bishnoi P, Singh S (2009) Nitrate contamination in groundwater of some rural areas of Rajasthan, India. J Hazard Mater 171(1–3):189–199. https://doi.org/10.1016/j.jhazmat.2009.05.111
Swaminathan S, Fonseca VA, Alam MG, Shah SV (2007) The role of iron in diabetes and its complications. Diabetes Care 30(7):1926–1933. https://doi.org/10.2337/dc06-2625
Tahera A, Jhohura FT, Akter F, Chowdhury TR, Mistry SK, Dey D, Barua MK, Akramul MI, Rahman M (2016) Water quality index for measuring drinking water quality in rural Bangladesh: a cross- sectional study. J Health Popul Nutr 35(4). https://doi.org/10.1186/s41043-016-0041-5
Tandogan B, Ulusu NN (2005) Importance of calcium. Turk J Med Sci 35:197–201. http://journals.tubitak.gov.tr/medical/issues/sag-05-35-4/sag-35-4-1-0504-11.pdf
Tanvir MATBR, Tirkey P, Bhattacharya T, Chakraborty S, Baraik S (2017) Assessment of groundwater quality and associated health risk: a case study of Ranchi City, Jharkhand, India. Groundwater Sustain Dev 5:85–100. https://doi.org/10.1016/j.gsd.2017.05.002
Tirkey P, Tanushree B, Sukalyan C (2013) Water quality indices- important tools for water quality assessment: a review. Int J Adv Chem 1(1):15–28
Todd DK (1980) Groundwater hydrology. Wiley, New York, pp 10–138. https://doi.org/10.1111/j.1349-7006.2008.01001.x

Toyokuni S (2009) Role of iron in carcinogenesis: cancer as a ferrotoxic disease. Cancer Sci 100(1):9–16. https://doi.org/10.1111/j.1349-7006.2008.01001.x

Tyagi S, Sharma B, Singh P, Dobhal R (2013) Water quality assessment in terms of water quality index. Am J Water Res 1(3):34–38. https://doi.org/10.12691/ajwr-1-3-3

UCCC (1974) Guidelines for interpretations of water quality for irrigation. University of California Committee of Consultants (UCCC), U.S.A. National field manual, US – Geological Survey

UKEVM (2003) Risk assessments: potassium. In: safe upper levels for vitamins and minerals. London, United Kingdom Food Standards Agency, expert group on vitamins and minerals, 299. http://cot.food.gov.uk/pdfs/vitmin2003.pdf

UNDP-BWDB (1982) Groundwater survey: the hydrogeological conditions of Bangladesh. UNDP and Bangladesh Water Development Board (BWDB) Technical Report DP/UN/BGD-74-009, 1, 113

UNESCO (2003) The UN World Water Development Report. Retrieved from: http://www.unesco.org/

UNICEF (2010) Towards an arsenic safe environment in Bangladesh. Executive summary: world water day, 22 March, 2010. www.unicef.org/infobycountry/files/

UN-water (2015) Wastewater Management – a UN-Water Analytical Brief. http://www.unwater.org/

US-CDC (2001) Centers for disease control and prevention. United States Department of Health and Human Services. https://en.wikipedia.org/wiki/Centers_for_Disease_Control_and_Prevention

US-DHEW (1962) Drinking water standards. US Department of Health, Education and Welfare, Public Health Service, Publication No. 956

US-EPA (1986) Quality Criteria for Water. U.S. Environmental Protection Agency, Office of Water Health and Ecological Criteria Division, Washington, DC. 20460, EPA: 44015-86-001. https://nepis.epa.gov/Exe/tiff2png.cgi

US-EPA (1989) Health and environmental effect document for 4-Aminopyridine. Environmental Protection Agency, Cincinnati, USA: PB91-216333, Report no. EPA/600/8-89/050. https://archive.epa.gov/osa/hsrb/web/pdf/epa-woe-document-4-aminopyridine.pdf

US-EPA (2003) Drinking water advisory: consumer acceptability advice and health effects analysis on sulfate. U.S. Environmental Protection Agency, Office of Water (4304T) Health and Ecological Criteria Division, Washington, DC. 20460. www.epa.gov/safewater/ccl/pdf/sulfate.pdf

US-EPA (2016a) Hydraulic fracturing for oil and gas: impacts from the hydraulic fracturing water cycle on drinking water resources in the United States – final report. EPA 600/R-16/236F, Washington, DC. https://cfpub.epa.gov/ncea/hfstudy/recordisplay.cfm?deid=332990

US-EPA (2016b) Development of a conceptual model to estimate pesticide concentrations for human health drinking water and guidance on conducting ecological risk assessments for the use of pesticides on Rice. Environmental Fate and Effects Division, Office of Pesticide Programs Office of Chemical Safety and Pollution Prevention, United States Environmental Protection Agency. https://www.epa.gov/sites/production/files/2016-10/documents/pfam-whitepaper.pdf

US-EPA (2018) Ground water rule. Drinking water requirements for states and public water systems. Environmental Protection Agencies, Washington, DC. https://www.epa.gov/dwreginfo/ground-water-rule

US-EPA (2019) Getting up to speed: ground water contamination. United State Environmental Protection Agency. https://www.epa.gov/sites/production/files/2015-08/documents/mgwc-gwc1.pdf

Uttam, K.C, Bhajan K. B., Tarit R. C., Samanta, S., Mandal, B.K., Basu, G.C., Chanda, C.R., Lodh, D., Saha, K.C., Mukherjee, S.K., Roy, S., Kabir, S., Quamruzzaman, Q., Chakraborti, D. 2000, Groundwater arsenic contamination in Bangladesh and West Bengal, India. Environ Health Perspect. 108(5), 393–397: https://doi.org/10.2307/3454378

Valavanidis A, Vlachogianni T (2010) Metal pollution in ecosystems. ecotoxicology studies and risk assessment in the marine environment. Science advances on Environment, Toxicity & Ecotoxicology issues. https://www.researchgate.net/publication/236623174

Valenzuela EF, Menezes HC, Cardeal ZL (2020) Passive and grab sampling methods to assess pesticide residues in water. A review. Environ Chem Lett 18:1019–1048. https://doi.org/10.1007/s10311-020-00998-8

Vallee BL, Wacker WE, Ulmer DD (1960) The magnesium-deficiency tetany syndrome in man. Boston Med Surg J 262(4):155–161. https://doi.org/10.1056/NEJM196001282620401

van Geen A, Ahmed K, Akita Y, Alam M, Culligan P, Feighery J, Ferguson A, Emch M, Escamilla V, Knappett P, Layton A, Mailloux B, Mckay L, Mey J, Serre M, Streatfield P, Yunus M (2011) Fecal contamination of shallow tube-wells in Bangladesh inversely related to arsenic. Environ Sci Technol 45(4):1199–1205. https://doi.org/10.1021/es103192b

Wallace JS, Batchelor CH (1997) Managing water resources for crop production. Philos Trans R Soc London Ser B 352:937–947

Waller RM (1982) Ground water and the rural homeowner, Pamphlet, U.S. Geological Survey. https://doi.org/10.3133/7000054

Wang F (2006) Water pollution, unneglectable point. Ecol Econ 7:142–145

Wang C, Yang Y, Wu N et al (2019) Combined toxicity of pyrethroid insecticides and heavy metals: a review. Environ Chem Lett 17:1693–1706. https://doi.org/10.1007/s10311-019-00905-w

WARPO (2000) Estimation of groundwater resources. Draft development strategy (DDS), Annex-C, Appendix-6, National Water Management Plan, Water Resources Planning Organization (WARPO), Dhaka, Bangladesh

Waste atlas (2012) Country Data: Bangladesh. http://www.atlas.d-waste.com/

WB Group (2019) Multi-Hazard groundwater risks to the drinking water supply in Bangladesh. Water Global Practice. 8922. https://doi.org/10.1007/s12403-019-00325-9

Weinberger DR, Berman KF, Zec RF (1986) Physiologic dysfunction of dorsolateral prefrontal cortex in schizophrenia: I. Regional cerebral blood flow evidence Archives of General Psychiatry 43:114–124

Whelton A, Dietrich A, Burlingame G, Schechs M, Duncan S (2007) Minerals in drinking water: impacts on taste and importance to consumer health. Water Sci Technol 55(5):283. https://doi.org/10.2166/wst.2007.190

WHO (1996a) Guidelines for drinking-water quality, Health criteria and other supporting Information, vol 2, 2nd edn. World Health Organization, Geneva, 1996, ISBN 92 4 154480 5. http://www.who.int/water_sanitation_health/publications/gdwq2v1/en/index1.html

WHO (1996b) Zinc in drinking-water. Background document for development of WHO guidelines for drinking-water quality, vol 2, 2nd ed. World Health Organization, Geneva (WHO/SDE/WSH/03.04/17). https://www.who.int/water_sanitation_health/dwq/chemicals/zinc.pdf

WHO (2004) Guidelines for drinking-water quality. World Health Organization, Geneva. http://www.who.int/iris/handle/10665/42852

WHO (2005) Nickel in Drinking-water. Background document for development of WHO Guidelines for drinking-water quality, World Health Organization, WHO/SDE/WSH/05.08/55. https://www.who.int/water_sanitation_health/gdwqrevision/nickel2005.pdf

WHO (2006) Section 4: approaches to drinking water source protection management. Protecting groundwater for health: managing the quality of drinking-water sources. IWA publishing for WHO. http://www.who.int/water_sanitation_health/publications/PGWsection4.pdf?ua=1

WHO (2009) Potassium in drinking-water. Background document for development of WHO guidelines for drinking-water Quality, WHO/HSE/WSH/09.01/7, Geneva, Switzerland. https://www.who.int/

WHO (2011) Guidelines for drinking-water quality, 4th edn. World Health Organization, Geneva. https://www.who.int/water_sanitation_health/

WHO (2018) Sustainable development and health environment. Retrieved from WHO Bangladesh. http://www.whoban.org/sust_dev_mental_env.html

WHO (2019) Chromium in drinking-water. Draft Background document for development of WHO Guidelines for Drinking-water Quality, WHO/SDE/WSH/xxxxxx. https://www.who.int/water_sanitation_health/

Wilcox LV (1955) Classification and use of irrigation water. U.S. Department of Agriculture, Circular No. 969, Washington D.C. U.S.A., 19. https://www.ars.usda.gov/arsuserfiles/20360500/pdf_pubs/P0192.pdf

Winton EF, Tardiff RG, McCabe LJ (1971) Nitrate in drinking water. J AWWA 63(2):95–98. https://doi.org/10.1002/j.1551-8833.1971.tb04035.x

Wolf L, Nick A, Cronin A (2015) How to keep your groundwater drinkable: safer siting of sanitation system. Sustainable sanitation alliance working group 11. http://www.susana.org/en/resources/library/details/2155

World Bank Group (2019) Multi-hazard groundwater risks to the drinking water supply in Bangladesh: challenges to achieving the sustainable development goals. Report No. WPS8922, Policy Research Working Paper, 8922, 3. http://documents.worldbank.org/curated/en/

Wosnie A, Wondie A (2014) Assessment of downstream impact of Bahir Dar tannery effluent on the head of Blue Nile River using macroinvertebrates as bioindicators. Int J Biodiver Conserv 6:342–350. https://doi.org/10.5897/IJBC2013.0646

Yidana SM, Yidana A (2010) Assessing water quality using water quality index and multivariate analysis. Environ Earth Sci 59:1461 1473. https://www.tib.eu/en/search/id/BLSE%3ARN264724957/

Zahid A (2008) Study on aquifer environment in the arsenic affected areas of Chandpur and Madaripur districts of southern Bangladesh (Ph.D. Thesis). Unpublished, University of Dhaka

Zahid A (2015) Groundwater management aspects in Bangladesh. Technical report, Center for Water and Environment, Bangladesh Water Development Board. https://www.researchgate.net/publication/320556522

Zahid A, Hassan MQ (2009) Flow paths and travel time of groundwater from arsenic contaminated zone to deeper aquifer units under different development stresses. In: Proceedings of the 2nd international conference on water and flood management. Bangladesh University of Engineering and Technology (BUET), 1, p 151–161. https://publications.iwmi.org/pdf/H039306.pdf

Zahid A, Hassan MQ, Nahar MNK (2010) Assessment of hydraulic connectivity of the multi-layered aquifers. Dhaka Uni J Earth Environ Sci 1(1):13–17

Zahid A, Hassan MQ, Ahmed KMU (2015) Simulation of flow paths and travel time of groundwater through arsenic-contaminated zone in the multi-layered aquifer system of Bengal Basin. Environ Earth Sci 73:979–991. https://doi.org/10.1007/s12665-014-3447-7

Zhang L, Wan L, Chang N, Liu J, Duan C, Zhou Q (2011) Removal of phosphate from water by activated carbon fiber loaded with lanthanum oxide. J Hazard Mater 190(1–3):848–855. https://doi.org/10.1016/j.jhazmat.2011.04.021

Zheng Y, Stute M, van Geen A, Gavrieli I, Dhar R, Simpson HJ, Schlosser P, Ahmed KM (2004) Redox control of arsenic mobilization in Bangladesh groundwater. Appl Geochem 19:201–214. https://doi.org/10.1016/j.apgeochem.2003.09.007

Zhou W, Long W, Xu T et al (2019) Organic ligands unexpectedly increase the toxicity of chromium(III) for luminescent bacteria. Environ Chem Lett 17:1849–1855. https://doi.org/10.1007/s10311-019-00892-y

Zuane JD (1996) Handbook of drinking water quality, 2nd edn. Wiley, New York. https://www.wiley.com/en-us/

Chapter 5
Influence of Organic Amendments on Soil Properties, Microflora and Plant Growth

Sanya Chaudhari, Abhidha Upadhyay, and Shweta Kulshreshtha

Abstract The demand for agricultural products is growing daily, thus exerting an unseen pressure on the land due to the absence of resting and fallow periods. The continuous use of land requires better soil management strategy in order to improve soil physical, chemical and biological status. For instance, plant growth can be improved by the application of organic soil amendments. This chapter focuses on the use of different types of organic amendments and their role in improving the soil quality, plant growth and microflora in soil. Biofertilizer increases nitrogen content in the soil up to 40–250 kg N_2/hectare/year and 50–80 kg N_2/hectare/year following the respective inoculation of *Rhizobium* and *Azotobacter*. Application of pearl millet green manure increases the soil bulk density by 13.3%. Legume green manure increased the average yield of rice grain by 1.7 Mg/hectare. Soil pathogens are reduced when plant waste and animal waste are applied in compost form. Moreover, compost prepared from animal manure possesses highest humus content of 11.8%, C:N ratio 13.5, and humic acid fraction 2.84. Bamboo biochar increases the activities of enzymes such as beta-glucosidase alkaline phosphatase and urease, while rice hull biochar increases the activity of ascorbate peroxidase and glutathione reductase enzyme. Biogas slurry and spent mushroom compost increases the nitrogen content by 35.98% and 2.17%, respectively. Kelp weed acts as biostimulant while biogas slurry acts as bactericide and fungicide. Mulching helps in maintaining the moisture of the soil.

Keywords Organic amendment · Green manure · Kelp weed amendment · Compost · Blood meal · Mulching · Plastic mulching · Biochar amendment · Biogas sludge · Soil properties · Soil quality · Soil fertility · Spent mushroom compost

S. Chaudhari · A. Upadhyay · S. Kulshreshtha (✉)
Amity Institute of Biotechnology, Amity University Rajasthan, Jaipur, Rajasthan, India

E. Lichtfouse (ed.), *Sustainable Agriculture Reviews 52*, Sustainable Agriculture Reviews 52, https://doi.org/10.1007/978-3-030-73245-5_5

Abbreviations

NRRL	Northern Regional Research Laboratory
C:N	Carbon:Nitrogen ratio
IPCC	Intergovernmental Panel on Climate Change
rRNA	Ribosomal ribonucleic acid
FAO	Food and Agriculture Organization

5.1 Introduction

Soil is the topmost fertile layer and an important part of terrestrial ecosystem. Its main function is to provide water and nutrients to the plants that grow in it. The world population is increasing at an alarming rate, with a total of 7.52 billion individuals till 2018. India constitutes of 17.15% of the world population with 1.29 billion individuals (United States Census Bureau, U.S. and World population clock). Of the total 13.4-billion-hectare land surface of the globe, 11% or 1.5-billion-hectare land is used for crop production worldwide. The annual worldwide crop production rate has substantially declined from 2.1% per annum to 1.4% per annum in the past 30 years (http://www.fao.org/docrep/005/y4252e/y4252e06.htm). With ever-increasing population, there is a need for increasing the crop production rate which further increases the demand of fertile land area. As the fertile land area is limited and cannot be expanded, it becomes more crucial to maintain high levels of fertility of the soil. The main cause of soil deterioration is soil erosion which is triggered by inappropriate agricultural practices, deforestation and overgrazing, and salinity. These activities degrade 38% of total agricultural land in the world, 21% pasture land and 18% forests and woodlands (Utuk and Daniel 2015). There are many other reasons for the degradation of soil quality including poor management of agricultural crops and cropland, over-exploitation of cropland, crop production and farming activities in fragile soil. Deterioration of soil quality poses a significant constraint on the crop production.

Crop farming is an extractive process in which as the plants grow, they absorb nutrients from the soil. Once the plants are harvested, the nutrients also go with them and leave the soil system nutrient deficient (www.sustainabletable.org/265/environment). This process is done incessantly without providing enough rest to the land which causes nutrient depletion in soil.

As soil characteristics and nutrient distribution in soil is the bigger driver of agricultural crops production, it is important to understand the nutrient distribution in different soil types in order to maintain soil fertility (Chen et al. 2017). Soil is fed with manure and chemical fertilizers to retain the fertility of soil, however; over-application of these soil additives causes the problem of soil deterioration and soil quality degradation.

Nitrogen (N), phosphorous (P), potassium (K) and iron (Fe) are the main nutrient elements required for soil replenishment and plant growth. These nutrients are also supplied to deficit soils artificially by the application of chemical fertilizers. According to Food and Agriculture Organization factsheet (FAO 2015), world fertilizer nutrient ($N+P_2O_5+K_2O$) utilization was estimated as 186,900,000 tonnes in 2014 and is expected to grow at the rate of 1.8% per annum from 2014 to 2018. The excessive application of chemical fertilizers to crop fields has led to reduced fertility of soil due to over accumulation of nutrients. Besides, it causes air pollution due to greenhouse gas emission and water pollution due to leaching of fertilizers in water bodies which leads to eutrophication (Rashid et al. 2016). Manufacturing the chemical fertilizers requires a huge input of non-renewable energy resources such as oil or natural gas. Therefore, there is a need to device an alternative method for soil amendment which is environment-friendly, sustainable and can overrule the use of costly and non-renewable chemical fertilizers.

The soil quality and growing potential can be improved by the use of organic amendments. Organic amendments are an important component of organic farming practices that exert positive or negative influence on the composition of soil and indigenous microbial communities that accomplish cycling of nutrient for the crop growth and health. Organic amendments can be provided in three forms to reclaim waste land or maintain the productivity of agricultural land (Figs. 5.1 and 5.2). First, application of organic amendment in the form of biofertilizer which provides essential nutrients such as phosphorous, nitrogen for the growth of plants. Second, use of

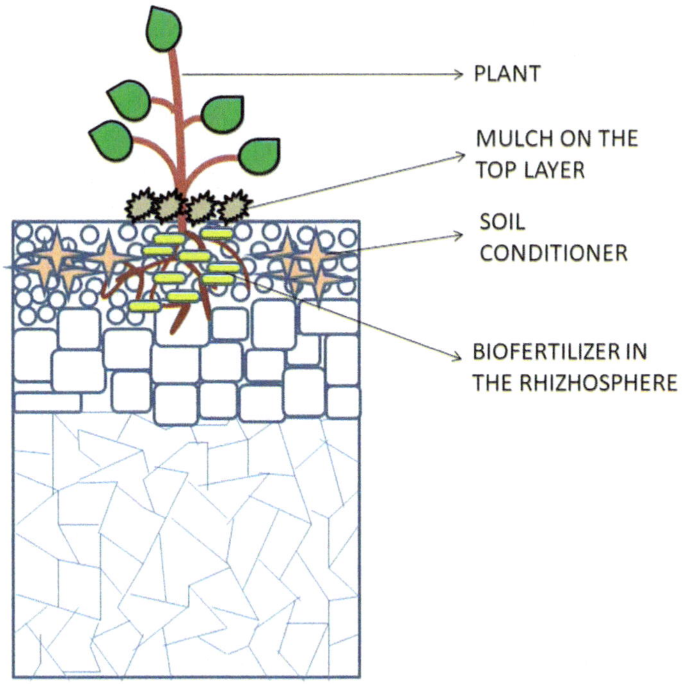

Fig. 5.1 Application of soil conditioner, biofertilizer and mulch as organic amendment

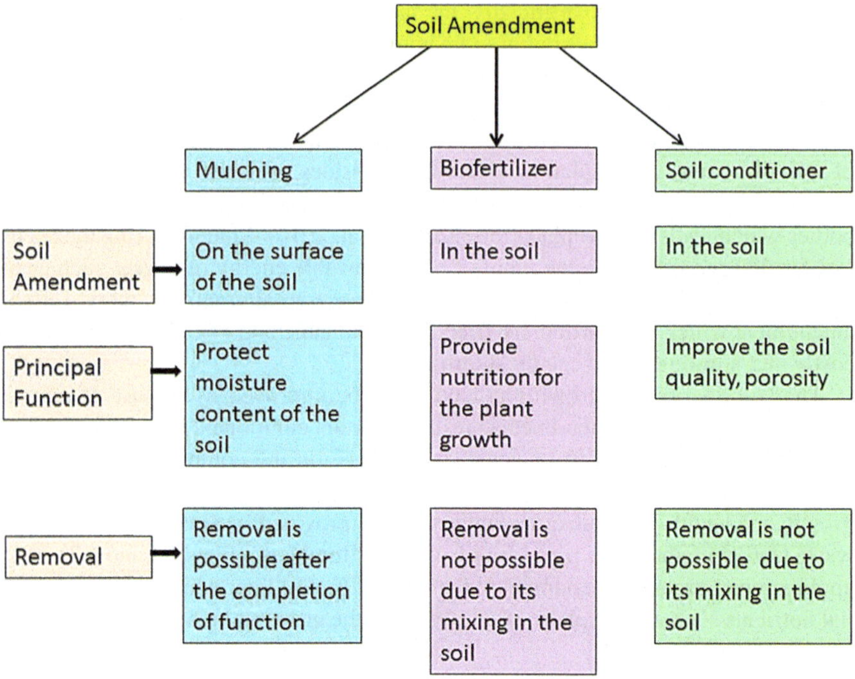

Fig. 5.2 Differences in soil conditioner, biofertilizer and mulch on the basis of soil amendment, principal function and removal

organic amendment as soil conditioner. Soil conditioners are the substances that incorporated in the soil to improve physical qualities of soil such as increase ability to retain soil water, increase oxygen-holding capacity; and provide space and nutrients for root growth. An organic material may act as either biofertilizer or soil conditioner, or both by providing appropriate pH for the growth of plant, by providing nutrients, improving the soil structure and stimulating the growth of beneficial microbes that support the growth of plant and its health, and third, application of mulch. Mulch is the substance which is placed on the soil surface (Fig. 5.1) in order to inhibit the growth of weeds, conserve water or moisture of soil and control soil temperature. Occasionally, it is allowed to degrade in soil to provide nutrients to plants. A comparison of soil conditioner, biofertilizer and mulch is presented in Fig. 5.2.

In this chapter, the focus will be given to the use of various organic soil amendments such as biofertilizers, green manure, spent mushroom compost and anaerobic sludge from biogas digesters, due to their high utility as carbon, nitrogen or phosphorous source. The complete analysis of soil will help us in selecting the appropriate organic source and method of amendment of soil. Organic amendments exert positive or negative effects on dynamic soil properties including physical, chemical and biological conditions which are discussed here in the chapter due to their important role in land reclamation decisions and improvement practices.

5.2 Soil Amendment by Biofertilizers

Biofertilizers are specific, selected and highly effective microbial strains with nitrogen fixing, phosphate solubilising or cellulolytic activity. Generally, bacterial and fungal strains and isolates are selected for biofertilizer preparation and applications. Biofertilizers ensure better and uniform supply of nutrients such as nitrogen, phosphorous and potassium along with some trace elements in the soil and make it accessible to plants (Mrkovački et al. 2012).

Biofertilizers enhance the number of microorganisms in the soil which further increase the amount of organic and inorganic compounds. These compounds are derived as a result of microbial metabolism. The dead microbes act as a substrate for the growth of new microorganisms to increase the microbial count in the soil (Jarak et al. 2010).

5.2.1 Bacterial Biofertilizers

Bacterial biofertilizers are generally nitrogen fixing bacteria. They enzymatically convert atmospheric nitrogen into ammonia by nitrogenase enzyme. They belong primarily to the genera *Rhizobium, Azorhizobium, Bradyrhizobium,* and *Sinorhizobium.* A number of Actinomycetes strains forms nodule-like structures on non-leguminous woody shrubs or trees like *Alnus* (alder) and *Malus* (apple).

The most important bacterial biofertilizers are *Rhizobium* and *Azotobacter*. *Rhizobium* sp. is a soil bacterium which can colonize the roots of leguminous plants. *Rhizobium* fixes atmospheric nitrogen in the form of ammonium ions during its symbiotic association with leguminous and certain non-leguminous plants like *Parasponia*. They form nodules in legume roots where they biologically convert elemental nitrogen into ammonia which is plant-accessible forms of nitrogen readily available to the plant for the synthesis of essential plant growth compounds like protein. Upon plant death, ammonia is released in the soil as nitrogen which is available for growth of other plants and for enhancing the soil fertility. It has been estimated that 40–250 kg nitrogen per hectare per year could be fixed by *Rhizobium* for various legume crops (Tamilkodi and Victoria 2018). The highest nitrogen fixation can be observed at initial planting stage and then decrease. The definite quantity of nitrogen fixed by symbiotic bacteria is affected by endogenous soil microbes, infectious microbial species present in soil and age of plant and its genotype, and agricultural practices. Besides this, physical, chemical and biological properties of soil also affect nitrogen fixation by leguminous crops.

Azotobacter sp. is a free living, aerobic nitrogen fixing bacteria which is used as a biofertilizer. *Azotobacter* is present in neutral or alkaline soils. It is a non-symbiotic nitrogen fixer which colonizes the root but do not form nodule. It is present in the rhizosphere of rice, maize, sugarcane, bajra, vegetables and plantation crops (Bhat et al. 2015). *Azotobacter* strains, in favourable environment, can fix 50–80 kg of nitrogen per hectare per year (Jarak et al. 2010) for the plant growth. *Azotobacter* sp. produces slime and causes soil aggregation which improves soil structure (Tamilkodi and Victoria 2018). The bacterium also produces anti-fungal antibiotics which helps in preventing seedlings' mortality due to inhibition of growth of pathogenic fungi.

5.2.2 *Fungal Biofertilizers*

The fungal biofertilizers mainly consist of phosphate solubilizers. Growth of these microorganisms causes the release of organic acids for example citric acid, fumaric acid, malic acid. These acids drop the pH and solubilize insoluble phosphates, chelate cations and compete with phosphate for adsorption sites in the soil. Thus, they increase the availability of phosphates which could be utilized for plant growth (Rathi and Gaur 2016).

Aspergillus is a filamentous fungus isolated from air, soil and plant debris. It accelerates phosphate solubilization of both organic and rock phosphate through the production of acids (citric, gluconic, glycolic, succinic, and oxalic acids) (Tamilkodi and Victoria 2018). Thus, it enhances the availability of phosphorous in soil and to plants. Another phosphate solubilizing biofertilizer is *Penicillium* which is isolated from air and soil. *Penicillium* produces gluconic, glycolic and malic acids which are responsible for rock phosphate solubilization (Sane and Mehta 2015). *Trichoderma harzianum* has the potential of phosphate solubilization after colonizing the cortex of the roots and hence promote plant growth (Mahato et al. 2018). *Trichoderma* sp. is found in soil which when enters the plant, can colonize the cortex of roots. They produce antibiotics and can parasitize other fungi and microorganisms in order to protect the host plant from various diseases. Furthermore, *Trichoderma* causes detoxification of toxins, mycoparasitism, inactivation of pathogenic enzymes, and increase in plant hormone to enhance rooting which are beneficial for plant growth.

5.2.3 *Arbuscular Mycorrhiza*

An arbuscular mycorrhiza is fungal symbiont which forms symbiotic association with the roots of plants and replenish phosphorous in soil by scavenging phosphorous from the soil and deliver it to the cortical cells of the root. Arbuscular mycorrhiza, belongs to the phylum Glomeromycota, penetrates the cortex of the root of vascular plant and form arbuscules. They provide a useful mean to replenish the phosphorous content of the soil. Thusly, it bypasses direct uptake mechanisms and provide phosphorous to the plants for improving growth. Pyrosequencing-based approach confirmed the presence of three largest families of arbuscular mycorrhiza fungi i.e. Glomeraceae, Gigasporaceae and Acaulosporaceae (Lin et al. 2012).

Long-term amendment with balanced fertilization helps in building-up the soil nutrients which consequently supports the growth of arbuscular mycorrhizal fungi. The self-regulation of plant-microbe system results in shifting of arbuscular mycorrhizal profile. This knowledge is important to understand the role of long-term balance fertilization on the development of arbuscular mycorrhizal fungal community. The application of arbuscular mycorrhizal fungi is environment-friendly and economic option to provide adequate amount of nutrients to the plant.

Organic Materials Review Institute (OMRI), a non-profit organization, reviewed organic biofertilizer and approved numerous varieties of organic soil amendments containing arbuscular mycorrhizal fungi which are listed in Table 5.1.

5.2.4 *Effects of Biofertilizers*

Biofertilizers amendment may improve the soil qualities which further, support the growth of beneficial microbes and helps in plant growth. The detailed description of the role of biofertilizer on plants, soil and soil microflora is discussed here.

5.2.4.1 Effect of Biofertilizer on Soil

Soil properties are related to plant health and soil microbial community. Naturally occurring and artificial application of microbes in soil affects its properties and plant growth. The physical characteristics and bio-chemical properties of soil influence the growth of microbes and their survival in the soil. The microbes, dwelling in the soil, play a crucial role in regulating and maintaining soil fertility (Wang et al.

Table 5.1 Organic Materials Review Institute (OMRI)-approved, commercially available, organic soil amendments having arbuscular mycorrhizal fungi

	Commercial product of mycorrhiza	Product content	Company name
1	Mycorise® ASP	Inoculants, endomycorrhize	Premier Tech Biotechnologies
2	MycoApply® Ecto Liquid Blend	Microbial products, ectomycorrhize	Mycorrhizal Applications, Inc.
3	Myco Seed™ Treat	Seed treatment, beneficial bacteria and fungi	AgriEnergy Resources®
4	Myke® Pro PS3	Seed treatment, mycorrhize	Premier Tech Biotechnologies
5	Nodulator® Liquid Inoculant for pea and lentil	Inoculants, rhizobia	Becker Underwood®, Inc.
6	Vault® NP Liquid Legume inoculant for soybeans	Inoculants, rhizobia	Becker Underwood®, Inc.
7	PRE-VAIL	Microbial inoculants, rhizobia	INTX Microbials, LLC

Source: OMRI Certification to the commercial product of mycorrhiza

2017). For instance, application of *Aspergillus* to the soil increased the phosphorous content of soil two times higher than the non-amended soil. Further, the potassium content increased up to 29 kg/Acre due to the soil amendment with *Aspergillus*. Moreover, the quantities of micronutrients increased due to biofertilizer application (Sane and Mehta 2015).

The difference in the species of microbes may show different effects in the soil and its properties. For instance, the application of *Penicillium spp.* showed differences in their ability to change the nutrient profile of the soil. When *Penicillium* sp. 1 was applied to the soil, it led to doubling of phosphorous content and increase in potassium content by 33 kg/Acre while with *Penicillium* sp. 2 the phosphorous content increased by 4 kg/Acre and potassium content increased by 20 kg/Acre (Sane and Mehta 2015). Therefore, different strains of same genus pose different effects on the soil.

Besides the nutrient content, the pH of the soil is also affected by supplementing the soil with microbes. The soil pH was reported to decrease to 6 by the application of *Trichoderma* in soil while the amount of phosphorous was 216.68 kg/ha and potassium was 534.8 kg/ha (Mahato et al. 2018). Application of competitive microbes in the soil can also solve the problem of bacterial wilt of soil. Therefore, the naturally abundant and artificially amended soil microbes exert positive effect on the soil quality, nutrient content, enzymatic and physicochemical properties of soil and plant growth (Wang et al. 2017).

5.2.4.2 Effect of Biofertilizer on Plant

The application of microbial inoculum as biofertilizer has subsequently increased the plant growth which could be measured by various parameters including leaf size, shoot size, number of leaves and roots produced as well as decrease in the incidence of diseases. Inoculation of *Rhizobium* in the soil improved nodulation, yield and seed quality of faba beans i.e. *Vicia faba*. It also decreased the incidences of diseases due to Bean Yellow Mosaic virus and Broad Bean Mottle Bromovirus in faba beans. Nitrogen fixation by *Rhizobium* further helped in reducing the salinity effects on faba beans (Elsalahi et al. 2016). The production of auxin, gibberellin, biotin, nicotinic acid and pyridoxine carried out by *Azotobacter* sp. which promoted the plant growth. Besides, it increased the bioavailability of phosphorous (Tošić et al. 2016). *Aspergillus* was applied to *Pennisetum glaucum* or bajra which gave 56% and 50% increase in root and shoot length, respectively (Sane and Mehta 2015). *Trichoderma* amendment to the wheat plants resulted in increasing plant height along with the growth in both root and shoot was observed (Mahato et al. 2018). The differences in the microbial strains also exert a variety of effects on the plant growth. Application of two strains of *Penicillium* i.e. *Penicillium* sp. 1 and *Penicillium* sp. 2 to the crop of *Pennisetum glaucum* or bajra showed diverse effects. The length of root and shoot and number of leaves was reported to be increased by 50% by Penicillium sp. 1 and 25–30% by Penicillium sp. 2 (Sane and Mehta 2015).

The role of arbuscular mycorrhizal fungi on plant diversity and ecosystem productivity has been well documented. However, the study on the different taxa of arbuscular mycorrhizal fungi and their effect on soil structure, plant diversity and survival and nutrient acquisition are limited (van der Heijden et al. 2006). Arbuscular mycorrhizal fungi help in nitrogen acquisition by plant but did not affect the total nitrogen uptake by them. Soil structure, plant nutrition and biomass depend on the different taxa of arbuscular mycorrhizal fungi. Zhu et al. (2003) used two cultivars i.e. Clipper and Sahara of *Hordeum vulgare* or barley to find out to the mycorrhizal responsiveness to the phosphorous deficiencies and reported that phosphorous mineralization and utilization efficiency of different strains of barley plants differs. Arbuscular mycorrhizal fungal diversity depends upon the availability of host plant. Any loss of dominant host plant species is associated with the loss of arbuscular mycorrhizal fungal diversity. This type of loss can be seen in the grassland with high grazing activity (Murray et al. 2010).

Occasionally, the negative impact of microbial fertilizer could be observed in terms of plant growth. This could be seen when *Trichoderma* is used as biofertilizer. *Trichoderma* when applied with chemical fertilizers had a negative impact on the leaf production, pinnacle number and pinnacle length of wheat due to its antagonistic effect with chemical fertilizers (Mahato et al. 2018).

5.2.4.3 Effect of Biofertilizer on Soil Microflora

One of the frontier topics of agricultural crop production is soil characterization and presence of microbes in the soil since it is associated with crop production and yield. Balanced application of chemical fertilizer exerts positive effect on the microbial population of soil (Su et al. 2015). However, long-term application of chemical fertilizer may result in loss of beneficial microbial communities. This loss can be avoided by the application of biofertilizers. Therefore, the importance of the short- and long-term application of biofertilizers cannot be ignored.

Biofertilizer application increases the microbial population in the soil. Microbial population in their live or dead form provide benefits to the soil. Live microbes support the microbial growth in soil by fixing nutrients and by synergistic or antagonistic effects. They increase organic matter in the soil; even the dead microbial cells provided organic matter for the growth of new microbial cells in order to increase microbial population (Jarak et al. 2007).

In the experiment conducted by Tošić et al. (2016), the treatment of soil with biofertilizers led to increase in the total number of bacterial cells from 37.02×10^6 to 53.15×10^6. Similarly, the total number of ammonifiers, that constitute bacteria, fungi and actinomycetes, also increased from 24.80×10^5 to 30.35×10^5. The number of *Azotobacter* after applying biofertilizer was found to increase from 57.03×10^2 to 75.24×10^2. However, the population of fungi was reported to decreased from 20.81×10^4 to 16.93×10^4 after biofertilizer application to the soil (Tošić et al. 2016). The possible reason for this is the availability and characteristic of organic matter in the soil. Moreover, many abiotic and biotic factors are responsible for this decrease in fungal population (Belanović et al. 2004).

5.3 Organic Manure

Organic manure, derived from animal waste and excreta, human excreta, plant waste, is nutrient rich substance for incorporating in soil in order to improve its physical, chemical and biological properties. The organic manure may exert positive or negative effect on soil qualities and microbial populations. To investigate the effect of organic manure on soil organic carbon content of paddy field soil of Zhejiang Province, China, an experiment was conducted by Mao et al. (2015) for 17-year long term fertilizer amended paddy field and results were compared with chemical fertilizers for both type of soil i.e. silt and clay. Result of this analysis revealed that plots amended with nitrogen-phosphorous-potassium fertilizer plus organic manure was more conductive to the stability of soil organic carbon as compared to nitrogen-phosphorous-potassium fertilizer alone. In both silt- and clay-soil, sequestered organic carbon can be stored back by the organic amendment. Moreover, high number of microbial products was recorded in the clay soil due to the formation of complex. In contrast, silt showed richness in plant-derived aromatic residues. Furthermore, the clay-fractions had relatively higher amount of alkyl-C, carbonyl-carbon, and lower amount of O-alkyl-carbon, aromatic-carbon as compared to silt fraction. Thus, long-term application of combination of organic manure and chemical fertilizer increased the accumulation of both carbon and recalcitrant compounds and decreased the decomposition of organic matter which further helps in accumulation of carbon in the paddy fields for facilitating crop growth (Mao et al. 2015).

Organic manure induced nutrient-shift in soil along with shift in the microbial communities of rice field which further affects the growth and productivity of rice (Su et al. 2015). Treatment of soil with nitrogen-phosphorous-potassium plus organic manure led to increase the microbial soil biomass significantly as compared to nitrogen-phosphorous-potassium amended field. Biolog's test indicated that nitrogen-phosphorous-potassium plus organic manure treatment enhanced the carbon utilization potential and functional diversity of soil microbial community as compared with the control and nitrogen-phosphorous-potassium amendment alone. Long-term amendment with organic manure resulted in the high nutrient-turnover and well-established carbon-utilization pattern of soil microbes (Hao et al. 2010).

5.3.1 *Green Manure*

Green manure is uprooted or sown crop parts that are left to wither on the fields. On its application to the soil, it acts as nutrient source for the growth of the subsequent crops. They have great prospects in enhancing the soil organic matter and crop productivity (Cherr et al. 2006). Green manure can be obtained either by growing green manure crops or by collecting leaves along with twigs of plant. Green manure plants belong to leguminous family and can be collected from forest, wastelands and field bunds. Green manure crops like sunnhemp, *Sesbania rostrata* and cluster beans are fast growing and can be applied to the soil after sufficient growth.

It has been observed that application of green manure in field improve soil condition with increase in water holding capacity. Besides, the declined weed growth, soil erosion decreased due to application of green manure. It further generates humic acid and acetic acid in the soil to reduce soil alkalinity. Many crops can be used as green manure due to their nutritional content as shown in Table 5.2. The selection of the crop depends on various factors such as climatic conditions, availability of seeds, cropping system practiced and local habitats (Meena et al. 2018).

The effects of green manure on soil properties, growth of plants and microorganisms have been accounted below.

5.3.1.1 Effect of Green Manure on Soil

Green manure application has generated overall positive effects for improving soil properties. The use of legume green manure has enhanced the physicochemical properties of soil, (Shukla et al. 2011). The organic substances bind the soil particles together for better soil aggregation, thus enhancing the hydraulic conductivity, infiltration and percolation, and water retention property (Boparai et al. 1992). For

Table 5.2 Application of plants and weeds as green manure due to their nitrogen, phosphorous and potassium content

			Nutrient content (%) on air dry basis		
	Plant	Scientific name	N	P205	K
1	Gliricidia	*Gliricidia sepium*	2.76	0.28	4.60
2	Pongania	*Pongamia glabra*	3.31	0.44	2.39
3	Neem	*Azadirachta indica*	2.83	0.28	0.35
4	Gulmohur	*Delonix regia*	2.76	0.46	0.50
5	Peltophorum	*Peltophorum ferrugenum*	2.63	0.37	0.50
6	Sunhemp	*Crotalaria juncea*	2.30	0.50	1.80
7	Dhaincha	*Sesbania aculeata*	3.50	0.60	1.20
8	Sesbania	*Sesbania speciosa*	2.71	0.53	2.21
Weeds					
1	Parthenium	*Parthenium hysterophorus*	2.68	0.68	1.45
2	Water hyacinth	*Eichhornia crassipes*	3.01	0.90	0.15
3	Trianthema	*Trianthema portulacastrum*	2.64	0.43	1.30
	Ipomoea	*Ipomoea*	2.01	0.33	0.40
	Calotrophis	*Calotropis gigantea*	2.06	0.54	0.31
	Cassia	*Cassia fistula*	1.60	0.24	1.20

Source: http://agritech.tnau.ac.in/org_farm/orgfarm_green manure.html instance, green manure when applied to fields of pearl millet and alfalfa, increased water retention property. Besides, the soil organic matter increased by approximately 1.2% for both the crops within a year. The nitrogen content of the soil was reported to be increased by 63.1%, potassium content by 33.5% and phosphorous content by 63.4% for pearl millet. While for alfa-alfa, the nitrogen, potassium and phosphorous content increased within a year by 66.9%, 31.6% and 63.6%, respectively (Ismail 2013). At emergence stage, the significant highest soil mineral nitrogen value recorded in 2011 was 69.4 mg/N kg while at harvest the highest value was recorded in 2012 for green manure treated samples (50.4 mg/N kg) (Ciaccia et al. 2017). Similarly, nitrogen, potassium and phosphorous content increased due to amendment of soil with sunflower and red clover plant derived green manure. This also increased the zinc mobility and its availability to the grains. The soil amendment with green manure decreased the pH from 7.75 to 7.50 for red clover green manure, and to 7.53 units for sunflower green manure (Aghili et al. 2014). However, ryegrass, pea and rape green manure amendment acidic with pH ranged from 5.10 to 5.76, higher for rape and lower for ryegrass. The concentrations of nitrogen, carbon and calcium were also higher for green manure treated soil samples (Yang et al. 2016). In sandy soil, green manuring improved the water holding capacity (Yadav et al. 2017).

Occasionally, the negative impact of green manure application on soil has also been observed. In the case of pearl millet and alfa-alfa plants, a decrease of soil bulk density by 13.3% for pearl millet and by 6.6% for alfa-alfa. There was a further 60% decrease of hydraulic conductivity of soil for both pearl millet and alfa-alfa (Ismail 2013). Similarly, the use of green gram and sesbania green manure reduced the bulk density of the soil to the extent of 0.03–0.07 Mg m^{-3} (Mandal et al. 2003).

In most of the cases, the soil amendment with green manure resulted in improving the soil characteristics which support the growth of plants.

5.3.1.2 Effect of Green Manure on Plant

Application of green manure makes the soil nutrient-rich to support and enhance plant growth. In pearl millet and alfa-alfa plants, the highest fresh and dry yields were obtained due to application of green manure (Ismail 2013). Application of green manure of red clover and sunflower, when amended the wheat fields, it significantly increased the nitrogen content of wheat grain by 43.3 mg/g dry weight and 58.4 mg/g dry weight, respectively. The zinc concentration of the grains was also increased by 45.6μg/g dry weight and 104.4μg/g dry weight for red clover and sunflower green manure, respectively which is a good enough amount to treat zinc deficiency diseases for people having wheat-based diets (Aghili et al. 2014).

Green manure treatment significantly enhanced maize crop total dry biomass at silking i.e. more than 14% and at harvest stages i.e. more than 8%. After the emergence, the stem elongation stages have had the highest total maize biomass for green manure amended samples. Nitrogen and phosphorous uptake of maize at grain ripening stage was reported to be 40% and 47%, respectively which was higher than non-amended samples (Ciaccia et al. 2017). Green manure of dhaincha i.e. *Sesbania aculeata* and sunhemp i.e. *Crotalaria juncea* was amended in combination with urea to the sugarcane (*Saccharum officinarum*) crop which exerted positive effect and increased the yield of sugarcane by 2–26% (Gafur and Rahman 2003). Similarly, legume green manure increased in the average yield of rice grain by 1.7 Mg ha^{-1} while higher yield i.e. 20 Mg ha^{-1} was recorded with Dhaincha green manure (Kumar et al. 2018).

Green manure of *Cassia spectsbilis* i.e. cassia was used alone and in combination with cow dung, when applied to the fields, the yield of maize grain increased. The highest nitrogen uptake i.e. 3.06–5.20 kg ha was reported for the cow dung and cassia combination at the 4–6 leaf stage. Potassium and magnesium increased significantly (Shadreck 1999). This showed that green manure in combination with cow dung employs positive effect on the yield of maize. Further, green manure helped in reduction of disease rate in certain plants. Ryegrass green manure treated soil samples, led to the reduction of tobacco disease occurrence i.e., 3.38% as compared to non-treated samples i.e., 80.45% (Yang et al. 2016).

5.3.1.3 Effect of Green Manure on Soil Microbes

Application of green manure helps in maintaining soil structure and its quality. Green manure amendment further enhanced the microbial population in the soil ecosystem. It supported the growth of cultivable bacterial colony, their appearance and growth. Green manure, derived from different plant, supports the growth of different microbial phyla in the soil. For instance, the bacterial phyla *Crenarchaeota* and *Acidobacteria* were found to be more abundant in Rape and pea green manure amended soil, respectively while *Chloroflexi* and *Planctomycetes* were found more abundant in Ryegrass green manure supplemented soil. Yang et al. (2016).

The application of green manure enhanced the mineral content of the soil which further produced positive or negative effect on the soil microbial flora. *Proteobacteria*, *Actinobacteria* were found to have negative correlation with soil calcium content, while *Acidobacteria* and *BRC1* had positive correlation with calcium content. *Proteobacteria*, *Actinobacteria* had negative correlation to nitrogen, while *Chloroflexi*, *Planctomycetes* had positive correlation to nitrogen.

Application of green manure further affects the population of soil microorganisms and their correlation to emergence of tobacco disease rate. The four major phyla- *Proteobacteria*, *Actinobacteria*, *Gemmatimonadetes* and *Nitrospira* and genera like *Dokdonella*, *Rhodanobacter* were positively correlated to tobacco disease rate whereas phyla like *Acidobacteria*, *Chloroflexi*, *Planctomycetes* and *WS3* and genera like *Acidobacteria Gp6* and *Acidobacteria Gp4* were negatively correlated to tobacco disease rate.

Soil amendment with manure compost and manure compost plus bacterial fertilizer led to increase in the amount of microbial biomass, especially, the number of cultivable bacteria. Further, increase in soil respiration and enzyme activities was reported. Besides, increase in microbial activities was reported for manure compost plus bacterial fertilizer (Zhen et al. 2014). Therefore, green manure poses positive effect on the microbial diversity of soil.

5.3.2 Farmyard Manure

All type of wastes generated on farm is used to prepare farmyard manure. This manure is prepared by using cow dung, cow urine, straw waste and dairy waste. Farmyard manure is prepared by dumping the cow dung, cow urine, straw material in a covered-pit in a form of uniform layer. The waste layer is made till it rises about 30 cm above the bottom of pit. The covered-pit protects the material from sunlight and rainfall. In the pit, water is sprinkled to prevent the drying of the matter. In approximately 6 months of time, nutrient rich farmyard manure is prepared which contains 0.32% nitrogen, 0.05% phosphorus (P), 0.25% potassium (K), 1.20% calcium (Ca) and 0.33% magnesium (Mg). Application of this manure in the fields provides nitrogen to the plant while some part decomposes slowly and provides nutrients to the plants as and when required.

5.3.2.1 Effect of Farmyard Manure on Soil

The method of preparation of farmyard manure poses a great influence on the properties of soil, diversity of soil organisms, enzymatic activities. To figure out the effect of traditionally composted farmyard manure and biodynamically composted farmyard manure on the chemical properties of soil, microbial survival, activity, biomass and respiration; enzymatic activities; and biomass of earthworms was assessed by Zaller and Köpke (2004). Their results showed significant increase in the soil pH by prepared and non-prepared farmyard manure. Phosphorous and potassium concentration, microbial biomass, dehydrogenase and saccharase activity and root growth was higher in the plots with prepared farmyard manure amendments compared to that of non-prepared farmyard manure or unamended farmyard manure plots. The biodynamically prepared farmyard manure decreased basal respiration rate of soil microbes and metabolic quotient but increased the rate of decomposition as compared to non-prepared farmyard manure or prepared farmyard manure with only Achillea plants. Soil amended with completely-prepared farmyard manure exerted positive effect on the population of earthworms. In the case of completely-prepared farmyard manure, significantly higher biomass and abundance of endogenic or anecic earthworms was reported as compared to non-prepared farmyard manure (Zaller and Köpke 2004). Therefore, the method of manure preparation greatly influences the characteristics of soil.

5.3.2.2 Effect of Farmyard Manure on Plants

Long-term continuous application of the farmyard manure on the crops affects its yield and soil parameters. In India and Nepal, a long-term study was conducted to investigate the role of farmyard manure on soil properties and growth of rice i.e. *Oryza sativa* L. and wheat i.e. *Triticum aestivum* L. Results of the study revealed that this organic amendment had have positive but variable effects on the soil and plants. The soil amended with farmyard manure led to increase in total carbon and nitrogen, permanganate-oxidizable C, and hot-water-extractable carbon by 40–70%. The phosphorous cation-exchange property, minerizable nitrogen and dehydrogenase activity increased, however, microbial biomass, basal respiration and metabolic quotients were not affected (Tirol-Padre et al. 2007).

Farmyard manure was amended to fodder beet plant in non-saline and saline sodic soil and effect of coated and non-coated seeds was observed. Non-coated seeds formed taller plant as compared to coated seeds. Moreover, there was significant increase in the number of leaves per plant, fresh and dry weight and increase in protein and chlorophyll content in those plants that were propagated from coated seed (Banaras et al. 2002). Hence, farmyard manure is a good option to enhance plant growth and yield. However, the effect of farmyard manure is not reported on the microbial community structure and microbial diversity so far.

5.4 Compost

Compost is an organic material prepared by the decomposition of organic matter by the microbial breakdown. This is mineral rich substrate introduced in the soil for increasing the growth of plants. It is recommended that the composition of compost must be analysed prior to its application by a well-established, reputed laboratory for the three parameters: physical and chemical properties, organic content and macro- or micronutrients. The testing accuracy and precision is required for better crop management in fields. As all agricultural testing labs are not specialized in compost testing, the search of compost testing laboratory is problematic. The same parameters must be considered while purchasing readymade compost. The choice and selection of right compost combination is critical step in obtaining a good quality crop. The application of compost in the soil can solve the problem of pathogenic microbe related incidences and protect the crop. Compost poses direct impact on soil quality, soil mineral content and support the growth of beneficial microbes which also reduce the incidence of microbial infection of crops (Panth et al. 2020). Compost amendment can control damping off disease of radish and *Phytophthora* diseases in vegetables and woody plants.

5.4.1 Plant Waste Compost

The shredded plant material is collected and 6 to 8 inches layer is formed which remain undisturbed till it reaches correct nitrogen balance. If plant material is dry, it can be mixed with livestock manure or after making few layers water can be sprinkled over it. Compost preparation does not require the addition of microbes. The plant material decomposers grow rapidly in the compost and accelerate the process of decomposition and compost formation. Besides providing nutrient and maintaining the soil microbial community, plant waste compost fight against soil-borne diseases and act as biocontrol agent.

5.4.1.1 Effect of Plant Waste Compost on Soil

According to Ren et al. (2017), compost extract is better material in improving the soil structure as compared to bio-organic manure. When applied in the soil, nitrogen, potassium and phosphorous content is higher than bioorganic manure.

5.4.1.2 Effect of Plant Waste Compost on Plants

Plant waste compost or its extract enhances the plant growth by improving the soil fertility and structure. In a study, the effect of compost extract and bio-organic manure on the growth of *Medicago sativa* i.e. lucerne plant was observed by Ren et al. (2017). The soil amendment with compost extract induced nodulation in the plant root likewise induced by *Rhizobium* sp. Compost extract alone was more effective in inducing plant growth as it stimulated the nodulation process which suppressed by the addition of bio-organic manure.

5.4.1.3 Effect of Plant Waste Compost on Microbes

The plant waste compost has had a variety of effects on soil microbial diversity and disease-causing population in soil. *Brassica juncea* seed meal and urea could reduce the *Rhizoctonia solani* AG-5 microbial population which infect apple plants. This is possibly due to increment of total microbial count in the soil and thereby, increased the production of antimicrobial substances which produces antagonistic effects on the pathogenic microbes (El-Sharouny, 2015). Therefore, soil amendment with compost reduces the pathogenic microbial population in the soil and exerted positive effect on the growth of apple plants.

Plant waste compost was prepared by El-Haddad et al. (2014) using rice straw alone and its combination with cattle dung, organic phosphorous, potassium and fungal accelerator i.e. *Trichoderma harzianum* NRRL 13019 and *Phanerorchaete chrysosporium* NRRL 6359. Similarly, other set of rice straw was prepared by using the same combinations but by adding earthworm *Eisenia fetida* to each of the plastic bin. The efficiency of the prepared compost and vermicompost was compared. The compost and vermicompost both were reported to be high quality products due to exerting positive influences on plants and soil. Moreover, this compost was devoid of total and fecal coliform and *Salmonella* sp. and *Shigella* sp. According to the Ren et al. (2017), compost extract can enrich indigenous flora of soil to carry-out nodulation and then, nitrogen fixation. It is more effective in inducing the plant growth as compared to the bio-organic manure.

5.4.2 Peat

Peat is fine amorphic, colloidal mass of incompletely decomposed plant material which has formed in the absence of oxygen and water saturated environment. It is heterogeneous blend of incompletely decomposed plant material i.e. humus. Wasteland reclamation is done by the amendment of soil with peat. Peat is added to plough layer. The impact of peat on the cultivation of different crops was assessed by Vepsäläinen et al. (2004). The physical, chemical and microbiological change in the soil due to amendment was compared with non-amended soil. Amendment of peat increased the soil moisture content which further affected arylsulphatase and β-xylosidase activities and plant growth. The microbial community structure was assessed by phospholipid fatty acid profiles which was stable for the period of 2 years. This stable profile revealed the stability in the microbial community structure. However, ATP content varied due to peat amendment (Vepsäläinen et al. 2004). Therefore, the effect of peat treatment cannot be explained clearly by analysing the microbial biomass and biological parameters.

5.4.3 Vermicompost

In small-scale agricultural farm, vermicomposting is a feasible waste management method. Agricultural and animal waste can be used for making vermicompost which further, increases the soil fertility and crop production. The poor management of agricultural and animal waste poses negative impact on the environment. Besides, it led to the spread of plant diseases. These problems of waste accumulation can be abated by the production of vermicompost. Vermicompost is a useful fertilizer for the crop which is produced by treating the animal or agricultural waste by using worms. In Kampala, Uganda, vermicompost system was established in which *Eudrilus eugeniae* species of earthworm was used to process cow manure and food waste. The waste-to-biomass conversion rate was 3.5% in the vermicomposting process on a total solids basis which increased on increasing the frequency of worms harvesting (Lalander et al. 2015). Thusly, vermicompost is technically and economically viable system to manage agricultural and animal waste.

The limitation associated with the production of vermicompost is increase in greenhouse gas emissions (Lubbers et al. 2013). Earthworms produce greenhouse gases during the production of vermicompost. Vermicompost is effective only when a healthy population of earthworms is maintained in the soil. The maintenance cost of earthworm makes it uneconomical, non-feasible approach.

5.4.4 *Animal Waste Compost*

Composting is a biological process to decompose organic material into nutrient rich soil like substances by the help of aerobic microbes. The microbes consume carbon and nitrogen from the organic substances and convert it into compost. The production of a good quality compost depends upon certain environmental conditions such as carbon to nitrogen ratio 25–30, 50–60% moisture content. Generally, pure manure is rich in nitrogen content and therefore, it needs to be mixed with carbon sources such as straw, leaves and wood residues. During decomposition by microbes, carbon and nitrogen content of animal and straw waste produce organic matter, carbon dioxide and heat and hence, pile need to overturn by manual or mechanical turning methods. This helps in aeration of compost which is required for the distribution of oxygen and survival of microbes and prevents them by removing excessive heat.

The composting process can be completed within 4–8 months and required post-compost period of 2–4 months for stabilization of compost. Completely prepared compost is of high quality and requires less time to spread. Besides, it possesses more humus as compared to unfinished compost i.e. prepared without proper composting. Incomplete composting affects the nutritional characteristics of compost. Application of compost on agricultural field is more beneficial than that of raw manure. The compost has soil like texture and therefore, it mixes with soil very easily, which improves the soil structure and health by providing nutrients. It provides immunity to plant diseases which supports the plant growth. Moreover, the animals can graze on compost amended pastures easily as opposed to the application of raw manure and therefore, it is important amendment for crop growth, managing animal manure and production of livestock.

5.4.4.1 Effect of Animal Manure and Compost on Soil

Animal manure may exert positive or negative effects on soil properties. In an experiment, conducted by Roy and Kashem (2014), agricultural fields were amended with cow dung and chicken manure and the combination of both at the rate of 10 ton per hectare. Different manure induced different effects depending upon their incubation period. The organic carbon content increased and reached peak value at 15th day of incubation and then, decreased in the control and cow dung amended fields on later days. In contrast to this, chicken manure and its combination with cow dung exerted no significant effect on soil properties (Roy and Kashem 2014). Therefore, cow dung was proved to be a better option in enhancing the soil properties as compared to chicken manure.

5.4.4.2 Effect of Animal Manure and Compost on Plants

Soil amendment with fish emulsion increased the tuber yield by 7–20%. In contrast to this, potato tuber yield remained unaffected in the absence of supplementation. Fish emulsion amendments decreased the percentage of *Verticillium dahliae*-infected potato plants and increased the yield of healthy marketable potatoes (Abbasi 2013).

5.4.4.3 Effect of Animal Compost on Soil Microbes

Application of animal compost is good option to control the spread and survival of *Verticillium dahliae* microsclerotia. Chicken manure reduced the incidences of scab and wilts and nematodes in both amended fields while amendment with swine manure reduced the incidences of scab diseases, wilt and nematodes. Contrast to previous application, in the third crop chicken manure amended field showed higher incidence of scab than that of control. In the third crop, swine manure reduced the incidence of scab disease in one plot only (Conn and Lazarovits 1999). This showed that *Verticillium dahliae* microsclerotia survive in the soil for longer time and its fate depends on the soil type and type of manure supplementation.

Animal wastes, when used in combination, decreased the carbon nitrogen ratio of wheat straw. The highest humus content of 11.8% and C:N ratio of 13.5, humic acid fraction of 2.84% and germination index of 59.66% was reported for compost prepared by grass clipping supplemented wheat straw-poultry dropping mixture (Gaind et al. 2009). Fish emulsion can be used in the replacement of inorganic fertilizer to improve the plant growth. The plant growth promoting rhizhobacteria utilize fish emulsion for nutrients acquisition and for the synthesis of plant growth regulators, auxins, gibberellins and cytokinins (Khaled et al. 2003). Fish emulsion amendment in soil can also suppress potato scab and verticillium wilt disease of potato. Besides, it increased the abundance of soil bacteria but there was no change in the number of fungi (Abbasi 2013).

5.5 Straw Amendment

Application of straw as organic amendment in the field has become a prevailing agricultural practice worldwide. Straw is rich in nitrogen, potassium and many other macronutrients. Straw can be amended as organic fertilizer or mulch in the agricultural field. Occasionally, it is burnt and then applied to the field. Burning of straw causes the loss of nitrogen and therefore, loses its potential to provide nutrients to the soil. It can also be supplied as compost after composting process. The effect of straw amendment on soil, microbes and plants is described below.

5.5.1 *Effect of Straw Amendment on Soil*

Soil amendment with straw deviates the physical, chemical and biological properties of soil and affects the plant growth, soil microbial community structure and abundance. In maize straw amended soil, abundance of actinomycetes decreased while abundance of fungi increased as compared to non-amended soil. Moreover, upsurged enzymatic activities β-glucosidase, β-D-cellobiosidase, and β-xylosidase was reported by Zhao and Zhang (2018). The microbial population and increased enzymatic activities led to the degradation of straw which improves the quality of maize straw amended soil. It improved soil organic carbon content and soil quality which further increased the yield nevertheless, depended on the type of soil. The formation of macro-aggregates and crop yield increased with the increasing soil organic carbon concentration (Liu et al. 2017). The straw amendment showed varied results in terms of soil organic carbon sequestration with the variation in soil quality.

Long-term supplementation with straw may increase the greenhouse gas emissions. In a laboratory experiment, conducted in China at different temperatures and soil moisture by Ding et al. (2007), it was revealed that long-term amendment of soil with straw led to significant increase in nitrous oxide gas emission. However, in the field experiment straw amendment in sandy loam soil depicted no significant difference in nitrous oxide emission as compared to organic manure and nitrogen-phosphorous-potassium supplemented soil. The possible reason of this is the coarse textured soil in which oxygen diffuses rapidly and diluted the concentration of nitrous oxide emission (Ding et al. 2007). Thus, straw amendment in loamy soil is good option to manage the straw waste and soil quality improvement because of its ability to fix nitrogen and sequestered atmospheric carbon.

5.5.2 *Effect of Straw Amendment on Plants*

Straw amendment in rice fields increases the production of methane gas. However, long-term amendment induces the population of methanotrophic bacteria in the rice fields. Straw amendment increases the oxygen transport in the soil which further helps in increasing the methane oxidation rates and reduces the effect of methane in the field. Jiang et al. (2019) reported 48% lower methane emissions from rice fields than that estimated by Intergovernmental Panel on Climate Change (IPCC). According to Tan et al. (2018), amendment of soil with hydrolysable amino acid, dilute-acid extractable carbohydrate produced methane gas with greater intensities as compared to lipid fractions. Moreover, methane emissions in the early phase occurred due to amendment with hydrolysable amino acid, dilute-acid extractable carbohydrate fraction, nevertheless, in late phase by lipid fractions. In contrast to this, acid-insoluble organic matter fraction released methane gas both in the early and late phase of incubation. This information will be helpful in alleviating the problem of methane emission by straw incorporation.

5.5.3 Effect of Straw Amendment on Microbes

Straw amendment to the rice field guided the development of microbial communities. However, the effect of long-term fertilization on development and shifting of microbial community structure after supplementation with rice straw and its combination with chemical fertilizers was analysed by Ding et al. (2018) using pyrosequencing and a microarray-based GeoChip. The composition of soil microbes was shifted significantly in the presence of the combination of rice straw with nitrogen, phosphorous, potassium and the combination of reduced rice straw plus reduced nitrogen, potassium and phosphorous. In this case, bradyrhizobiaceae and rhodospirillaceae families dominated and benefitted the soil with higher productivity as compared to other treatment and rice straw alone. Besides, this treatment increased the microbial diversity and abundance of genes involved in carbon and nitrogen cycling. This diversity enriched the soil with nutrients and therefore, exerted positive effect on yield. The combined application of rice straw and balanced fertilizers increased the microbial diversity and improved the productivity of crops as compared to the application of rice straw alone (Ding et al. 2018). Hence, this is sustainable way of reducing straw waste and nutrient cycling in soil.

In an experiment conducted by Zhao et al. (2017), straw was amended with and without decomposer along with nitrogen, phosphorous and potassium nutrients. A shift in bacterial composition was reported (Zhao et al. 2017). The nutrient content such as total nitrogen, available nitrogen and available phosphorous stimulated the development of copiotrophic bacteria (proteobacteria, betaproteobacteria, and actinobacteria) and increased their number in the straw plus decomposer amended soil. This change in the microbial community was related with the soil biological activity and productivity of crops. Therefore, straw amendment with decomposer is a viable option for increasing the yield of the crops and productivity in sustainable and economic way.

In rapeseed-rice rotation system, long-term straw amendment supported the growth of Nitrospira-like bacterial community which was analysed by the nitrite oxidoreductase gene-nxrB (Luo et al. 2017). Further, the available nutrients like nitrate, phosphorous, ammonium and potassium in the soil also guide the development of microbial communities.

5.6 Biochar

Biochar is carboniferous substance formed by incomplete burning of organic waste generated from agricultural waste. It is applied to the soil to manage soil nutrients and characteristics since historic period. The production parameters of biochar affect its potential in changing the soil structure and environmental conditions. Further, nutritional composition of soil and its amendment with biochar posed positive effect on the root biomass which helped in increasing aboveground biomass and plant nutrient uptake (Backer et al. 2018).

5.6.1 Bamboo biochar

Bamboo biochar is prepared from bamboo wood by incomplete burning at high temperature in the absence of oxygen. Application of bamboo biochar in the soil helps in reducing the heavy metals toxicity by their immobilization. This further leads to the reduction of phytotoxicity. Besides, it imposes positive effect on the plant growth, soil properties and microflora of soil which is discussed below.

5.6.1.1 Effect of Bamboo Biochar on Soil

Soil amendment with bamboo biochar improved the soil health of heavy metal contaminated soil and increased the enzymatic activities of β-glucosidase, alkaline phosphatase and urease Application of biochar in mine polluted soil led to increase in soil pH and electrical conductivity. The bioavailability of heavy metals zinc and cadmium decreased in the soil by bamboo biochar (Ali et al. 2017a, b). It is comparatively less effective than the rice-straw biochar in decreasing the acid extractable metal fractions (Lu et al. 2017).

5.6.1.2 Effect of Bamboo Biochar on Plant Growth

Ali et al. (2017a) observed the effects of biochar amendment on *Brassica juncea* and reported that heavy metal uptake in *Brassica juncea* plants decreased which produced profound positive effect on the growth of plant as revealed by increase in antioxidant activities. After amending the soil with bamboo biochar, root and shoot growth of plant and dry biomass, chlorophyll a and b, and carotenoids increased (Ali et al. 2017a). Therefore, in the mine site area the bamboo biochar can be used for the cultivation of *Brassica juncea.*

5.6.1.3 Effect of Bamboo Biochar on Microbes

Bamboo biochar was prepared by the pyrolysis of Moso-Bamboo chips at 600 °C. According to Li et al. (2018), application of this biochar to the soil led to increase in microbial biomass carbon but decrease in microbial diversity of soil. This microbial diversity is largely affected by biochar amendment rates and nitrogen deposition. The effect of bamboo biochar amendment has not been not reported so far for fungal diversity in the soil. The bacterial diversity was analysed by high-throughput sequencing of the 16S rRNA gene. Amendment of soil with 20 tonnes bamboo biochar significantly increased the microbial biomass carbon and bacterial diversity. However, increasing the dose of bamboo biochar to 40 ton per hectare decreased the soil microbial biomass carbon but increased the bacterial diversity. Thus, soil amendment with bamboo biochar can mitigate the effect of nitrogen supplementation and soil microbial biomass and diversity. Initially, the amendment of bamboo biochar up to 20 years changed the microbial diversity and abundance where after no change was observed up to 40 years (Han et al. 2019). This finding suggested the role of bamboo biochar amendment in managing the soil microbial biomass, diversity and abundance.

5.6.2 Biomass Biochar

Organic waste can be applied indirectly after converting it to green compost or biochar to the field. Preparation of biochar, its concentration, long-term or short-term amendment, affects soil properties, pollutants removal potential and plant growth.

5.6.2.1 Effect of Biomass Biochar on Soil

Biochar produced from oak at 650 °C was applied to silty-loam soil and results of soybean plant growth were compared with the fields amended with humic acid and residuals of water treatment. These amendments increased the pH of soil. Oakwood biochar increased the soil carbon by 7% but reduced soil bulk density by 13%. In biochar amended soil, cumulative nitrous oxide emission reduced however, methane and carbon dioxide emission remain unaffected. All amendments were effective in reducing the global warming potential of the amended soil, nevertheless, biochar addition showed a net negative global warming potential (Mukherjee et al. 2014). In this way, biochar amendment not only improved the soil qualities but also reduced the emission of greenhouse gases.

Biochar, compost and co-composted biochar when applied to banana and papaya planted soil, the soil quality improved due to rise in the soil water content and soil carbon content, and increased nutrient like potassium, calcium, nitrate and ammonium content (Bass et al. 2016). Biochar amendment on drought-stressed soil enhanced its water holding capacity and improved the physical and biological properties of soil. In salt-stressed soil, biochar amendment led to decrease the sodium uptake and increased the potassium uptake by the plant (Ali et al. 2017b). Biochar may also ameliorate the effect of salinity in soil. When soil was amended with biochar along with wheat straw biochar and poultry manure compost and diluted pyroligneous solution for the period of 2-year in a field experiment, the effect of salinity stress ameliorated and the productivity of plants improved. Furthermore, this combination reduced the emission of greenhouse gases and global climate change (Lashari et al. 2015). In the drought conditions, biochar amendment to the maize cropped soil exerted positive effect by increasing water availability and water retention ability of the soil. Compared to the other treatments, aged biochar increased the soil organic carbon and improved the condition of drought (Paetsch et al. 2018). Similarly, Rice hull biochar amendment on reclaimed tidal land soil exerted positive effect on soil properties and therefore, increased plant yield. This amendment increased the water retention ability and raised the level of phosphate in the soil. Besides, there was increase in the number of exchangeable cations, cation exchange capacity and soil organic carbon content (Kim et al. 2016). Hence, biochar amendment increased the soil properties of tidal land soil. Biochars prepared from palm kernel and rice husk when amended in the soil, increase in pH, moisture content was recorded (Simarani et al. 2018). Moreover, organic carbon and labile carbon were reported to be higher in the amended soil that that of non-amended soil. Biochar amendment resulted in the higher activities of enzymes β-glucosidase and xylanase (Simarani et al. 2018). Therefore, biochar supplementation presented a good and feasible option in the soil reclamation by ameliorating the problem of salinity and drought.

5.6.2.2 Effect of Biomass Biochar on Plants

Besides, there was increase in mineral uptake by the plants and production of phytohormones. Biochar amendment is helpful in regulating the stomatal conductance (Ali et al. 2017b). Biochar, compost and co-composted biochar employed positive effect on soil properties and nutrient content. However, the yield of plant could not be correlated with these properties. For instance, the yield of banana decreased and yield of papaya was not affected significantly by the biochar amendment (Bass et al. 2016). When wheat straw biochar, poultry manure compost and diluted pyroligneous solution was amended to maize plants the problem of salinity ameliorated and improved the osmotic stress related activity of leaf. The leaf area index, plant growth, yield of maize increased however, leakage of leaf electrolyte decreased. Besides, potassium, nitrogen and phosphorous content increased in leaf sap of maize plant, nevertheless, sodium and chloride decreased (Lashari et al. 2015).

Rice hull derived biochar amendment on the reclaimed tidal land soil increased the maize yield due to improvement in soil properties. Reclaimed tidal land soil contains exchangeable sodium content and high levels of sodium salt. Biochar amendment increased the potassium content of the soil and therefore, reduced the sodium uptake by maize plants. This amendment reduced the antioxidant activities which were carried out by ascorbate peroxidase and glutathione reductase enzymes (Kim et al. 2016). The rice hull derived amendment at 5% (w/w) level, ameliorated the negative effect of reclaimed tidal land soil.

5.6.2.3 Effect of Biomass Biochar on Microbes

The effect of biochar in the microbial diversity may be positive or negative. According to Simarani et al. (2018), any shift in the microbial population was not observed on supplementing the loam soil using palm kernel and rice husk biochar. Microbial biomass carbon was higher in the rice husk biochar and rice husk with fertilizer amended soil as compared to control while microbial biomass nitrogen was higher in the combination of fertilizer with rice husk and palm kernel biochar than alone.

Biochar amendment also increased the growth and survival of plant by increasing the soil fertility and inducing the microbial population to provide resistance to pathogens. In a study, conducted by Kolton et al. (2011) on sweet pepper plant i.e. *Capsicum annum* L. to find out root-associated bacterial community composition. There was increase in the growth in the relative abundance of members of bacteroidetes phylum from 12–30% but decreased the level of proteobacteria from 71–47%. The most dominating bacterium due to biochar amendment was *Flavobacterium* followed by the population of chitin degrader *Chitinophaga*, cellulose degraders *Cellvibrio* and aromatic compound degraders *Hydrogenophaga* and *Dechloromonas* genus (Kolton et al. 2011). All these microbes may bring out positive change on the plant growth. Further, research is required to figure out the effect of microbial diversity on the growth of plants. Contrast to this, Zhang et al. (2014) reported the ineffectiveness of biochar amendment in improving the soil fertility and microbial diversity. Biochar prepared at 700 °C from spent coffee grounds, wood pellets, and horse bedding compost separately and results of amendment showed that it did not provide fruitful results in terms of microbial activity in soil irrespective of the raw substrate used in biomass preparation (Zhang et al. 2014) which is possibly due to the use of unfavourable temperature for biochar preparation. This shows the effect of biochar preparation protocol on the properties of biochar and further microbial activities.

Biochar amendment to the soil results in improving the soil quality, carbon sequestration and plant growth and thus, increases the crop productivity. Besides, it poses positive impact on the microbial community. However, it depends on the protocol used for biochar preparation. Biochar amendment to soil has positive environmental and agricultural potential.

5.6.3 *Torrefied Grass Fibres*

Torrefied grass fibres are produced by thermal process in which high temperature converts the biomass into a coal-like substance. The torrefaction process led to the generation of phytotoxic compounds namely phenol, 2-methoxyphenol, benzopyran-2-one, and tetrahydro-5,6,7,7a-benzofuranone which were extracted from torrefied grass fibres and analysed by gas chromatography mass spectrometry by Trifonova et al. (2009).

5.6.3.1 Effect of Torrefied Grass Fibres on Soil

The application of torrefied biomass as soil amendment is not as popular as the soil amendment with biochar. The torrefied biomass acts as fertilizer and a substitute for peat for growing the crops. Torrefied biomass has lower carbon content as compared to biochar and therefore, lower amount of carbon is added to the soil. Furthermore, they stimulate the microbial activity in soil and cause the degradation of soil organic carbon. In a greenhouse experiment, conducted by Backer et al. (2018), the potential of pyrogenic carbonaceous biosolids in increasing the growth of maize i.e. *Zea mays* seedling was observed under controlled environmental conditions. On increasing the temperature from 270 to 320 °C, pyrogenic carbonaceous biosolid reduced from this soil amendment and therefore, reducing the phytotoxicity. At 270 °C water soluble components of pyrogenic carbonaceous biosolids were produced which decreased the productivity of maize seedlings. At 320 °C, pyrogenic carbonaceous biosolids increased the mineralization of soil and uptake of nitrogen despite of smaller root system of maize plant. Thus, this supplementation improves the productivity of maize plant. However, it needs to be tested in the field prior to its implementation.

5.6.3.2 Effect of Torrefied Grass Fibres on Plants

Trifonova et al. (2009) reported that on amending the soil with torrefied grass fibres along with certain microbes, production of phytotoxic compounds decreased. Besides, it suppressed the disease outbreak in the field. The combination of torrefied grass fibre with *Coniochaeta ligniaria* F/TGF15 fungi and bacteria was reported to be best combination to reduce greenhouse gas emission and phytotoxicity as compared to soil amendment with the torrefied fibres alone (Trifonova et al. 2009). Conclusively, the soil amendment with torrefied grass fibre is good option to treat the soil and control the disease outbreak in plants.

5.6.3.3 Effect of Torrefied Grass Fibres on Microbes

Torrefied grass fibre and its extracts, when amended to the soil, induce a specific population of microbes which can be analysed by polymerase chain reaction-denaturing gradient gel electrophoresis. The microbial community structure was analysed by cultivable or non-cultivable based approaches. Analysis of partial sequencing of the 16S ribosomal RNA gene from isolates and bands from denaturing gradient gel electrophoresis gels revealed the presence of alpha-proteobacteria (*Methylobacterium radiotolerans*, *Rhizobium radiobacter*), gamma-proteobacteria (*Serratia plymuthica, Pseudomonas putida*), *Cytophaga-Flavobacterium-Bacteroides* (CFB) group (*Flavobacterium denitrificans*), beta-proteobacteria (*Ralstonia campinensis*), actinobacteria (*Cellulomonas parahominis, Leifsonia poae*, *L. xyli* subsp. *xyli*, and *Mycobacterium anthracenicum*), and the firmicutes (*Bacillus megaterium*) (Trifonova et al. 2009). These microbes plays an important role in the reduction of toxicity of torrefied grass fibers and its extract.

5.7 Spent Mushroom Compost

Spent mushroom compost, is the left-over post-harvest waste of mushroom cultivation (Fasidi et al. 2008). The waste from various agriculture-based industries could be used to prepare the mushroom substrate which is later discharged as waste. The compost could be made from different waste materials such as sawdust, rice and wheat straw, bedded horse manure, cotton wastes, paper wastes, cocoa shells, maize husks and various other wastes (Jonathan 2002) which when discarded becomes the part of spent mushroom compost. Additional materials such as rice bran, calcium carbonate or wheat bran may be added to the mushroom substrate in order to increase mushroom growth. The residues of these supplemented materials may be present in spent mushroom substrate after harvesting mushroom fruit bodies (Gbolagade 2006).

Spent mushroom compost is thought to be a good nutrient source for agriculture of crops because it is rich in nutrients, with organic matter as high as 66.6%, total nitrogen to be 2.17%, total available phosphorous to be 25 mg/kg (Polat et al. 2009) and has high cation exchange capacity along with its slow rate of mineralization or degradation (Shivaji 2012).

5.7.1 Effect of Spent Mushroom Compost on Soil

The application of spent mushroom compost enhances the soil properties, including its nutritional quality, and aggregate formation and stability. Generally, positive results were obtained by the application of spent mushroom compost.

The organic matter of the soil has a huge impact on the aggregate stability, which in turn affects the fertility of the soil. The increase in aggregate stability has been reported in case of soil amendment with spent mushroom compost. The highest of 16% aggregate stability was reported after 42 days incubation at 8% spent mushroom compost supplementation. Hence, application of spent mushroom compost led to the enhancement of organic matter of the soil along with its fertility. Moreover, the soil organic carbon content also increased by spent mushroom compost application, with the highest of around 2.5% after 21 days incubation at 8% spent mushroom compost amendment. The soil organic carbon has a major impact on the maintenance of soil properties and hence, its supplementation to the soil further leads to betterment of soil properties. Soil amendment with 8% spent mushroom compost led to increase the nitrogen content to a maximum of 0.16% after 62 days incubation (Gümüş and Şeker 2017). Thus, nitrogen content of soil enhanced with the application of spent mushroom compost

5.7.2 Effect of Spent Mushroom Compost on Plants

Treatment of soil with spent mushroom compost generated both positive and negative results. Generally, supplementation of spent mushroom compost in the soil exerts positive effect on plant growth. A significant impact on the total yield of cucumber i.e. 14.40 kg m^{-2} was reported by applying 40 tonnes ha^{-1} spent mushroom compost (Polat et al. 2009). Both macronutrients and micronutrients of the fruit increased with the increase in amount of applied spent mushroom compost. 600 g spent mushroom compost, when applied to *Abelmoschus esculentus, Lycopersicon esculentum, Capsicum annum* and *Capsicum chinense,* 100 cm, 71 cm, 27 cm and 41 cm growth in height was observed for the respective crops after 14 weeks of planting (Jonathan et al. 2011). Similarly, 5–30% increase in the height of the tomato saplings was observed as compared to non-amended normal loamy soil, with the maximum height of 69.9 cm at 10% spent mushroom compost amendment (Wiafe-Kwagyan and Odamtten 2018). Further, the application of spent mushroom compost with cow dung and chemical fertilizers both in soil used for cultivation of brinjal gave the maximum plant height of BARI Begun-6 of 80.11 cm and BARI Begun-8 of 75.56 cm. The highest yield per plant and the maximum fibre content of BARI Begun-6 and BARI Begun-8 observed to be 1.938 kg and 1.452 kg and 2.15% and 2.34%, respectively after the soil amendment with spent mushroom compost in addition to cow dung and chemical fertilizer (Rahman et al. 2016). Therefore, spent mushroom compost exerts positive effect on plant growth, fibre content and fruits yield.

Occasionally, application of spent mushroom compost may impose negative impact on plant growth. The negative impacts of spent mushroom compost treatment have been majorly related to the concentration of spent mushroom compost. Use of spent mushroom compost in high quantity led to decrease in yield. For instance, reduction in yield of cucumber was observed possibly due to higher salt concentration. The salt concentration increased due to increase in the amount of spent mushroom compost (Polat et al. 2009). Another drawback of spent mushroom compost application is the decreased chlorophyll content in leaves (Wiafe-Kwagyan and Odamtten 2018). Thus, spent mushroom compost imposes both positive or negative impact on plant growth.

5.7.3 Effect of Spent Mushroom Compost on Soil Microbes

Various fungal, bacterial and *Actinomycetes* colonies increases in the soil when it is amended with spent mushroom compost. There was no significant variation in the microbial colonies appeared on the un-treated soil sample as compared to spent mushroom compost treated sample, with 4.2–4.3 $\log_{10}$ CFU/g for non-treated soil and 4.3–4.5 $\log_{10}$ CFU/g for spent mushroom compost treated soil. However, fungi *Aspergillus flavus, A. niger, Penicillium citrinum, Rhizopus stolonifer* and *Trichoderma harzianum* were dominant residents (Wiafe-Kwagyan and Odamtten 2018). According to Shivaji (2012), maximum population of bacteria was 36 × 10^6 per gram soil after 30 days incubation and 43 × 10^6 per gram soil after 60 days incubation after applying 15 tonnes spent mushroom compost per hectare along with 15 tonnes farm yard manure plus recommended doses of fertilizer treatment.

Therefore, spent mushroom compost not only shows positive effects on plants but also on microbes. Moreover, spent mushroom compost is useful for improving the soil quality. Therefore, it offers a complete package to improve the soil quality and plant growth.

5.8 Biogas Sludge

Biogas sludge or biogas slurry or bio-slurry is the by-product of the anaerobic digestion for biogas production. It consists of 93% water and 7% of dry matter, of which 4.5% is organic matter and 2.5% inorganic matter (Kumar et al. 2015). It contains sufficient amounts of both macronutrients such as nitrogen, potassium and phosphorous as well as micronutrients like zinc, manganese, boron that are essential for plant growth (Alam 2006). Biogas sludge can be used as a fertilizer directly in the fields, or it can be used along with organic materials and synthetic fertilizers.

Anaerobically digested animal waste is rich in nitrogen content and can be used to improve soil quality and plant growth. Addition of urine during anaerobic digestion enhances the nitrogen content of the animal waste bio-slurry. Moreover, this can increase the process of making compost in a short time interval and increases the carbon/nitrogen (C/N) ratio in the sludge (Kumar et al. 2015). The average nutrient content of biogas sludge is as follows- 2.14% nitrogen, 1.36% phosphorous and 0.70% potassium (Jared et al. 2016). Further, biogas sludge repels termites and pests, and reduces weed growth in the soil. It is contaminant-free and also enhances water holding capacity of the soil.

The effects of biogas sludge on soil properties, growth of plants and microorganisms have been accounted below.

5.8.1 Effect of Biogas Sludge on Soil

The use of biogas sludge for soil conditioning has generated positive results. The nutrient qualities and quantities of biogas sludge could also be used for increasing beneficial soil properties. Improvement in soil structure, water holding capacity, aeration and drainage was observed for soil treated with biogas sludge (Ahmad and Jabeen 2009).

Biogas sludge has the highest composition of nitrogen, phosphorous and potassium compared to farmyard manure and compost. In the biogas sludge, nitrogen is 35.98% and 69.15%, phosphorous is 48.52% and 64.70% and potassium is 4.28% and 12.85% higher than compost and farmyard manure, respectively (Jared et al. 2016). Thus, due to high nutritional quality of biogas sludge, it is a better alternative to be used as organic manure. The biogas slurry is used in its fresh and liquid form directly from the biogas reactor. All of these minerals further increase the quality of the soil in which it is applied.

5.8.2 Effect of Biogas Sludge on Plants

Biogas sludge application has significantly enhanced the plant growth and productivity. The height of plants, number of leaves produced by the plants and crop yield also reported to increase. An increase in the plant growth and yield of sunflower was observed upon bio-slurry application due to better availability of essential nutrients and plant growth promoting material such as auxin, amino acids and vitamins. These essential nutrients are produced by their decay of biogas slurry which enhances the plant growth by increasing 16.66% height, 27.27% stem diameter and 24.13% leaf area of the plant grown in non-saline soil, respectively. The application of biogas sludge also reduced the adverse effects of salinity on vegetative growth as well as on the yield of sunflower by approximately 19.23% and 26.6% for fresh biomass and dry biomass of the plant, which further prevented sodium induced toxicity in them (Ahmad and Jabeen 2009).

Application of bio-slurry on four different varieties of boro rice namely BRRI dhan28, BRRI dhan29, Binadhan-8 and Binadhan-10 resulted in maximum plant height, number of tillers per hill, pinnacle length, number of grains per pinnacle, weight of 1000 grains and harvest index of 79.15 cm, 13.76, 20.60 cm, 85.32, 25.94 g and 46.75%, respectively (Hossain et al. 2018). The use of dry biogas slurry has considerably enhanced the yields of cotton, wheat, maize and rice by 154.5 kg/

ha, 450 kg/ha, 555.9 kg/ha and 634.4 kg/ha, respectively (Kumar et al. 2015). Application of biogas sludge with chemical fertilizers in potato cultivation fields had a significant effect on potato stem length, number of tubers and their weight. Tallest plants were obtained for biogas sludge-chemical fertilizer amendment with a height of 52.5 cm. The number of tubers and weight of tubers were found to be highest for biogas sludge-chemical fertilizer amendment in the year 2011. The mean tuber yield for the year 2011–12 of 23.44 t/ha was maximum for biogas sludge-chemical fertilizer amended soil as compared to non-treated (Haque et al. 2015). Occasionally, negative impacts of application of bio-slurry can be observed. The toxicity in seed filling could not be controlled by biogas slurry (Ahmad and Jabeen 2009).

5.8.3 Effect of Biogas Sludge on Soil Microbes

Biogas sludge could be used as a bactericide and fungicide. Anaerobic digestion of bio-waste is carried out at high temperature i.e. thermophilic range (53–58 °C) or mesophilic (30–42 °C) conditions which significantly reduces the pathogens like *Clostridium perfringens, Listeria monocytogenes, Salmonella sp.* In many cases, bio-slurry contains significant amounts of pathogens such as bacteria, nematodes or viruses. Hitherto, biogas slurry has lesser amount of microbial population as compared to the undigested manure. Biogas sludge is also reported to control *Phyllosticta citricarpa* and *Fusarium graminearum* infection in plant. These microbes are pathogenic and cause diseases to plant. Biogas sludge application reduced the disease-causing ability of fungi (FAO 2018).

5.9 Seaweed or Kelp Weed Amendment

Seaweed extracts have potential to improve the plant growth by inducing seed germination, root development, leaf quality and yield and therefore, it is used as soil amendments. This amendment provides nutrition to the soil and plant by regulating gene expression for nutrient uptake, improves the soil structure and ability of plant to resist disease (Arioli et al. 2015) and biotic and abiotic stresses. Besides, it increases shelf-life after the harvesting of crops. It can be applied to vegetable, fruit, flower and ornamental crops.

One of the popular seaweed amendments is kelp weed i.e. *Laminaria japonica* Aresch. which is a rich source of nutrients and minerals. Kelp weed also increases the amount of total soluble protein, phenolics, flavonoids and antioxidants. Norwegian kelp i.e. *Ascophyllum nodosum* can be used as soil amendment after drying due to its richness in potassium salt along with many other trace minerals, vitamins, amino acids and plant hormones. It is only source of mannitol, a natural sugar that helps the plant to break down the micronutrients and utilize them for their growth.

5.9.1 Effect of Kelp Meal on Soil

Kelp meal can be applied as liquid fertilizer and an organic stimulant in the soil. In the organic farming, it is applied in the field as foliar spray and soil drench. It is biodegradable, non-toxic and therefore, considered as eco-friendly option and safe choice for human and animals. Maxicorp Kelp meal is a commercially available preparation which is used as soil conditioner. The potting soil, seed and transplant beds, seeds and composting material can be amended with kelp meal in order to reduce the stress and improve the plant health. Besides, the mineral and amino acid of the kelp meal induces the growth of microbe in the field.

5.9.2 Effect of Kelp Meal on Plants

Kelp meal supports the plant growth at every step like growing, budding, flowering and fruiting along with providing resistance to diseases, pests and weather. Dried and composted kelp meal is chemical free, non-hazardous, non-toxic, nutrients rich which can be used for amending all types of indoor and outdoor plants. Indoor plants cultivated in pots and hence, have very limited space to grow. When amended the potting soil, kelp meal increases not only the efficiency of soil but also the size and storage life of harvest. Outdoor plants have sufficient soil to grow. Kelp amendment increases root mass and further, the growth of plants.

Kelp weed extract exerted a variety of effect on the plants depending on its amended amount in the soil. The main problem of pakchoi or *Brassica chinensis* L. is the production and quality which can be solved by the use of kelp weed amendment. Kelp waste can also be used as biostimulant or biofertilizer for improving the soil quality and microbial diversity and the crops yield. It possesses a variety of mineral nutrients and organic substances. It contains the highest amount of nitrogen and high amount of calcium. Kelp waste is extracted by enzymolysis procedure which makes it rich in soluble sugars, amino acids, nitrogen, phosphorous, calcium, magnesium. It also possesses quaternary ammonium molecules such as proline and betaines (Battacharyya et al. 2015). Therefore, seaweed treated plants showed the presence of phytohormones, unique polysaccharides and polyphenols and mineral elements.

5.9.3 *Effect of Kelp Meal on Microbes*

Kelp meal enhances the microbial growth in the soil. However, the effect of kelp waste amendment on the microbial diversity and its abundance in the soil is meagrely reported. Further, research in this direction will provide the information on the microbial composition of the soil which will be helpful in increasing the efficiency of microbes to stimulate plant growth.

5.10 Blood Meal

Blood meal is powdered form of blood prepared by drying of any animals' blood especially, the cow's blood. The blood can be obtained from meat packaging units and by sacrificing animals from the slaughter house. The hemoglobin of the blood contains iron as prosthetic group which increases the availability of iron during the incubation of blood in soil. This is possibly due to the progressive degradation of prosthetic group which further helps in the chelation of iron from the humic substances. Blood meal acts as the rich source of nitrogen which is required in optimum amount for the healthy growth of plants. Excessive application of blood meal burns the plants and kills them due to undue amount of nitrogen amendment in the soil. The blood meal nitrogen can increase the acidity of the soil which is beneficial to the plants having potential to survive in the soil with low pH.

Besides the use of blood meal as nitrogen source, it can be used as source of iron. When amended to the soil, blood meal has the potential to take iron from iron hydroxides and incorporate it into porphyrin ring of hematin (Yunta et al. 2013). It improves the structure of soil by making soil aggregates.

The major limitation of blood meal amendment is its ability to attract dogs, raccoons and other carnivorous or omnivorous animals nevertheless, repels the moles, squirrels and deer. Second limitation is related with the use of higher dose to amend the field as higher dose may kill the plant. This can be solved by using the balanced dose of blood meal amendment. Overall, blood meal is good source of nitrogen to improve the fertility of soil and plant growth, if amended in balanced amount.

5.11 Soil Amendment as Mulch

The water-saving and management strategies are crucial component of agriculture in rain-fed, dry and sub-humid areas for sustainable agriculture. Mulch is an option to save water and maintain moisture content of the soil for better agricultural practices. Any material which is placed over the soil to protect the moisture content and reduce the effect of direct sunlight is called as mulch. The advantageous function of mulch includes decrease in incidences of plant injury caused by sun heat, weed growth suppressor, soil moisture retaining ability, reduction of soil erosion. Organic mulches can provide soil nutrients and encourage the growth of beneficial soil organisms and worm activity which improves the soil quality. Mulches save water, money, time and plants. A 4-inch layer of mulch is sufficient to protect the plant. Newspaper mulch is ideal for protecting the crops against weeds.

Mulching can be done any time of the year however; certain points must be considered prior to mulching. In the case of heavy rain, soil retains excessive amount of water therefore, mulching should be removed to support natural evaporation process. Second major issue related to mulch is its ability to block the penetration of sunlight which inhibits seed germination process. To overcome this problem, mulching can be done after the growth of plant. Third issue is related to the development to pests and worms which poses a risk to plant growth. This problem can be solved by making a thin layer of mulch and keep it away from the plant bases. Mulching material, mulch amendment in on-ridge and on-furrow pattern has a great influence on the crop production. A number of mulches and their effect on soil, plant and microbes are mentioned in Table 5.3. The selection of appropriate mulching technique helps in managing the crops and their yield.

5.11.1 Effect of Mulching on Soil

The water and nitrogen content of soil are the critical factors for crop production. The increase in maize grains yield and positive effects of mulching with fertilizer and top dressing on maize root-zone was reported (Wang et al. 2016). Therefore, mulching of maize crop increases the fertilizer use efficiency which supports the plant growth.

In another experiment, Shen et al. (2016) compared the efficiency of film mulching and gravel mulching on crop production. The crop production was higher in film mulching as compared to that obtained by gravel mulching due to better quality of soil. This experiment proved the positive effect of mulching on soil and plants and focuses on the use of mulching in rain-fed areas.

5.11.2 Effect of Mulching on Plants

Mulching practices have profound effect on the soil moisture content which may further affect the yield of crops. The ridge-furrow mulching pattern was introduced to the wheat crop and yield was reported to be 27.4% higher compared to flat mulching (Wang et al. 2015). Therefore, ridge-furrow mulching pattern is the most suitable pattern to increase the wheat yield. Mulching has positive effect on the growth of plants.

Table 5.3 Mulching materials and their effect on plant growth, properties of soil and microbial communities

S. no.	Mulches	Effects on soil properties	Plant	Microbial communities	References
1.	Sewage Sludge	Significantly increases hydraulic conductivity, final infiltration rate. Significantly decreases soil bulk density. Sewage sludge can help to improve soil physical conditions which persist over long periods.	Highest plant height Significantly highest dry matter production. Maximum number of leaves. Maximum number of spikes per plant were recorded with 100% sewage sludge amendment.	Increased bacterial community tolerance	Asagi et al. (2007)
2.	Rice Straw	Increases the per cent of organic matter and organic carbon in soil. Reduces soil salinity. Improves the crop yield. The surface soil depth was deepened; soil hardness and bulk density decreases while porosity increases.	Increases the number of spikelets per square meter and the percentage of ripeness. Increases infiltration and better retention. Facilitates a better crop growth and development and resulted in higher yield.	Signicantly increases the amount phospholipid fatty acid analysis.	Rezk et al. (2015)
3.	Sawdust	The addition of peat increases soil matric potential and available water over all other treatments. It had higher N concentrations. Well suited organic soil amendment for the enhancement of soil properties.	Increases the plant heights, number of leaves, and number of branches Seedlings grow tallest, with the largest stem diameter.		Koll (2009)

(continued)

Table 5.3 (continued)

S. no.	Mulches	Effects on soil properties	Plant	Microbial communities	References
4.	Bark	Bark amendment increased metal bioavailability. Bark amendment decreases the soil pH and significantly changes the nutrient status of soil.	Increases the plant growth and yields were increased. The growth response is likely to have been principally related to improved physical conditions	Bark amendment significantly increases the microbial growth rates, but with distinct differences between fungal and bacterial groups. Practices increasing the amount of soil organic matter should be priorities for soil reclamation.	Pérez-Esteban et al. (2012)
5	Cocoa	Save moisture of soil.	Cocoa mulch contains a naturally occurring stimulant called theobromine. This stimulant is toxic to dogs and cats.		https://www.chainsawjournal.com/types-of-mulch

5.11.3 *Effect of Mulching on Microbes*

A detailed analysis of microbial diversity of soil improves our understanding to different cropping pattern, crop rotation and yield of crop which is important to develop sustainable management strategies and for the application of appropriate mulching process.

Shen et al. (2016) observed the effect of plastic-film mulching with inorganic nitrogen fertilizer and organic manure addition on *Zea mays* L. i.e. spring maize and reported a positive impact on the dehydrogenase activity, microbial activity and Shannon diversity index. Moreover, with this amendment, the highest value of relative use of amino acids and carbohydrates by microbes was obtained. These changes were not present in non-mulched soil (Shen et al. 2016). Therefore, mulching techniques supports the growth of beneficial microorganism in soil which further, enhances the crop production. According to Qin et al. (2017), the soil where on-ridge planting of potato crop with half mulch was done, the dominant genus was *Mortierella*, whereas 1–2% population of *Scutellinia*, *Cryphonectria*, *Acremonium*, and *Alternaria* was present. In on-ridge planting with full mulch amended soil, potato tuber yield was highest which was correlated with the highest fungal diversity (Qin et al. 2017).

5.12 Plastic Film Mulching

Plastic mulch is a plastic sheet used to suppress weeds and conserve water in the soil which is beneficial for crop production. Plastic mulching technology modifies the environment for the crop growth and crop production which is mainly affected by cropping system and methods, the climatic conditions and spatial distribution characteristics. Amendment of plastic mulch depicted a variety of effect on soil quality and plant growth which are discussed below.

5.12.1 *Effect of Plastic Mulching on Soil*

Mulching not only improved the soil temperature and moisture in the dry land, but also improved the soybean plant health by improving seed mass, plant height, branching in plant and pod's yield. In 0 to 200 cm layer of soil, soil moisture content increased by 1.2–1.4% and water storage by 62.7–70.3 mm (Yang et al. 2013).

The plastic film mulch exhibited more significant and positive effects on maize yield as compared to straw mulch in ridge and furrow pattern. Furthermore, the combined effect of mulching and nitrogen fertilizer on maize yield showed significant increase in water use efficiency and residual soil nitrate accumulation (Liu et al. 2017). Mulching for maize plant imposed a little effect on the fate of nitrogen fertilizer. Nitrogen amendment along with mulch prevented the nitrogen leaching and its accumulation in the deeper layer of soil (Dang et al. 2003). The soil-water storage capacity of soil increased from 0 to 60 cm and rate of soil temperature increase by ridge-furrow with plastic film-mulched ridge and straw-mulched furrow method as compared to conventional flat planting method. Therefore, this ridge-furrow with plastic film-mulched ridge and straw-mulched furrow method has potential to increase the yield of wheat crop in winter season (Liu et al. 2018).

5.12.2 Effect of Plastic Mulching on Plants

Plastic mulching performance, crop yield and water utilizing efficiency is influenced by mulching method, colour of plastic mulch sheet and type of crop. The plastic film mulch exhibited more significant and positive effects on maize yield as compared to straw mulch. Furthermore, the combined effect of mulching and nitrogen fertilizer on maize yield showed significant increase in water use efficiency and residual soil nitrate accumulation (Liu et al. 2017). Similarly, Wang et al. (2018) showed the synergistic effect of plastic mulch and nitrogen fertilizer on the maize plant. According to Yang et al. (2013), the effect of plastic film mulching was positive as compared to the non-mulched soil in dry land in soybean cultivation irrespective of mulching method.

The most influencing factor in the case of plastic mulching is its application pattern in the soil which helps in increasing the water availability to the crops for increasing production. The completely mulched alternating narrow and wide ridges with ridge and furrow planting showed better results to save water and high productivity of potato in semiarid areas (Qin et al. 2017). Second, influencing factor is the colour of plastic film mulch which may affect the planting density, soil moisture content and dry matter accumulation and yield in the rain-fed areas and therefore, need to be considered prior to implement mulching techniques. For instance, the plastic film mulching with black sheet not only increased the dry matter accumulation but also increased the water use efficiency in maize while transparent plastic film increased the mean soil and contributed to the yield improvement of maize crop (Zhang et al. 2018). Similar observations were made by Gao et al. (2019) on the efficiency of black plastic film mulch on the potato, maize, wheat, and cotton crops.

The only limitation of applying plastic mulch is its non-biodegradable nature. Unlike sludges, paper wastes, leaf wastes, the plastic residues are not degraded in the crop field. Therefore, these residues accumulated in the field and poses threat to environment and soil. There is need to develop the methods for the recovery of residues of plastic film to protect the environment.

5.13 Conclusion

A variety of organic amendment are available to improve the soil quality and increases the growth of the plants by exerting positive effects. For instance, green manure of red clover plant increased the nitrogen content i.e. 43.3 mg/g nitrogen content/dry weight of wheat grain. Waste conversion to compost and its application to the fields to reduce the population of pathogens. Spent mushroom compost provided organic matter 66.6%, total nitrogen 2.17%, and increased the total cation exchange properties of the soil. Further application involved the conversion of waste into biochar. Biochar and rice hull biochar increased the enzymatic activities of beta-glucosidase, alkaline phosphatase and urease and glutathione reductase activity (Duminda et al. 2015). Organic waste can be applied to the field after anaerobic digestion and conversion in the form of biogas slurry which can reduce the number of pathogens and acts as bactericide and fungicide. Application of biofertilizers improved the plant growth, nevertheless, its application with chemical fertilizer reduced plant growth, pinnacle number and yield (Mahato et al. 2018). An effective application of organic amendment depends on its quality and proper application. Inappropriate selection may produce adverse effects on soil and crops. Therefore, the most critical step in the organic amendment is soil characterization and application of appropriate doses of organic amendments. Recently, simulation model and RothC model are developed for modifying and optimizing the application of organic amendments which helps in finding out the organic matter mineralization curve (Mondini et al. 2017). Besides, simulation models will be helpful in maintaining the carbon and nitrogen dynamics of soil (De Rosa 2017). In future, organic amendment will be great tool to enhance productivity, yield of plants and soil quality if applied properly after soil characterization and simulation modelling.

Acknowledgement The authors of this chapter are thankful to Prof. Vinay Sharma, Director-AIB, for providing constant support and guidance.

References

Abbasi PA (2013) Establishing suppressive conditions against soil borne potato diseases with low rates of fish emulsion applied serially as a pre-plant soil amendment. Can J Plant Pathol 35(1). https://doi.org/10.1080/07060661.2012.740076

Aghili F, Gamper HA, Eikenberg J, Khoshgoftarmanesh AH, Afyuni M, Schulin R, Jansa J, Frossard E (2014) Green manure addition to soil increases grain zinc concentration in bread wheat. PLoS ONE 9(7):1–13. https://doi.org/10.1371/journal.pone.0101487

Ahmad R, Jabeen N (2009) Demonstration of growth improvement in sunflower (*Helianthus annus* L.) by the use of organic fertilizers under saline conditions. Pak J Bot 41(3):1373–1384

Alam S (2006) Production of organic manure in Bangladesh. Bangladesh Livestock Research Institute's Report, Savar

Ali A, Guo D, Zhang Y, Sun X, Jiang S, Guo Z, Huang H, Liang W, Li R, Zhang Z (2017a) Using bamboo biochar with compost for the stabilization and phytotoxicity reduction of

heavy metals in mine-contaminated soils of China. Sci Rep 7(1):2690. https://doi.org/10.1038/s41598-017-03045-9
Ali S, Rizwan M, Qayyum MF, Ok YS, Ibrahim M, Riaz M, Arif MS, Hafeez F, Al-Wabel MI, Shahzad AN (2017b) Biochar soil amendment on alleviation of drought and salt stress in plants: a critical review. Environ Sci Pollut Res Int 24(14):12700–12712. https://doi.org/10.1007/s11356-017-8904-x
Arioli T, Mattner SW, Winberg PC (2015) Applications of seaweed extracts in Australian agriculture: past, present and future. J Appl Phycol 27(5). https://doi.org/10.1007/s10811-015-0574-9
Asagi N, Ueno H, Ebid A (2007) Effects of sewage sludge application on rice growth, soil properties, and n fate in low fertile paddy soil. Int J Soil Sci 2:171–181. https://doi.org/10.3923/ijss.2007.171.181
Backer R, Ghidotti M, Schwinghamer T, Saeed W, Grenier C, Dion-Laplante C, Fabbri D, Dutilleul P, Segiun P, Smith DL (2018) Getting to the root of the matter: water-soluble and volatile components in thermally-treated biosolids and biochar differentially regulate maize (*Zea mays*) seedling growth. PLoS ONE 13(11). https://doi.org/10.1371/journal.pone.0206924
Banaras NH, Rozema J, Badr-uz-Zaman SM (2002) Improvement of saline soil productivity through farm yard manure, amendment and coated seeds for fodderbeet cultivation. Asian J Plant Sci 1:82–84. https://doi.org/10.3923/ajps.2002.82.84
Bass AM, Bird MI, Kay G, Muirhead B (2016) Soil properties, greenhouse gas emissions and crop yield under compost, biochar and co-composted biochar in two tropical agronomic systems. Sci Total Environ 550:459–470. https://doi.org/10.1016/j.scitotenv.2016.01.143
Battacharyya D, Babgohari MZ, Rathor P, Prithiviraj B (2015) Seaweed extracts as biostimulants in horticulture. Sci Hortic 196:39–48. https://doi.org/10.1016/j.scienta.2015.09.012
Belanović S, Knežević M, Miličić B, Đorović M (2004) Contents of heavy metals and microflora in some soil of Mt. Stara planina. Bull Fac Fores Univ Belgrade 89:53–61. https://doi.org/10.2298/GSF0489053B
Bhat TA, Ahmad L, Ganai MA, Shams-Ul-Haq KOA (2015) Nitrogen fixing biofertilizers; mechanism and growth promotion: a review. J Pure Appl Microbiol 9(2):1675–1690
Boparai BS, Singh Y, Sharma BD (1992) Effect of green manure (*Sesbania aculeata*) on physical properties of soil and growth of rice-wheat and maize-wheat cropping system. Int Agrophys 6:95–101
Chen J, Lu F, Zou Y, Shen Z, Li G, Zhou X (2017) Evaluation of soil nutrients about typical economic forest lands of low hilly areas in eastern part of Zhejiang Province. Environ Earth Sci 61:1–6. https://doi.org/10.1088/1755-1315/61/1/012053
Cherr C, Scholberg J, McSorley R (2006) Green manure approaches to crop production. Agron J 98(2):302–319
Ciaccia C, Ceglie F, Tittarelli F, Antichi D, Carlesi S, Testani E, Canali S (2017) Green manure and compost effects on N-P dynamics in Mediterranean organic stockless systems. J Soil Sci Plant Nutr 17(3):751–769
Conn KL, Lazarovits G (1999) Impact of animal manures on *Verticillium* wilt, potato scab, and soil microbial populations. Can J Plant Pathol 21(1):81–92. https://doi.org/10.1080/07060661.1999.10600089
Dang T, Hao M, Guo S, Cai G (2003) Effect of plastic-film mulch on water and nitrogen use by spring maize and on fate of applied nitrogen in the southern Loess Plateau. Ying Yong Sheng Tai Xue Bao 14(11):1901–1905
De Rosa D (2017) Optimizing the use of organic amendments in sub-tropical vegetable cropping systems for improved environmental and agronomic outcomes. PhD by Publication, Queensland University of Technology
Ding WX, Meng L, Cai ZC, Han FX (2007) Effects of long-term amendment of organic manure and nitrogen fertilizer on nitrous oxide emission in a sandy loam soil. J Environ Sci (China) 19(2):185–193. https://doi.org/10.1016/s1001-0742(07)60030-8

Ding LJ, Su JQ, Sun GX, Wu JS, Wei WX (2018) Increased microbial functional diversity under long-term organic and integrated fertilization in a paddy soil. Appl Microbiol Biotechnol 102(4):1969–1982. https://doi.org/10.1007/s00253-017-8704-8

Duminda VG, Ranjith M, Randombage D, Asim B (2015) Effect of rice husk biochar on selected soil properties in tropical Alfisols. Soil Res 54. https://doi.org/10.1071/SR15102

El-Haddad ME, Zayed MS, El-Sayed GAM, Hassanein MK, El-Satarb AMA (2014) Evaluation of compost, vermicompost and their teas produced from rice straw as affected by addition of supplements. Ann Agric Sci 59(2):243–251. https://doi.org/10.1016/j.aoas.2014.11.013

Elsalahi R, Mohamed SS, Sherif AM, Osman AG (2016) *Rhizobium* biofertilizer (Okadin) production and future prospects in Sudan. Environ Nat Resour J 1(1):1–12. https://doi.org/10.12692/ijb/11.5.277-285

El-Sharouny EE (2015) Effect of different soil amendments on the microbial count correlated with resistance of apple plants towards pathogenic *Rhizoctonia solani* AG-5. Biotechnol Biotechnol Equip 29(3):1–7. https://doi.org/10.1080/13102818.2014.1002285

FAO (2015) Food and Agriculture Organization of the United Nations

FAO (2018) Food and Agriculture Organization of the United Nations, World agriculture; towards 2015/2030- an FAO perspective http://www.fao.org/docrep/005/y4252e/y4252e06.htm

Fasidi IO, Kadiri M, Jonathan SG, Adenipekun CO, Kuforiji OO (2008) Cultivation of tropical mushrooms. Ibadan University Press, Ibadan

Gafur MA, Rahman ABMM (2003) Effects of Crotalaria and Sesbania aculeata green manures and N fertilizer on soil fertility and the productivity of sugarcane. J Agric Sci 140(03):305–309. https://doi.org/10.1017/S0021859603003022

Gaind S, Nain L, Patel VB (2009) Quality evaluation of co-composted wheat straw, poultry droppings and oil seed cakes. Biodegradation 20(3):307–317. https://doi.org/10.1007/s10532-008-9223-1

Gao H, Yan C, Liu Q, Ding W, Chen B, Li Z (2019) Effects of plastic mulching and plastic residue on agricultural production: a meta-analysis. Sci Total Environ 651(Pt 1):484–492. https://doi.org/10.1016/j.scitotenv.2018.09.105

Gbolagade JS (2006) Bacteria associated with compost used for cultivation of Nigerian edible mushrooms: *Pleurotus tuberregium* (Fr.) Singer, and *Lentinus squarrosulus* (Berk.). Afr J Biotechnol 5(4):338–342

Gümüş I, Şeker C (2017) Effects of spent mushroom compost on physicochemical properties of degraded soil. Solid Earth Discuss. https://doi.org/10.5194/se-2016-122

Han G, Chen Q, Zhang S, Li G, Yi X, Feng C, Wang X, Yu C, Lan J (2019) Biochar effects on bacterial community and metabolic pathways in continuously cotton-cropped soil. J Soil Sci Plant Nutr:1–13. https://doi.org/10.1007/s42729-019-0014-z

Hao XH, Hu RG, Wu JS, Tang SR, Luo XQ (2010) Effects of long-term fertilization on paddy soils organic nitrogen, microbial biomass, and microbial functional diversity. Ying Yong Sheng Tai Xue Bao 21(6):1477–1484

Haque MA, Jahiruddin M, Rahman MM, Saleque MA (2015) Usability of bioslurry to improve system productivity and ecomomic return under potato-rice cropping system. Res Agric Livest Fish 2(1):27–33. https://doi.org/10.3329/ralf.v2i1.23026

Hossain MN, Sarker UK, Uddin MR, Rehana S, Hoque MML, Islam MA (2018) Effects of bioslurry with chemical fertilizer on the performance of some high yielding varieties of boro rice (*Oryza sativa* L.). Arch Agric Environ Sci 3(2):109–115. https://doi.org/10.26832/24566632.2018.030202

Ismail SM (2013) Influence of effective microorganisms and green manure on soil properties and productivity of pearl millet and alfalfa grown on sandy loam in Saudi Arabia. Afr J Microbiol Res 7(5):375–382. https://doi.org/10.5897/AJMR12.1693

Jarak M, Đurić S, Simikić M, Savin L, Vasin J (2007) Microbiological activity in soil under wheat. Traktors and power machines 12(3):49–53

Jarak MN, Đurić SS, Đorđević BD (2010) Benefits of inoculation with *Azotobacter* in the growth and production of tomato and peppers. J Nat Sci 119:71–76. https://doi.org/10.2298/ZMSPN1019071J

Jared N, Erastus G, Christopher N, Steve A (2016) Evaluation of biogas slurry as an alternative organic fertilizer: a case study in Kenya. Int J Extensive Res 9:10–14

Jiang Y, Qian H, Huang S, Zhang X, Wang L, Zhang L, Shen M, Xiao X, Chen F, Zhang H, Lu C, Li C, Zhang J, Deng A, van Groenigen KJ, Zhang W (2019) Acclimation of methane emissions from rice paddy fields to straw addition. Sci Adv 5(1):eaau9038. https://doi.org/10.1126/sciadv.aau9038

Jonathan SG (2002) Vegetative growth requirements and antimicrobial activities of some higher fungi in Nigeria. Dissertation, University of Ibadan

Jonathan SG, Lawal MM, Oyetunji OJ (2011) Effect of spent mushroom compost of *Pleurotus pulmonarius* on growth performance of four Nigerian vegetables. Mycobiology 39(3):164–169. https://doi.org/10.5941/MYCO.2011.39.3.164

Kim HS, Kim KR, Yang JE, Ok YS, Owens G, Nehls T, Wessolek G, Kim KH (2016) Effect of biochar on reclaimed tidal land soil properties and maize (*Zea mays* L.) response. Chemosphere 142:153–159. https://doi.org/10.1016/j.chemosphere.2015.06.041

Koll PJ (2009) Effects of conifer sawdust, hardwood sawdust, and peat on soil properties and a barefoot conifer seedling development. Master's thesis, Michigan Technological University

Kolton M, Harel YM, Pasternak Z, Graber ER, Elad Y, Cytryn E (2011) Impact of biochar application to soil on the root-associated bacterial community structure of fully developed greenhouse pepper plants. Appl Environ Microbiol 77(14):4924–4930. https://doi.org/10.1128/AEM.00148-11

Kumar S, Malav LC, Malav MK, Khan SA (2015) Biogas slurry: source of nutrients for eco-friendly agriculture. Int J Extensive Res 2:42–46

Kumar A, Meena RN, Singh RK, Gilotia YK (2018) Effect of organic and inorganic sources of nutrient on yield, yield attributes and nutrient uptake of rice CV. PRH-10. Biscan 9(2):595–597. https://doi.org/10.13140/RG.2.2.16540.39044

Lalander CH, Komakech AJ, Vinnerås B (2015) Vermicomposting as manure management strategy for urban small-holder animal farms- Kampala case study. Waste Manag 39:96–103. https://doi.org/10.1016/j.wasman.2015.02.009

Lashari MS, Ye Y, Ji H, Li L, Kibue GW, Lu H, Zheng J, Pan G (2015) Biochar-manure compost in conjunction with pyroligneous solution alleviated salt stress and improved leaf bioactivity of maize in a saline soil from central China: a 2-year field experiment. J Sci Food Agric 95(6):1321–1327. https://doi.org/10.1002/jsfa.6825

Li Q, Lei Z, Song X, Zhang Z, Ying Y, Peng C (2018) Biochar amendment decreases soil microbial biomass and increases bacterial diversity in Moso bamboo (*Phyllostachys edulis*) plantations under simulated nitrogen deposition. Environ Res Lett 13. https://doi.org/10.1088/1748-9326/aab53a

Lin X, Feng Y, Zhang H, Chen R, Wang J, Zhang J, Chu H (2012) Long-term balanced fertilization decreases arbuscular mycorrhizal fungal diversity in an arable soil in North China revealed by 454 pyrosequencing. Environ Sci Technol 46(11):5764–5771. https://doi.org/10.1021/es3001695

Liu Z, Meng Y, Cai M, Zhou J (2017) Coupled effects of mulching and nitrogen fertilization on crop yield, residual soil nitrate, and water use efficiency of summer maize in the Chinese Loess Plateau. Environ Sci Pollut Res Int 24(33):25849–25860. https://doi.org/10.1007/s11356-017-0194-9

Liu G, Zuo Y, Zhang Q, Yang L, Zhao E, Liang L, Tong YA (2018) Ridge-furrow with plastic film and straw mulch increases water availability and wheat production on the Loess Plateau. Sci Rep 8(1):6503. https://doi.org/10.1038/s41598-018-24864-4

Lu K, Yang X, Gielen G, Bolan N, Ok YS, Niazi NK, Xu S, Yuan G, Chen X, Zhang X, Liu D, Song Z, Liu X, Wang H (2017) Effect of bamboo and rice straw biochars on the mobility and redistribution of heavy metals (Cd, Cu, Pb and Zn) in contaminated soil. J Environ Manag 186(2):285–292. https://doi.org/10.1016/j.jenvman.2016.05.068

Lubbers IM, van Groenigen KJ, Fonte SJ, Six J, Brussaard L, van Groenigen JW (2013) Greenhouse-gas emissions from soils increased by earthworms. Nat Clim Chang 3(3):187–194. https://doi.org/10.1038/NCLIMATE1692

Luo X, Han S, Lai S, Huang Q, Chen W (2017) Long-term straw returning affects *Nitrospira*-like nitrite oxidizing bacterial community in a rapeseed-rice rotation soil. J Basic Microbiol 57(4):309–315. https://doi.org/10.1002/jobm.201600400

Mahato S, Bhuju S, Shrestha J (2018) Effect of Trichoderma Viride as biofertilizer on growth and yield of wheat. Malaysian J Sustain Agric (MJSA) 2(2):1–5. https://doi.org/10.26480/mjsa.02.2018.01.05

Mandal UK, Singh G, Victor US, Sharma KL (2003) Green manuring: its effect on soil properties and crop growth under rice-wheat cropping system. Eur J Agron 19:225–237. https://doi.org/10.1016/S1161-0301(02)00037-0

Mao XL, Lu KP, Sun T, Zhang XK, He LZ, Wang HL (2015) Effect of long-term fertilizer application on the stability of organic carbon in particle size fractions of a paddy soil in Zhejiang Province, China. Huan Jing Ke Xue 36(5):1827–1835

Meena BL, Fagodiya RK, Prajapat K, Dotaniya ML, Kaledhonkar MJ, Sharma PC, Meena RS, Mitran T, Kumar S (2018) Legume green manuring: an option for soil sustainability. In: Meena R, Das A, Yadav G, Lal R (eds) Legumes for soil health and sustainable management. Springer, Singapore, pp 387–408

Mondini C, Cayuela ML, Sinicco T, Fornasier F, Galvez A, Sánchez-Monedero MA (2017) Modification of the RothC model to simulate soil C mineralization of exogenous organic matter. Biogeosciences 14(13):3253–3274. https://doi.org/10.5194/bg-14-3253-2017

Mukherjee A, Lal R, Zimmerman AR (2014) Effects of biochar and other amendments on the physical properties and greenhouse gas emissions of an artificially degraded soil. Sci Total Environ 487:26–36. https://doi.org/10.1016/j.scitotenv.2014.03.141

Murray TR, Frank DA, Gehring CA (2010) Ungulate and topographic control of arbuscular mycorrhizal fungal spore community composition in a temperate grassland. Ecology 91(3):815–827. https://doi.org/10.1890/09-0209.1

Paetsch L, Mueller CW, Kögel-Knabner I, von Lützow M, Girardin C, Rumpel C (2018) Effect of in-situ aged and fresh biochar on soil hydraulic conditions and microbial C use under drought conditions. Sci Rep 8:6852. https://doi.org/10.1038/s41598-018-25039-x

Panth M, Hassler SC, Baysal-Gurel F (2020) Methods for Management of Soilborne Diseases in Crop Production. Agriculture 10(1):16. https://doi.org/10.3390/agriculture10010016

Pérez-Esteban J, Escolástico C, Masaguer A, Moliner A (2012) Effects of sheep and horse manure and pine bark amendments on metal distribution and chemical properties of contaminated mine soils. Eur J Soil Sci. https://doi.org/10.1111/j.1365-2389.2012.01468.x

Polat E, Uzun HB, Topçuolu B, Önal K, Onus AN, Mehmet Karaca M (2009) Effects of spent mushroom compost on quality and productivity of cucumber (*Cucumis sativus* L.) grown in greenhouses. Afr J Biotechnol 8(2):176–180

Qin S, Yeboah S, Xu X, Liu Y, Yu B (2017) Analysis on fungal diversity in rhizosphere soil of continuous cropping potato subjected to different furrow-ridge mulching managements. Front Microbiol 8:845. https://doi.org/10.3389/fmicb.2017.00845

Rahman MS, Rahman MH, Chowdhary MFN, Sultana MS, Ahmed KU (2016) Effect of spent mushroom substrate and cow dung on growth, yield and proximate composition of brinjal. Int J Sci Res 6(10):468–475

Rashid MI, Mujawar LH, Shahzad T, Almeelbi T, Ismail MI, Oves M (2016) Bacteria and fungi can contribute to nutrients bioavailability and aggregate formation in degraded soils. Microbiol Res 183:26–41. https://doi.org/10.1016/j.micres.2015.11.007

Rathi M, Gaur N (2016) Phosphate solubilizing bacteria as biofertilizer and its applications. J Pharm Res 10(3):146–148

Ren H, Hu J, Hu Y, Yang G, Zhang Y (2017) Divergence of compost extract and bio-organic manure effects on Lucerne plant and soil. PeerJ 5:e3775. https://doi.org/10.7717/peerj.3775

Rezk E, Ibrahmin M, Robin P, Nourya AC, El-Saka M (2015) Rice straw composting and its effect on soil properties. Compost Sci Utilization 17(3):146–150. https://doi.org/10.13140/RG.2.1.4089.3283

Roy S, Kashem MA (2014) Effects of organic manures in changes of some soil properties at different incubation periods. Open J Soil Sci 4(3):6. https://doi.org/10.4236/ojss.2014.43011

Sane SA, Mehta SK (2015) Isolation and evaluation of rock phosphate solubilizing fungi as potential biofertilizer. J Fertil Pest 6(2):1–6

Shadreck B (1999) The effect of green manure, cowdung and their combinations on growth and grain yield of maize (*Zea mays* L.) in low input systems. Thesis, The University of Zambia, Lusaka

Shen Y, Chen Y, Li S (2016) Microbial functional diversity, biomass and activity as affected by soil surface mulching in a semiarid farmland. PLoS One 11(7):e0159144. https://doi.org/10.1371/journal.pone.0159144

Shivaji RS (2012) Studies on effect of spent mushroom compost on growth parameter of cabbage (*Brassica oleracea* var. *capitata* L.) and soil microflora. Thesis, Mahatma Phule Krishi Vidyapeeth, Mumbai

Shukla KP, Sharma S, Singh NK, Singh V, Tiwari K, Singh S (2011) Nature and role of root exudates: efficacy in bioremediation. Afr J Biotechnol 10(48):9717–9724. https://doi.org/10.5897/AJB10.2552

Simarani K, Halmi MFA, Abdullah R (2018) Short-term effects of biochar amendment on soil microbial community in humid tropics. Arch Agron Soil Sci. https://doi.org/10.1080/03650340.2018.1464149

Su JQ, Ding LJ, Xue K, Yao HY, Quensen J, Bai SJ, Wei WX, Wu JS, Zhou J, Tiedje JM, Zhu YG (2015) Long-term balanced fertilization increases the soil microbial functional diversity in a phosphorus-limited paddy soil. Mol Ecol 24(1):136–150. https://doi.org/10.1111/mec.13010

Tamilkodi R, Victoria J (2018) Liquid Microbial Biofertilizers (LMF) for enhancing soil fertility- a review. Int J Trend Sci Res Dev (IJTSRD) 2(3):673–678

Tan W, Yu H, Huang C, Li D, Zhang H, Jia Y, Wang G, Xi B (2018) Discrepant responses of methane emissions to additions with different organic compound classes of rice straw in paddy soil. Sci Total Environ 630:141–145. https://doi.org/10.1016/j.scitotenv.2018.02.230

Tirol-Padre A, Ladha JK, Regmi AP, Bhandari AL, Inubushi K (2007) Organic amendments affect soil parameters in two long-term rice-wheat experiments. Soil Sci Soc Am J 71(2):442–452. https://doi.org/10.2136/sssaj2006.0141

Tošić I, Golić Z, Radosavac A (2016) Effects of the application of biofertilizers on the microflora and yield of lettuce (*Lactuca sativa* L.). Acta Agric Serb 21(42):91–98

Trifonova R, Postma J, Schilder MT, van Elsas JD (2009) Microbial enrichment of a novel growing substrate and its effect on plant growth. Microb Ecol 58(3):632–641. https://doi.org/10.1007/s00248-009-9518-8

United States Census Bureau, U.S. and World population clock. https://www.census.gov/popclock/. Accessed Oct 2018

Utuk IO, Daniel EE (2015) Land degradation: a threat to food security: a global assessment. Environ Earth Sci 5(8):13–21

van der Heijden MG, Streitwolf-Engel R, Riedl R, Siegrist S, Neudecker A, Ineichen K, Boller T, Wiemken A, Sanders IR (2006) The mycorrhizal contribution to plant productivity, plant nutrition and soil structure in experimental grassland. New Phytol 172(4):739–752. https://doi.org/10.1111/j.1469-8137.2006.01862.x

Vepsäläinen M, Erkomaa K, Kukkonen VSM, Wallenius K, Niemi RM (2004) The impact of crop plant cultivation and peat amendment on soil microbial activity and structure. Plant Soil 264(1):273–286. https://doi.org/10.1023/B:PLSO.0000047763.46795.cb

Wang C, Wu N, Hou HJ, Tang YF, Shen JL, Qin HL (2016) Effects of fertilization on soil microbial abundance and community structure at DNA and cDNA levels in paddy soils. Huan Jing Ke Xue 37(11):4372–4379. https://doi.org/10.13227/j.hjkx.201604220

Wang R, Zhang H, Sun L, Qi G, Chen S, Zhao X (2017) Microbial community composition is related to soil biological and chemical properties and bacterial wilt outbreak. Sci Rep 7:343. https://doi.org/10.1038/s41598-017-00472-6

Wang X, Wang N, Xing Y, Ben El Caid M (2018) Synergetic effects of plastic mulching and nitrogen application rates on grain yield, nitrogen uptake and translocation of maize planted in the Loess Plateau of China. Sci Rep 8(1):14319. https://doi.org/10.1038/s41598-018-32749-9

Wiafe-Kwagyan M, Odamtten GT (2018) Use of *Pleurotus eous* Strain P-31 spent mushroom compost (smc) as soil conditioner on the growth and yield performance of *Capsicum annuum* L. and *Solanum lycopersicon* L. Seedlings under greenhouse conditions in Ghana. Trop Life Sci Res 29(1):173–194. https://doi.org/10.21315/tlsr2018.29.1.12

Yadav GS, Lal R, Meena RS, Babu S, Das A, Bhomik SN, Datta M, Layak J, Saha P (2017) Conservation tillage and nutrient management effects on productivity and soil carbon sequestration under double cropping of rice in North Eastern Region of India. Ecol India Ecol Indicators. https://doi.org/10.1016/j.ecolind.2017.08.071

Yang FK, Wang LM, Zhang GH (2013) Effects of plastic film mulching with double ridges and furrow planting on soil moisture and temperature and soybean yield on a semiarid dryland of Gansu Province, Northwest China. Ying Yong Sheng Tai Xue Bao 24(11):3145–3152

Yang H, Niu J, Tao J, Gu Y, Zhang C, She S, Chen W, Yang H, Yin H (2016) The impacts of different green manure on soil microbial communities and crop health. Preprints 1:1–12. https://doi.org/10.20944/preprints201609.0056.v1

Yunta F, Di Foggia M, Bellido-Díaz V, Morales-Calderón M, Tessarin P, López-Rayo S, Tinti A, Kovács K, Klencsár Z, Fodor F, Rombolà AD (2013) Blood meal-based compound. Good choice as iron fertilizer for organic farming. J Agric Food Chem 61(17):3995–4003. https://doi.org/10.1021/jf305563b

Zaller JG, Köpke U (2004) Effects of traditional and biodynamic farmyard manure amendment on yields, soil chemical, biochemical and biological properties in a long-term field experiment. Biol Fertil Soils 40(4):222–229

Zhang H, Voroney P, Price G (2014) Effects of biochar amendments on soil microbial biomass and activity. J Environ Qual 43(6):2104–2114. https://doi.org/10.2134/jeq2014.03.0132

Zhang LL, Sun SJ, Chen ZJ, Jiang H, Zhang XD, Chi DC (2018) Effects of different colored plastic film mulching and planting density on dry matter accumulation and yield of spring maize. Ying Yong Sheng Tai Xue Bao 29(1):113–124. https://doi.org/10.13287/j.1001-9332.201801.019

Zhao S, Zhang S (2018) Linkages between straw decomposition rate and the change in microbial fractions and extracellular enzyme activities in soils under different long-term fertilization treatments. PLoS One 13(9):e0202660. https://doi.org/10.1371/journal.pone.0202660

Zhao J, Ni T, Xun W, Huang X, Huang Q, Ran W, Shen B, Zhang R, Shen Q (2017) Influence of straw incorporation with and without straw decomposer on soil bacterial community structure and function in a rice-wheat cropping system. Appl Microbiol Biotechnol 101(11):4761–4773. https://doi.org/10.1007/s00253-017-8170-3

Zhen Z, Liu H, Wang N, Guo L, Meng J, Ding N, Wu G, Jiang G (2014) Effects of manure compost application on soil microbial community diversity and soil microenvironments in a temperate cropland in China. PLoS One 9(10):e108555. https://doi.org/10.1371/journal.pone.0108555

Zhu YG, Smith FA, Smith SE (2003) Phosphorus efficiencies and responses of barley (Hordeum vulgare L.) to arbuscular mycorrhizal fungi grown in highly calcareous soil. Mycorrhiza 13(2):93–100

Mrkovački N, Jarak M, Đalović I, Jocković Đ (2012) Importance of PGPR application and its effect on microbial activity in maize rhizosphere. Proc Inst Field Veg Crops Novi Sad 49(3):335–344. https://doi.org/10.5937/ratpov49-1915

Chapter 6
Basics of Waste Management and Disposal Practices in the Indian Context

Kalpna Varshney

Abstract Accelerated growth in population, rapid industrialization and urbanization have increased the amount of solid waste, calling for efficient strategies of waste management. Waste management is an ancient practice dating back to 2000 BC. Different methods are used for different types of waste, including thermal methods, biological methods and recycling. In this chapter I review the type of wastes, the history of waste disposal, and methods of waste disposal such as composting, landfills, incineration, gasification, waste-to-energy, recycling and waste minimization.

Keywords Solid waste management · Waste treatment · Incineration · Composting · Landfills · Waste to energy · Recycling · Integrated solid waste management

6.1 Introduction

Urbanization and present growth rate of the world's population is resulting in approximately 2.5 billion additional people to the urban population by 2050 in Asia. The rising health level, and the high level of resource usage patterns have had an unintended and adverse impact on the urban environment. The other growing problem with the urban as well as rural areas is generation of high levels of waste (Gutberlet 2017). Various technologies and methods of disposal are involved from ancient times. Now new techniques and methodology are required to tackle with this huge quantity of waste. A lot of effort is required from public and the private sectors (Ladhe et al. 2014). Over the past decades, the production and creation of solid waste has become a growing problem in many countries (Fig. 6.1). (Kaza et al. 2018).Waste treatment and recycling reduces the environmental damage prior to dispose it through landfills. Policymakers and stakeholders have seen the benefits of

K. Varshney (✉)
Faculty of Engineering and Technology, Manav Rachna International Institute of Research and Studies, Faridabad, Haryana, India
e-mail: kalpna.fet@mriu.edu.in

E. Lichtfouse (ed.), *Sustainable Agriculture Reviews 52*, Sustainable Agriculture Reviews 52, https://doi.org/10.1007/978-3-030-73245-5_6

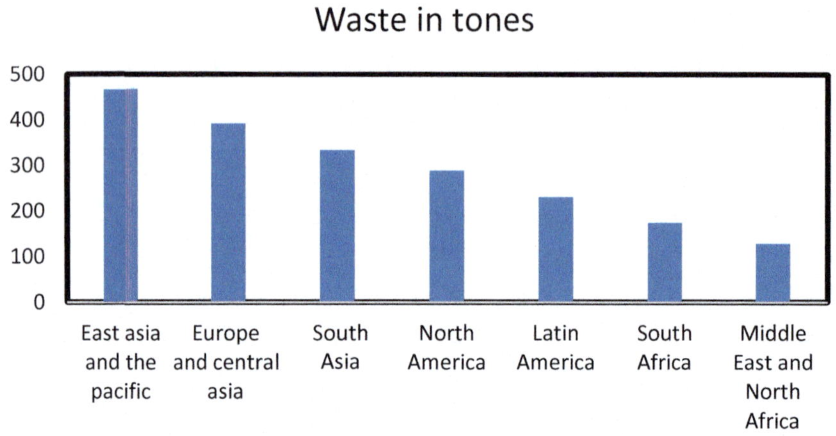

Fig. 6.1 Global waste generation continent wise. Source: world bank

waste management and recycling and therefore always willing to update with new rules and regulations. Solid waste disposal that covers a wide range of issues including the reduction, recycling, separation, conversion, treatment and dumping of various levels of complexity (USDE 1999).

6.1.1 Categories of Waste

Waste or refuse is either biodegradable or non-biodegradable (Table 6.1). It can be categorized into five types which are frequently originate in and around us by most of our daily activities. These include liquid waste, solid rubbish, organic waste, recyclable rubbish and hazardous waste (Abdel-Shafya and Mansour 2018). Liquid waste generally comes from industrial effluent and sewage.

6.1.2 Sustainable Disposal of Waste

Disposal of waste into the surrounding locality is the general practice and no one is really concerned about the environment while doing this. The disposal or treatment of waste should be in proper way so that the earth must be sustainable for the future generations (Hamer 2004). Most industries and general public find easy to dispose of waste either in landfills, oceans or unrestrained burning of waste. All these practices are unhealthy and hazardous in nature. The main aim of waste treatment is to remove toxic substances and stabilization, so that the ultimate residues do not get accumulated after further change. Therefore, the complete mineralization is a necessary step in waste treatment. After the mineralization the degraded and simpler minerals can easily pass through

Table 6.1 Various categories of waste can be partly degraded in soils, but the rest are non-degradable

Biodegradable waste	Non-biodegradable waste
Municipal solid waste from household, commercial wastes	Plastic, paper, cardboard and cartons
Food industry waste, food stalls, restraint waste including bones and eggshells	Bottles, flasks and vessels and all types of packing materials like foil, tetra packs etc.
Juice, beverage, agriculture and agroindustry waste	Glass and metallic waste
Domestic food (fruits and vegetable)	Rags/shreds, rubber
Wood/forestry industry waste, garden waste	Building industry and their destruction waste
Animal manures from intensive cattle farming	Petroleum and energy sector waste
Sewage waste	Ash and dust from different sources like construction
Butchery waste and contaminated skeletons and shells	Discarded electronic items from offices, residences. Cassettes, computer diskettes, printer cartridges and electronic parts and other electrical and electronic appliances
Farmyard waste containing green and dry leaves	Used and discarded clothes, equipment, instruments and furniture

the various biogeochemical cycles that maintain the materials cycling in the environment (Bindschedler et al. 2016). The main objective of a sustainable waste management system is to maximize the materials and energy recovery without impacting the environment and social cost (Stringfellow 2014). Although it is difficult to decide that which treatment process is used for which kind of waste, yet the sustainability objectives should always be maintained. Therefore it is imperative to evaluate the various waste treatment and management methods time to time (Seadon 2010).

6.2 Historical Aspects of Waste Management

Historically, waste management has been introduced as a treatment method for the disposal of waste since ancient times. The disposal characteristics of the waste are often altered due to degradation. Degradability is a factor that greatly increases the ability of many pollutants to be disposed of. Degradation of waste is governed by many factors like environmental conditions, the nature of the material for degradation and the accessibility of appropriate consortia of microbes (Barles 2014). In early civilization, waste was mainly consisting fire ash of bones, wood and crop residues (Vishwakarma 2012). The vegetable waste was used to feed the faunas and the rest was discarded where it would rot. Archaeological excavations made by archaeologists reported only a small amount of ash, soil and broken tools, which tell us that these early civilizations used the reused and repair techniques (Theodore and Theodore 2010).

6.2.1 Milestones of Waste Management

As population growth in cities increased, waste treatment techniques became essential to manage the burden of waste. Athens, and Rome in ancient civilization were the first that started waste management. Rome established a waste collection group to collect waste from the streets. The waste then transported in trucks to mines outside the city. The history of management of solid wastes is not available thoroughly as customary no tradition of academic research into the history of solid waste disposal. This history timeline is the extract of various articles (Erhard 1964; Hosch 1967; Priestly 1968; Anonymous 1969; Wilson 1976) (Table 6.2).

India has a unique history of waste treatment. At each stage of immigration, either in the times of Aryans who were one of the ancient rulers in the world, or during the time of Mughal invaders and finally the 300-year rule of England, environmental laws have always carried the stage according to the administrative wishes of the inhabitants of the country. Since ancient times, humans and animals have used land resources to support health and dispose of waste. In the first community, waste disposal of human and other waste did not present any major problems, because of less population and less land available for waste removal. The challenge of nature today, is the most difficult and the most important that the world's people have ever faced, and yet it is new. Our ancestors knew the threats of pollution and waste. They therefore worshiped and serve the natural resources to maintain a healthy environment. Ancient Indian texts have enough to reveal about the protection and preservation of ecological balance. All religions that find their densities in India have a more natural view of the observance of the laws of nature and therefore show respect towards nature and creation (Agarwal and Mathur 2008; Ahmeda and Alib 2004; Ahsan 1999).

6.3 Methods of Waste Disposal

Waste is comprising of two components. One component is consisting of the useful substances which we can again reuse or recycle. The other one is the products which we cannot reuse, and their recycling is also either not possible or very costly. Therefore, this component is the unusable substances. While treating the waste, segregation of the valuable part of the waste and dispose of the unusable component is a significant step. The final disposal should not be harmful for environment and living organisms. Let's go through the methods for waste disposal.

Table 6.2 The detailed chronology of waste management practices and rules

Year	Major timeline event
2000 BC	Bronze scrap recovery systems were in place in Europe. Early records from the Han dynasty suggest that composting was a part of life in China "fertilizer recipes" listing human excreta, animal waste, straw, plant ash.
1500 BC	The Minoan people created dump sites where waste was placed in large pits and covered with earth
500 BC	Waste can be deposited no less than one mile from the city and banning the dumping of refuse in city streets of Athens
250 AD	Mayan Indians of Central America had dumps, they also recycled their inorganic waste into fill for building projects.
200 AD	The first sanitation force is created by the Romans, they also recycled their inorganic waste into fill for building projects
1297	A law is passed in Britain requiring householders to keep the front of their house clear from refuse
1340	The Black Death spread to Western Europe and North Africa during the 1340s, resulting in an estimated 75 million deaths worldwide.
1354	Earliest English garbage men generally loaded the waste into carts and deposited it in the Thames or Fleet rivers.
1388	The English Parliament bans dumping of waste in ditches and public waterways
1400	Garbage piles build up so high outside the gates of Paris that they interfere with the city's defenses. The preliminary attempt of English govt to manage waste was not principally successful, but paved the way for further regulation
1500	A recycling Spanish technique in copper mines to use scrap iron for the cementation of copper, which is relevant even today.
1657	A regulation against discarding waste passed in the New York.
1690	First paper mill opened in America and prepared paper from used cotton, linen, and old paper.
1757	Benjamin Franklin starts the first American municipal street-cleaning operation in Philadelphia, Pennsylvania
1776	The first metal recycling in America
1848	The first regulation for the process of waste regulation in Britain was formed as Public health act of 1848
1874	First Incinerator designed in Britain and energy from waste was transformed
1885	In NY first American garbage incinerator was designed and in the next two decades, approximately 200 garbage incinerators were built in US
1890	First paper mill opened in Britain and prepared paper from recycled materials
1895	First U.S. comprehensive system of public-sector garbage management by cart off garbage to dumps, incinerators and in the Atlantic Ocean. First U.S. waste sorting plant for recycling
1902	127 U.S. cities surveyed in MIT project provided regular collection of refuse.
1904	US first two major aluminum recycling plants open in Cleveland and Chicago
1905	First garbage incinerator was used in New York to produce electricity.
1917	Waste reclamation service was started in US due to deficiencies of raw materials during world war I
1918	Motorized waste collection units began in US which earlier were drawn by horses

(continued)

Table 6.2 (continued)

Year	Major timeline event
1932	Hydraulic rear loader compactor trucks began to use in Europe.
1942	Americans started collect rubber, paper, scrap metal, fats, and tin cans for recycling after world war I
1962	The national solid wastes management association founded in US
1965	First solid waste management act
1970	First earth day celebration, EPA was formed
1972	The first buy-back centers for recyclables are opened in Washington state. They accept beer bottles, aluminum cans and newspapers
1984	Hazardous and solid waste act amendments and reauthorization to the resource conservation and recovery act
1989	Hazardous waste management rules in India
1990	Both Coca-Cola and Pepsi begin using a bottle made of about 25 percent recycled plastic resin. Mc Donald announced to stop the use of polystyrene plastic packaging
1998	Biomedical waste management rule in India
2000	Municipal solid waste handling and management in India
2002	NY first landfill closed and converted into a park
2007	According to the U.S. EPA, the recycle and compost in US was 33.4 percent of the municipal solid waste stream, as compared to a 6.4 percent recycling rate in 1960

6.3.1 Biological Treatment Methods

6.3.1.1 Composting and Vermicomposting

Composting is a very important and effective method to dispose of biodegradable waste. A diverse kind of degradable waste materials can be deposited down in the dumping pits for rotting by microorganisms present in the soil. By composting the waste materials are transformed into valuable manure. In this method the soil is sanitized as well because all harmful chemicals and other organisms are killed by the microorganisms. Composting can be executed in aerobic as well as anaerobic environments (Tanugur 2009) Composting usually takes a longer duration for the process of putrefaction around 2–3 months. Vermicomposting is a special kind of composting method. A specific kind of earthworms namely *Eisenia foetida* and *Lumbricus rubellis* are used to enhance the rate of composting and thus reduce the time for degradation. The composting is transformed into vermicomposting (Fig. 6.2). Vermicompost is better manure than the normal compost (Naikwade et al. 2012). There are nearly 3600 types of earthworms which are divided into burrowing and non-burrowing types. Red earthworm species, like *Eisenia foetida,* and are the most efficient in compost making. The non-burrowing earthworms eat 10 percent soil and 90 percent organic waste materials (Manyuchi and Anthony 2013). Composting can be considered as one of recycling method as well because this mimic the natural process of converting dead plants and animal waste into humus

Fig. 6.2 Composting (left) is used to convert the household domestic degradable waste into manure, and vermi-composting (right) is bioprocessing method to convert waste into manure using earthworms

(Sharholy et al. 2008). Rate and fate of composting is dependent on many factors. The important factors are C/N ratio, humidity, temperature (Guo et al. 2012).

6.3.1.2 Advantages of Composting

Waste amount is reduced in composting. Harmful insect pests are killed by the heat produced during the putrefaction process. The composting not only increases the nutrients and the productivity of soil but also convert the invaluable waste into valuable manure. This is one of the best methods to dispose of the biodegradable waste easily (Argun et al. 2017).

6.3.1.3 Types of Compost System

There are several methods for converting waste into compost.

Small Scale Composting

(a) **Home and farmyards composting**

Just like other industrial waste food and agriculture waste is among the ancient human practices (Shilev et al. 2006). Maximum home compost systems can be constructed with foraged materials, few need nothing but the soil, but others may involve much cost. Composting systems can be managed by the type of wastes like farm wastes utilize large space and tractors while vegetative kitchen wastes are composted either through soil pits or by vermicomposting. Generally, food wastes should be composted in closed systems separately from farm wastes to keep rodents and other pests from becoming a problem in the open (Fig. 6.3). The type of composting depends on what materials are to be recycled, how much space is available, when compost is needed, and what it will be used for.

Fig. 6.3 Home composting can be done in the garden where all domestic waste is dumped into a pit then converted into manure easily (left). Farmyard composting is mixed with cow dung and urine and some dairy waste

Large Scale Composting

(a) **Open windrow**

The windrow composting method consists of linear rows of compost materials (rice straw and cow manure), which are placed one on each other and mechanically turned periodically. The air contained in the interspaces of the composting mass varies in composition. The CO_2 content gradually increases and the O_2 level falls during composting process. The concentration of O_2 for composting varies from 15 to 20%. (Diaz and Savage 2007).Trash is kept in heaps (windrows) of1.5 m height and 2.5 m width and 60% moisture is also maintained in these heaps (Fig. 6.4). Temperature may go upto 70 °C because of the heat generated during this process. After 7–10 weeks of repeated cycles, temperature down to atmospheric temperature and this signify the stabilization of composting.

(b) **Mechanized composting**

Machineries are used to mix and turn the waste in the windrows to enhance the rate of composting for the fast stabilization (Fig. 6.5). The duration is around 1–2 weeks only for transforming organic waste into to manure. The refuse and the other material are turned over after every 5 days. The manure quality can be increased by adding cow dung and night soil with the waste. This process is generally done in compost pits. The extra moisture content can be drained out of the pits. The pit consist of the layer of ash, powdered calcium carbonate (lime stone), or loamy soil to counterbalance the acidity in the manure these alkaline substances may be beneficial for the microorganisms also which are degrading the organic waste into manure. The compost pit is having different layers of organic/food waste/ degradable waste and night soil (human excreta) or cow dung in alternate manner. After about 4 weeks the manure is ready for use.

Fig. 6.4 Open windrow composting is used in farmyards where biodegradable waste is collected in the form of heaps with some moisture

Fig. 6.5 Mechanized composting requires machineries to turn over the layers of the degradable waste, cow dung and soil

Fig. 6.6 A typical landfill site where all types of waste are dumped

6.3.1.4 Landfills

One of the most accepted methods of solid waste disposal is landfill. Solid waste is filled in multiple layers inside the pits, compressed and enclosed in a landfill which is a specially designed pit. Historically the places where the rubbish can be disposed was known as landfills that is also known as rubbish dump or dumping site (Meegoda et al. 2016) (Fig. 6.6). Many times the landfill sites are old quarries, unused mines, gravel pits or marshy areas. The bottom of the pit is coated with limestone to avoid groundwater pollution (Gupta et al. 2007). The designs of a landfill should be proficient, and the skilled workforce is required to operate there (Bagchi 2004). For proper management, usually the landfills are at the outskirts of the city and are low lying areas where garbage is collected. it is covered with layers of soil. After some time, the landfill becomes settle and microorganisms start degrading the organic matter and convert them to carbon dioxide and water under aerobic conditions like composting. In recent years, there has been a shift in philosophy of landfill design from the dry storage concept towards the bioreactor approach. In the bioreactor approach, the moisture content of the solid waste is increased by recirculation of leachate to enhance the biodegradation (Reddy et al. 2009). As a new approach the government agencies are trying to convert the inactive landfill sites into a park or a playground in India.

Limitations of Landfills

Though Landfills are the most efficient method for waste management, but few limitations are there. Fire hazards are there if excess of waste is buried. The Moisture should be maintained (60%) for decent biodegradation. The waste layer deepness should normally be up to 2 m. Combustion starts at a temperature of 70 degree C in the initial stages of decomposition and then it drops remnants open for a longer duration of time, where flies come easily and may cause various diseases.

Types of Landfills

There are four types of landfills.

1. **Sanitary landfills** – since ancient times, the waste is dumped into open dumps. The toxic chemicals and gases contaminate the surrounding atmosphere, soil and groundwater. These landfills were considered as disease causing areas as mosquitoes and other pathological pests may grow there. The face of open old landfill sites is now changing, and only illegal dumping is now found in near surrounding. In a sanitary landfill, waste is segregated from the neighboring atmosphere by a multilayered system that allow waste to decay securely and the last layer is a clay line to isolate the trash from the surroundings (Fig. 6.7) (Terlecky 1980). The putrefaction of waste generates methane which is a greenhouse gas. Methane is accumulated and collected in the sanitary landfills and reuse as fuel or electricity generator.
2. **Municipal solid waste landfills** – This is having a plastic lining at the bottom of the landfill to separate the landfill from the surrounding.

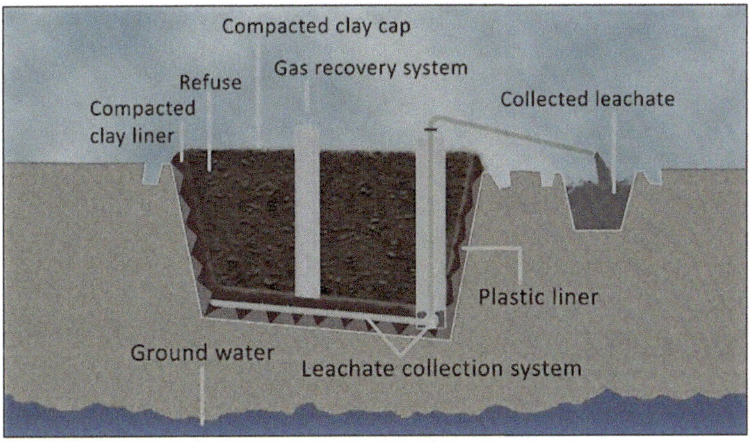

Fig. 6.7 A sanitary landfill provides an organized way for management of hazardous as well as non-hazardous waste

3. **Construction waste landfills** – Comprising of the remains of the construction, renovation, and demolition of buildings waste and roads, and bridges construction waste.
4. **Industrial waste landfills** – Consisting non-hazardous debris associated with manufacturing and other industrial applications.

Structure of a Modern Landfill

Scientifically a modern landfill (Fig. 6.8) should have the following elements. A modern landfill is smartly designed and constructed so that there are minimum adverse effects from the by- products of the landfill, on the environment, both during its construction and operation and after closure (Reinhart and Townsend 2018).

- **Bottom lining System** – The bottom liner separates and prevents buried waste from coming into contact with the natural bottom soil and groundwater. In municipal waste disposal facilities, suburbs are usually constructed using some form of durability, made of plastic HDPE (High Density Polyethylene) from 30 to 100 mils thick. Plastic liners can also be made of a mixture of compacted clay, and synthetic plastic.
- **Compartments** – There are a number compartments of different sizes depending on the total amount of waste generated daily in landfills from a few acres to 20 acres. They are in large numbers so that one day trash is under some cells and rests are empty and then they are replaced as per the demand.
- **Leachate collection system** – The lowermost layer of each landfill is designed in such a way that the liquid under the soil is absorbed into the ground, called a

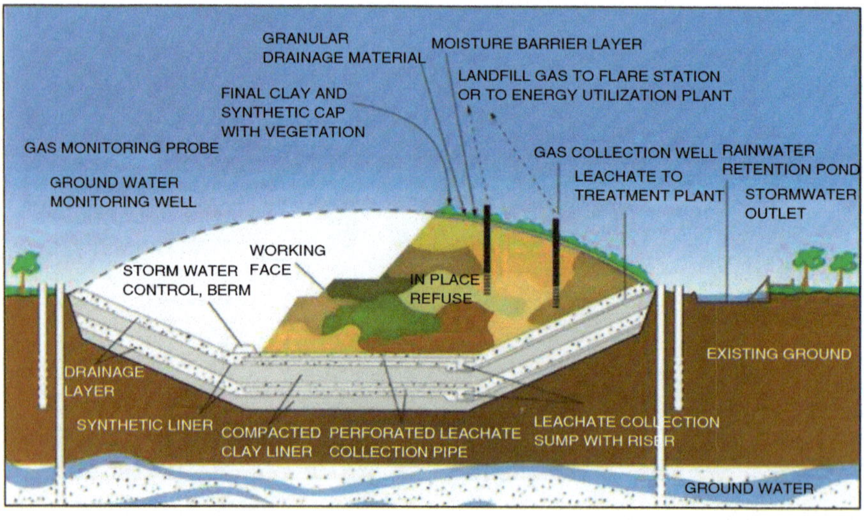

Fig. 6.8 A modern landfill is safer for the environment as waste is not degraded in open condition

cone. The liquid trapped inside the soil dump – known as the wastewater – is collected and removed from the landfill. When the leachate is removed from the dump, it is usually drained or flowed into a tank or catching pond, where it is treated on site or hauled to a public area where waste disposal is discussed.

- **Drainage channel**- This is software driven and designed to control the flow of water during rain or storms. This is done by guiding the rider through a series of holes in the holding areas such as clay ponds. In these ponds the water flow is reduced or held long enough for the suspended particles of soil to function before the water is released into the area.
- **Methane collection unit** -The major gas released from landfills is methane, the gas is highly inflammable and explosive. It must be removed from the landfill site. For the removal of methane various pipe are placed in succession inside the landfill to collect methane gas. After the collection either it can be used or disposed of.
- **Coverings** – A thick layer of approximately 6 inches compacted soil or foam or metal-retardant fiber layer may be used to cover the waste present in a landfill. These are positioned over the drain at the end of each day and removed the day before the waste is deposited. Capping is designed to isolate air debris, insects, birds and rodents and control odor. Once the landfill stage is completed, it is permanently covered with plastic, covers composted soil and a surface layer that supports the growth of the vegetation to prevent erosion.
- **Groundwater stations** – These stations are planned for assessment of ground water around the land availability of leachate chemicals.

6.3.2 Thermal Treatment Methods

6.3.2.1 Incineration

Once the valuable components from the waste are treated by different techniques to convert them into manure or other valuable things the rest unusable and non-recyclable waste like, e-waste, hazardous and medical waste, are treated by incineration. In this technology, rubbish is combusted at a high temperature in a unique furnace called *Incinerator*. Incineration also decreases the quantity of waste product as after the burning the waste product is converted into ash which can then be dumped into the landfills. The capacity of different incinerators may vary from 50 to 1000 tons per day to about 100–3000 tons per day (Nie 2008).

Advantages of Incineration

No or little pretreatment is required for this process. All types of refuse and rubbish can be combusted through this technique. It is the most extensively used and effective treatment method. Volume of waste is considerably reduced and energy recovery is maximum. The only disadvantage is that it involves a large amount of cost into its operations.

Types of Incinerators

1. **Mass-burn or rotary kiln incinerator**

Solid refuse is dropped down in the incinerator by trucks where it is dropped into a hopper. The waste is gradually slide down into the furnace at a high temperature of about 750 °C. The heat produced in this process is used to generate electricity through turbine generator. The remaining fly ash is then run over on to an electromagnet to take out any metallic refuse for recycling and the other gases are passed through the electrostatic precipitators or scrubbers for the treatment before these are released into the atmosphere through chimneys (Chen and Chen 2001; Ryan et al. 1996; Zjup et al. 2019; Lombardi et al. 2013; Zhang et al. 2019).

2. **Fluidized-bed incinerator**

Fluidized bed incinerators are used both the industrial and sewage sludge incineration processes. The principle of fluidization is the process where a granular material in a solid state is turned into a fluid-like state by passing a fluid (liquid or gas) through it. As the fluid passes through the granular material, drag forces will overcome the force of gravity and the material will expand as the particles are moved away from each other. Eventually, as the fluid and material spread farther apart, the velocity slows and the force of gravity overcomes the drag forces, causing the particles to fall back down until they begin to fluidize again. A line of limestone or sand is layered in this incinerator to operate at high temperatures, a perfect air circulation is increase the heating that results in the bubble formation and all the solid matter is converted into fluid like structure and that introduce the term 'fluidized' (Malerius et al. 2003). These are of two types, bubbling bed fluidized incinerator. The difference between the two is in air flow and bed material, types of waste treated, and the energy recovery system. They are widely used in Japan and have a capacity between 50 and 150 tons per day (Caneghem et al. 2012; Chandel and Alappat 2002).

3. **Liquid injection incinerators**

Liquid waste or fuel oil is sprayed using an atomizing medium (compressed air or steam), and the liquid is sprayed to form a mist with an average droplet diameter of 100–150 μm. The atomization of liquid wastes by means of a high-performance injector is the most important factor for the oxidative decomposition of liquid wastes in an incinerator. Liquid injection incineration is an easy and approachable means for removing organic liquid waste completely (Galek et al. 2018; Kramlich 1990).

4. **Sludge incineration or multiple hearth incinerator**

This incinerator is used for the treatment of all types of sewage waste and sludge from wastewater treatment plants. Nontoxic and nonpathogenic residues are resulted after the treatment process. Multiple hearth Incinerator is very much useful if less space is available and composting is not possible (Lewis 2004).

5. **Catalytic combustion incinerator**

These incinerators deploy a line of some catalyst bed to enhance the total combustion process. By using the catalyst, the combustion can be done at lower temperature than in other thermal incinerators. The general waste treated by catalytic combustion are volatile organic chemicals (Ojala et al. 2005; Liab et al. 2009a).

6.3.2.2 Plasma Gasification

Plasma gasification is an additional form of solid waste management. Plasma is a highly ionized gaseous state which are electrically charged, this may produce the temperature upto 12,000–12,600 °F. in this technique, a container is having the typical plasma sources that can generate a temperature of about 10,000 °F that produce a gasification zone till 3000 °F (Zhovtyansky and Valinčius 2018) by this gasification any solid or liquid refuse material is transformed into a syngas., the product's chemical bonds cleaved because of extensive heat in the container. Hazardous waste can be disposed of easily and safely by this method. Waste to energy is also recovered effectively during this process. Expired medicines, barred insecticides, plastic waste, biomedical waste, etc. are treated effectively by plasma pyrolysis. It is already working fine in several countries across the globe: like Australia, USA, in Europe (France, Japan, Germany, Switzerland,) Asia (India, China, Belarus, etc.). We can convert the all recyclable plastic waste into fuel by using plasma gasification (Byun et al. 2012).

6.4 Disposal of Hazardous Waste

Hazardous wastes are those refuse which are harmful to the environment and the health of living beings. These materials are noxious, highly combustible, explosive, may act as carcinogens, teratogenic and mutagenic agents. The sources of hazardous refuse are mines, household and small industrial wastes. Nuclear weapons, nuclear power plants, radioactive materials and hospitals are also potential sources of hazardous waste (Wolbeck 1983). Effective management of hazardous trash is necessary to save the environment and human health.

The other category of hazardous waste is E-waste, all household electrical and electronic items are classified as e-waste for example unused mobile phones, cameras, television sets, radio sets, fax units, photocopy machines, compressed disc players, printers and its accessories. Electrochemical batteries, single use batteries, watches, computers and its accessories, other heavy electronic items of household like washing machines, fridge, oven. Several harmful and toxic heavy metals, acids, noxious chemicals and nonbiodegradable plastics are other hazardous waste materials that need to be managed wisely and safely (Sivaramanan 2013).

Safe and effective disposal of hazardous waste is a major concern in any country and various rules and regulations are required for the management. The complete management of hazardous waste starts with the collection and identification then separation and segregation of the radioactive wastes/e-waste is done. Their handling, treatment and disposal, storage, transport, and final disposal are the basic steps. Many toxic materials are discarded, burnt or sold to recyclers. The best methods for the management of hazardous waste is incineration, landfills and recycling (Saleh 2016).

6.5 Waste Minimization and Recycling

The best and easiest method for solid waste management is to minimize the generation of trash materials form the source itself, so that minimum waste going to landfills. Reduction in the quantity of waste is possible by recycling and reusing of old and unused materials as an alternate to buy a new product. We should not consume the single use products like plastic bags, redesigning of single used products is a necessary step. Environmental Protection Agency, U.S. EPA has given a standard hierarchy of waste management process (Nandan et al. 2017).

6.5.1 Recycling

One of the best methods in waste management is recycling and composting. In comparison to composting, recycling is extensively popular globally (Jackson et al. 1998). Plastic, metal, glass, cloth and paper are the most recyclable items. 5R principle for waste disposal and recycling is a guiding principle in waste management.

6.5.2 Reduce

We should minimize the use of goods and services. Irrational use of electricity should be avoided by saving electricity should be avoided so that our precious natural resource coal can be preserved. The production of waste can be minimized by distinct individual efforts and its originating place like unnecessary purchasing, shopping and storage must be reduced and avoided.

6.5.3 Reuse

One thing may be used again and again for various purposes like carry bags may be reused, old clothes must be shared with the needy one, plastic vessels can be reused for storage of various things etc. old toys and stationery must be donated to NGOs. Used cards can be converted into beautiful envelopes etc. By this we can lessen the demand for new goods and the natural resources can be saved sustainably.

6.5.4 Refuse

We should refuse unnecessary goods and services. People should be encouraged to carry a cloth bag instead of using the plastic bags provided by shopkeepers. Reduction in demand will lead to less production.

6.5.5 Repair

Repairing old broken goods and using them will reduce the demand for new goods. Repairing old furniture is ecofriendly as well as cost effective.

6.5.6 Advanced Recycling Processes

6.5.6.1 Recycling of Used Concrete and Iron Bars

Whenever an old building is collapsed concrete and iron bars can be recycled and converted into the new one and then it can be used a fresh. Though this process requires a lot of energy but that is worth and less as compare to create the new materials (Jassim 2017).

6.5.6.2 Recycling of Consumed Cigarettes

Though smoking is injurious to health, yet it crates ample of rubbish. Currently, used cigarette butts can be recycled to convert them into plastic shipping pallets. One company Terracycle in US has collaborated with urban local bodies in the collection of used butts and to recycle them (Terracycle).

6.5.6.3 Recycling of Plastics by 3-D Printing

3-D printing technology is a new innovative technique in the engineering and that is helping the world a lot in managing the waste. By this technology plastic can be recycled and designed in the desired way. Old toys, plastic goods all can be remoulded by 3-D printing (Kreiger et al. 2014).

6.5.6.4 E-Waste Recycling Innovations

E-waste is a huge problem across the globe because of its toxic, nondegradable and harmful nature to living beings (Aldierie et al. 2019). The best way is to manage e-waste is to re-cycle them. Apple's new recycling robot "Liam" is an innovation that specialized in new dismantling technologies (apple.com). It separates all individual parts, plastic and chips of all e-waste in scientific manner and send them for recycling in efficient way fully automatically, therefore no harmful side effect on any human.

6.5.6.5 Prohibitions on Nonrecyclable Goods

All non-recyclable and single used materials and goods must be banned for use, like plastic carry bags and single use plastic bottles. These regulations compel the consumer to use more recyclable and reusable goods so that the environment can be safe, and the natural resources remain sustainable.

6.5.6.6 Bioplastics and Conversion of Plastic into Liquid Fuel

Various types of natural products like starch, banana peels, sugar cane bagasse etc. are using to convert them into usable and harmless plastic that can be degraded easily into the nature (Mehlika et al. 2017). The other technology which is gaining its popularity day by day is the conversion of used polythene and plastic polypropylene, polyethylene and polyethylene ethylene terephthalate into usable liquid fuel. This can be done by catalytic pyrolysis (Tushar et al. 2013; Chandran et al. 2020).

6.5.6.7 General Methods

The other methods we can apply for waste minimization are All the organic contents can be converted into sugar or proteins. Food waste must be reduced by using better storage and transport systems so that the minimum food is spoiled. The food waste can be stopped by providing organic waste management programs education to the consumers. Maximum construction waste and paper should be reused for saving natural resources and energy requirement. The used plastic and other construction waste may be used in road construction and all used refuse is also using to convert them into bricks. Similarly fly ash from the power plants is also converting into bricks. Better policies and practices must be formalized to use more and more waste management techniques may be used effectively. Developing countries must be provided with financial help to grow and utilize the waste management systems. By using the new recycling technologies use of green practices will be more in the world and the world can be safer and with less waste.

6.6 Waste to Energy: Recovery of Energy

This is a very much necessary step in waste management. In this process non-recyclable substances when incinerated are converted into usable form of energy, which is either electricity or fuel or usable gases. This is a kind of non-conventional energy as non-recyclable waste can be used again and again for long duration of time to generate energy. This is also helpful in lowering the carbon emission. As per EPA, it is one of the cleanest sources of energy (Psomopoulos et al. 2009). Waste to energy can be captured through several processes such as incineration, gasification, pyrolysis, anaerobic digestion, and landfill gas recovery (Figs. 6.9, 6.10, 6.11, 6.12, 6.13, and 6.14).

6.7 Waste Generation in India and Disposal Practices

India is a giant country divided into 29 States and 7 Union Territories (UTs). Around 31.2% inhabitants are residing in metro cities and towns. More than 377 million town people are residing in 7935 cities. There are three mega cities -Greater Mumbai, Delhi, and Kolkata having population of more than 10 million, 53 cities have more than one million population, and 415 cities having population 100,000 or more (Census 2011) India produces more than 160 thousand metric tons of solid waste every day (source: Swachhta Status Report 2016, NSSO). The metro cities produce 200 gm to 600 gm of refuse on per capita basis every day. As per FICCI data, 37% (out of 22 major cities surveyed) generates over 1000 tons per day (Ahmedabad, Delhi, Mumbai, Jaipur, Kanpur, Lucknow) (Pune, and Surat) 13.6% (3 of 22) Cities produce waste between 500–1000 tons per day (Indore, Ludhiana and Vadodara). Rest cities 63.6% (14 of 22) throw their rubbish at landfill sites, including Ahmedabad and Surat. Urbanization is another factor that is straightly related to waste production (Vaish et al. 2016; Kumar and Chakrabarti 2015). The waste production in town areas of India will be 700 gm individually per day in 2025 (Fig. 6.15b). The amount of solid waste generated in cities of India is over 62 million tonnes annually, and that may increase with a growth rate of 5% per year. By 2041, waste production tends to be increased up to 161 lakh tonnes with the population growth (Pappu et al. 2007; Arora 2019). Solid waste is one of the chief

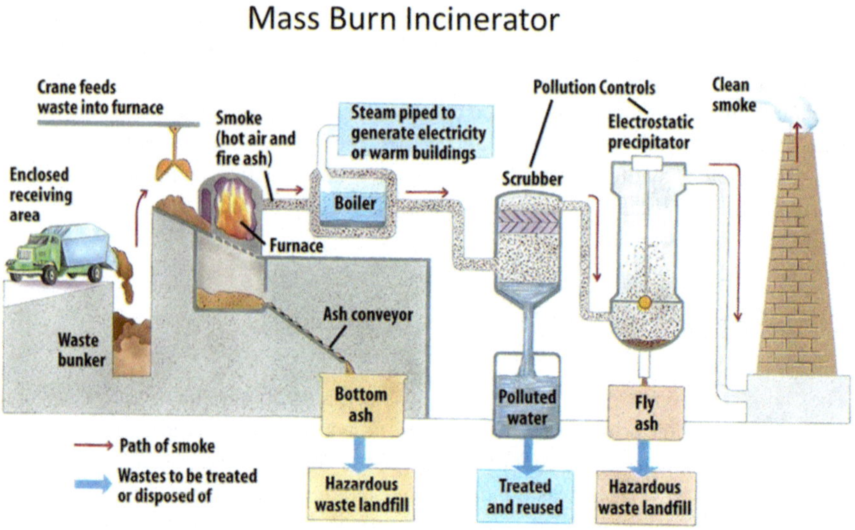

Fig. 6.9 Rotary kiln incinerator
It is used to dispose non hazardous waste

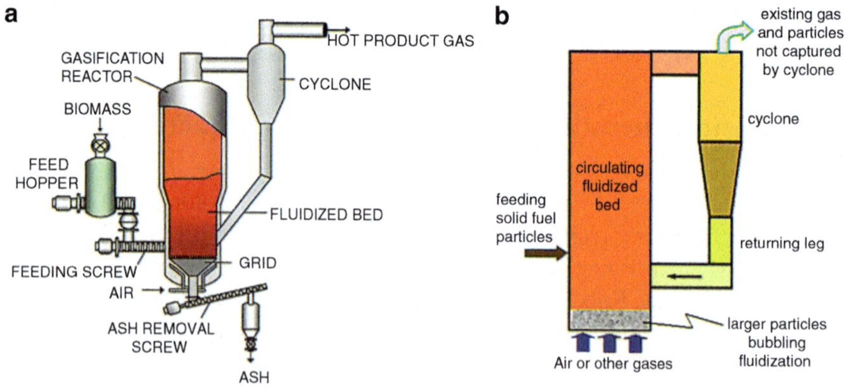

Fig. 6.10 (**a**) Bubbling bed fluidized Incinerator: Fluidized bed combustion (FBC) is utilized to combust solid fuels/waste like petroleum coke, various oil shales, coals and other wastes coming out of fuels (**b**) Circulating bed fluidized Incinerator

environmental problem of Indian metro cities (Fig. 6.16b). Urbanization and industrialization rise are having their side effects and one them is large amount of waste generation (Agarwal et al. 2015). The solid waste is not only harmful to the environment, but it also affects the health of all living beings. The data provided by CPCB (Central Pollution Control Board 2016). Currently, 1.30 lakh of MSW is generating in India (Fig. 6.16a) every day. Domestic, industrial and commercial activities generate many types of waste every day (Fig. 6.17) (Gupta et al. 1998).

Data shows that solid waste from Indian cities comprises around 40–65% organic waste (Fig. 6.17), which can easily be degraded and converted into compost or can

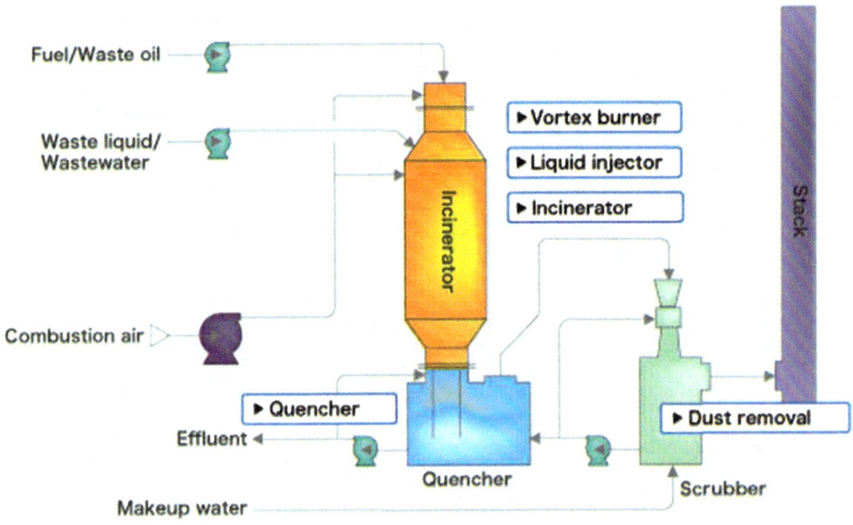

Fig. 6.11 Liquid injection incineration
It is a reliable and friendly means for removing organic liquid waste

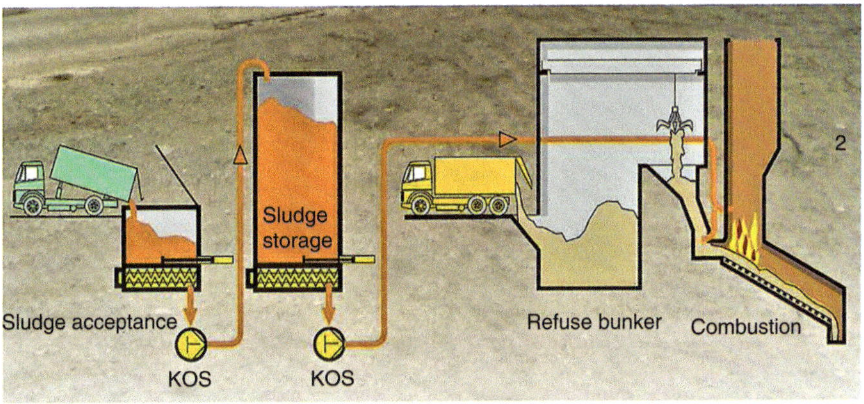

Fig. 6.12 Sludge incinerator
It treats all types of sewage waste and sludge from wastewater treatment plants

be utilized for producing biogas (Bag et al. 2015; Devi et al. 2016; Kalyani and Pandey 2014; Kaushal et al. 2012; Kumar and Samadder 2017; Mani and Singh 2016; Narayana 2009). Recycling and reuse practices must be increased as these are the major challenges in India to follow. Segregation of waste is another challenge in India. The color-coded containers must be provided to all residential societies and publicized to encourage segregation and recycling practices. The role of urban local bodies is also solicited for the effective waste management in India (Fig. 6.18).

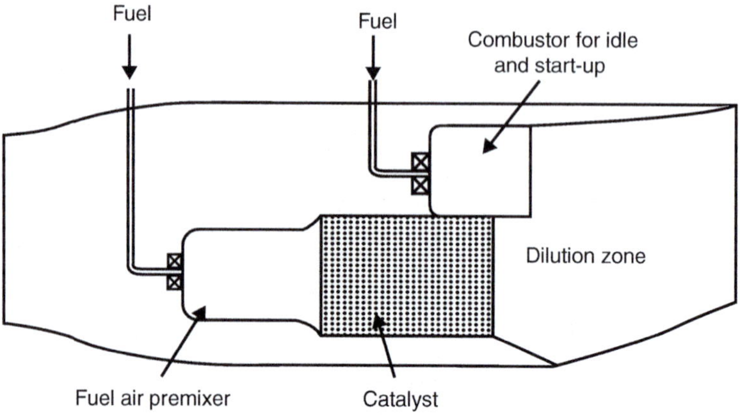

Fig. 6.13 Catalytic combustion incinerator
It uses a lining of catalyst that simplifies the overall combustion reaction for solid wastes

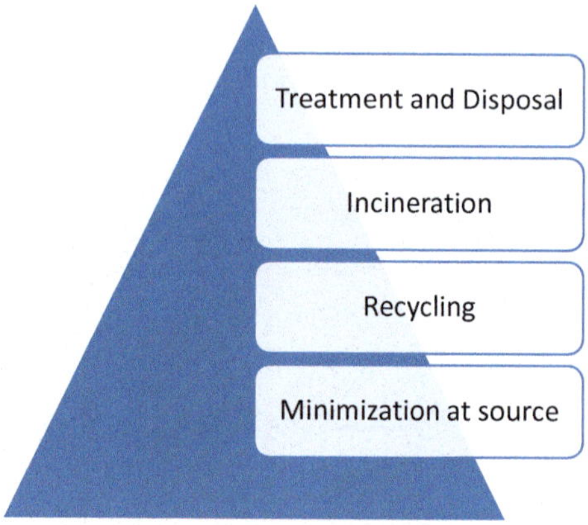

Fig. 6.14 Triangle of waste management
This shows the standard hierarchy of waste management that shows the minimised use of products for less waste generation

6.7.1 Role of Ragpickers and Kabadiwalas in Waste Disposal

Rag picking is one of the business in India for many people who earn their bread and butter from this practice. They have a significant role in the solid waste management in India. These rag-pickers are converting wealth to waste by collecting the materials from the garbage which are having resale value. They collect the dumped plastic, metal and paper waste and sell these items to scrap dealers or directly to recycling units. The rapicking is not only useful in waste management but also a helping process for the environment and society (Chandramohan et al. 2010). Apart from rag pickers one more informal segment working in India known as "kabadi system" or scrap dealers. They also help in managing the solid waste and help the municipalities in true sense without any extra cost. The collection and segregation are done at their end and the waste is finally sold to recycling units. This system performs over

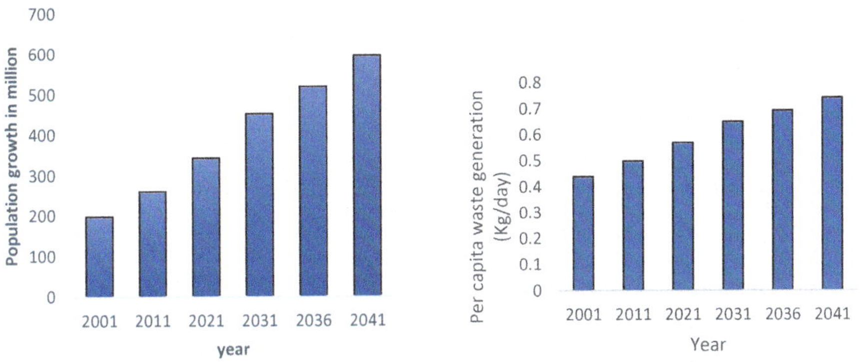

Fig. 6.15 Growth rate of the Indian population in million and per capita waste generation in last forty years

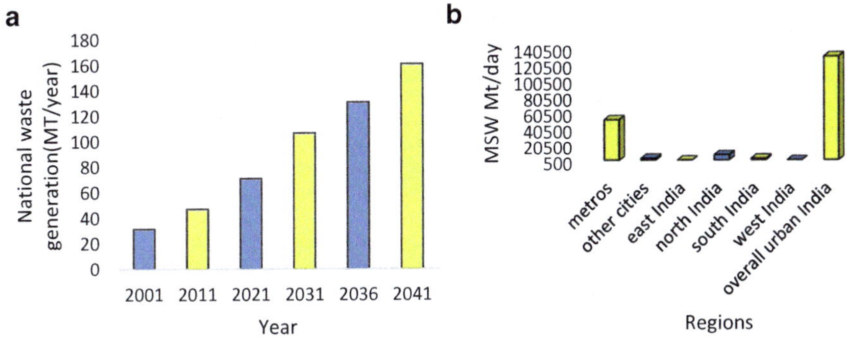

Fig. 6.16 (**a**) National waste generation growth in forty years annually since 2001: PIB 2016. (**b**) Total waste (Mt/day) generated per day

70 to 75% of recycling of goods in. by a rough estimate, ragpickers and scrap dealers save approximately 14% of the civic budget annually (Pooja and Shelar 2019).

6.8 Predictions on Future Waste Growth

The worldwide production of waste is over 2.01 billion metric tons presently. As per world bank global waste generation will rise to 3.40 billion metric tons by 2050. The main contributing countries are China and India and other countries in Asia. Approximately 40% of global waste generated is not managed properly worldwide and this is directly thrown into the open landfills or burnt in open. The estimate for 2025 is that solid waste will be increased to 4.3 billion urban residents, generating about 1.42 kg/capita/day of municipal solid waste (2.2 billion tonnes per year) The World Bank, What a waste, A global review of Solid Waste Management. Urban

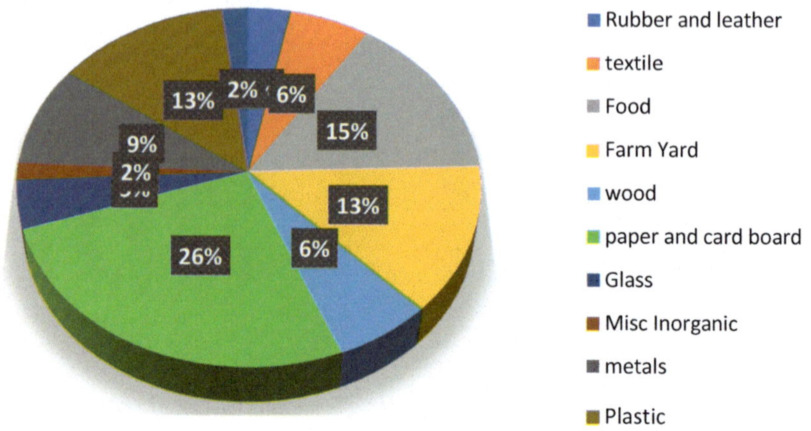

Fig. 6.17 In India organic waste is approximately 65%, and the rest is non-degradable

Fig. 6.18 ISWM: (Integrated solid waste management)
This figure is showing the current waste management approaches which all can be integrated together to give a more sustainable waste management approach

Development Series, No.15 (March 2012). Hoornweg, Bhada-Tata, and Chris Kennedy in 2013 predicted that if traditional industrialization practice as in normal pace as of now, solid waste production will be triple as of today to reach over 11 million tonnes per day by 2100. Few cities in the world are setting an example like San Francisco has aimed "zero waste" by 2020 by using all waste management techniques specially using the recycling vigorously [EPA]. Globally around 55% of its waste is recycled or reused today. Japan is another example where the waste management system is running in very sophisticated manner and that is the best example for the rest world. All kind of waste from polystyrene to all small packaging is segregated and then recycled efficiently (Weforum 2019).

6.9 Integrated Solid Waste Management

Integrated Solid Waste Management is referred to integrate various strategies to manage and minimize the waste sustainably (Marshall and Farahbakhsh 2013). Integrated solid waste management started in the 1970s can be well-defined as technical problems with technical solutions, the endeavour of Integrated solid waste management is to make a balance between three segments of waste management: these are financial affordability, eco-friendly approach and social acceptability redrawn from (van de Klundert and Anschutz 2001; McDougall et al. 2001). United Nations Environmental Programme(UNEP) has defined integrated waste management as a framework of reference for designing and implementing new waste management systems and for analysing and optimising existing systems. The chief objective of integrated solid waste management is to recover maximum energy and materials from the waste management technologies in numerous sectors for technical, ecological, community health and financial sustainability. Integrated solid waste management approach is the waste reduction, collection and recycling, composting and disposal system. Integrating solid waste management is the need of the hour in today's scenario to maintain the energy and natural resources in a sustainable approach. Many countries, including Germany, Japan, Austria, Denmark, Netherlands, and Singapore, have almost abolished landfilling by a blend of composting, recycling, and waste to energy techniques in integrated solid waste management system.

6.10 Conclusion

The effective management of waste is a challenging process for our society. It is imperative to manage the waste in such a way so that it does not cause any harm to any of the environmental components. The ideal steps of waste management are collection, sorting, treatment and disposal must be followed without the creating the other problems such as foul odour, noise, and pollution. Waste is accumulated due the inappropriate methods of handling, transportation or storage. Improper or informal disposal of solid waste may cause adverse effects on the environment as well as on the human health. The management of solid waste necessitates the use of suitable infrastructure, strict regulations, regular maintenance and commitment of working staff. The major challenges in the waste management is usually the cash crunches in the local urban bodies to manage the waste properly. A well-organized solid waste management scheme is currently as significant and vital as other essential services such as food, electricity, construction, markets, airports, and highways. Various techniques like reuse, recycling, composting and waste to energy from incineration are the favoured and efficient techniques than the landfills. Advanced and innovative recycling approaches can save the earth for further loading from waste.

References

Abdel-Shafya HI, Mansour MSM (2018) Solid waste issue: sources, composition, disposal, recycling, and valorization. Egypt J Pet 27(4):1275–1290

Agarwal V, Mathur V (2008) Urban solid waste management in India- 'Swachta Doot Aple Dari' a case study of Nagpur City. J Waste Soc 4 Context EBSCO Publishing, pp X1–X10, 10p. http://search.ebscohost.com

Agarwal R, Chaudhary M, Singh J (2015) Waste management initiatives in India for human well being. Eur Scientif J Spl Ed ISSN: 1857–7881

Ahmeda S, Alib M (2004) Partnerships for solid waste management in developing countries: linking theories to realities. J Habit Intl 28:467–479. www.elsevier.com

Ahsan N (1999) Solid waste management plan for Indian mega cities. Ind J Environ Protect 19(2):90–95

Aldieri L, Ioppolo G, Vinci CP, Yigitcanlar T (2019) Waste recycling patents and environmental innovations: an economic analysis of policy instruments in the USA, Japan and Europe. Waste Manage 95:612–619. https://doi.org/10.1016/j.wasman.2019.06.045

Anonymous (1969) Refuse disposal in Greater London (anon.) published by the Greater London Council, London

Argun Y, Tunç A, Çalışır U, Kılınç N (2017) Composting as a waste management method. J Intl Environ Appl Sci 12:244–255

Arora PK (2019) Microbial technology for the welfare of society. Book Springer Singapore, Singapore

Bag S, Dubey R, Mondal N (2015) Solid waste to energy status in India: a short review. Int Daily J 2(11):283–289

Bagchi A (2004) Design of landfills and integrated solid waste management. ISBN 0471254991. Wiley & Sons, Incorporated, John (Book)

Barles S (2014) History of waste management and the social and cultural representations of waste. In: Agnoletti M, Neri Serneri S (eds) The basic environmental history, Environmental history, vol 4. Springer, Cham

Bindschedler S, Cailleau G, Verrecchia E (2016) Role of Fungi in the biomineralization of calcite. Minerals 6(2):41. https://doi.org/10.3390/min6020041

Byun Y, Cho M, Hwang S-M, Chung J (2012) Thermal plasma gasification of municipal solid waste (MSW). Gasification for practical applications. Intech Open. https://doi.org/10.5772/48537

Caneghem J, Brems A, Lievens P, Block C, Billen P, Vermeulen I, Dewil R, Baeyens J, Vandecasteele C (2012) Fluidized bed waste incinerators: Design, operational and environmental issues. Prog Energy Combust Sci 38:551–582. https://doi.org/10.1016/j.pecs.2012.03.001

Census (2011) India-Census of India Website: Office of the Registrar General & Census Commissioner, India (censusindia.gov.in)

Chandel M, Alappat BJ (2002) Fluidized bed incineration of hospital waste. Sustainable environmental sanitation and water services. In: Proceedings of the 28th WEDC conference

Chandramohan A, Ravichandran C, Sivasankar V (2010) Solid waste, its health impairments and role of rag pickers in Tiruchirappalli city, Tamil Nadu. Southern India. Waste Manag Res 28(10):951–958. https://doi.org/10.1177/0734242X09352705

Chandran M, Tamilkolundu S, Murugesan C (2020) Conversion of plastic waste to fuel. Plastic Waste Recycl:385–399. https://doi.org/10.1016/b978-0-12-817880-5.00014-1

Chen W-H, Chen J-C (2001) Combustion characteristics and energy recovery of a small mass burn incinerator. Intl Commun Heat Mass Transfer 28:299–310

CPCB (Central Pollution Control Board) (2016) Central pollution control board (CPCB) bulletin, Government of India. Accessed 3 Dec 2016. http://cpcb.nic.in/openpdffile.php?id=TGF0ZXN0RmlsZS9MYXRlc3RfMTIzX1NVTU1BUllfQk9PS19GUy5wZGY

Devi K, Swamy AVV, Nailofer S (2016) Municipal solid waste management in India: an overview. Asia Pacific J Res 1(XXXIX):118–126

Diaz LF, Savage GM (2007) Chapter 4: Factors that affect the process. In: Diaz LF, de Bertoldi M, Bidlingmaier W, Stentiford E (eds) Waste management series, vol 8. Elsevier, pp 49–65, ISSN 1478-7482, ISBN 9780080439600. https://doi.org/10.1016/S1478-7482(07)80007-8

Erhard H (1964) Refuse and waste disposal, history of city sanitation. Druckhaus Lichterfeld GMBH, Berlin

Galek V, Hrbek J, Hadrava J, Stoklasa J (2018) The process and optimization of the liquid waste incineration in mso technology. Conference: 6th international conference on chemical technology, Mikulov

Guo R, Li G, Jiang T, Schuchardt F, Chen T, Zhao Y, Shen Y (2012 May) Effect of aeration rate, C/N ratio and moisture content on the stability and maturity of compost. Bioresour Technol 112:171–178. https://doi.org/10.1016/j.biortech.2012.02.099

Gupta S, Mohan K, Gupta S, Kansal A, Prasad R (1998) Solid waste Management in India: options and opportunities. J Resour Conserv Recycl EBSCO Publishing 24(2):1370. http://search.ebscohost.com

Gupta S, Choudhary N, Alappat BJ (2007) Bioreactor landfill for MSW disposal in Delhi. In: Proceedings of the international conference on sustainable solid waste management, Chennai, India, pp 474–481

Gutberlet J (2017) Waste in the city: challenges and opportunities for urban agglomerations. https://doi.org/10.5772/intechopen.72047. Book Chapter in Intech Open

Hamer G (2004) Solid waste treatment and disposal: effects on public health and environmental safety. Biotechnol Adv 22:71–79. https://doi.org/10.1016/j.biotechadv.2003.08.007

Hoornweg D, Bhada-Tata P, Kennedy C (2013) Environment: waste production must peak this century Nature 502:615. https://doi.org/10.1038/502615a

Hosch K (1967) From donkey-drawn wheelbarrows to sanitary landfills. Stadthygiene 18(10):228–231. Uelzen/Hamburg

https://www.apple.com/environment/pdf/Liam_white_paper_Sept2016.pdf

https://www.epa.gov/transforming-waste-tool/zero-waste-case-study-san-francisco

https://www.terracycle.com/en-US/brigades/cigarette-waste-recycling

https://www.weforum.org/agenda/2019/08/the-japanese-have-a-word-to-help-them-be-less-wasteful-mottainai/

Jackson SR, Shepherd P, Dixon M (1998) The recycling of waste. De Montfort University, Flexible Learning Systems, Leicester

Jassim AK (2017) Sustainable solid waste recycling. In: Skills development for sustainable manufacturing. Intech Open Book Chapter. https://doi.org/10.5772/intechopen.70046

Kalyani KA, Pandey KK (2014) Waste to energy status in India: a short review. Renew Sustain Energy Rev 31:113–120. https://doi.org/10.1016/j.rser.2013.11.020

Kaushal RK, Varghese GK, Chabukdhara M (2012) Municipal solid waste management in India—current state and future challenges: a review. Int J Eng Sci Technol 4(4):1473–1489

Kaza S, Yao LC, Bhada-Tata P, Van Woerden F (2018) What a Waste 2.0: a global snapshot of solid waste management to 2050 urban development. World Bank © World Bank, Washington, DC. https://openknowledge.worldbank.org/handle/10986/30317

Kramlich JC (1990) Observations on waste destruction in liquid injection incinerators. Combust Sci Technol 74(1–6):17–30. https://doi.org/10.1080/00102209008951677

Kreiger M, Mulder M, Glover A, Pearce J (2014) Life cycle analysis of distributed recycling of post-consumer high density polyethylene for 3-D printing filament. J Clean Prod 70:90–96. https://doi.org/10.1016/j.jclepro.2014.02.009

Kumar S, Chakrabarti T (2015) Effective municipal solid waste management in India. Waste Management Book. Intech open 125–128. https://doi.org/10.5772/8459
Kumar A, Samadder SR (2017) An empirical model for prediction of household solid waste generation rate: a case study of Dhanbad, India. Waste Manag 68:3–15. https://doi.org/10.1016/j.wasman.2017.07.034
Ladhe T, Magnusson J, Nilsson A (2014) From trash to cash: a case of waste management business model formation. In: Rocha Á, Correia A, Tan F, Stroetmann K (eds) New perspectives in information systems and technologies, Volume 1: Advances in intelligent systems and computing, vol 275. Springer, Cham
Lewis F (2004) Operating strategies to reduce fuel usage in multiple hearth and fluid bed sludge incinerators. Air & Waste Management Association(13) (PDF) Operating Strategies to Reduce Fuel Usage in Multiple Hearth and Fluid Bed Sludge Incinerators (researchgate.net)
Liab WB, Wanga JX, Gonga H (2009a) Catalytic combustion of VOCs on non-noble metal catalysts. Catalysis Today 148(1–2):81–87
Liab WB, Wanga JX, Gonga H (2009b) Catalytic combustion of VOCs on non-noble metal catalysts. Catalysis Today 148(1–2):81–87
Lombardi F, Lategano E, Cordiner S, Torretta V (2013) Waste incineration in rotary kilns: a new simulation combustion tool to support design and technical change. Waste Manage Res ISWA:31. https://doi.org/10.1177/0734242X13484187
Malerius O, Werther J, Mineur M (2003) Mercury removal in the flue gas scrubber of a fluidized-bed sewage sludge incinerator. 17th international conference on fluidized bed combustion
Mani S, Singh S (2016) Sustainable municipal solid waste management in India: a policy agenda. Procedia Environ Sci 35:150–157. https://doi.org/10.1016/j.proenv.2016.07.064
Manyuchi, Anthony (2013) Vermicomposting in solid waste management: a review. Intl J Scientif Eng Technol 2:1234–1242
Marshall RE, Farahbakhsh K (2013) Systems approaches to integrated solid waste management in developing countries. Waste Manage 33(4):988–1003
McDougall F, White P, Franke M, Hindle P (2001) Integrated solid waste management: a life cycle inventory. Intl J Life Cycle Assess 6. https://doi.org/10.1007/BF02978794
Meegoda J, Hettiarachchi H, Hettiaratchi J (2016) Landfill design and operation. https://doi.org/10.1061/9780784414101.ch18
Mehlika K, Richard L, Geoffrey P, Robson D (2017) Abiotic and biotic environmental degradation of the bioplastic polymer poly(lactic acid): a review, ISSN: 0141-3910. Polymer Degrad Stab 137:122–130
Naikwade P, Sankpal S, Jadhav B (2012) Management of waste by composting, vermicomposting and it's use for improvement of growth, yield and quality of fodder maize. Arpn J Sci Technol 2:184–194
Nandan A, Yadav B, Baksi S, Bose D (2017) Recent Scenario of Solid Waste Management in India. World Scientific News 66:56–74
Narayana T (2009) Municipal solid waste management in India: from waste disposal to recovery of resources? Waste Manag 29(3):1163–1166. https://doi.org/10.1016/j.wasman.2008.06.038
National sample survey office (NSSO), Ministry of statistics and programme implementation, Government of India. Swachhta status report (2016) Available at. http://mospi.nic.in/sites/default/files/publication_reports/Swachhta_Status_Report%202016_17apr17.pdf
Nie Y (2008) Development and prospects of municipal solid waste (MSW) incineration in China. Front Environ Sci Eng China 2:1–7. https://doi.org/10.1007/s11783-008-0028-6
Ojala S, Lassi U, Ylönen R, Karjalainen H, Keiski RL (2005) Influence of the heat exchangers' construction material on catalytic incineration. Catal Today 100:367–372
Pappu A, Saxena M, Asolekar S (2007) Solid wastes generation in India and their recycling potential in building materials. Build Environ 42:2311–2320. https://doi.org/10.1016/j.buildenv.2006.04.015
Pooja SD, Shelar A (2019) Role of rag pickers in solid waste management in Nashik City. Intl J Eng Res Technol (IJERT) 8(5)
Priestly JJ (1968) Civilization, water and wastes. Chem Indust:353–363

Psomopoulos C, Bourka A, Themelis N (2009) Waste-to-energy: a review of the status and benefits in USA. Waste Manag 29:1718–1724. https://doi.org/10.1016/j.wasman.2008.11.020
Reddy KR, Hettiarachchi H, Gangathulasi J, Parakalla N, Bogner J, Lagier T (2009) Compressibility and shear strength of landfillled municipal solid waste at Orchard Hills landfill. Waste Manage Res 27(6):578–587
Reinhart DR, Townsend TG (2018) Modern landfill fundamentals. Landfill Bioreactor Design Operat:7–23
Ryan J, Lemieux P, Lutes C, Tabor D (1996) Development of pic target analyte list for hazardous waste incineration processes. U.S. Environmental Protection Agency, Washington, DC
Saleh HE-DM (2016) Introductory chapter: Introduction to Hazardous Waste Management. Management of Hazardous Wastes. Intech Open. https://doi.org/10.5772/64245
Seadon JK (2010) Sustainable waste management systems. J Cleaner Prod 18(16–17):1643–1651
Sharholy M, Ahmad K, Mahmood G, Trivedi R (2008) Municipal solid waste management in Indian cities – a review. Waste Manage (New York, NY) 28:459–467. https://doi.org/10.1016/j.wasman.2007.02.008
Shilev S, Naydenov M, Vancheva V, Aladjadjiyan A (2006) Composting of food and agricultural wastes. https://doi.org/10.1007/978-0-387-35766-9_15
Sivaramanan S (2013) E-waste management, disposal and its impacts on the environment. UJERT 2:531–537. https://doi.org/10.13140/2.1.2978.0489
Stringfellow T (2014) An independent engineering evaluation of waste-to-energy technologies. Renew Energy World. https://www.renewableenergyworld.com/2014/01/13/an-independent-engineering-evaluation-of-waste-to-energy-technologies/#gref
Tanugur I (2009) Aerobic composting of broiler waste with different amendments. M.Sc. thesis. Institute of Science and Technology, Istanbul Technical University. Istanbul Turkey
Terlecky PM (1980) Sanitary landfill. In: Wang LK, Pereira NC (eds) Solid waste processing and resource recovery, Handbook of environmental engineering, vol 2. Humana Press, Totowa
The State of Development of Waste Forms for Mixed Wastes: U.S. Department of Energy's Office of Environmental Management (1999) Chapter: 4. Waste treatment and stabilization
Theodore MK, Theodore L (2010) Introduction to environmental management, 1st edn. CRC Press, Boca Raton
Tushar R, Khan S, Varshney K (2013) Polythene Pollution Management by Converting them into Fuel. Intl J Chem Chem Eng 3:223–224. ISSN 2248-9924
Vaish B, Sarkar A, Singh P, Singh PK, Sengupta C, Singh RP (2016) Prospects of biomethanation in Indian urban solid waste: stepping towards a sustainable future. In: Karthikeyan O, Heimann K, Muthu S (eds) Recycling of solid waste for biofuels and bio-chemicals. Environmental footprints and eco-design of products and processes. Springer, Singapore
van de Klundert A, Anschutz J (2001) Integrated sustainable waste management – the concept. ISBN: 9076639027. The Waste, Netherlands
Vishwakarma M (2012) Recognition of environmental rights as human rights: a sine qua non for peaceful existence of human society. CPJ Law J:II
Wilson DG (1976) A brief history of solid-waste management. Intl J Environ Stud 9(2):123–129. https://doi.org/10.1080/00207237608737618. Published online in 2008
Wolbeck B (1983) Political dimensions and implications of hazardous waste disposal. Hazard Waste Disposal. Challenges of Modern Society, vol 4. Springer, Boston, MA, pp 7–17. https://doi.org/10.1007/978-1-4613-3602-0_3
Zhang B, He J, Hu C, Chen W (2019) Experimental and numerical simulation study on co-incineration of solid and liquid wastes for green production of pesticides. PRO 7:649. https://doi.org/10.3390/pr7100649
Zhovtyansky V, Valinčius V (2018) Efficiency of plasma gasification technologies for hazardous waste treatment. Gasification for low-grade Feedstock. IntechOpen
Zjup W, Jiang X, Li Y, Yan J (2019) Hazardous waste incineration in a rotary kiln: a review. Waste Disp Sustain Energy 1:3–37. https://doi.org/10.1007/s42768-019-00001-3

Chapter 7
Impact of the Invasive *Prosopis juliflora* on Terrestrial Ecosystems

M. Iftikhar Hussain, Ross Shackleton, Ali El-Keblawy, Luís González, and M. Mar Trigo

Abstract Invasive non-native plants are changing ecosystems and native biodiversity, and modifying soil microbial feedback. The invasive species *Prosopis juliflora* (Sw.) D.C. (mesquite) has been introduced into several ecosystems, especially in tropical and subtropical regions, causing economic, ecological and health problems. This article reviews *P. juliflora* ecophysiological and reproductive attributes, such as phenology, vegetative, seed germination and dispersal, allelopathy and invasion mechanisms. We found that *P. juliflora* invasion has negative impacts on native biodiversity, ecosystem structure and function, bulk soil, seed bank, and hydrological cycle. We discuss *P. juliflora* as a ruminant food and for human use, and new management techniques. The easy naturalization *P. juliflora* in tropical regions has been explained by allelopathy, repeated flowering, vegetative propagation, production and dispersal of huge viable seeds. In particular, *P. juliflora* produces allelochemicals that are not produced in close relatives. Such chemicals have facilitative effects on associated vegetation in the native range, but have detrimental effects in the introduced range. Management strategies are presented to control *P. juliflora* invasion.

M. I. Hussain (✉)
Department of Plant Biology & Soil Science, Universidade de Vigo, Vigo, Spain

Research Institute of Science and Engineering, University of Sharjah, Sharjah, United Arab Emirates

R. Shackleton
Centre for Invasion Biology (CIB), Department of Botany and Zoology, Natural Sciences Building, Private Bag X1, Stellenbosch University, Matieland, South Africa

Institute of Geography and Sustainability, University of Lausanne, Lausanne, Switzerland

A. El-Keblawy
Department of Applied Biology, University of Sharjah, Sharjah, UAE

L. González
Department of Plant Biology & Soil Science, Universidade de Vigo, Vigo, Spain

M. M. Trigo
Department of Plant Biology, Universidad de Malaga, Malaga, Spain

E. Lichtfouse (ed.), *Sustainable Agriculture Reviews 52*, Sustainable Agriculture Reviews 52, https://doi.org/10.1007/978-3-030-73245-5_7

Keywords Ecological constrains · Reproductive features · Invasing mechanisms · Prosopis · Adverse effects

7.1 Introduction

Many human activities, such as agriculture, recreation, global trade and transportation have promoted both the intentional and accidental spread of species across their natural dispersal barriers (Kolar and Lodge 2001). A number of introduced species have been established and spread in the new range and are considered pests for agricultural, landscape and horticultural sector as they pose economic threats to these industries (Barbier et al. 2013). Several authors have considered the increasing number of introduced or invasive species as a major component of global change because of their potential to alter social-ecological systems (Richardson et al. 2011; Moodley et al. 2013; Potgieter et al. 2013; Shackleton et al. 2014). Biological invasions are recognized as one of the most important causes of ecosystem degradation, community structure, local species and biodiversity loss worldwide (Wardle et al 2001; Pyšek et al. 2012).

Prosopis L. is a genus with 44 species of medium-sized trees and shrubs belonging to family Fabaceae, subfamily Mimosoideae (Burkart 1976). Naranjo et al. (1984) and Hunziker et al. (1986) have reported the possibility of interspecific hybridization between *Prosopis juliflora* (Sw.) D.C. (Fabaceae) and *P. pallida* (Humb. & Bonpl. ex Willd.) Kunth, which led to the difficulty in differentiating the two taxa. Despite molecular studies were useful in differentiating the two species (Landeras et al. 2006; Sherry et al. 2011), such morphological similarities led to misidentification of *Prosopis* species and consequently all records of introduction of *Prosopis* need to be reevaluated. To avoid this problem, it is often referred to as the *P. juliflora – P. pallida* complex (Pasiecznik et al. 2001). *Prosopis* was found to occur as a native or introduced species in 129 countries worldwide mainly in the hot arid and semi-arid climatic regions of the world (Shackleton et al. 2014). The numerous goods and services provided by *Prosopis* have led to global introductions and have made some species important for local communities (Pasiecznik et al. 2001; Shackleton et al. 2014). At least 19 (invasive and weedy) of the 44 species in the genus *Prosopis* are known to generate benefits and costs, with the rest being primarily beneficial in their native ranges (Shackleton et al. 2014). It has been introduced to different parts of the world for different purposes, but it has become aggressive invader in most introduced ranges. For example, it was introduced to Sudan in 1917 to combat desertification and to provide fuelwood (Elfadl and Luukkanen 2006). Similarly, this species was brought to Lake Baringo, Kenya, in the 1980s to alleviate fuelwood shortage (Mwangi and Swallow 2005). *Prosopis juliflora* (Sw.) D.C. was introduced to India during late nineteenth century for the rehabilitation of sodic lands and to supply of fuelwood, fodder, timber, and fiber (Mishra et al. 2003; Sharma and Dakshini 1996) and into Ethiopia in the 1970s and

1980s mainly for soil and water conservation (Tegegn 2008). In fact, *P. juliflora* is present in IUCN's new list of 100 world's worst invasive alien species (Luque et al. 2014).

Management of the invaded species is inefficient in many areas due to lack of knowledge on key aspects of the invasive species. For proper management of *P. juliflora*, it is important to understand the reasons for introductions, benefits, costs, ecology and scales of invasions (Shackleton et al. 2014; Wilson et al. 2014). All of the used approaches for controlling mesquite were mainly to maximize its benefits and minimize its negative impacts (Wise et al. 2012). Patnaik et al. (2017) highlights the dual role of *P. juliflora* and itemizes the facts that make it a blessing in some contexts and a bane in other contexts with a more realistic view. Previous and recent reviews on genus *Prosopis* have focused on the special attributes that give plants competitive advantage, origin and systematics, spread and distribution, impacts, benefits and management (Patnaik et al. 2017 and Shackleton et al. 2014). This review article synthesizes all aspects of the taxonomy, current distribution, history of introduction and spread, ecological constrains (including preferred climate, substratum and habitats), ecophysiological responses to biotic and abiotic factors, biology (including phenology, vegetative and reproductive biology, seed germination and dispersal). Although, we review why *Prosopis juliflora* is invasive in many arid regions, where it was introduced while it is not invasive in its native distribution environment and areas. This will particularly help to find clues of its invasion mechanism (congeneric and biogeographic approach). Indicating potential trade-offs between economic benefits (ruminants feed and human uses) from biomass use and ecological impacts (aboveground, belowground, water and soil fertility) within a defined landscape, whilst also considering interventions in other habitats, will aid decision making, planners and risk managers about its better management in arid and semi-arid regions including Middle East and North Africa in a sustainable way.

7.2 Current Distribution and Status

Most species of the genus *Prosopis* are native to the Americas, ranging from the southwestern USA, through Mexico and Central America into South America, as far as Argentina (Morgan et al. 2017). It was then introduced to Ethiopia, Kenya, Sudan, Eritrea, Iraq, Pakistan, India, Australia, South Africa, the Caribbean, the Atlantic Islands, Venezuela, Bolivia, Brazil, the Dominican Republic, El Salvador, Nicaragua, the United States (USA), and Uruguay (Abdulahi et al. 2017) where *P. juliflora* is an invasive weed. At present, *Prosopis juliflora* are distributed worldwide, they are more frequently invasive in Mediterranean, tropical and arid climates (Fig. 7.1). Furthermore, Table 7.1, demonstrate the Global distribution of *Prosopis* species.

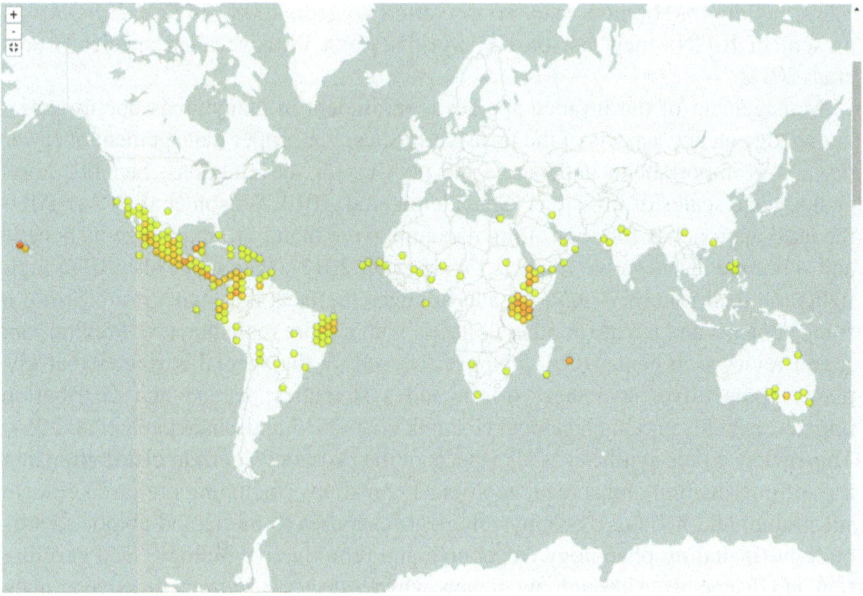

Fig. 7.1 Worldwide distribution of reported *Prosopis juliflora* in different countries

7.2.1 Native Range: Climate, Soil and Habitats

P. juliflora is a shrub native to Mexico, South America and the Caribbean (Burkart 1976; Pasiecznik et al. 2001). It require at least 250 mm annual rainfall, but some have been found in areas with <100 mm (Morgan et al. 2017). Kaur et al. (2012) identified Venezuela as one of the native ranges of *P. juliflora*. Detrimental and beneficial mechanisms do not act in isolation from each other in nature. The relative importance of these two processes in a particular plant community determines the structure of that community (Callaway and Walker 1997). Many *Prosopis* species, in their native ranges, contribute with higher concentrations of organic matter, nitrogen, phosphorus and potassium beneath their canopies and behave as strong facilitators of other species (Tiedemann and Klemmedson 1973). In its native range in Venzuela, *P. juliflora* probably has much stronger facilitative effects on neighbours than other leguminous tree species (Larrea-Alcázar and Soriano 2008).

However, in the introduced range, *P. juliflora* has both facilitaive and allelopathic effects. However, the allelopathic effects of the litters of *P. juliflora* override its potential positive effects on soil fertility. For example, El-Keblawy and Abdelfatah (2013) assessed the impact of *P. juliflora* on the soil and associated flora in the natural habitats of the United Arab Emirates and concluded that this species has facilitative effects in relation to the available nutrients (i.e., most important macro-nutrients K, N and P were increased). In addition, *P. juliflora* increased the organic matter content, which would also increase the water holding capacity that would improve soil texture and increase soil moistures. However, the negative effect of *P. juliflora*

Table 7.1 Global distribution of *Prosopis* species

Country	*Prosopis* species status	References
Afghanistan	*P. cineraria* (NA); *P. farcta* (W) *P. juliflora* (N)	Sohrabi et al. (2011) and GBIF (2013)
Algeria	*P. farcta* (NA), *P. juliflora* (I)	Habit et al. (1990), Pasiecznik et al. (2001), and ISSG (2005)
Angola	*Prosopis* spp. (N)	Poynton (2009)
Antigua and Barbuda	*P. juliflora* (N)	Johnston (1962)
Argentina*	*P. abbreviata* (NA), *P. affinis* (W), *P. alba* (NA), *P. algarobilla* (NA), *P. alpataco* (NA), *P. argentina* (NA) *P. caldenia* (W), *P. calingastana* (NA), *P. camperstris* (W), *P. castellanosii* (NA), *P. chilensis* (NA), *P. denudans* (NA), *P. elata* (NA), *P. ferox* (NA), *P. fiebrigii* (NA), *P. flexuosa* (NA),*P. glandulosa* (N), *P. hasslei* (W), *P. humilis* (W), *P. juliflora* (N), *P. kuntzei* (W), *P. laevigata* (NA), *P. nigra* (W), *P. pugionata* (NA), *P. reptans* (NA), *P. rojasiana* (NA), *P. ruizleali* (NA), *P. ruscifolia* (W), *P. sericantha* (NA), *P. strombulifera* (NA), *P. torquata* (NA), *P. tamarugo* (N), *P. vinalilla* (NA), *P. hybrids* (I/W)	CABI (2005) and GBIF (2013)
Aruba	*P. juliflora* (N)	D.C.B (2009)
Ascension Island*	*P. juliflora* (I)	Pickup (1999) and Belton (2008);
Australia*	*P. glandulosa* (I), *P. juliflora* (I), *P. pallida* (I), *P. velutina* (I), *P.* hybrids (I)	Panetta and Carstairs (1989), Osmond (2003), and van Klinken et al. (2006)
Azerbaijan	*P. farcta* (NA)	CABI (2005)
Bahamas	*P. juliflora* (N)	CABI (2005)
Bahrain	*P. farcta* (NA), *P. glandulosa* (U), *P. juliflora* (N)	Pasiecznik et al. (2001) and AFTD (2009)
Bangladesh	*P. juliflora* (N)	CABI (2005)
Barbados	*P. juliflora* (N)	Jonston (1962)
Belize	*P. juliflora* (NA)	AFTD (2009)
Benin	*P. africana* (NA) *P. juliflora* (U)	AFTD (2009) and CABI (2009)
Bermuda	*P. juliflora* (U)	CABI (2009)
Bolivia	*P. alba* (NA), *P. algarobilla* (NA), *P. alpataco* (NA), *P. chilensis* (NA), *P. denudans* (NA), *P. elata* (NA), *P. ferox* (NA), *P. flexuosa* (NA), *P. juliflora* (NA), *P. kuntzei* (NA), *P. laevigata* (NA), *P. nigra* (NA), *P. pallida*(NA), *P. ruscifolia* (NA), *P. vinalillo* (NA)	Pasiecznik et al. (2001), Grandtner (2005), and AFTD (2009)

(continued)

Table 7.1 (continued)

Country	*Prosopis* species status	References
Botswana	*P. chilensis* (I), *P. glandulosa* (I), *P. pallida* (N), *P. juliflora* (N), *P. velutina* (I), *P.* hybrids (I)	Botswana Gov (2009), Poynton (2009), and Muzila et al. (2011)
Brazil	*P. affinis*(N/W), *P. alba* (N), *P. flexuosa* (NA), *P. juliflora* (I), *P. laevigata* (NA), *P. pallida* (N), *P.rubriflora* (NA), *P. ruscifolia* (N)	Pasiecznik et al. (2001), Leão et al. (2011), and de Olivera et al. (2012)
Brunei Darussalam	*P. juliflora* (U)	Pasiecznik et al. (2001) and AFTD (2009)
Burkina Faso	*P. africana* (NA), *P. juliflora* (I)	Ræbild et al. (2003), Weber et al. (2008)
Cambodia	*P. juliflora* (U)	AFTD (2009)
Cameroon	*P. africana* (NA)	Pasiecznik et al (2001) and AFTD (2009)
Cape Verde	*P. glandulosa* (N), *P. juliflora* (I), *P. pallida* (I) *P. velutina* (N), *P.* hybrids (I)	FAO (2006), AFTD (2009), and Cienciala et al. (2013)
Cayman Islands	*P. juliflora* (N)	CABI (2005)
Central African Republic	*P. africana* (NA)	AFTD (2009)
Chad	*P. africana* (NA), *chilensis* (N), *P. juliflora* (I)	Pasiecznik et al. (2001), ISSG (2005), Geesing et al. (2004)
Chile	*P. alba* (NA), *P. alpataco* (NA), *P. burkartii* (NA), *P. chilensis* (W),*P. flexuosa* (NA), *P. fruticosa* (NA),*P. laevigata* (NA), *P. strombulifera* (NA), *P. tamarugo* (W)	Pasiecznik et al. (2001), AFTD (2009), and GBIF (2013)
Colombia	*P. flexuosa* (NA), *P. juliflora* (W), *P. nigra* (W), *P. pallida* (NA)	Grandtner (2005), Vallejo et al. (2012), and GBIF (2013)
Costa Rica	*P. juliflora* (NA)	Burkart (1976)
Cote d'Ivore	*P. africana* (NA), *P. juliflora* (U)	CABI (2005)
Cuba	*P. juliflora* (N), *P. glandulosa* (N)	Johnston (1962), Pasiecznik et al. (2001)
Curacao	*P. juliflora* (N)	Pasiecznik et al. (2001)
Cyprus	*P. farcta* (NA)	Pasiecznik et al. (2004)
Djibouti*	*P. cineraria* (I), *P. juliflora* (I), P. pallida (I),	Pasiecznik et al. (2001)

(continued)

Table 7.1 (continued)

Country	*Prosopis* species status	References
Dominican Republic	*P. juliflora* (N)	ISSG (2010), and Pasiecznik et al. (2001)
Ecuador	*P. chilensis* (NA), *P. juliflora* (NA), *P. pallida* (NA)	Grandtner, (2005), GBIF (2013)
Egypt*	*P. africana* (NA), *P. chilensis* (N), *P. farcta* (NA), *P. glandulosa* (I), *P. juliflora* (I), *P. velutina* (I)	Pasiecznik et al. (2001), Ghazali (2006), Weber et al. (2008)
El Salvador	*P. juliflora* (NA)	Burkart (1976)
Eritrea*	*P. juliflora* (I)	Zimmermman (1991) ISSG (2005), Bokerezion (2008)
Ethiopia*	*P. africana* (NA), *P. chilensis* (I), *P. juliflora* (I), *P. pallida* (I)	ISSG (2005), Berhanu and Tesfay (2006), Shiferaw et al. (2004), FARM-Africa (2008)
Fiji	*P. pallida* (U)	Gallaber and Merlin (2010)
Galapagos	*P. juliflora* (NA), *Prosopis spp.* (NA)	Wiggins et al. (1971)
Gambia	*P. africana* (NA), *P. juliflora* (I)	Pasiecznik et al. (2001)
Georgia	*P. farcta* (W)	Pasiecznik et al. (2001)
Ghana	*P. africana* (NA), *P. juliflora* (I)	Pasiecznik et al. (2001)
Guam	*P. pallida* (N)	Fosberg et al. (2013)
Guatemala	*P. juliflora* (NA)	Pasiecznik et al. (2001)
Guinea	*P. africana* (NA)	Pasiecznik et al. 2001
Guinea-Bissau	*P. africana* (NA), *P. juliflora* (I)	AFTD (2009)
Haiti	*P. chilensis* (U), *P. flexuosa* (IP), *P. juliflora* (I), *P. velutina* (N)	Burkart (1976), Lee et al. (1992), and Timyan (1996)
Hawaii*	*P. juliflora* (I), *P. pallida* (I), *P. hybrids* (I)	Kaur et al. (2012)
Honduras	*P. juliflora* (NA)	Burkart (1976)
India*	*P. cineraria* (W), *P. farcta* (NA), *P. glandulosa* (I), *P. juliflora* (I), *P. pallida* (N), *P. velutina* (N), *P.* hybrids (I)	Pasiecznick et al. (2004), CABI (2005), FAO (2006), AFTD (2009), Kaur et al. (2012);
Indonesia	*P. juliflora* (U)	Pasiecznik et al. (2001), CABI (2005)

(continued)

Table 7.1 (continued)

Country	*Prosopis* species status	References
Iran*	*P. cineraria* (W), *P. farcta* (W), *P. juliflora* (I), *P. koelziana* (NA)	Pasiecznik et al. (2001), AFTD (2009), Sohribi et al. (2011), Sajad and Sefidi (2012)
Iraq	*P. farcta* (W), *P. juliflora* (I)	Burkart (1976), Berhanu and Tesfay (2006)
Israel	*P. farcta* (W), *P. glandulosa* (N), *P. juliflora* (I), *P. pallida* (N)	Pasiecznik et al. (2001), CABI (2005)
Jamaica	*P. juliflora* (N)	Pasiecznik et al. (2001)
Jordan*	*P. farcta* (W), *P. glandulosa* (N), *P. juliflora* (I), *P. pallida* (N)	Pasiecznik et al. (2001), and Quasem (2007)
Kenya*	*P. africana* (NA), *P. chilensis* (I), *P. juliflora* (I), *P. pallida* (N)	Choge et al. (2002); Mwangi and Swallow (2005), FAO (2006), and Maturi (2013)
Kiribati	*P. pallida* (U)	Fosberg et al. (1976), GISD (2013)
Kuwait	*P. glandulosa* (N), *P. juliflora* (I)	Burkart (1976)
Laos	*P. juliflora* (N)	AFTD (2009)
Lesotho	*Prosopis spp.* (N)	Poynton (2009)
Liberia	*P. africana* (NA), *P. juliflora* (N)	CABI (2005)
Libya	*P. farcta* (NA), *P. juliflora* (I)	CABI (2005)
Madagascar	*P. juliflora* (N)	Du Puy (1990), CABI (2005)
Malaysia	*P. juliflora* (U)	CABI (2005)
Malawi	*P. glandulosa* (I)	Chikuni et al. (2004)
Mali	*P. africana*(NA), *P. juliflora* (I)	Weber et al. (2008), Djoudi et al. (2011)
Mauritania	*P. chilensis*(U), *P. juliflora* (I), *P. pallida*(N)	Gritzner (1979), Jensen and Hajej (2001)
Mariana Islands	*P. pallida* (N)	Fosberg et al. (1976)
Marquesas Islands	*P. juliflora* (I), *P. pallida* (N)	Gallaber and Merlin (2010), and Lorence and Wagner (2013)
Mauritius	*P. pallida* (N)	Gallaber and Merlin (2010)
Mexico	*P. articulata* (NA), *P. glandulosa* (W), *P. juliflora* (NA), *P. laevigata* (NA), *P. nigra* (N), *P. palmeri* (NA), *P. pubescens* (NA), *P. reptans (NA)*, *P. tamaulipana* (NA), *P. velutina* (W), *P.* hybrids (I/W)	Grandtner (2005) and GBIF (2013)

(continued)

Table 7.1 (continued)

Country	*Prosopis* species status	References
Montserrat	*P. juliflora* (N)	Gallaber and Merlin (2010)
Morocco	*P. juliflora* (N)	Benata et al. (2008)
Mozambique	*P. juliflora* (I)	Witt (2013, personal communication)
Myanmar	*P. glandulosa* (N)	Gallaber and Merlin (2010)
Namibia*	*P. chilensis*(N), *P. glandulosa*(I), *P. pallida*(I), *P. juliflora* (I), P. velutina (I), *P.* hybrids (I)	Zimmerman (1991), Smit (2004), and Poynton (2009)
Netherlands Antilles	*P. juliflora* (N)	Grandtner (2005)
New Caledonia	*P. pallida* (U)	MacKee (1994)
Nicaragua	*P. juliflora* (NA)	AFTD (2009)
Niger*	*P. africana*(NA), P. juliflora (I)	Geesing et al. (2004), FAO (2006), Weber et al. (2008), and GBIF (2013)
Nigeria	*P. africana*(NA), P. juliflora (I)	Burkart (1976), Borokini and Babalola (2012)
Oman*	*P. cineraria* (NA), *P. juliflora* (I),	Ghazanfar (1996), Al Rawahy et al. 2003 and Al Abri et al. (2004)
Pakistan*	*P. cineraria* (W), *P. glandulosa* (N), *P. juliflora* (I), *P. koelziana* (NA)	Pasiecznik et al. (2004), AFTD (2009), Hussain et al. (2010), Khan et al., (2011)
Panama	*P. juliflora* (NA)	AFTD (2009)
Papua New Guinea	*P. pallida* (N), *P. juliflora* (I)	CABI (2005), and AFTD (2009)
Paraguay	*P. affinis* (NA), *P. algarobilla* (NA), *P. argentina* (NA), *P. alba* (NA), *P. campestris* (W), *P. chilensis* (NA),*P. elata* (NA), *P. fiebrigii* (NA), *P. hassleri* (NA), *P. humilis* (NA), *P. kuntzei* (NA), *P. juliflora* (NA), *P. nigra* (NA), *P. P. rojasiana* (NA), *P. ruscifolia* (W), *P. rubriflora* (NA), *P. sericantha* (NA), *P. vinalillo* (NA)	GBIF (2013)
Peru	*P. alba* (NA), *P. chilensis* (NA), *P. juliflora* (NA), *P. laevigata* (NA), *P. pallida* (NA), *P. reptans* (NA)	Grandtner (2005), GBIF (2013)
Philippines	*P. juliflora* (N)	Burkart (1976), Gallaber and Merlin (2010)

(continued)

Table 7.1 (continued)

Country	*Prosopis* species status	References
Polynesia	*P. pallida* (N)	Burkart (1976), ISSU (2005), Lorence and Wagner (2013)
Puerto Rico	*P. glandulosa* (N), *P. juliflora* (I), *P. pallida* (N)	Little and Wadsworth (1964), GBIF (2013)
Qatar	*P. cineraria* (NA), *P. glandulosa* (N), *P. juliflora* (N)	Pasiecznik et al. (2001)
Reunion	*P. glandulosa* (N), *P. juliflora* (I)	Kueffer and Lavergne (2004), CABI (2005)
Saint Helena	*P. juliflora*(N)	GISD (2013)
Saudi Arabia*	*P. africana* (U), *P. cineraria* (NA), *P. farcta* (W), *P. glandulosa* (N), *P. juliflora* (I), *P. koelziana* (NA)	Pasiecznik et al. (2001) and Hall et al. (2010)
Senegal	*P. africana* (NA), *P. alba* (N), *P. cineraria* (N), *P. pallida* (N) *P. juliflora* (I)	Diagne (1992) and Pasiecznik et al. (2001)
Sierra Leone	*P. africana* (NA), *P. chilensis* (N)	CABI (2005) and GBIF (2013)
Society Islands	*P. juliflora* (U), *P. pallida* (U)	Fosberg (1997)
Somalia	*P. chilensis* (I), *P. juliflora* (I)	Zollner (1986) and CABI (2005)
South Africa*	*P. alba* (I), *P. chilensis* (I), *P. glandulosa* (I), *P. juliflora* (I), *P. laevigata* (I), *P. pubescens* (N), *P. velutina* (I), *P. hybrids* (I)	Zimmermann (1991), Poynton (2009), Van den Berg (2010), Mazibuko (2012), and Wise et al. (2012)
Spain	*P. chilensis* (N), *P. velutina* (N)	Tilstone et al. (1998), Pasiecznik and Peñalvo López (in review)
Sri Lanka	*P. cineraria* (N), *P. juliflora* (I)	Parera and Pasiecznik (2005), AFTD (2009)
St Lucia	*P. juliflora* (N)	AFTD (2009)
St Vincent	*P. juliflora* (N)	AFTD (2009)
Sudan*	*P. africana* (NA), *P. chilensis* (I), *P. glandulosa* (I), *P. pallida* (N), *P. juliflora* (I), *P. velutina* (I)	Burkart (1976), El Fadl (1997), CABI (2005), FAO (2006), ISSG (2006), and Bokrezion (2008)
Syria	*P. farcta* (W)	ISSU (2005)
Tajikistan	*P. farcta* (NA)	GBIF (2013)
Tanzania*	*P. africana* (NA), *P. chilensis* (N), *P. juliflora* (I)	AFTD (2009) and Witt (2013, personal communication
Thailand	*P. juliflora* (U)	Pasiecznik et al. (2001)

(continued)

Table 7.1 (continued)

Country	*Prosopis* species status	References
Togo	*P. africana* (NA), *P. chilensis* (N)	AFTD (2009)
Trinidad and Tobago	*P. juliflora* (N)	ISSG (2013)
Tunisia	*P. chilensis* (N), *P. cineraria* (N), *P. farcta* (NA), *P. glandulosa* (N), *P. juliflora* (I), *P. laevigata* (N), *P. velutina* (N)	Habit and Saavedra (1990)
Turkey	*P. farcta* (W)	Pasiecznik et al. (2001); ISSU (2005)
Turkmenistan	*P. farcta* (NA)	GBIF (2013)
Uganda	*P. africana* (NA), *P. juliflora* (N)	Pasiecznik et al. (2001)
Ukraine	*P. farcta* (W)	Pasiecznik et al. (2001)
United Arab Emirates*	*P. cineraria* (N), *P. farcta* (NA), *P. glandulosa* (N), *P. juliflora* (I)	El-Keblawy and Al-Rawai (2007) and AFTD (2009)
United States*	*P. alba* (N), *P. articulata* (NA), *P. chilensis* (W), *P. cineraria* (N), *P. cinerascens* (NA), *P. farcta* (I), *P. glandulosa* (W), *P. juliflora* (I), *P. laevigata* (NA), *P. pallida* (I), *P. pubesens* (W), *P. strombulifera* (W), *P. velutina* (W), *P.* hybrids (I/W)	Johnston (1962), Grandtner (2005), and GBIF (2013)
Uruguay	*P. affinis* (NA), *P. alba* (NA), *P. caldenia* (W), *P. chilensis* (NA), *P. nigra* (NA), *P. ruscifolia* (NA)	GBIF (2013)
Venezuela	*P. flexuosa* (NA), *P. juliflora* (W)	Grandtner (2005) and Burkart (1976)
Vietnam	*P. juliflora* (N)	Pasiecznik et al. (2001)
Virgin Islands	*P. juliflora* (N), *P. pallida* (N)	Little and Wadsworth (1964) and Burkart (1976)
Western Sahara	*P. juliflora* (N)	Habit et al. (1990) and Witt (2013, personal communication)
Yemen*	*P. chilensis* (I), *P. cineraria* (NA), *P. farcta* (NA), *P. glandulosa* (N), *P. juliflora* (I), *P. koelziana* (NA)	FAO (2006) and Geesing et al. (2004)
Zanzibar	*P. juliflora* (N)	Nahonyo et al. (2005)
Zimbabwe	*P. juliflora* (N), *P. pallida* (U)	White (1962) and Poynton (2009)

Status codes (*sensu* Pyšek et al. 2004 with additional category weedy to describe native species that are invasive in their native rages) are given in brackets: *N* naturalised, *I* invasive, *NA* native, *W* weedy, *U* unknown. Countries partaking in management of Prosopis species are marked with an asterisk on the associated flora indicates that the allelopathic effects of the litters may override its potential positive effects on soil fertility (El-Keblawy and Abdelfatah 2013). In India, Kaur et al. (2012) showed that *P. juliflora* form resource islands by accumulating total organic N and organic carbon in their rhizosphere soil. However, the allelochemicals produced by this species outweigh the facilitative by nutrient enrichment (Kaur et al. 2012).

The history and range of distribution of the invasive *P. juliflora* – *P. pallida* complex has been covered by several authors including Burkart (1976), Habit and Saavedra (1990a, b, c), Poynton (1990), Felker and Moss (1996) and Perry (1998). In their monograph about *P. juliflora*, Pasiecznik et al. (2001) described the environmental conditions where this complex grows and creates environmental problems. According to Pasiecznik et al. (2001), the *P. juliflora* – *P. pallida* complex has been recorded in most of the African countries.

7.2.2 Introduced Range: Climate, Soil and Habitats

In the introduced range *P. juliflora* can grow in a wide range of conditions ranging from sand dunes to clay soils; from saline to alkaline soils; from areas below 200 to more than 1500 m above sea level; and from 50 to 1500 mm mean annual rain fall (Pasiecznik et al. 2004; Zeila et al. 2004). It can also withstand and survive temperatures from as high as 50 °C (air temperature) and 70 °C (soil temperature) (Pasiecznik et al. 2004). The introduction of *Prosopis* species from the Americas started to Senegal in 1822, and then to Australia, Hawaii, India, Philippines, South Africa, Sri Lanka and Sudan in the late 1800s and early 1900s (Pasiecznik et al. 2001). However, reforestation programmes after major droughts in the Sahel encouraged the widespread introductions into Africa and Asia between the 1970s and 1990s (Shackleton et al. 2014). The reasons of the introductions of *Prosopis* include providing fodder, fuelwood and shade in the arid areas of South Africa and Australia. In addition, the introductions were for dune stabilization, afforestation and fuel wood supply, provision of and fodder in several African countries. In India and Middle Eastern countries, the introductions were for rehabilitating degraded soil, local greening, ornamental cultivation and soil stabilization (Shackleton et al. 2014).

Prosopis juliflora is a major invasive species in India, and has also invaded other regions throughout the world including Saharan and southern Africa, the Middle East, Pakistan, and Hawaii (USA) (Pasiecznik et al. 2001), where it appears to strongly suppress the species native to those regions (Kaur et al. 2012). *Prosopis juliflora* forms pure stands in its invaded range in India and Arabia (Figs. 7.1 and 7.2), and occurs in forests, wastelands and at the boundaries of crop fields. *P. juliflora* also occurs in saline habitats in Hawaii USA. Above all, the worst thing is its negative impacts on the ecosystem like forming impenetrable shrubby thickets, invading water courses, lowering the water-table and thus indirectly starving plants of other species of moisture and nutrients, creating what are known as 'green deserts', largely devoid of life, instead of meeting the stated objective (Abdulahi et al.

Fig. 7.2 Satellite images show the invasion of *P. juliflora* to sand dunes next to Sharjah Airport, United Arab Emirates. Before the invasion of *P. juliflora*, these sandy dunes were almost barren of vegetation

2017). Even in the arid lands of the Arabia, *P. juliflora* dominates areas with shallower water table (El-Keblawy et al. 2015). These functional properties of Prosopis and otherfoster its adaptability and support the invasion of thespecies across various agro-ecosystems includingwetlands, dry lands, and irrigated agricultural lands (Shiferaw et al. 2004).

The *P. juliflora* have typically invaded areas in Arabian Peninsula where water table is close to the ground, such as gravel deserts, open plains, sand sheets, wadis and edges of farms (El-Keblawy and Al-Rawai 2007; El-Keblawy and Abdelfatah 2013; El-Keblawy et al. 2015; Thomas et al. 2016). In Saudi Arabia, *P. juliflora*was reported as a ruderal plant, mostly found inside the urban and suburban areas and has rarely seen among the native plant populations (Thomas et al. 2016). In the UAE, this species was reported in many natural habitats that have shallower water table. In addition, several farms of Ras Al Khaimah have been invaded and, consequently, ruined during the last 30 years (Ali El-Keblawy, unpublished data). *Prosopis juliflora* was identified as the most prominent invasive species in majority of lowlands of Saudi Arabia (Thomas et al. 2016).

This species was found as a troublesome exotic in the main inhabited island (Farasan Al-Kabir) in the Red Sea; approximately 12.5% of the total cover of inland vegetation is represented by *P. juliflora*. In addition, this species was found to be the most troublesome one in Najd, Central region of Saudi Arabia, which is considered as the most arid part of country; average annual rainfall is less than 100 mm. In this region, *P. juliflora* was reported as a ruderal plant, mostly found inside the urban and suburban areas and rarely seen among the native plant populations (Thomas et al. 2016). Careful examination of the *P. juliflora* plants showed the presence of *P. juliflora* growing mixed with *P. pallida* in most of the invaded habitat of the UAE and this complex is highly aggressive and coppices so well that it crowds out native vegetation (El-Keblawy and Al-Rawai 2007).The suitable habitat of *P. juliflora* is roadsides, but it is also planted in close spacing at an area of few hectares on sandy desert. Similarly, this species has been introduced to Qatar during 1950s and it grows around farmland as well as in depressions (Ahmad and Ismail 1996).

7.2.3 History of Introduction and Spread

Prosopis juliflora (Sw.) D.C. is native to south-west United States and north-west Mexico, and it is invasive in many tropical and subtropical regions, including North Africa and the Arab Gulf region (Abbas et al. 2016). *P. juliflora* was introduced to the UAE in the 1970s for greening deserts (El-Keblawy and Al-Rawai 2007; Tourenq and Shuriqi 2010). During the last four decades, this species has escaped the forests and currently is considered as a serious weed (Figs. 7.1 and 7.2). In Bahrain, scattered patches of young *P. juliflora* plants were recorded on coastal lowland. The presence of a huge single isolated *P. juliflora*tree in the central plateau of Bahrain indicates the probability of its early introduction. In Ethiopia, *Prosopis* was introduced in the late 1970's, or early 1980's through collaborative efforts of governments and international development organizations to rehabilitate degraded soils, to supply firewood and fodder and to combat desertification (Berhanu and Tesfaye 2006; Rettberg and Müller-mahn 2012; Shackleton et al. 2014). Now a day, around one million hectares of Ethiopia are already covered by *Prosopis* (Abdulahi et al. 2017). In Egypt, *P. juliflora* was introduced intentionally in the Gebel-Elba National Park by local people of Old-Hala'ib. village for agroforestry purposes during 1980s. However, they found it as highly invasive, spreading rapidely, destroying the local flora (Abbas et al. 2016). At the moment, now, it has been spread to 116 Km Northwest of its introduced range in the wadis of Shalal, Mericowan, Sarara, Shab and Aibib (Shiferaw et al. 2004; Hundessa 2016).

In the arid and semi-arid areas of Kenya, the *Prosopis* species were first introduced to rehabilitate quarries near the coastal town of Mombasa with seed sourced from Brazil and Hawaii in the early 1970's as a source of fuelwood and later, in the rehabilitation of degraded lands (Oduor and Githiomi 2013). In the Sudan, *Prosopis* spp. were introduced in many semi-arid areas, to combat desertification and provide fuel wood and fodder (Elfadl 1997a, b). However, this species have spread rapidly into fertile, productive areas, and irrigation and drainage channels (Morgan et al. 2017) (Fig. 7.3).

Fig. 7.3 *Prosopis juliflora* and its congeneric species *Prosopis cineraria* are co-occurring in many habitats of the United Arab Emirates. A. *Prosopis cineraria* provided the shade for domestic and wild animals that left big amount of *Prosopis juliflora* seeds resulted in dense growing of their seedlings under the crown of native *Prosopis cineraria* (b) the dense competition between the two congeneric species, especially for water resources (c and d) the intense competition between the two species resulted in death of the native *Prosopis cineraria*

The highly invasive nature of *P. juliflora* is a result of several factors; it is extremely drought resistant, a quality mainly attributed to its deep taproot (Samuel et al. 2013). Riversand water canals also play a significant role in the dissemination of seeds to different areas. Swampareas, road sides and irrigation canals are highly-invaded by *Prosopis*. This shows *Prosopis* establisheswell in areas where water is available and also wheresurface runoff water present for its seed dispersal. In Kenya, The banks of the Tana river, which is the largest river in Kenya have been invaded by Prosopis. Seeds could be dispersed by water, but the main dipsersal agent is the digestive system of most herbivores (Maundu et al. 2009)

7.3 Ecological Constraints

In the native range *Prosopis juliflora* exhibits different physiological responses to environmental variables, mainly temperature, rainfall and day length (Pasiecznik et al. 2001). *P. juliflora* can grow in arid and semi-arid regions because of its resistance to drought and heat. It thrives on all soil types under variable climatic conditions (Sawal 2004).

7.3.1 Climate

Prosopis juliflora is reported to tolerate an annual precipitation of 150 to 1670 mm, annual temperature of 20.3 °C to 28.5 °C, and pH around neutral (Ruskin 1980 and Larrea-Alcázar and Soriano 2008) but it can growth in areas with annual rainfall less than 100 mm as in the Central region of Saudi Arabia (Thomas et al. 2016) and UAE (El-Keblawy and Abdelfatah 2013). Given the fact that *P. juliflora* is evergreen and has fast growth aloo over the year, this is support the hypothesis that it relies manily on underground water as main source for life. *P. juliflora* is frost sensitive (Felker et al. 1982) and severe frost can cause stem and tree mortality in countries with cold weather. Consequently, there is doubt about the presence of this species in the Mediterranean climates of countries such as Morocco, Algeria, Tunisia, Libya and Egypt. For example, winter frosts occur over much of Egypt making it unsuitable for the *P. juliflora* – *P. pallida* complex unless it grow in sheltered or coastal locations (Burkart 1976). Similarly, Akrimi (1990) stated that cold sensitive *P. juliflora* was killed outright in Tunisia. Most of Prosopis species that are present in these countries are more likely be *P. Glandulosa* Torr., *P. velutina* Wooton, P. chilensis (Molina) Stuntz or hybrid forms from these species (Burkart 1976). In northern India, *P. juliflora* grows in regions that have high temperatures and high light intensities for most of the year. However, during the winter (December–January), the nocturnal temperatures fall to 2 °C, while during the summer (May–June), the temperatures of the day often exceed 45 °C (Shirke and Pathre 2004). *P. juliflora* shows a high capacity to adapt to theseextreme temperatures, especially, to the high temperaturesin summer.Similarly, in the UAE, where temperatures in summer can reach around 50 °C, *P. juliflora* grews nicely allover the year (El-Keblawy et al. 2015).

7.3.2 *Substratum*

*Prosopis juliflora*is a facultative phreatophyte that has a higher tolerance to droughts and ability to access declining groundwater tables through its deeper root system.It survives and flourishes in heavy or sandy soils, as well as saline dry flats and it can tolerate grazing (Morgan et al. 2017). In the Lagunillas semiarid enclave in the Venezuelan Andes *P. Juliflora* is present on loamy clay soils with high concentrations of organic matter (4–5%), nitrogen (0.15–0.30%), potassium (0.4–1.1 ppm), magnesium (2.0–5.5 me/l) and calcium (30–45 me/1) (Larrea-Alcázar and Soriano 2008). *Prosopis juliflora* usually grows in non-saline soils in most of its native and introduced range, but it has been recorded to occur in saline habitats in Hawaii USA and Gujarat, India (Kaur et al. 2012). In the UAE, it is reported in both saline and non-saline hábitats as well as farms and disturbied sites (El-Keblawy and Al-Rawai 2007).

7.4 New Insights into Traits That Promote Invasion

7.4.1 *Morphological Characteristics*

The life history traits of invasive plants demonstrate an important feature that can help themselves in the invasion, adaptation and establishment in a new habitat or introduced ones. The morphological and agro-physiological characteristics include plant height, dispersal efficiency, compition potential and sprouting capacity (Gibson et al. 2011). However, the human and environmental factors are also a important contributor in the invasion and dispersal of exotic plant species into a new habitat (Castro-Díez et al. 2011). *Prosopis juliflora* (mesquite, vilayati babul, algarrobo) is a xerophytic evergreen and fast growing tree, native to frost-free tropical regions of Peru, Central America and the Caribbean, with long lateral shallow and deep taproot systems that are also able to fix nitrogen (Burkart 1976; Pasiecznik et al. 2001). Stem is green-brown, sinuous and twisted, up to 5–10 m in height, with strong axial thorns situated on both sides of the nodes, the bark is rough and light red in colour. Leaves are compound, bipinnate with one or sometimes two pairs of rachis, each having 12 to 25 pairs of green folioles (Silva 1986). Flowers are small (4–6 mm long, 1 cm long), produced in inflorescences of various sizes and shapes but generally in spike-like inflorescences (racemes) 9.5–16.5 cm long (Díaz 1995). *Prosopis* flowers are arranged in cylindrical racemes of about 300–400 flowers

(Zaitoun et al. 2009), in clusters of 2–5, at the end of the branches. Flowers are hermaphrodite, sometimes sterile, actinomorphic and pentamerous (Burkart 1976). Fruit is a non-dehiscent pod, curved and about 4 mm thick, 1 cm wide and up to 15 cm in length made up of light yellow hardened epicarp, fleshy mesocarp and woody endocarp which contains seed (Silva 1986).

7.4.2 Reproductive Features

7.4.2.1 Phenology and Vegetative Growth Attributes

Prosopis trees generally initiate flowering and fruiting in the third or fourth year after germination. First flowering is dependent on optimal conditions and under drought conditions or very poor soils it can be considerable later (Pasiecznik et al. 2001). *Prosopis* species produce flowers every year increasing gradually up to 10–20 years and may be expected to continue at this high level for several decades. In the native range *P. juliflora* exhibits different physiological responses to environmental variables, mainly temperature, rainfall and day length (Pasiecznik et al. 2001), so variation in the onset of flowering is present between different populations due to climatic variation. Flowering is also variable within and between trees of the same population. Despite, flowering times are genetically controlled (Graham 1960; Pasiecznik et al. 2001). *P. juliflora* flowering season at the native range varies, with one or two periods of main flower production. It generally coincides with the wet season, from December to February and it is delayed from March to April and from July to September when there are two periods of flowering. Therefore, legume production generally overlaps with the end of the wet season improving seedling establishment or partly cover the dry season, ensuring pod consumption and seed dispersal by wild animals.

In the invaded area *P. juliflora* present a different multiple possibilities of flowering, associated not only to climatological conditions, but to evolutionary associations, extremely fast, with pollinator insects. In Punjab (India) *P. juliflora* showed one flowering period from January to May with earlier flowering time period with the increase in the temperature (Kaur et al. 2013) but in the ridge of Delhi (India) *P. juliflora* flowering occurs all year round and reaches its flower peak in April and September (Thakur 1989). Flowering almost continuous year round was observed in some invaded areas as Brasil or Haiti (Silva 1988; Timyan 1996). In the Arabian Peninsula, El-Keblawy and Al Rawai (2006) reported two episodes for flowering: autumn (November – Dec.) and (April – May). However, in moist year, the plants was observed in the flowering stage of the year around. There is controversy about time development of flowers in *P. juliflora*. Styles emerge from most flowers prior to anthesis and flowers remain in this state for some days (Pasiecznik et al. 2001). Anthesis occurs when flowers are fully open and accessible to pollinators. Flower maturation is variable in the same tree, often lower flowers in the raceme are fully developed than the upper ones that are still immature (acropetal succession). In general, the main pollen season took place in spring, the highest concentrations being recorded mainly in April, with a secondary peak in autumn, if any (Davis 1995 and Simpson et al. 1977). On the other hand, Perveen et al. (2014) studied the intradiurnal behaviour followed by *P. juliflora* pollen at Khairpur (Pakistan), where the maximum peak occurred near midday, at 11:00.

7.4.2.2 Prosopis juliflora and Pollinators

Pollinator assemblage structure can have important influences on floral evolution and reproductive interactions among plant species (Moeller and Tiffin 2005). *Prosopis* anthers have a glandular appendage that release an exudate containing protein and carbohydrates (Chaudhry and Vijayaraghavan 1992). Then, flowers produce copious amounts of nectar, as a nutritious reward for potential insect pollinators especially bee species (Simpson et al. 1977). In a geographical context, the spatial structure of variation in pollinator abundance and community composition can also have important implications for plant reproductive performance and ultimately floral evolution (Gomez et al. 2007). This entomophilia is extremely generalized, 77 insect visitors being reported in India by the former authors, Hymenoptera (bees and wasps) and Diptera (flies) being the main visitors.

P. juliflora is self-incompatible and receives insect pollinators and entomological visitors from a wide type of insect orders (Ward et al. 1977). *P. juliflora* has small size of flowers that ensures the contact of small insects with reproductive organs and there will be less chances of nectar robbing. Despite *P. juliflora* produces large number of inflorescences; the pollination success rate is significantly less. Based on inflorescence number/tree, according to reports of DeOliveira & Pires (1990), *P. juliflora* exhibited 29% pollination efficiency while it was dropped to 1.48%. Flowers of *Prosopis* are visited by different types of insects and more specifically from Hymenoptera (bees, wasps and Pepsis spp.), Coleoptera (Bruchidae, Lycidae, Tenebrionidae, and Scarabaeidae), Diptera (Syrphidae), and Lepidoptera (Noctuidae, Geometridae and various butterflies) (Simpson et al. 1977; Keys 1993).

Many plant species produces mass-flowers and mostly for floral display but usually they do not have enough food resources to transport to these points that help then to develop into fruit (Trueman & Wallace 1999). However, several plants produces significant number of seeds that are enough for that particular species to survive (Karron & Mitchell 2011). Other researchers demonstrated that there are several factors responsible for productivity and includes stigma reception, pollen release time, period and pollen viability. In this regard, ovary abortion, flower sterility and few pollen visitors are others major reasons for low pollination and productivity DeOliveira & Pires (1990). In case of *P. juliflora* that mostly grown in arid and semi arid regions with high summer temperature are another reason of low pollination. The insect pollinators are less mobile during midday (above 40 °C) that conincided with higher pollen production at midday (Goel and Behl 1995). There was also significant spatial variation observed among locations in pollinator assemblages in terms of species richness, diversity and dominance. The abundance-richness relationship is frequent in pollinator assemblages (Steffan-Dewenter et al. 2002),

According to studies of Sajjad et al. (2012), number of pods and germination rate of seeds rise with incease in its flower visitors. They concluded that flower visitor abundance was a better predictor of plant reproductive performance than species richness and the Shannon-Wiener index. We also found a significant spatial variation among locations in pollinator assemblages in terms of species richness, diversity and dominance. Diversity and identity (Gomez et al. 2010), and relative abundance of floral visitors (Sahli and Conner 2006) have been reported as the important predictors of plant reproductive success. The sensitivity of pollinator assemblages to predict plant reproductive success may vary with plant species and the pollination effectiveness of available pollinator species (Talavera et al. 2001).

7.4.2.3 Seed Production, Dispersal and Germination

The racemes work as reproduction units and the small size of the flower ensures contact of small insects with reproductive organs. Different maturation period in the reproductive organs and flowers development make *P. juliflora* plants self-incompatibles, depending of insects for seed setting (Sajjad et al. 2012). Obligate outcrossing leading to high genetic variability as an evolutionary mechanism for survival in zones with a high variability in rainfall, temperature and soil types. This trait, among others, allows to *P. juliflora* being an effective plant invader. By other hand, as was mentioned, long periods of asynchronous flower production would assume a long period of pollen release. However, due to *P. juliflora* is a tree and its flowers exhibit exerted stamens, its pollen grains present a secondary anemophily and can easily pass into the atmosphere and become airborne. Coat imposed dormancy is of wide occurrence in legumes and has been reported as a characteristic governing the weediness of *P. juliflora* (Morgan et al. 2017). In the field germination is favoured because of *P. juliflora* seeds pass through the digestive system of grazing sheep, goats and cows that might help in seed dissemination for long distances. In Ethiopia, it was observed that cattles are major source of seed disoersals followed by camels and goats (Shiferaw et al. 2004). Morgan et al. (2017) revealed that seeds collected from sheep's droppings were free from the pericarp and displayed 65% germination and 60% emergence suggesting that soil seed bank contain seeds of *P. juliflora* ready to germinate when conditions are optimal or moved to a shallower depth by agricultural implements.

The seed traits such as seed coat, endosperm, pericarp and extrafloral organs prevent germination until they are removed or damaged (Finkelstein 2006). On average, fruits of *P. juliflora* (15–20 cm width, 10–40 cm length) possess 20 seeds/pod. The endocarp is divided into rigid, leathery segments with one brown, elliptical seed in each segment (Meyer et al. 1971). The seeds frequently have hard, impermeable seminal integuments (Manga and Sen 1995), which ensure dormancy and are well adapted to long-term survival (Tschirley & Martin 1960) and dispersal strategies (Bewley 1997). Several Environmental factors such as fire, temperature, salinity and light as well as their interaction might stimilate or retard the seed germination of *Prosopis juliflora.* Salinity stress of 400 mM NaCl at 40 °C and in 600 mM NaCl at 25 °C decreased the seed germination. Furthermore, light also stimulated the germination at low salinity (El-Keblawy et al. 2015). Autotoxicity due to dry leaves of *P. juliflora* also caused significant inhibition in germination, radicle and hypocotyl growth of mesquite sedes (Warrag 1995). Due to climate change and increase in soil temperature will results in inhibition or stimulation of certain plant seed germination (Ooi et al. 2009). Al-Rawahy and co-workers (2003), demonstrated that germination of both *Prosopis juliflora* and *P. cineraria* seeds were unaffected following exposure to 90 °C for 6 h.

Senthilkumar et al. (2005), documented that prosopis has also capacity to bioacumulate heavy and toxic metals in various plant organs. Prosopis accumulated significant quantity of Cd in roots than shoots. While, concentration of Cu was not much distinguished between roots or shoots. They concluded that animal grazing on heavy metal contaminated soil vegetation should be avoided because the plant foliage and pods can retain more toxic metals than other parts. Several diverse animanl species such as camels, goats, sheep, deer, horses, cattles, rodents, feral pigs, warthogs are also responsible for dispersal of Prosopis seeds from one place to other through their feces (Kneuper et al. 2003, Shiferaw et al. 2004). Shiferaw et al. (2004) reported germination rates for *P. juliflora* of 100% with mechanical scarification, 97–99% germination with sulfuric acid, 37% after passage through goats, 47% through warthogs, 15% through camels, 4% through cattle, and 21% with no treatment. However, El-Keblawy and Al-Rawai (2005), documented more than 80% germination success in unscarified seeds. A study conducted on anther glands of P. juliflora demonstrated that an exudates (composed of carbohydrates and proteins) was extreated through cuticular openings to exterior of glands and have function like insect and pollinators attractance (Chaudhry and Vijayaraghavan 1992).

The process of Endozoochory might be responsible for enhancing the germination through releasing the seeds from fruit and ruminants feces that can also serve as source of water and nutrients for seedling establishments. Local animals, cattles, goats, camals, mules and wild fauna (gazelles and coyotes) can play a significant role in dispersal of seeds from one to other places (Wronski et al. 2012; Alvarez et al. 2017). However, exposure f the seeds to animal gut microenvironment, stomach juices, chewing, and herbivores species also affect the quality of endozoochorous (Schupp (1993; Kneuper et al. 2003; Jaganathan et al. 2016). In this context, determining whether endozoochory is advantageous to plant germination requires examining the cumulative effects of gut passage, deposition in faeces and seed dispersal patterns on germination and viability. Thus, understanding the functional role of domestic and wild animal species in seed dispersal of invasive species is central to determining how biotic interactions could be affected by anthropogenic drivers.

7.5 Possible Mechanisms explaining the Invasive Ability of *P. juliflora*

The traditional, congeneric, and bio-geographical approaches have been used to examine the mechanism of invasive species success (Inderjit et al. 2008).

7.5.1 *Traditional Approach*

The traditional approach focuses on the fate, dose, replenishment, and effect of chemicals produced by invaders in the soil environment (Inderjit 2001; Hierro and Callaway 2003).

Allelopathy is mediated through the release of secondary metabolites that directly affects many physiological and biochemical reactions and thereby, influence the growth and development of neighboring plants (Lara-Núñez et al. 2006; Hussain et al. 2011; Hussain and Reigosa 2011). Allelopathy has been suggested as one of the mechanisms driving *P. juliflora* to become more abundant and competitively dominant in their introduced range than in their native range (Elfadl and Luukkanen 2006; El-Keblawy and Al-Rawai 2007; Inderjit et al. 2008 and Kaur et al. 2012) reducing significantly the number of annual plants under the canopy of *P. juliflora*. The plant has little or no autoallelopathic effect under field condition (EI-Keblawy and AI-Rawai 2006). This mechanism, combined with drought condition can inhibit other species and eliminate any kind of competition (Abdulahi et al. 2017).

7.5.2 *Congeneric Approach*

The congeneric, or phylogenetic, approach involves comparative studies of exotic species with natives in the same genus (Inderjit et al. 2008). Native plants typically do not share a co-evolutionary history with the exotic invasive species, and therefore greater allelopathic effects of the alien invasive, as compared the native congeneric in such ecosystems. The allelochemicals produced by the invaders are new to the native plant communities (i.e., novel weapons, Bais et al. 2003; Callaway and Ridenour 2004). Assessment of the impact of allelopathy of two *Prosopis* species on the germination and existing of the associated species helped in understating the role of allelopathy as a mechanism for invasion of *P. juliflora* in its new ranges (El-Keblawy and Abdelfatah 2013; Kaur et al. 2012). In both Arabia and India, the native *P. cineraria* is a slow growing tree and is beneficial for the growth and development of other species (Abdel Bari et al. 2007). It is rarely, if ever, seen as a weedy species and has not been successfully introduced into other parts of the world (Pasiecznik et al. 2001). In addition, *P. cineraria* had a facilitative rther tan determinetal effect on the asscoated flora (El-Keblawy and Abdelfatah 2013; Kaur et al. 2012). In the two regions, however,the introduced *P. juliflora* has invaded many habitats and residential area and significantly reduced the native flora diversity (El-Keblawy and Abdelfatah 2013; Kaur et al. 2012). In the introduced range of the UAE and India, both species produced allelochemcials. El-Keblawy and Abdelfatah (2015) assessed the impacts of allelopathy produced by *P. juliflora* and *P. cineraria* and soil properties on understory native plants in the arid deserts of the UAE. They found inhibitory effect for the aqueous extracts of fresh and old leaves of *P. juliflora* on the associated flora, but *P. cineraria* leaves and litter had positive effects on other native species. The quantity of the allelochemicals produced by the two species could help understaning their defferential effect on the assocted flora. Kaur et al. (2012, 2014) found that the amounts of phenolics and tryptophan produced by *P. juliflora* in India were significantly higher than those of *P. cineraria*. For example, Kaur et al. (2012) detected L-tryptophan in leaf leachates of both *P. juliflora* and *P. cineraria*, but the amounts were 73% higher in leaf leachate of the former than that of the latter. Similarly, Inderjit et al. (2008) compared soils collected from the rhizospheres of the two *Prosopis* species and found that soils beneath the exotic *P. juliflora* contained 63.2% higher concentrations of total phenolics than soil beneath the native *P. cineraria*. This result could explain the greater effect of *P. juliflora* litterson mortality of native Indian species, compared to litters from *P. cineraria* (Kaur et al. 2012). Similarly, aqueous extract of *P. juliflora* leaves showed negative effects on root growth of three common crop species of north-west India, whereas *P. cineraria* leaf leachate had positive effects (Kaur et al. 2012). Furthermore, Goel et al. (1989) reported greater allelopathic potential of *P. juliflora* leaf leachate and decomposing litter residues compared with *P. cineraria*.

7.5.3 Biogeographic Approach

The biogeographic approach studies ecological traits of species and ecological processes in native and non-native ranges. Exotic species bring chemicals novel for invaded communities that has potential to exhibit allelopathic effects due to naïve soil communities and sensitive neighbors (Callaway and Ridenour 2004, Inderjit et al. 2011). The 'novel weapons hypothesis' was posed by Callaway and Aschehoug 2000; also see Rabotnov 1982; Malik and Pellisier 2000; Ridenour and Callaway 2001) to study role of plant chemicals in ecological processes and evolutionary context (Inderjit et al. 2006, 2011). Several plant secondary metabolites (phenolics, flavonoids, glycosides, chalcones, cinnamic acid derivatives, Terpenoids, Coumarin, Saponins, and alkaloids) were reported from bark, leaves, stems, flowers, pods and sedes of *P. juliflora* (Table 7.2). Nakano et al. (2003), demlonstrated the presence of phenolics, tryptophan and juliflorine from *P. juliflora* foliage. Whereas *P. juliflora* has facilitative effect in its native range in Venezuela, it has an inhibitory effect in its introduced range.

Table 7.2 Secondary metabolites isolated from different *Prosopis juliflora* organs: leaves, stems, flowers, fruit, pods, seeds, and bark

Plant part	Metabolite class	Compounds	Extract type	References
Pollen	Flavonoids	Apigenin derivative-7-O-R	Identified in ethanol–water (50% v/v) extract by high-performance liquid chromatography/diode array detector (HPLC/DAD)	Almaraz-Abarca et al. (2007)
	Cinnamic acids	Luteolin derivative		
		Flavonol glycoside		
		Quercetin-3-glycoside		
		Genistein glycoside or dihidroquercetin		
		Isorhamnetin-3-O-R		
		Chalcone		
		Cinnamic acid derivative		
Fruits	Flavonoids	Patulitrin (flavonoid)	Fruit ethanol extract (flavonoid identification) and aqueous extract (free sugar identification)	Wassel et al. (1972)
	Free sugars	Glucose and sucrose (free sugars		
Bark	Flavonol glycoside	Kaempferide 3-O-b-dgalactopyranoside	Characterized in ether soluble fraction from acetone extract	Nee' Shukla and Misra (1981)
	Isoflavone glycoside	Retusin 7-O-neohesperoside	Characterized in etanol extract	
Leaves	Flavonoid	–	Phytochemical screening reactions analyzed by gravimetric and spectrophotometric methods	Ibrahim et al. (2013)
	Alkaloids			
	Saponins			
	Phenols			
	Tannins			
	Fibers			
	Pectic substances			
Leaves	Flavonoids	Apigenin	Identified in standard extracts and chromatographic procedures	Bragg et al. (1978)
		Luteolin		
		Apigenin-6,8-di-C-glycoside		
		Chrysoeriol 7-O-glucoside		

(continued)

Table 7.2 (continued)

Plant part	Metabolite class	Compounds	Extract type	References
		Luteolin 7-O-glucoside		
		Kaempferol 3-O-methyl ether		
		Quercitin 3-O-methyl ether		
		Isoharmentin 3-O-glucoside		
		Isoharmentin 3-O-rutinoside		
		Quercitin 3-O- rutinoside		
		Quercitin 3-O-diglycoside (glucose and arabinose)		
Leaves	Terpenoids	–	Phytochemical screening reactions on methanol, ethanol, chloroform and benzene extracts	Sharmila et al. (2013)
	Phenol			
	Flavanoid			
	Coumarin			
	Glycoside			
	Carbonyl			
	Saponins			
Leaves	Pigments	Carotene, xanthophylls and pheophytin	Identificated in etanol extracts	Tesoriere et al. (2005)
Pods	Ellagic acid glycoside	Ellagic acid 4-O-a-L-rhamnosylgentiobioside	Isolated in ethanol extract	Malhotra and Misra (1981c)
Pods	Ellagic acid glycoside	Ellagic acid 4-O-rutinoside	Isolated in ethanol extract	Malhotra and Misra (1981b)
Pods	Tannins	Tannins	Different solvents extracts as acetone and methanol	Makkar et al. (1990)
Roots	Flavanone glycosides	30,40-dihydroxy 5-methoxy 6-methyl flavanone 7–0-b-D-glucopyranoside	Isolated in benzene and ethyl acetate fractions of ethanolic	Malhotra and Misra (1983)
Roots	Ellagic acid glycoside	3,30-di-O-methyl ellagic acid 4-O-a-Lrhamnopyranoside	Isolated in acetone extraction	Malhotra and Misra (1981a)
Seed	Amino acids	Essential amino acids except lysine, methionine and cysteine	Hydrothermically processed seed meal	Bhatt et al. (2011)

Costa et al. (2018), reported the ecological damage to ecosystems, stand stabilization and settlement of *P. juliflora*. The leaves, barks and roots (aqueous extracts) (125 g plant/500 ml of distilled water) inhibited the emergence and seedling growth of jurema-preta, ***Mimosa tenuiflora***, native species of Caatinga and 100% root extract proved to be deleterious than all other extracts. Shah et al. (2018), in a two-year field experiments demonstrated the phytotoxicity of *P. juliflora* on weed control and yield of wheat. They found that aqueous extracts (0, 10, 20, 30 and 40% concentration of leaves, stems, and roots) reduced the weed density, biomass, leaf area index, leaf area duration, crop growth rate, net assimilation rate, chlorophyll contents, plant height, number of tillers, spike length, number of grains, 1000-grain weight, grain yield, biological yield, harvest index and grain protein content.

Prosopis juliflora is known to inhibit germination of seeds of other plants species that lie in its vicinity (Muturi et al. 2017; Shaik & Mehar 2015). It also discourages other species of plants to grow near it. It releases allelochemicals from its leaves, roots, as well as fruits to achieve this (Noor et al. 1995).Goel et al. (1989) found that leaf extracts as well as leaf leachates of *P. juliflora* carried allelochemicals, so did decaying leaves. These authors, as well as Chellamuthu et al. (1997), who studied the influence of *P. juliflora* leaf litter on the germination of seeds of other species, attributed the allelopathy to phenolic compounds present in *P. juliflora*. Al-Humaid and Warrag (1998) recorded suppression of seed germination and early growth of bermuda grass *(Cynodon dactylon)*, and Kaur et al. (2014) of *Brassica campestris*, by aqueous extracts of *Prosopis* leaves. These studies indicate that *P. juliflora* foliage may contain water-soluble allelochemicals, which get leached to the ground as rain water falls on them and tikles down (Abbasi and Abbasi 2011). These chemicals were isolated by Nakano et al. (2002, 2003, 2004) and identified as syringin, (-) – lariciresinol, L- tryptophan, juliprosopine, juliprosine, and juliprosopinal. Among these, juloprosine derivatives exhibited the most pronounced allelopathy.

This indicates that allelopathy plays a significant role in shaping plant community structure. In its native range in Venezuela, *P. juliflora* appear to coexist with and facilitate large numbers of other native species (Kaur et al. 2012). In its non-native ranges, *P. juliflora* strongly suppress species native to those regions (Pasiecznik et al. 2001) and forms pure stands in India, Hawaii (USA) (Kaur et al. 2012) and the UAE (El-Keblawy and Al-Rawai 2007). Competition is another mechanism that would enable *P. juliflora* to replace native flora. Root density of *P. juliflora* was 3 cm of root/cm^3 of soil in the upper 15 cm of the soil profile, dropping to less than 0.5 cm root/cm^3 of soil at below 45 cm depth, and less than 0.2 cm root/cm^3 of soil at 1.8 m depth (Jones et al. 1998). Hoshino et al. (2011) indicated that *P. juliflora* can detect even very tiny soil moisture and grow to various conditions. Some of the many adaptive abilities that allow *P. juliflora* to thrive under such conditions include ability of roots to adapt to a wide variety of soil conditions (Hoshino et al. 2011). Roots can grow upwards towards the soil surface to capitalize on little rainfall, but can also grow to depths of 80 m and extend laterally more than 30 m (Thorp and Lynch, 2001). Such high density of the superficial roots could enhance the competitive ability of *P. juliflora*to extract the limited nutrients and water resources of the arid deserts (El-Keblawy and Abdelfatah 2014). This could explain the high aggressive ability and how it could crowd out native vegetation in most invaded sites. It has been noticed that farmers in many places of the UAE just ruin their farms once they are invaded with this species.

Shirke et al. (2018) demonstrated that *Prosopis juliflora* has shown optimal physiological characteristics to adapt in monsoon season through its leaf architecture that exhibit maximum carbon fixation under moderate temperatures and a wide range of photosynthetic photon flux density. The leaves produced in spring were sensitive to very high temperature and others that develop during monsoon were sensitive to low temperatures causingsenescence in winter. Reinoso et al. (2004) found that salinity induced anatomical changes in roots (young and mature zones), hypocotyls, young stems, and leaflets, with small leaves, reduced cortex and nascular system. Salt stress was also lethal for stem and in hypocotyl size, diameter was reduced along with a reduction in secondary phloem. However, salinity stimulated the production of tannins in stem and leaflets of treated plants. They concluded that anatomical modifications in this species are related to metabolic adaptations, such as an early development of the endodermal barrier for ion exclusion, to allow survival in high salinity.

7.6 Impacts of Prosopis Invasions on Biophysical Features and Ecosystem Services

Propsopis juliflora can grow in different arid climates and substrates because of its resistance to drought and heat and it has many potential uses and impacts (Mendes 1986).

7.6.1 Uses and Positive Impacts

In some areas the benefits from *Prosopis* are regarded as a key income source for many households. For example, 44% of people in a village in Malawi replied on *Prosopis* products as a primary or supplementary source of income (Chikuni et al. 2004). Similarly, sale of charcoal and *Prosopis* pods for fodder have enhanced the local economy in some areas in Kenya by US$1.5 million per year (Choge et al. 2012). In India, *Prosopis* provides up to 70% of fuel wood needs for local households in some dry region villages (Pasiecznik et al. 2001). In Senegal, the positive aspects of *P. juliflora* are predominantly associated with regulatory services, such as soil erosion regulation, rehabilitation of sodic soils, flow regulation, and water purification (Tripathi and Singh 2010). The dense shrubs of *P. juliflora* stabilize the soil, regulate the flow of water, and promote the infiltration of water into the soil. The dense shrubs also provide physical protection to soils against wind erosion during dry periods and against heavy rains during the rainy seasons (Ayanu et al. 2015).

Prosopis juliflora increase organic matter and nutrients on soil beneath its canopy, and commonly used to improve soil physiochemical and biological properties (Vallejo et al. 2012).

Secondary products from *P. juliflora* includes honey, edible exudates gums, fibres, tannins, foliage for fodder, mulch, biopesticides and medicines, and other uses for wood and particle board, wood chips for energy generation, pods for ethanol production, galactomannan gums from the seeds and other specialist products (Wise et al. 2012; Oduor and Githiomi 2013; Haji and Mohammed 2013).

7.6.1.1 Pods Can Be Used as a Livestock Feed and for Making Human Foods

The pods of *P. juliflora* contain high levels of protein and are available for a minimum of 4–5 months in the UAE. This could be a good source for proteins for livestock. However, the low digestibility of the leaves and pods of *P. juliflora* is mainly associated with the presence of harmful substances, such as tannins, glucosinolate, cyanogens, alkaloids, and nitrates (Chaturvedi and Sahoo 2013; Leonard 2011). In some cases, some ruminants are spontaneously poisoned and intoxicated by pods of P. juliflora (Camara et al. 2011). In addition, the dried leaves of this species suppress feed intake and nutrient availability and shouldn't be included in the feed of livestock (Chaturvedi and Sahoo 2013). An overview of use of *Prosopis juliflora* and its different organs in livestock feed was elaborated in Table 7.3.

P. juliflora was assayed as biosorbent due to their extremely rapid proliferation, massive growth, renewability, high biosorption capacity and low cost. Seed powder of the *P. juliflora* exhibit an ability to absorb Pb (II) from the contaminated environment (Jayaram and Prasad 2009). Biomass contributes a significant share of global primary energy consumption because liquid fuels produced from biomass contain no sulphur, thus avoiding SO_2 emissions and also reducing emission of NOx (Kumar and Kotiya 2004). The *P. juliflora* was used as a potential renewable energy source and their large scale utilisation can represent one of the best strategies for their management. The fermentation of both acid and enzymatic hydrolysates, containing 18.24 g/L and 37.47 g/L sugars from *P. juliflora*, with *Pichia stipitis* and *Saccharomyces cerevisiae* produced 7.13 g/L and 18.52 g/L of ethanol with corresponding yield of 0.39 g/g and 0.49 g/g, respectively (Gupta et al. 2009).*P. juliflora* was used for extraction of energy precursors in the form of volatile fatty acids (VFAs). Patnaik et al. (2018) in a single one-pot step, were able to convert up to 10.7% of the total solids present in the *Prosopis* leaves to VFAs.

7.6.2 *Negative Impacts*

7.6.2.1 Aboveground Effects

Maundu et al. 2009 studied the negative aspects of *P. juliflora* onlocal livelihoods of three sites in Eastern Province of Kenya and reported 24 negative effects. The magnitude varied according to people's sources of livelihood. The aspect that caught most attention in all sites was the thorn problems on humans. Several problems are related to animals that depended heavily on the pods. In addition, diarrhoea is

Table 7.3 Use of *Prosopis juliflora* and its different organs in livestock feed

Type of study	Key findings	References
Digestibility of globulins from mesquite pods and cowpea by mammalian digestive enzymes.	Pods were difficult to digest due to the presence of globulins, in comparison to immature cowpea.	Araujo et al. (2002)
Pods in the diets of laying hens	The pods can be included upto 13.6% of the diets without adversely affecting the performance of the laying hens.	Vilar Da Silva et al. (2002)
Prosopis flour in diets of rats	Due to the protein content present in the flour, there was no adverse effect during pregnancy and lactation phases.	Da Silva et al. (2003)
General feed for livestock.	Pods could replace costlier feed ingredient such as grain and bran, contributing 10–50% of the diet. If the percentage increases to 20%, then phosphorous supplements need to be added	Sawal et al. (2004)
Pod meal in the diet of growing crossbred heifers along with wheat straw.	Complete feed with 30% wheat straw and 20% Prosopis pods replacing rice polish, can be given as feed to heifers, without an adverse effect on growth and reproduction.	Pandya et al. (2005)
Pods as a feed for goats	Pods can be utilized as feed up to a proportion of 200 g–1 kg of rhodegrass hay. Higher pod fraction affects the carcass yield and quality.	Mahgoub et al. 2005
Pods as a meal for horses.	Pod meal can be used in the diets of horses even though a decrease in dietary fiber digestibility is seen.	Da silva stein et al. (2005)
Pods as a feed for Awassi lambs	Upto 200 kg–1 replacement of barely grains with pods is feasible, and is also cost effective. More than that will affect the growth, digestibility, and quality of the meat of the lambs.	Obeidat et al. (2008)
Pods as a feed for Nile tilapia fries.	Diet supplemented with 60 g kg–1 pods improved the growth, nutrient utilization and whole body composition in Nile tilapia fry.	Mabrouk et al. (2008)
Pod meal in diets of lactating goats	Substitution of corn meal by pod meal had no significant effect on the ruminal parameters but showed a linear negative response for microbial efficiency synthesis	Argolo et al. (2010)
Pods as a feed ingredient in the diets of broiler chicks	Pods can replace corn by 20% in the diet of broiler chickens.	Al-Beitawi et al. (2010)
Prosopis seed meal for Labeo rohita fingerlings	Processed seed meal can be incorporated into the carp diet at an inclusion level of not more than 20%.	Bhatt et al. (2011)
Ground pods as feed ingredient in poultry diet (broilers).	About 20% of the broilers' diet can be replaced by ground pods which will reduce feed cost without any negative effect. More at 20%, will affect the feed intake, growth and carcass quality	Girma et al. (2011)

(continued)

Table 7.3 (continued)

Type of study	Key findings	References
Ground pods as feed ingredient in poultry diet (layers).	Upto 20% ground pod in layers' ration is recommended, even though 10% is better suited. But more than 20% results in reduction of egg production and egg mass.	Girma et al. (2011)
Pods as feed for sheep and goats	Pods can be used without restriction in the feed of sheep but goats may be kept in Prosopis invaded areas for no more than one fructification period	Riet-Correa et al. (2012)
Pods and leaves as feed for lambs.	Pods can be used as a supplement in the diets of lambs without any adverse effect. However, leaves are unpalatable.	Ali et al. (2012)
Ground pods as feed ingredient broilers	Upto 30% inclusion of ground pods in broilers diet did not alter the chemical and fatty acid composition and sensory test of the meat. But at high levels the immune response to parasitic infection was impaired	Girma et al. (2012)
Pods as feed for cattle and horses	Despite their toxicity, pods can be used as cattle feed at concentrations of 30% of the food. Horses can be given pods as diet in confined and semi-confined systems. Horses are not recommended to be kept in grazing areas where Prosopis is in fructificating whereas cattle can be kept, butfor no more than 30 days.	Medeiros et al. (2012)
Assessment of nutritive value of pods of Prosopis along with some other plants.	Prosopis along with leucaena and blue panic plant might be promising alternative feed supplement for ruminants to replace alfalfa. Prosopis had the highest volatile fatty acids concentration.	Allam et al. (2012)
Prosopis pod meal in the diet of sheep	Pod meal can replace grass upto 45% in the diet	Pereira et al. (2013)
Prosopis pod meal in the diet of lactating goats.	Pod meal can replace corn not exceeding 40.5% of the total diet.	Pereira et al. (2013)
Pods as feed along with cenchrus grass for sheep.	Pods can replace 40% of the feed mixture in sheep without any adverse effect	Chaturvedi and Sahoo (2013)
Prosopis seeds in the diets of broiler chickens	Upto 2% of the diet of broilers can have Prosopis seeds. More than that will have adverse effect	Mohammadi et al. (2013)
Partial substitution of barley grain with Prosopis pods in lactating ewes's diets	Pods can be included in the diet of nursing ewes and their lambs, to reduce the cost of feed and improve milk production, upto 250 g kg–1.	Obeidat and Shdaifat (2013)
Replacement of cottonseed meal with ground Prosopis pods as a supplementary feed for sheep	Compared to feeding hay alone, supplementing the feed with cottonseed meal and Prosopis pods is a better feeding strategy	Yasin and Animut (2014)
Substitution of corn for pod meal in lambs' diet.	At a substitution level of 47.5%, maximum weight gain is seen in the lambs	Pereira et al. (2014)

(continued)

Table 7.3 (continued)

Type of study	Key findings	References
Preference and ingestive behavior of sheep feed on tropical tree fruits	The fruits of Prosopiswere 'highly preferred	Pinto-Ruiz et al. (2014)
Prosopis pods as a partial replacement of corn in the diet of growing broiler chicken.	Prosopis pods can be included at levels of 5% in broiler diets without affecting performance	Al-Marzooqi et al. (2015)
Prosopis pod meal in the diet of sheep	Diet consisting 30–45% of pod meal can be given to sheep.	dos Santos et al. (2015)
Pods as a replacement to concentrate feed forgoats	Upto 40% of the concentrate feed can be replaced with Prosopis pods without any adverse effect	Hintsa et al. (2015)
Prosopis pods as a replacement of corn in the diet of juvenile Nile tilapia.	Prosopis meal can completely replace corn in Nile tilapia juvenile diets during periods of low water temperature	Silva et al. (2015)
Milled mature Prosopis pods as a replacement of maize in the diet of broiler chicken	Milled Prosopis pods negatively affected the performance of the broiler chickens.	Odero-Waitituh et al. (2016)
Prosopis pod meal as a total replacement of corn in the diet of Holstein-Zebu crossbred dairy sterrs.	Prosopis pod meal can totally replace corn in the diet of Holstein-zebu crossbred dairy steer.	De Oliveira Moraes et al. (2016)

happen among inexperienced goats that feed on the leaves for the first time. Furthermore, diarrhoea in goats was reported mainly in Loiyangalani town due to influx of pastoralists and their goats escaping drought and insecurity caused by raids for stealing animals. Encroachment was most serious in Baringo and Garissa sites where the species had displaced farmers from their crop farms, invaded areas used for grazing and browsing. Moreover, encroachment to paths, dwellings, water sources, farms and pastureland, constraining movement and other activities were reported as problems asscoaited with the introduction of the species.

Vegetation and Community Structure

Several reviews have shown deserts to be among the least-invaded ecosystems worldwide, at least in terms of the number of naturalized and invasive species (Lonsdale 1999). However, this species is expanding its range at an alarming rate and damaging native diversity and ecosystem health of the arid and hyper-arid regions. It discourages abundance, richness and growth of native species. *P. juliflora* often observed to form pure stands, and does not allow other species to grow beneath or around its canopies. Despite *P. juliflora* is a strong facilitator in the native range (Tiedemann and Klemmedson 1973; Kaur et al. 2012); it was found that the canopies of the invasive *P. juliflora* had far fewer understory species in the invaded areas than native congeners (Aggarwal et al. 1976). In the Arabian Peninsula the invasive has strong negative impacts on native species: number, richness, evenness, density and frequency of the associated native speciesdespite increases in the concentrations of some nutrients in subcanopy soil (El-Keblawy and Al-Rawai 2007). Especially important is that this depressive effect extended beyond the canopy-covered ground for dense sites. Old and dense sites of *P. juliflora*resulted in significantly lower density, frequency and diversity for most associated annual species (El-Keblawy and Al-Rawai 2007). Similarly, the growing of *P. juliflora* shrubs, as well as exotic *Eucalyptus*, in the forests of the UAE has also resulted in significant reductions in species diversity and abundance of understory species, compared to the native *P. cineraria* and *Acacia arabica* (El-Keblawy and Ksiksi 2005).

7.6.2.2 Belowground Effects

Soil Microorganisms

Soil microbial communities play an important role in the invasion success of exotic plant species (Inderjit and van der Putten 2010). Plant-soil feedbacks (PSFs) include plant-mediated changes in soil communities that affect the establishment and growth of plant species (van der Putten et al. 2013, Inderjit and Cahill 2015). Exotic invaders are known to culture soil biota that exert neutral or positive impacts on the invader compared to the negative impacts of soil biota cultured by native species on themselves (Callaway et al. 2004). The effect of *P. juliflora* on soil properties could be mediated through soil microbial flora. *P. juliflora* trees have the potential to establish a symbiotic relationship with N-fixing Rhizobium bacteria, which increase the N levels in the rhizosphere with root and nodule turnover (Reyes-Reyes et al. 2002; Perroni-Ventura et al. 2010). For example, a total of 150 bacterial strains were isolated from the root nodules of *P. juliflora* growing in soils collected from Marigat area of Kenya (Otieno et al. 2017). In addition, the microbial activity of *P. juliflora* as witnessed by the emission of CO_2 was larger for soil sampled under canopy than outside it (Herrera-Arreola et al. 2007). Those authors indicated that addition of leaves increased production of CO_2 and between 40% and 50% of the organic C of the leaves was mineralized. Prosopis has been recognized as a fertility island tree that significantly increase soil nutrient concentration under its canopy, and improve soil microbial activity and diversity (Abril et al. 2009; Vallejo et al. 2012). The organic matter inputs from leaf litter, fruit, and root exudates returned to the soil all could increase organic C and nutrients under canopy trees. Therefore, soil beneath the *P. juliflora* canopies had higher organic C, total N, nitrate, available P and lower bulk density than soil outside, improving chemical and physical soil quality, even in the arid deserts in which water is a limiting factor for decomposition process (Vallejo et al. 2012; El-Keblawy and Abdelfatah 2013; Kaur et al. 2012). This was attributed to organic matter build-up, higher biological activity, and improved soil structure (aggregation and porosity) favored by tree roots as well as fungal and actinomycetes hyphae. As *P. juliflora* produce allelochemcials that interfer with understory native plants and at the same time enhance soil and microbial activities, the relative importance of these two processes determines the structure of the plant community under and around the native and exotic trees (Callaway and Walker 1997).

Water and Soil Fertility

Several authors have considered the increasing number of introduced or invasive species as a major component of global change because of their potential to alter primary productivity, decomposition, hydrology, nutrient cycling, and natural disturbance regimes (Pyšek et al. 2012). The conversion of barren sand dunes in the arid deserts into dense thickets of *P. juliflora* can generate many environmental consequences through altering water and nutrient cycling, changing carbon storage and enhancing climate variability.

Considering that water is the limiting resource for the ecosystem in semi-arid and arid regions, hydrological response to *P. juliflora* invasionmay directly impact the availability of ecosystem services and consequently human wellbeing (Nie et al. 2012; Vaz et al. 2017). Groundwater, which is the main water source in semiarid regions, is severely and increasingly threatened. Consequently, the greater ability of *P. juliflora* to deplete the ground water would threaten the groundwater resource and consequently affect native plants. Nie et al. (2012) assessed hydrological consequences of *P. juliflora* invasionin the upper San Pedro watershed (U.S./Mexico) and found that the simulated average annual evapotranspiration increases with *P. juliflora* encroachment, leading to the decrease of annual water yield and percolation by 9.8% and 9.7%, respectively. Therefore, *P. juliflora* increase water demand from soil, especially in the arid regions, where the limited non-renewable groundwater is depleted and/or salinated (Murad et al. 2007, AlRukaibi 2010).The deep taproot system of Prosopis has been implicated in declining ground water tables in Hawaii (Richmond and Mueller-Dombois 1972). Similarly, the decline in the groundwater level in the island of Kahoolawe was attributed to the spread of P. pallida (Stearns 1940).

Dzikiti et al. (2017), documented the impact of removing *Prosopis*, co-occuring with indigenous trees, *Vachellia karroo* (Hayne) Banfi & Galasso on the groundwater characteristics in the Northern Cape (South Africa). They measured the water consumption of both tested species through stem sap flows and found that *Prosopis*transpire 5 times more water than *V. karroo*. Following the removal of invasive *Prosopis* from the area demonstrated that water table decline was significantly slow down. The *P. juliflora* has largely invaded alluvial floodplains in North Cape, uplands, and shrub lands in South Africa and competing with native species (Mucina and Rutherford 2006). The study suggests that clearing of invasive *Prosopis* would conserve groundwater in the arid parts of South Africa (Dzikiti et al. 2013). It has been observed the *P. juliflora* is mainly invading the lands that have shallower water tables in the UAE (El-Keblawy and Al-Rawai 2007). This tree relies mainly on ground water as a main source in the absence of rainfalls. After rainfall, the extensive superficial dense root system enables it to extract the available water, which should be at the expense of the associated native plants (Pasiecznik et al. 2001).

Several studies have assessed soil physical and chemical characters under *P. juliflora* canopies and in bare ground next to them and concluded that the growth of this species improves some soil physical and chemical properties. For example, Menezes et al. (2002) indicated that *P. juliflora*in semiarid northeastern Brazil significantly affected microclimate and the dynamics of litter and soil nutrients, and may contribute to increases in the cycling rate of nutrients in these systems. In addition, Garg and Singh (2003) have shown that nutrient concentrations (N, P, K, Ca and Mg) of *P. juliflora*stand were significantly greater than that of other woody species. Furthermore, Goel and Behl (1999) reported that *P. juliflora* plantation resulted in a marked decrease in soil pH and sodium content, and improved organic carbon, N, K and P concentrations of the soil. In the arid climate of the UAE, canopies of *P. juliflora*increased the most important macro-nutrients K, N and P and the organic matter contents (El-Keblawy and Al-Rawai 2007; El-Keblawy and Abdelfatah 2014). The increase in organic content could increase the water holding capacity that would improve soil texture and increase soil moistures (El-Keblawy and Abdelfatah 2014). The effect of *P. juliflora*on soil properties could be mediated through soil microbial flora. For example, *P. juliflora*trees have the potential to establish a symbiotic relationship with N-fixing Rhizobium bacteria, which increase the N levels in the rhizosphere with root and nodule turnover (Reyes-Reyes et al. 2002; Perroni-Ventura et al. 2010). The organic matter inputs from leaf litter, fruit, and root exudates returned to the soil all could increase organic C and nutrients under canopy trees. Thwerefore, soil beneath the *P. juliflora* canopies had higher organic C, total N, nitrate, available P and lower bulk density than soil outside, improving chemical and physical soil quality, even in the arid deserts in which water is a limiting factor for decomposition process (Vallejo et al. 2012; El-Keblawy and Abdelfatah 2013; Kaur et al. 2012). This was attributed to organic matter build-up, higher biological activity, and improved soil structure (aggregation and porosity) favored by tree roots as well as fungal and actinobacteria hyphae.

7.7 Control Through Utilization

The high costs of *P. juliflora* eradication have argued several countries to follow a new and less expensive approach known as control through utilization. It has been argued that the negative impacts of *P. juliflora* invasion in Ethiopia are partially offset by provisioning of firewood and charcoal production. Wakie et al. (2016) have assessed the economic feasibility of selected *P. juliflora* eradication and utilization approaches that are currently practiced in Ethiopia. Their results showed that conversion of the infested area of *P. juliflora* to irrigated cotton reduces the spread of *P. juliflora* on farmlands and is economically feasible. In addition, managing *P. juliflora* infested lands for charcoal production with a four-year harvest cycle is also profitable (Wakie et al. 2016). However, the difficulties to control its rapid spread indicate that the threats it poses to ecosystem services, people's livelihoods and lifestyles may exceed its benefits (Ayanu et al. 2015).

The plant biomass that is rich with high lignin content results in high biochar yields (Sohi et al. 2010). The major component of mesquite wood was lignin (63.96%), which makes it a potential source for biochar. Li et al. (2016) used the mesquite wood to produce three biochars with different pyrolysis conditions (no pyrolysis, 450 °C, or 750 °C). Among the three pyrolysis conditions, biochar pyrolyzed at 450 °C showed a moderate yield (40 wt%), the highest surface area (500 m^2 g^{-1}), the highest pore volume (1.02 cm^3 g^{-1}) and the highest CO_2 uptake (26 mmol g^{-1}) at 30 bar and 25 °C. When the pyrolysis was done at 700 °C, the yield of the activation reaction (63 wt%) was considerably higher. However, the CO_2 uptake performance (13.9 mmol g^{-1}) was much lower (Li et al. 2016). The yield of biochar produced by mesquite was 44.4% with pyrolysis temperature of 400 °C (Liu et al. 2016). The C content of the biochar depended on mesquite feedstock particle sizes; small particle sizes (<0.853 mm) produced higher C (68%), compared to larger particle sizes (1.70–2.00 mm) that produced (73%) (Liu et al. 2016). In addition, pH depended on particle sizes; small particle sizes (<0.853 mm) produced pH = 7.95, but larger particle sizes (1.70–2.00 mm) produced lower pH (7.41, Liu et al. 2016).The acidic nature of biochar produced from *P. juliflora* makes it perfect for nutrelizing the calcareous alkaline soils of the arid deserts.In addition, biochar can improve the physical and chemical properties of low quality sandy and marginal soils that devoid of nutrients and have very low water holding capacity (Cao et al. 2009). Many studies have shown that biochar, in different soils, is a useful resource to improve the physicochemical properties of soil, effectively maintain soil organic matter levels, increase fertilizer-use efficiency and increase crop production, particularly for long-term cultivated soils in subtropical and tropical regions (Lehmann et al. 2003).As biochar is a carbon-rich solid product of thermal stabilization of organic matter, it could be stored safely as a carbon source in soil, which could control Co2 emission and consequently is good solution for global warming.

7.8 Conclusion

The exotic invasive *P. juliflora* is threatening the ecosystem services and human well-being. The species has introduced to combat desertification in different places around the world, but became one of the factors causing land degradation. In the native range, it has very limited effects on ecosystem components and native plants, but came with their new weapons, such as new allelochemicals and may be some endophytic bacteria that helped them to have serious impacts on the native flora of the introduced range. *P. juliflora* invasions have led to changes in ecosystem services, ground water depletion, plant biodiversity loss and alteration in soil physicochemical properties. The widespread of *P. juliflora* and modification of its phenology to flower more than one episode per year enable it to cause serious health threat, such as allergies. A change in hydrological cycles associated with *P. juliflora* invasion has aggravated the drought episodes that will be further exacerbated due to future climate change scenarios. However, *P. juliflora* introduced several economic and environmental benefits in the introduced range, such as soil conservation, fuel-wood, livestock fodder, timber, and fiber. It is important to calculate the beneficial and harmful impacts of *P. juliflora* and consequently take the decision whether to eradicate or conserve it. Allelochemicals produced by this species would affect the associated soil microbial communities, which in turn might have negative feedbacks for nutrient cycling, ecosystem processes and native vegetation. Further studies are needed to assess the impact of the introduced allelochemicals on local microbial communities. In the introduced range, where the plant has now become naturalized and became part of the flora, the threats posed by *P. juliflora* have been identified and control actions are currently implemented at both national and regional levels to reduce the deleterious effect of this species on the environment and human health. Regional regulations, prevention of expansion, knowledge platforms and environment protection laws are some of the crucial steps that needs to be undertaken for exotic species control measures and further spread and naturalization. In stipulations of the appraisal of native plant communities, functional traits should be assessed for understanding different mechanisms of controlling this plant.

Contribution of the Co-authors M. Iftikhar Hussain: conceived the paper, obtained information, coordinate the review, wrote the first draft and edited the manuscript. Ross Shackleton: obtained information, wrote different sections and contributed to significant revision and organize the manuscript. Ali El-Keblawy: obtained information and contributed to draft and revise different sections. Luís González: contributed to draft, revise and organize the manuscript. M. Mar Trigo: obtained information and contributed to draft and revise different sections.

Conflict of Interest The authors declare that they have no conflict of interest.

References

Abbas AM, Soliman WS, Mansour A et al (2016) Predicting the spatial spread of invasive Prosopis juliflora (SW.) D.C along environmental gradients in Gabel Elba National Park, Egypt. Int J Sci Eng Res 7:596–599

Abdel Bari E, Fahmy G, Al Thani N, Al Thani R, Abdel-Dayem M (2007) The Ghaf Tree Prosopis cineraria in Qatar. Qatar University Environmental Studies Centre, Doha

Abdillahi A, Aboud P, Kisoyan K, Coppock DL (2005) Agro-Pastoralists' Wrath for the *P. juliflora* tree: the case of the Il Chamus of Baringo District. Pastoral risk Management project, Kenya

Abdulahi MM, Ute JA, Regasa T (2017) Prosopis juliflora L: distribution, impacts and available control methods in Ethiopia. Trop Subtrop Agroecosyst 20(1)

Admasu D (2008) Invasive plants and food security: the case of *Prosopis juliflora* in the Afar region of Ethiopia. FARM-Africa and IUCN, Ethiopia, pp 1–13

Aggarwal RK, Gupta JP, Saxena SK, Muthana KD (1976) Studies on soil physico-chemical and ecological changes under twelve years old desert tree species of Western Rajasthan. Indian For 102:863–872

Agroforestree Database (AFTD) (2009) Prosopis. World Agroforestry Centre, Kenya

Ahmad R, Ismail S (1996) Use of Prosopis in Arab/Gulf states including possible cultivation with saline water in deserts. In: Felker P, Moss J (eds) Prosopis: semi-arid fuelwood and forage tree, vol 1. Building Consensus for the Disenfranchised. Center Semi-Arid Forest Resources, Kingsville, pp 41–1.52

Akrimi N (1990) *Prosopis* in Tunisia. In: Habit MA, Saavedra JC (eds) The current state of knowledge on *Prosopis juliflora* (Eds.) . FAO, Rome, pp 147–151

Al Abri AS, Al Ajmi DS, Al Halhali AS, Al Saqry NM, Forsberg NE, Kadim IT, Mahgoub O, Richie AR (2004) In: Salem B, Nefzaoui A, Morand-Fehr P (eds) Nutrition and feeding strategies of sheep and goats under harsh climates. CIHEAM, Zaragoza

Al Mutairi K, El-Bana M, Mansor M, Al-Rowaily S, Mansor A (2012) Floristic diversity, composition and environmental correlates on the arid Coralline Islands of the Farasan Archipelago, Red Sea, Saudi Arabia. Arid Land Res Manag 26:137–150

Al Rukaibi D (2010) Water resources of GCC countries. The Sustainable Resources Research Institute, Al-Riyadh

Al-Beitawi NA, Awawdeh FT, Khwaileh MM (2010) Preliminary study on *Prosopis juliflora* pods as unconventional feed ingredient in diets of broiler chicks. Anim Nutr Feed Technol 10:51–60

Al-Frayh A, Hasnain SM, Gad-elRab MO, Al-Turki T, Al-Mobeireek K, Al-Sedairy ST (1999) Human sensitization to *Prosopis juliflora* antigen in Saudi Arabia. Annals Saudi Med 19:331–336

Al-Humaid AI, Warrag MOA (1998) Allelopathic effects of mesquite (Prosopis juliflora) foliage on seed germination and seedling growth of bermudagrass (Cynodon dactylon). J Arid Environ 38(2):237–243

Ali AS, Tudsri S, Rungmekarat S, Kaewtrakulpong K (2012) Effect of feeding *Prosopis juliflora* pods and leaves on performance and carcass characteristics of Afar sheep. Kasetsart J Nat Sci 46:871–881

Al-Keblawy (2012) Impacts of native and exotic Prosopis species on native plants in aridlands of the UAE. International Conference on Ecology, Agriculture and Chemical Engineering December, Phuket, pp 18–19

Allam AM, Bakhashwain AA, Nagadi SA, Sallam SM (2012) Nutritive value assessment of some sub-tropical browses grown in arid region using in vitro gas production technique. J Food Agricult Environ 10:1339–1343

Almanza SG, Moya EG (1986) The uses of mesquite (*Prosopis* spp.) in the highlands of San Luis Potosi, Mexico. For Ecol Manag 16:49–56

Almaraz-Abarca N, da Graça Campos M, Ávila-Reyes J et al (2007) Antioxidant activity of polyphenolic extract of monofloral honeybee-collected pollen from mesquite (Prosopis juliflora, Leguminosae). J Food Compos Anal 20:119–124

Al-Marzooqi W, Al-Kharousi K, Kadim IT, Mahgoub O, Zekri S, Al-Maqbaly R, Al-Busaidi M (2015) Effects of feeding *Prosopis juliflora* pods with and without exogenous enzyme on performance, meat quality and health of broiler chickens. Int J Poult Sci 14:76–88

Al-Rawahy SH, Al-Dhafri KS, Al-Bahlany SS (2003) Germination, growth and drought resistance of native and alien plant species of the genus Prosopis in the sultanate of Oman. Asian J Plant Sci 2:1020–1023

Al-Soqeèr AA, Alsubaie QD, Motawei MI, Mousa HM, Ahmed M, Abdel-Salam AM (2017) Isolation and identification of allergens and biogenic amines of Prosopis juliflora genotypes. Electron J Biotechnol 30:24–32

Alvarez M, Leparmarai P, Heller G, Becker M (2017) Recovery and germination of Prosopis juliflora (Sw.) D.C. seeds after ingestion by goats and cattle. Arid Land Res Manag 31:71–80

Anderson S (2005) Spread of the introduced tree species *Prosopis juliflora* (Sw.) D.C. Sveriges lantbruksuniversitet: Institutionen för skoglig vegetationsekologi

Anderson LJ, van Klinken RD, Parr RJ, Climas R, Barton D (2006) Integrated management of hybrid mesquite: a collaborative fight against one of Australia's worst woody weeds. Fifteenth Australian Weeds Conference, Adelaide

Ansley RJ, Pinchak WE, Teague WR, Kramp BA, Jones DL, Jacoby PW (2004) Long-term grass yield following chemical control of honey mesquite. J Range Manag 57:49–57

Ansley RJ, Boutton TW, Mirik M, Castellano MJ, Kramp BA (2010) Restoration of C4 grasses with seasonal fires in a C3/C4 grassland invaded by *Prosopis glandulosa*, a fire-resistant shrub. Appl Veg Sci 13:520–530

Araüjo AH, Cardoso PC, Pereira RA, Lima LM, Oliveira AS, Miranda MRA, Xavier-Filho J, Sales MP (2002) In vitro digestibility of globulins from cowpea (*Vigna unguiculata*) and xerophitic algaroba (*Prosopis juliflora*) seeds by mammalian digestive proteinases: a comparative study. Food Chem 78:143–147

Archer S (1989) Have southern Texas savannas been converted into woodlands in recent history? Am Nat 134:545–561

Argôlo LDS, Pereira MLA, Dias JCT, Cruz JFD, Del Rei AJ, Oliveira CASD (2010) Mesquite pod meal in diets of lactating goats: ruminal parameters and microbial efficiency synthesis. Rev Bras Zootec 39:541–548

Assarehzadegan MA, Khodadadi A, Amini A, Shakurnia AH, Marashi SS, Ali-Sadeghi H, Zarinhadideh F, Sepahi N (2015) Immunochemical Characterization of *Prosopis Juliflora* Pollen Allergens and Evaluation of Cross-Reactivity Pattern with the Most Allergenic Pollens in Tropical Areas. Iran J Allergy Asthma Immunol 14(1):74–82

Badano E, Pugnaire FI (2004) Invasion of Agave species (Agavaceae) in south-east Spain: invader demo- graphic parameters and impacts on native species. Divers Distrib 10:493–500

Bahre CJ, Shelton ML (1993) Historic vegetation change, mesquite increases, and climate in southeastern Arizona. J Biogeogr 20:489–504

Barbier EB, Knowler D, Gwatipedza J, Reichard SH (2013) An economic analysis of the invasive plant problem associated with the horticulture industry in North America. Invas Plant Ecol:259

Beggs PJ (2015) Environmental allergens: from asthma to hay fever and beyond. Curr Clim Change Rep 1:176–184

Beggs PJ, Bambrick HJ (2005) Is the global rise of asthma an early impact of anthropogenic climate change? Environ Health Perspect 113:915–919. https://doi.org/10.1289/ehp.7724

Behrendt H, Ring J (2012) Climate change, environment and allergy. Chem Immunol Allergy 96:7–14

Belton T (2008) Management strategy for Mexican thorn (Prosopis juliflora) on Ascension Island: an assessment of this species, and recommendations for management. RSPB, Bedfordshire

Benata H, Mohammed O, Noureddine B, Abdelbasset B, Abdelmoumen H, Muresu R, Squartini A, El Idrissi MM (2008) Diversity of bacteria that nodulate Prosopis juliflora in the eastern area of Morocco. Syst Appl Microbiol 31(5):378–386

Bener A, Safa W, Abdulhaik S, Lestringant GG (2002) An analysis of skin prick test reaction in asthmatics in a hot climate and desert environment. Allery Immunol 34:281–286

Berhanu A, Tesfaye G (2006) The *Prosopis* dilemma, impacts on drylands biodiversity and some controlling methods. J Drylands 1(2):158–164

Bestelmeyer BT (2005) Does desertification diminish biodiversity? Enhancement of ant diversity by shrub invasions in south-western USA. Divers Distrib 11:45–55

Bethune S, Griffin M, Joubert D (2004) National review of invasive alien species Namibia. Windhoek, Directorate of Environmental Affairs, Ministry of Environment and Tourism

Bewley JD (1997) Seed germination and dormancy. Plant Cell 9:1055–1066

Bhatt SS, Chovatiya SG, Shah AR (2011) Evaluation of raw and hydrothermically processed *Prosopis juliflora* seed meal as supplementary feed for the growth of *Labeo rohita* fingerlings. Aquac Nutr 17:e164–e173

Bieberdord FW, Swinny B (1952) Mesquite and related plants in allergy. Ann Allergy 10:720

Bielory L, Lyons K, Goldberg R (2012) Climate change and allergic disease. Curr Allergy Asthma Rep 12:485–494

Bock CE, Kennedy L, Bock JH, Jones ZF (2007) Effects of fire frequency and intensity on velvet mesquite in an Arizona grassland. Rangel Ecol Manag 60:508–514

Bogino SM, Villalba R (2008) Radical growth and biological rotation age of *Prosopis caldenia* Burkart in Central Argentina. J Arid Environ 72:16–23
Bokrezion H (2008) The ecological and socio-economic role of Prosopis juliflora in Eritrea: an analytical assessment within the context of rural development in the Horn of Africa. PhD thesis, Johannes Gutenberg University, Mainz
Borokini TI, Babalola FD (2012) Management of invasive plant species in Nigeria through economic exploitation: lessons from other countries. Manage Biol Inv 3(1):45–55
Botswana Government (2009) Botswana fourth national report to the convention of biological diversity. Botswana Government, Gaborone
Bradley BA, Curtis CA, Chambers JC (2016) Bromus Response to Climate and Projected Changes with Climate Change. M.J. Germino et al. (eds.), Exotic Brome-Grasses in Arid and Semiarid Ecosystems of the Western US, Springer Series on Environmental Management. https://doi.org/10.1007/978-3-319-24930-8_9
Bragg LH, Bacon JD, McMillan C et al (1978) Flavonoid patterns in the Prosopis juliflora complex. Biochem Syst Ecol 6:113–116
Brown K (1991) Biology of *Prosopis cineraria* (Leguminosae) in the Sultanate of Oman. Phd in Science. Durham University, Durham
Brown JR, Archer S (1989) Woody plant invasion of grasslands: establishment of honey mesquite (*Prosopis gnaldulosa* var. *glandulosa*) on sites differing in herbaceous biomass and grazing history. Oecologia 80:19–26
Brown AF, Massey RE (1929) Flora of the Sudan. Thomas Murby and CO, p 376
Brown CJ, Macdonald IAW, Brown SE (1985) Invasive alien organisms in South West Africa. Foundation for Research Development, Namibia
Browning DM, Archer S, Asner G, McClaran MP, Wessman CA (2008) Woody plants in grasslands post-encroachment and stand dynamics. Ecol Appl 18:928–944
Burkart A (1976) A monograph of the genus *Prosopis* (Leguminosae subfam. Mimosoideae). (Part 1 and 2). Catalogue of the recognized species of *Prosopis*. J Arnold Arborat 57: 219–249; 450–525
Busch M, Knight C, Hodara CNMK, Chaneton EME (2012) Rodent seed predation on tree invader species in grassland habitats of the inland Pampa. Ecol Res 27:369–376
Cadman A (1991) Airspora of Johannesburg and Pretoria, South Africa,1987/88. II. Meteorological relationships. Grana 30:181–183
Callaway RM, Ridenour WM (2004) Novel weapons: invasive success and the evolution of increased competitive ability. Front Ecol Environ 2:436–443
Cao X, Ma L, Gao B, Harris W (2009) Dairy-manure derived biochar effectively sorbs lead and atrazine. Environ Sci Technol 43(9):3285–3291
Cardoso MB, Ladio AH, Lozada M (2012) The use of firewood in a Mapuche community in a semi-arid region of Patagonia, Argentina. Biomass Bioenergy 46:155–164
Cecchi L, D'Amato G, Ayres JG, Galan C, Forastiere F, Forsberg B, Gerritsen J, Nunes C, Behrendt H, Akdis C, Dahl R, Annesi-Maesano I (2010) Projections of the effects of climate change on allergic asthma: the contribution of aerobiology. Allergy 65:1073–1081
Chandrasekaran S, Swamy PS (Undated) Ecological and socio-economic impacts of *Prosopis juliflora* invasion into the semi-arid ecosystems in selected villages of Ramand district in Tamilnadu. Madurai Kamaraj University, Madurai
Chaturvedi OH, Sahoo A (2013) Nutrient utilization and rumen metabolism in sheep fed *Prosopis juliflora* pods and *Cenchrus* grass. Springer Plus 2:1–7
Chaudry B, Vijayaraghavan MR (1992) Structure and function of the anther gland in Prosopis juliflora (Leguminosae, Mimosoideae): a histochemical analysis. Phyton (Horn) 32:01–07
Chikuni MF, Dudley CO, Sambo EY (2004) *Prosopis glandulosa* Torry (Leguminosae-Mimosoidae) at Swang'oma, Lake Chilwa plain: a blessing in disguise. Malawi J Sci Technol 7:10–16
Choge SK, Chikami BN (2004) Experiences of Prosopis utilization and management from outside Kenya. In: Proceedings of the Workshop on Integrated Management of Prosopis Species in Kenya. KEFRI, Nairobi, Kenya

Choge SK, Ngujiri FD, Kuria MN, Busaka EA, Muthondeki JK (2002) The status and impact of Prosopis spp. in Kenya. KEFRI, Nairobi

Choge SK, Harvey M, Chesang S, Pasiecznik NM (2006) Cooking with Prosopis flour. Recipes tried and tested in Bango District. Kenya. Nairobi and Coventry, KEFRI and HDRA

Choge SK, Pasiecznik NM, Harvey M, Wright J, Awan SZ, Harris PJC (2007) *Prosopis* pods as human food, with special reference to Kenya. Water South Afr 33:419–424

Choge SK, Clement N, Gitonga M, Okuye J (2012) Status report on commercialization of Prosopis tree resources in Kenya, Technical report for the KEFRI/KFS Technical Forest Management and Research Liaison Committee. KEFRI, Nairobi

Cibichakravarthy B, Preetha R, Sundaram SP, Kumar K, Balachandar D (2012) Diazothrophic diversity in the rhizosphere of two exotic weed plants, *Prosopis juliflora* and *Parthenium hysterophorus*. World J Microbiol Biotechnol 28:605–613

Cienciala E, Centeio A, Balazek P, da Cruz Gomes Soares M, Russ R (2013) Estimation of stem and tree level biomass models for *Prosopi juliflora/pallida* applicable to multi-stemmed tree species. Trees 27:1061–1070

Coetzer W, Hoffmann JH (1997) Establishment of *Neltumius arizonensis* (Coleoptera: Bruchidae) on Mesquite (*Prosopis* species: Mimosaceae) in South Africa. Biol Control 10:187–182

Costa AC, Souchie EL, Teixeira MB, Resende O, Soares FAL, Saléh BB, Faria DJG, Soares MA, Ceballos DCJ, Martins EM (2018) Allelopathic effect of aqueous extracts of Prosopis juliflora (SW.) D.C. in the emergence and initial growth of Mimosa tenuiflora (WILLD.) Poiret seedlings. Conteúdo da revista. Capa 11(1)

Creamer CA, Filley TR, Olk DC, Stott DE, Dooling V, Boutton TW (2013) Changes to soil organic N dynamics with leguminous woody plant encroachment into grasslands. Biogeochemistry 113:307–321

Csurhes S (ed) (1996) Mesquite (Prosopis spp) in Queensland. Department of Natural Resources and Mines, Australia

da Silva CGM, Stamford TLM, de Andrade SAC, de Souza EL, de Araújo JM (2010) Production of ethanol from mesquite (*Prosopis juliflora* (SW) D.C.) pods mash by *Zymomonas mobilis* and *Saccharomyces cerevisiae*. Electron J Biotechnol 13:12–13

Davis OK (1995) Climate and vegetation patterns in surface samples from arid western U.S.A.: application to Holocene climatic reconstructions. Palynology 19:97–120

Dayal V (2007) Social diversity and ecological complexity: how an invasive tree could affect diverse agents in the land of the tiger. Environ Develop Eonom 12:553–571

De Andrade AA, Fabricante JR, de Oliveira FX (2010) Impactos da invasao de *Prosopis juliflora* (sw.) D.C.. (Fabaceae) sobre o estrato arbustivo-arboreo em areas de Caatinga no Estafo da Paraiba, Braazil. Maringá 32:249–255

de Dios V, Weltzin JF, Huxman TE, Williams DG (2012) Windows of opportunity for *Prosopis velutina* seeding establishment and encroachment in a semiarid grassland. Perspect Plant Ecolo Evol System 14:275–282

de Oliveira Moraes GS, de Souza EJ, Véras AS, de Paula Almeida M, da Cunha MV, Torres TR, Pereira GFC (2016) Total replacement of corn by mesquite pod meal considering nutritional value, performance, feeding behavior, nitrogen balance, and microbial protein synthesis of Holstein-Zebu crossbred dairy steers. Trop Anim Health Prod 48:1415–1420

De Oliveira VR, Pires IE (1990) Pollination efficiency of P. juliflora (Sw) D.C. in Petrolina, Pernambuco. In: Habit MA, Saavedra JC (eds) The current state of knowledge on P. juliflora. FAO, Rome

de Oliveira LSB, de Andrade LA, Fabricante JR, Gonçalves GS (2012) Structure of a *Prosopis juliflora* (Sw.) D.C.. Population established in a temporary riverbed in the microregion of Cariri in the Stat of Paraiba. Semina Ciências Agrárias Londrina 33(5):1769–1778

de Villalobos AE, Paláez DV, Elia OR (2005a) Factors related to establishment of *Prosopis caldenia* Burk. Seedlings in central rangelands of Argentina. Acta Oecol 27:99–106

de Villalobos AE, Paláez DV, Elia OR (2005b) Growth of *Prosopis cladenia* Burk: seedlings in central semi-arid rangelands of Argentina. J Arid Environ 61:345–346

de Villalobos AE, Peláez DV, Bóo RM, Mayor MD, Elia OR (2007) Effect of a postfire environment on the establishment of *Prosopis caldenia* seedlings in central semiarid Argentina. Austral Ecology 32:581–591
Dean WRJ, Anderson MD, Milton SJ, Anderson TA (2002) Avian assemblages in native Acacia and alien Prosopis drainage line woodland in the Kalahari, South Africa. J Arid Environ 51:1–19
DeLoach CJ (1984) Conflicts of interest over beneficial and undesirable aspects of Prosopis (Prosopis spp.) in the United States as related to biological control. Vancouver, Canada: 6th international symposium on biological control, 19–25 August 1984, pp 301–340
Demissie H (2009) Invasion of *Prosopis juliflora* (Sw.) D.C.. into Awash National Park and its impacts on plant species diversity and soil characteristics. Masters in Science. Addis Ababa University, Addis Ababa
Dhyani A, Arora N, Gaur SN, Jain VK, Sridhara S, Singh BP (2006) Analysis of IgE binding proteins of mesquite (Prosopis juliflora)pollencross-reactivitywithpredominant tree pollens. Immunobiology 211:733–740
Dhyani A, Arora N, Jain VK, Sridhara S, Singh BP (2007) Immunoglobulin E (IgE)-mediated cross-reactivity between mesquite pollen proteins and lima bean, an edible legume. Clin Exp Immunol 149:517–524
Diagne O (1996) Utilization and nitrogen fixation of *Prosopis juliflora* in Senegal. In: Felker P, Moss J (eds) Prosopis: Semiarid fuelwood and forage tree; Building consensus for the disenfranchised. Kingsville, Centre for Semi-Arid Forest Resources
Diaz A (1995) Los Algarrobos. Consejo Nacional de Ciencia y Tecnología (CONCYTEC). Lima, Perú, 207 pp
Díaz Celis A (1995) Los Algarrobos. CONCYTEC, Lima, Peru
Distel RA, Peláez DV, Bóo RM, Mayor MD, Elia OR (1996) Growth of *Prosopis caldenia* seedlings in the fields as related to grazing history of the site and in a greenhouse as related to different levels of competition from *Stipa tenuis*. J Arid Environ 32:251–257
Djoudi H, Brockhaus M, Locatelli B (2011) Once there was a lake: vulnerability to environmental changes in northern Mali. Reg Environ Chang. https://doi.org/10.1007/s10113-011-0262-5
dos Santos E, Pereira ML, Almeida PJ, Moreira JV, Souza AC, Pereira CA (2015) Mesquite pod meal in sheep diet: intake, apparent digestibility of nutrients and nitrogen balance. Acta Scientiarum. Anim Sci 37:55–59
Du Puy (1990) *Prosopis juliflora* (Sw.) D.C.. From Madagascar. Royal Botanic Gardens, Kew
Dudley BD, Flint Hughes R, Ostertag R (2014) Groundwater availability mediates the ecosystem effects of an invasion of Prosopis pallida. Ecol Appl 24(8):1954–1971
Dufgas WA, Mayeux HS (1991) Evaporation from rangeland with and without honey mesquite. J Range Manag 44:161–170
Dussart E, Peinetti R, Boninesegra JA (1997) Análisis del crecimiento de *Prosopis caldenia* (L) Burk. En relación con parametros ambientales y fuego. Reunión Argentina de Ecologia, Buenos Aires
Dutch Caribbean Biodiversity Explorer (D.C.BE) (2009) *Prosopis*. Government of the Netherlands Antilles, Netherlands Antilles
Dzikiti S, Schachtschneider K, Naiken V, Gush M, Moses G, Le Maitre DC (2013) Water relations and the effects of clearing invasive *Prosopis* trees on groundwater in an arid environment in the Northern Cape, South Africa. J Arid Environ 90:103–113
Eggemeyer KD, Schwinning S (2009) Biogeography of woody encroachment: why is mesquite excluded from shallow soils? Ecohydrology 2:81–87
Ehrenfeld JG, Kourtev P, Huang WZ (2001) Changes in soil functions following invasions of exotic understory plants in deciduous forests. Ecol Appl 11:1287–1300
El Fadl MA (1997) Management of *Prosopis juliflora* for use in agroforestry systems in the Sudan, Tropical forestry reports 16. University of Helsinki, Helsinki
Elfadl MA (1997a) Management of *Prosopis juliflora* for use in agroforestry systems in the Sudan, Tropical forestry reports 16. University of Helsinki, Helsinki

Elfadl MA (1997b) Management of *Prosopis juliflora* for muse in agroforestry in the Sudan. Doctoral dissertation, PhD thesis. Department of forest ecology, Tropical Silviculture Unit, University of Helsinki, Finland

Elfadl MA, Luukkanen O (2006) Field studies on the ecological strategies of Prosopis juliflora in a dryland ecosystem 1: a leaf gas exchange approach. J Arid Environ 66(1):1–15

El-Keblawy A, Abdelfatah MA (2014) Impacts of native and invasive exotic Prosopis congeners on soil properties and associated flora in the arid United Arab Emirates 100–101:1–8

El-Keblawy A, Al-Rawai A (2005) Effect of salinity, temperature and light on germination of invasive *Prosopis juliflor* (Sw.) D.C. J Arid Environ 61(4):555–565

El-Keblawy A, Al-Rawai A (2007) Impacts of the invasive exotic *Prosopis juliflora* (Sw.) D.C. on the native flora and soils of the UAE. Plant Ecol 190:23–35

El-Keblawy A, Ksiksi T (2005) Artificial forests as conservation sites for native flora of the UAE. For Ecol Manag 213:288–296

Emlan JT (1986) Land-bird densities in matched habitats on six Hawaiian Islands: a test of resource regulation theory. Am Naturalist 127:125–139

Ethiopia Government (2002) Ethiopia-national plan for *Prosopis* – final draft. Ethiopia Government

Ethridge DE, Dahl BE, Sosebee RE (1984) Economic evaluation of chemical mesquite control using 2,4,5-T. J Range Manag 37:152–156

FAO (2004) The introduction of *Prosopis* spp. In the drylands of the world ledge of plight for biodiversity. Food and Agricultural Organization of the United Nations, Rome

FAO (2006) Problems posed by the introduction of Prosopis spp. in selected countries. Plant Production and Protection Division, Food and Agricultural Organization of the United Nations, Rome, pp 233–239

FARM-Africa (2008) Experiences on Prosopis management case of Afar region. FARM-Africa, Ethiopia

Felker P (1987) Recommendations for development of *Prosopis* in Pakistan. Centre for Semi-Arid Forest Resources, Texas

Felker P (2003) Management, use and control of *Prosopis* in Yemen. Yemen: Project Number: TCP/YEM/0169(A)

Felker P, Clark PR, Nash P, Osborn JF, Cannell GH (1982) Screening Prosopis (mesquite) for cold tolerance. For Sci 28:556–562

Finkelstein R (2006) Abscisic acid: a seed maturation and antistress signal. In: Taiz L, Zeiger E (eds) Plant physiology, 4th edn. Sinauer Associates, Sunderland, pp 539–558

Foroughbakhch R, Parra AC, Pinero JLH, Vázquez MAA, Estrada AR, Cardenas ML (2012) Wood volume production ans use of 10 woody species in semiarid zones of northeastern Mexico. Intl J For Res. https://doi.org/10.1155/2012/529829

Fosberg FR, Sachet M, Oliver R (1976) A geographical checklist of the Micronesian dicotyledonae. Micronesia 15:1–295

Gairola S, Mahmoud T, Shabana H, El-Keblawy A (2016) Growing knowledge about the floral diversity of United Arab Emirates: new additions and conservation through seed banking. Tribulus 24:136–143

Gallaber T, Merlin M (2010) Biology and impacts of Pacific Island invasive species. 6. *Prosopis pallida* and *Prosopis juliflora* (Algarroba, Mesquite, Kiawe) (Fabaceae). Pac Sci 64:489–526

Garg VK, Singh B (2003) Macronutrient dynamics and use efficiency in three species of short rotation forestry developed on sodic soils in North India. J Trop For Sci 15:289–302

Geesing D, Al-Khawlani M, Abba ML (2004) Management of introduced *Prosopis* species: can economic exploitation control and invasive species? Unasylva 217:36–44

Ghazaly UF (2006) Community-based management of invasive *Prosopis juliflora* in Egypt. GISP, Washington, DC

Ghazanfar SA (1996) Invasive *Prosopis* in the Sultanate of Oman. Alien 3:10

Girma M, Urge M, Animut G (2012) Ground *Prosopis juliflora* pods as feed ingredient in poultry diet: effects on nutrient intake, muscle fatty acid composition, sensory quality and hematology of broilers. Pak J Nutr 11:1014–1022

Global Biodiversity Information Facility (GBIF) (2013) *Prosopis species lists*. GBIF, Copenhagen
Global Invasive Species Database (ISSG) (2005) *Prosopis*. IUCN
Goel VL, Behl HM (1995) Propagation of Prosopis juliflora from rooted stem cuttings. Intl Tree Crops J 8:193–201
Goel U, Saxena DB, Kumar B (1989) Comparative study of allelopathy as exhibited by *Prosopis juliflora* swartz and *Prosopis cineraria* (L) Druce. J Chem Ecol 15:591–600
Gomez JM, Bosch J, Perfectti F, Fernandez J, Abdelaziz M (2007) Pollinator diversity affects plant reproduction and recruitments: the tradeoffs of generalization. Oecologia 153:597–605
Goslee SC, Havstad KM, Peters DP, Rango A, Schlesinger WH (2003) High-resolution images reveal rate and pattern of shrub encroachment over six decades in New Mexico, U.S.A. J Arid Environ 54:755–767
Graham JD (1960) Morphological variation in mesquite (*Prosopis*, Leguminosae) in the lowlands of northeastern Mexico. Southwest Nat:187–193
Grandtner MM (2005) *Elsevier's directory of trees*. ELSEVIER, Amsterdam
Gritzner JA (1979) Environmental degradation in Mauritania: staff report. Board on Science and Technology for International Development, Commission on International Relations and National research Council, Mauritania
Gupta R, Sharma KK, Kuhad RC (2009) Separate hydrolysis and fermentation (SHF) of Prosopis juliflora, a woody substrate, for the production of cellulosic ethanol by Saccharomyces cerevisiae and Pichia stipitis-NCIM 3498. Bioresour Technol 100(3):1214–1220
Habit MA, Saavedra JA (1990a) The current state of knowledge on *Prosopis juliflora*. II international conference on *Prosopis*. Brazil 25–29 August. FAO, Rome
Habit MA, Saavedra JC (1990b) The current state of knowledge on *Prosopis juliflora*. FAO, Rome
Habit MA, Saavedra JC (eds) (1990c) The current state of knowledge on *Prosopis juliflora*. FAO, Rome
Haji J, Mohammed A (2013) Economic impact of Prosopis juliflora on agropastoral households of Dire Dawa Administration, Ethiopia. Afr J Agric Res 8(9):768–779
Hall M, Llewellyn OA, Miller AG, Al-Abbasi TM, Al-Wetaid AH, Al-Harbi RJ, Al-Shammari KF (2010) Important plant areas in the Arabian Peninsula: 2. Farasan Archipelago. Edinb J Bot 67(2):189–208
Harding GB (1987) The status of Prosopis as a weed. Plant Protection Research Institute, Uitenhage
Harding GB, Bate GC (1991) The occurrence of invasive Prosopis species in the north- western cape, South Africa. S Afr J Sci 87:188–192
Harfi HH, Alsaeed AH (2011) Allergenicity to allergins like *Prosopis juliflora* and date tree pollens in Saudi Arabia. Kuwait Med J 43:109–112
Hasan MK, Alam AKAA (2006) Land degradation situation in Bangladesh and the role of Agroforestry. J Agricult Rural Develop 4:19–25
Havstad KM, James D (2010) Prescribed burning to affect a state transition in a shrub-encroached desert grassland. J Arid Environ 74:1324–1328
Hawaii Ecosystem at Risk Project (1998) Prosopis pallida (Fabaceae). Map of estimated distribution on Molokai. Hawaii Ecosystem at Risk Project, Hawaii
Hawke PR, Meadows ME (1989) Winter airspora spectra and meteorological conditions in Cape Town, South Africa. Grana 28:187–192
Heiss S, Fischer S, Muller WD, Weber B, Hirschwehr R, Spitzauer S et al (1996) Identification of a 60 kD cross-reactive allergen in pollen and plant derived food. J Allergy Clin Immunol 98:938–947
Heitschmidt RK, Ansley RJ, Dowhower SL, Jacoby PW, Price DL (1988) Some observations from the excavation of honey mesquite root systems. J Range Manag 41:227–231
Hennessy JT, Gibbens RP, Tromble JM, Cardenas M (1983) Vegetation changes from 1935 to 1980 in mesquite dunelands and former grasslands of southern New Mexico. J Range Manag:370–374
Henschel JR, Parr T (2010) Population changes of alien invasive plants in the Lower Kuiseb River. DINTERIA No. 31, Windhoek, pp 5–17

Hierro JL, Callaway RM (2003) Allelopathy and exotic plant invasion. Plant Soil 256:29–39
Hintsa K, Balehegn M, Birhane E (2015) Utilization of pods of *Prosopis juliflora*, an invasive tree, as a replacement to concentrate feed for goats in Ethiopia. Livest Res Rural Dev 27:9
Hoffman MT, Sonnenberg D, Hurford JL, Jagger BW (1995) Ecology and management of Riemvasmaak's natural resources. National Botanical Institute, Cape Town
Hoffmann JH, Impson FAC, Moran VC (1993) Competitive interactions between two Burchid species (*Algarobius* spp.) introduced into South Africa for biological control of mesquite weeds 9*Prosopis* spp. Biol Control 3:215–220
Hollister EB, Schadt CW, Palumbo AV, Ansley JR, Boutton TW (2010) Structural and functional diversity of soil bacteria and fungal communities following woody plant encroachment in the southern Great Planes. Soil Biol Biochem 42:1816–1824
Hoshino B, Yonemori M, Manayeva K, Karamalla A, Yoda K, Suliman M, Elgamri M, Nawata H, Mori Y, Yabuki S, Aida S (2011, July). Remote sensing methods for the evaluation of the mesquite tree (*Prosopis juliflora*) environmental adaptation to semi-arid Africa. In: Geoscience and Remote Sensing Symposium (IGARSS), 2011 IEEE International. IEEE, pp 1910–1913
Hoshino B, Karamalla A, Abd Elbasit MAM, Manayeva K, Yoda K, Suliman M, Elgamri M, Nawata H, Yasuda H (2012) Evaluating the invasion strategic of Mesquite (*Prosopis juliflora*) in eastern Sudan using remotely sensed technique. J Arid Land Stud 22:1–4
Hundessa N (2016) Distribution and socio-economic impacts of Prosopis juliflora in East Shewa and West Arsi zones, Ethiopia. Intl J Afr Asian Stud 24:31–41
Hunziker J-H, Saidman BO, Naranjo CA, Palacios RA, Poggio L, Burghardt AD (1986) Hybridization and genetic variation of Argentina species of Prosopis. For Ecol Manag 16:301–316
Hussain MI, Reigosa MJ (2011) Allelochemical stress inhibits growth, leaf water relations, PSII photochemistry, non-photochemical fluorescence quenching and heat energy dissipation in three C_3 perennial species. J Exp Bot 62:4533–4545
Hussain SS, Ahmed M, Siddiqui MF, Wahab M (2010) Threatened and endangered native plants of Karachi. Int J Biol Biotechnol 7(3):259–266
Hussain MI, González L, Souto C, Reigosa MJ (2011) Ecophysiological responses of native plants to phytotoxic effect of *Acacia melanoxylon* R. Br. Agrofor Syst 83:149–166. https://doi.org/10.1007/s10457-011-9433-0
Ibrahim F (2004) No place like home: history, politics and mobility among a pastoral nomadic community in western India. Nomadic Peoples 8:168–190
Ibrahim M, Nadir M, Ali A et al (2013) Phytochemical analyses of Prosopis juliflora Swartz D.C. Pak J Bot 45:2101–2104
Immerzeel P, Schols HA, Voragen AGJ, de Vries SC (2004) Different arabinogalactan proteins are present in carrot (Daucus carota) cell culture medium and in seeds. Physiol Plant 122:181–189
Inderjit (2001) Soils: environmental effect on allelochemical activity. Agron J 93:79–84
Inderjit S, Callaway TR, Pollock RM, Kaur JL (2008) Allelopathy and plant invasions: traditional, congeneric, and biogeographical approaches. Biol Invasions 10:875e890
Inderjit EH, Crocoll C, Bajpai D, Kaur R, Feng Y, Silva C, Carreo'n JT, Valiente-BanuetA GJ, Callaway RM (2011) Volatile chemicals from leaf litter are associated with invasiveness of a neotropical weed in Asia. Ecology 92:316–324
Invasive Species Compendium (CABI) (2005) *Prosopis*. CABI, Oxfordshire
IPCC. Summary for policymakers. In: Parry ML, Canziani OF, Palutikof JP, van der Linden
Issa S, Dohai B (2008) GIS analysis of invasive *Prosopis juliflora* dynamics in two selected sites from the United Arab Emirates. Can J Pure Appl Sci 2:235–242
Jackson RB, Banner JL, Jobbágy EG, Pockman WT, Wall DH (2002) Ecosystem carbon loss with woody plant invasion of grasslands. Nature 418(6898):623–626
Jaganathan GK, Yule K, Liu B (2016) On the evolutionary and ecological value of breaking physical dormancy by endozoochory. Perspect Plant Ecolo Evol System 22:11–22
Jayaram K, Prasad MNV (2009) Removal of Pb (II) from aqueous solution by seed powder of Prosopis juliflora D.C. J Hazard Mater 169(1–3):991–997

Jensen AM, Hajej MS (2001) The road of hope: control of moving sand dunes in Mauritania. Unasylva 207:31–36

Jhala YV (1993) Predation on blackbuck by wolves in the Velavadar National Park, Gujarat, India. Conserv Biol 7:874–881

Johns RE, Lee JS, Agahian B, Gibbons HL, Reading JC (1986) Respiratory effects of mesquite broiling. J Occup Med 28:1181–1184

Johnston MC (1962) The North American mesquites Prosopis Sect. Algarobia (Leguminosae). Brittonia 14(1):72–89

Jones M, Sinclair F, Grime V (1998) Effect of tree species and crown pruning on root length and soil water content in semi-arid agroforestry. Plant Soil 201:-197

Joshi EB, Joshi PN, Jain BK (2012) Phytosociological study and management action for the invasive weed : *Prosopis juliflora* (SW.) D.C.. in Tapkeshwari Hill Ranges in the Kachchh Island, Gujarat, India. J Non-Timber Forest Prod 19:29–36

Jyrkama MI, Sykes JF (2007) The impact of climate change on spatially varying groundwater recharge in the grand river watershed (Ontario). J Hydrol 338(3):237–250

Kahi CH, Ngugi RK, Mureithi SM, Ng'ethe JC (2009) The canopy effects of *Prosopis juliflora* (D.C..) and *Acacia tortillis* (hayne) trees on herbaceous plants species and soil physio-chemical properties in Njemps Flats, Kenya. Trop Subtrop Agroecosyst 10:441–449

Karron JD, Mitchell RJ (2011) Effects of floral display size on male and female reproductive success in Mimulus ringens. Annals Bot. https://doi.org/10.1093/aob/mcr193

Kaur R, Gonzáles WL, Llambi LD, Soriano PJ, Callaway RM, Rout ME et al (2012) Community impacts of *Prosopis juliflora* invasion: biogeographic and congeneric comparisons. PLoS One 7(9):e44966

Kaur G, Singh BP, Nagpal AK (2013) Phenology of some phanerogams (trees and shrubs) of Northwestern Punjab, India. J Bot 2013

Kazmi JH (2009) Ecological and Socio-economic evaluation of the use of *Prosopis juliflora* for bio-char production in Pakistan. Drynet, Pakistan

Kedzie-Webb SA, Sheley RL, Borkowski JJ, Jacobs JJ (2001) Relationships between Centaurea maculosa and indigenous plant assemblages. Western North Am Nat 61:43–49

Keys RN (1993) Mating systems and pollination biology of velvet mesquite (Prosopis velutina Wooton). PhD dissertation, The University of Arizona. Tucson, AZ

Khan I, Marwat KB, Khan IA, Ali H, Dawar K, Khan H (2011) Invasive weeds of southern districts of Khyber Pakhtunkhwa -Pakistan. Pakistani J Weed Sci Res 17(2):161–174

Khider MM, Al Abjar ZA, Eltigani A (2011) Studies on ecology and biology of some insect pests of the mesquite trees *Prosopis juliflora* in Sudan. Egypt J Biol Pest Contr 21:353–359

Killian S, McMichael J (2004) The human allergens of mesquite (Prosopis juliflora). Clini Mol All 2(1):8

Kim SH, Park HS, Jang JY (2011) Impact of meteorological variation on hospital visits of patients with tree pollen allergy. BMC Public Health 11:890

Kneuper CL, Scott CB, Pinchak WE (2003) Consumption and dispersion of Mesquite seeds by ruminants. J Range Manag 56:255–259

Knox RB, Suphioglu C, Taylor P, Desai R, Watson HC, Peng JL, Bursill LA (1997) Major grass pollen allergen Lol p 1 binds to diesel exhaust particles: implications for asthma and air pollution. Clin Exp Allergy 27(3):246–251

Kolar CS, Lodge DM (2001) Progress in invasion biology: predicting invaders. Trends Ecol Evol 16(4):199–204

Kreuter UP, Amestoy HE, Kothmann MM, Ueckert DN, McGinty WA, Cummings SR (2005) The use of brush management methods: a Texas landowner survey. Rangel Ecol Manag 58:284–291

Kueffer C, Lavergne C (2004) Case studies on the status of invasive woody plant species in the western Indian ocean. FAO, Rome

Kumar ASHWANI, Kotiya AMIT (2004) Some Potential Plants for Bio-energy. Biomass for energy industry and climate protection. WIP-Munich, pp 180–183

Landeras G, Alfonso M, Pasiecznik NM, Harris PJC, Ramirez L (2006) Identification of Prosopis juliflora and Prosopis pallida accessions using molecular markers. Biodivers Conserv 15(5):1829–1844
Lara-Núñez A, Romero-Romero T, Blancas V, Ventura JL, Anaya AL, Cruz-Ortega R (2006) Allelochemical stress cause inhibition of growth and oxidative damage in *Lycopersicon esculentum*. Plant Cell Environ 29:2009–2016
Larrea-Alcázar DM, Soriano PJ (2008) Columnar cacti-shrub relationships in an Andean semiarid valley in Western Venezuela. Plant Ecol 196:153–161
Lasalle JC (1962) El increment de la masa forestall del caldén (*Prosopis caldenia* Burk). Rev Forestal Argentina 6:44–50
Latorre F, Perez CF (1997) One year of airborne pollen sampling in Mar del Plata (Argentina). Grana 36:49–53
Lauenstein DAL, Fernanández ME, Verga AR (2013) Drought stress tolerance of *Prosopis chilensis* and *Prosopis flexuosa* species and their hybrids. Trees 27:285–296
Laxén J (2007) Is Prosopis a curse of a blessing? – an ecological-economic analysis of the invasive alien tree species in Sudan. Doctorate of Science dissertation, Faculty of Agriculture and Forestry, University of Helsinki, Helsinki
Leão TCC, de Almeida WR, de Sá Dechoum M, Ziller SR (2011) Espécies exóticas invasoras: no Nordeste do Brasil. CEPAN Instituto Hórus, Brazil
Lee SG, Russel EJ, Bingham RL, Felker P (1992) Discovery of thornless, non-browsed, erect tropical *Prosopis* in 3-year-old Hatian progeny trials. For Ecol Manag 48:1–13
Lemlem N (2003) Mapping of Prosopis juliflora using remote sensing and GIS – case srudy of Marigat Division, Baringo District. Bachelor of Science in Surveying. University of Nairobi, Nairobi
Lewis WH (1986) Airborne Pollen of the Neotropics. Grana 25:75–83
Lewis WH, Elvin Lewis MPF (2004) Medical Botany. Wiley, Plants Affecting Human Health. Economic Botany
Liao C, Peng R, Luo Y, Zhou X, Wu X, Fang C, Li B (2008) Altered ecosystem carbon and nitrogen cycles by plant invasion: A meta-analysis. New Phytol 177(3):706–714
Little EL, Wadsworth FH (1964) Common trees of Porto Rico and the Virgin Islands. Department of Agriculture, Forest Services, Washington, DC
Lloyd J, Mannan RW, Destefano S, Kirkpatrick C (1998) The effects of mesquite invasion on a southeastern Arizona grassland bird community. Wilson Bull 110:403–408
López-Portillo J, Montana C (1999) Spatiall distribution of *Prosopis glandulosa* var. *torreyana* in vegetation strips of the southern Chihuahuan Desert. Acta Oecol 20:197–208
Lorence DH, Wagner WL (2013) Flora of the Marquesas Islands. National Tropical Botanical Garden and the Smithsonian Institution
Lucas SK, Buckley CE (1989) Quantitative studies of cutaneous hypersensitivity. The prevalence of epicutaneous flare reactions to allergenic pollen extracts. J Allergy Clin Immunol 84:465–474
Luque GM, Bellard C, Bertelsmeier C, Bonnaud E, Genovesi P, Simberloff D, Courchamp F (2014) The 100th of the world's worst invasive alien species. Biol Invasions 16(5):981–985
Lynes BC, Campell SD (2000) Germination and viability of mesquite (*Prosopis pallida*) seed following ingestion and excretion by feral pigs (*Sus scrofa*). Tropical Grasslands 34:125–128
Mabrouk H, Hilmi E, Abdullah M (2008) Nutritional value of *Prosopis juliflora* pods in feeding nile tilapia (*Oreochromis niloticus*) fries. Arab Gulf J Scientific Res 26:49–62
MacKee HS (1994) Catalogue des plantes introduites et cultivées en Nouvelle-Calédonie. Muséum Natural d'Historie, Paris
Mahgoub O, Kadim IT, Johnston EH, Srikandakumar A, Al-Saqri NM, Al-Abri AS, Ritchie A (2005a) The use of concentrate containing Meskit (*Prosopis juliflora*) pods and date palm by-products to replace commercial concentrate in diets of Omani sheep. Anim Feed Sci Technol 120:33–41

Mahgoub O, Kadim IT, Forsberg NE, Al-Ajmi DS, Al-Saqry NM, Al-Abri AS, Annamalai K (2005b) Evaluation of meskit (*Prosopis juliflora*) pods as a feed for goats. Anim Feed Sci Technol 121:319–327

Makkar HPS, Singh B, Negi SS (1990) Tannin levels and their degree of polymerisation and specific activity in some agro-industrial by-products. Biol Wastes 31:137–144

Malhotra S, Misra K (1981a) 3,30-di-O-methylellagic acid 4-Orhamnoside from the roots of Prosopis juliflora. Phytochemistry 20:2043–2044

Malhotra S, Misra K (1981b) Ellagic acid 4-O-rutinoside from pods of Prosopis juliflora. Phytochemistry 20:2439–2440

Malhotra S, Misra K (1981c) An ellagic acid glycoside from the pods of Prosopis juliflora. Phytochemistry 20:860–861

Malhotra S, Misra K (1983) New flavanones from Prosopis juliflora roots. Planta Med 47:46–48

Malik P, Singh AB, Babu CR, Gangal SV (1990) Head-high, airborne pollen grains from different areas of metropolitan Delhi. Allergy 45:298–305

Mandu P, Kibet S, Morimoto Y, Imbumi M, Adeka R (2009) Impact of Prosopis juliflora on Kenya's semi-arid and arid ecosystems and local livelihoods. Biodiversity 2(3):33–50

Manga VK, Sen DN (1995) Influence of seed traits on germination in *Prosopis cineraria* (L.) MacBride. *J Arid Environ* **31**:371–375

Mann HS, Shankarnarayan KA (1980) The role of Prosopis cineraria in an agropastoral system in Western Rajasthan. Browse in Africa. International Livestock Centre for Africa, Addis Ababa, pp 437–442

Mari A (2008) When does a protein become an allergen? Searching for a dynamic definition based on most advanced technology tools. Clin Exp Allergy 38:1089–1094. PMID: 18477011; Bergmann KC, Ring J. (2014) History of allergy. Karger AG, Basel, pp 172–192

Maundu P, Kibet S, Morimoto Y, Imbumi M, Adeka R (2009) Impact of *Prosopis juliflora*on Kenya's semi-arid and arid ecosystems and local livelihoods. Biodiversity 10:33–50

Maundu P, Kibet S, Morimoto Y, Imbumi M, Adeka R Impact of *Prosopis juliflora* on Kenya's semi-arid and arid ecosystems and local livelihoods. Biodiversity 10:33–50

Mazibuko DM. 2012. Phylogenetic relationships of Prosopis in South Africa: an assessment of the extent of hybridization, and the role of genome size and seed size in the invasion dynamics. MSc thesis, Stellenbosch University, Stellenbosch

Mc Kay F, Gandolfo D, Witt ABR (2012) Biology and host range of *Coelocephalapion gandolfoi* Kissinger (Brentidae), a promosing candidate for the biological control of invasive *Prosopis* species (Legumiosae) in South Africa. Afr Entomol 20:281–291

McClaran MP, Angell DL (2006) Long-term vegetation response to mesquite removal in Desert Grassland. J Arid Environ 66:686–697

Medeiros MA, Riet-Correa F, Dantas FP, Santos JR, Medeiros RM (2014) Teratogenic effects of *Prosopis juliflora* in rats and analysis of the toxicity of the pods. Pesqui Vet Bras 34:1089–1093

Medina AA, Dussart EG, Estelrich HD, Morici EA (2000) Reconstrucci6n de la historia del fuego en un bosque de *Prosopis caldenia* (Burk.) de Arizona, south of San Luis Province. Multequina 9:91–98

Mehar SK (2011) Assessment of effect of Prosopis juliflora litter extract on seed germination and growth of rice. Food Sci Qual Manage 2:9–18

Mendes BV (1986) Potential offered by *Prosopis juliflora* (SW) D.C. pods in the Brazilian semi arid region. The current state of knowledge onProsopis juliflora. II International Conference on Prosopis recife, 25–29 August, Brazil, pp 61–62

Menezes RSC, Salcedo IH, Elliott ET (2002) Microclimate and nutrient dynamics in a silvopastoral system of semi- arid northeastern Brazil. Agrofor Syst 56

Merga Y (2012) Assessing the impact of *Prosopis juliflora* (SW.) D.C.. invasion on livelihood of pastoralists: the case of fentale woreda, east showa zone of Oromia. MSc thesis, Haramaya University, Haramaya, Ethiopia

Meyer RE, Morton HL, Haas RH, Robison ED, Riley TE (1971) Morphology and anatomy of honey mesquite. USDA Technology Bulletin, Washington, DC

Miller C, Darlow A (2008) A reiview of the Ascension Island action plan for SAIS Project. RSPB

Miranda RQ, Oliveira MTP, Correia RM, Almedia-Cortez JS, Pompelli MF (2011) Germination of *Prosopis juliflora*(Sw) D.C. seeds after scarification treatments. Plant Species Biol 26:186–192

Mishra A, Sharma SD, Gupta MK (2003) Soil rehabilitation through afforestation: evaluation of the performance of Prosopis julifl- ora, Dalbergia sissoo and Eucalyptus tereticornis plantations in a sodic environment. Arid Land Res Manag 17(3):257–269

Moeller DA, Tiffin P (2005) Genetic diversity and the evolutionary history of plant immunity genes in two species of Zea. Mol Biol Evol 22:2480–2490

Mohammadi A, Nasr J, Rahmatnejad E, Dashtizadeh M, Golshahi A (2013) Performance and carcass quality of broiler chickens in response to *Prosopis juliflora* seed (PJS) as a by-product. J Agricult Technol 9:317–322

Moodley D, Geerts S, Richardson DM, Wilson JRU (2013) Different traits determine introduction, naturalization and invasion success in woody plants: Proteaceae as a test case. PLoS One 8:75–78

More D, Whisman L, Whisman B, Jordan-Wagner D (2002) Identification of specific IgE to mesquite wood smoke in individuals with mesquite pollen allergy. J Allergy Clin Immunol 110:814–816

Morgan AM, Hamdoun AM, Bashir NH (2017) Studies on Seed Germination and Seedling Emergence of Mesquite, Prosopis juliflora (Swartz) D.C.. in Sudan Gezira. Univ J Agricult Res 5(2):159–163

Mosweu S, Munyati C, Kabnda T, Setshongo M, Muzila M (2013) *Prosopis* L. invasion in the south-western region of Botswana: the perceptions of rural communities and management options. Natural Resour 4:496–505

Muller, G.C., Junnila, A., Traore, M.M., Traore, S.F., Doumbia, S., Sissoko, F., Dembele, S.M., Schlein, Y., Arheart, K.L.., Revay, E.E., Kravchenko, V.D., Witt, A. and Beier, J.C. 2017. The invasive shrub *Prosopis juliflora*enhances the malaria parasite transmission capacity of *Anopheles* mosquitos: a habitat manipulation experiment

Munzbergova Z, Ward D (2002) Acacia trees as keystone species in Negev desert ecosystems. J Veg Sci 13:227e236

Murad AA, Al Nuaimi H, Al Hammadi M (2007) Comprehensive assessment of water resources in the United Arab Emirates (UAE). Water Resour Manag 21(9):1449–1463

Muturi GM, Mohren GMJ, Kimani JN. (2009) Prediction of *Prosopis* species invasion in Kenya using geographical information system techniques

Muturi GM, Poorter L, Mohren GMJ, Kigomo BN (2013) Ecological impacts of *Prosopis* species invasion in Turkwel riverine forest, Kenya. J Arid Environ 92:89–97

Muzila M, Setshogo MP, Moskei B, Morapedi R (2011) An assessment of *Prosopis* L. in the Bokspits area south-western Botswana, based on morphology. Afr J Plant Sci Biotechnol 5:75–80

Mwangi M, Swallow B (2005) Invasion of *Prosopis juliflora* and local livelihoods: case study from the Lake Baringo area of Kenya. ICRAF Working Paper – no. 3. World Agroforestry Centre, Nairobi

Mwangi E, Swallow B (2008) *Prosopis juliflora* invasion and rural livelihoods in the Lake Baringo Area of Kenya. Conserv Soc 6(2):130–140

Mworia JK, Kinyamario JL, Omari JK, Wambua JK (2011) Patterns of seed dispersal and establishment of invader *Prosopis juliflora* in the upper floodplain of Tana River, Kenya. Afr J Range Forage Sci 28:35–41

Nahonyo CL, Mwasumbi LB, Msuya CA, Masao CA, Suya TB, Shing'weda C (2005) NGEZI – vumawimbi forest reserves biodiversity inventory report. Department of Zoology and Marin Biology, University of Dar es Salam, Dar es Salam

Nakano H, Fujii Y, Suzuki T, Yamada K, Kosemura S, Yamamura S, Hasegawa K (2001) A growth-inhibi- tory substance exuded from freeze-dried mesquite (Prosopis juliflora (Sw.) D.C..) leaves. Plant Growth Regul 33:165–168

Naranjo CA, Poggio L, Enus-Zeiger S (1984) Phenol chromatography, morphology and cytogenetics in three species and natural hybrids of Prosopis (Leguminosae, Mimosoideae). Plant Syst Evol 144:257–276

Naseeruddin S, Yadav KS, Sateesh L, Manikyam A, Desai S, Rao LV (2013) Selection of the best chemical pretreatment for lignocellulosic substance *Prosopis juliflora*. Bioresour Technol 136:542–549

National Weeds Strategy Executive Committee (2001) Mesquite (Prosopis Speceis) strategic plan. National Weeds Strategy Executive Committee, Launceston

Naylor RL (1996) Invasions in agriculture: assessing the cost of the golden apple snail in Asia. Ambio 25:443–448

Ndhlovu T, Milton-Dean SJ, Esler KJ (2011) Impact of Prosopis (mesquite) invasion and clearing on the grazing capacity of semiarid Nama Karoo rangeland, South Africa. Afr J Range Forage Sci 28:129–137

Nee' Shukla RV, Misra K (1981) Two flavonoid glycosides from the bark of Prosopis juliflora. Phytochemistry 20:339–340

Neudecker P, Schweimer K, Nerkamp J, Scheurer S, Vieths S, Sticht H et al (2001) Allergic cross-reactivity made visible: solution structure of the major cherry allergen Pru av 1. J Biol Chem 276:22756–22763

Nie W, Yuan Y, Kepner W, Erickson C, Jackson M (2012) Hydrological impacts of mesquite encroachment in the upper San Pedro watershed. J Arid Environ 82:147–155

Niguse H, Amare F (2016) Distribution and Socio-economic Impacts of Prosopis juliflora in East Shewa and West Arsi Zones, Ethiopia. Intl J Afr Asian Stud 24:31–41

Njoroge E, Sirmah P, Mburu F, Koech E, Mware M, Chepkwony J (2012) Preference and adoption of farmer field school (FFS) *Prosopis juliflora* management practices: experiences in Baringo District, Kenya. Forest Stud China 14(4):283–290

Nolte KR, Fulbright TE (1997) Plant, small mammal and avian diversity following control of honey mesquite. J Range Manag 50:205–212

Northern Territory Government (2012) Weed management plan for mesquite (Prosopis species). Northern Territory Government, Palmeton

Novey HA, Roth M, Wells ID (1977) Mesquite pollen: an aeroallergen in asthma and allergic rhinitis. J Allergy Clin Immun 59:359–363

Obeidat BS, Shdaifat MM (2013) Partial substitution of barley grain with *Prosopis juliflora* pods in lactating Awassi ewes' diets: effect on intake, digestibility, and nursing performance. Small Rumin Res 111:50–55

Obeidat BS, Abdullah AY, Al-Lataifeh FA (2008) The effect of partial replacement of barley grains by *Prosopis juliflora* pods on growth performance, nutrient intake, digestibility, and carcass characteristics of Awassi lambs fed finishing diets. Anim Feed Sci Technol 146:42–54

Odero-Waitituh JA, King'ori AM, Guliye AY (2016) Effect of replacing maize with milled mature pods of *Prosopis juliflora* on performance of finishing broiler chicken. Livestock Res Rural Develop 28:Article No. 22

Oduor NM, Githiomi JK (2013) Fuel-wood energy properties of Prosopis juliflora and Prosopis pallida grown in Baringo District, Kenya. Afr J Agric Res 8(21):2476–2481

Ojaveer H, Galil BS, Minchin D, Olenin S, Amorim A, Canning-Clode J et al (2014) Ten recommendations for advancing the assessment and management of non-indigenous species in marine ecosystems. Mar Policy 44:160–165

Ooi MKJ, Auld TD, Denham AJ (2009) Climate change and bet-hedging: interactions between increased soil temperatures and seed bank persistence. Glob Chang Biol 15:2375e2386

Ordman D (1950) *Prosopis* tree as a cause of seasonal hay fever and asthma in South Africa. S Afr Med J 33:12

Osmond R (2003) Best practice manual, mesquite: control and management options for mesquite (Prosopis spp.) in Australia. National Weeds Programme and Queensland Department of Natural Resources and Mines, Queensland

Page AR, Lacey KL (2006) Economic impact assessment of Australia weed biological control. CRC for Australian Weed Management, Australia

Palacios RA (2006) Los mezquites Mexicanos: biodiversidad y distrabucion geografica. Bull Bot Soc Argentina 41:99–121

Pandey CN, Pandey R, Bhatt JR. (2012) *Prosopis juliflora (Swartz)* D.C.: management dilemmas and regulatory issues in Gujarat. In: Bhatt et al (eds) Invasive alien plants: an ecological appraisal for the Indian subcontinent. CAB International

Panetta FD, Carstairs SA (1989) Isozymic discrimination of tropical Australian populations of mesquite (*Prosopis* spp.): implications for biological control. Weed Res 29:157–165

Pasiecznik NM, Peñalvo López E (In review) 25 year results from an arid zone tree species elimination trial in Almeria, Spain, and an invasive risk assessment of the exotic species introduced. Biodiversity and Conservation

Pasiecznik NM, Felker P, Harris PJ, Harsh L, Cruz G, Tewari JC, Cadoretm K, Maldonado LJ (2001) The *'Prosopis Juliflora'-'Prosopis Pallida'* complex: a monograph, vol 172. HDRA, Coventry

Pasiecznik NM, Harris PJC, Smith SJ (2004) Identifying tropical Prosopis species: a field guide. HDRA, Coventry

Pasiecznik NM, Choge SK, Muthike GM, Chesang S, Fehr C, Bakewell-Stone P, Wright J, PJC H (2006) Putting knowledge on Prosopis into use in Kenya. Pioneering advances in 2006. KEFRI and HDRA, Nairobi/Coventry

Patnaik P, Abbasi T, Abbasi SA (2017) Prosopis (Prosopis juliflora): blessing and bane. Trop Ecol 58(3):455–483

Patnaik P, Abbasi T, Abbasi SA (2018) Extraction of energy precursors in the form of volatile fatty acids (VFAs) from the xerophyte prosopis (Prosopis juliflora), In Advances in Health and Environment Safety (pp. 255–261). Springer, Singapore

Pauli G (2000) Evolution in the understanding of cross-reactivities of respiratory allergens: the role of recombinant allergens. Int Arch Allergy Immunol 123:183–195

Peinetti R, Sosa A, Kin A, Cerqueira E (1997) Modelo de simulación del banco de semillas del caldén (*Prosopis caldenia*) Reunión Argentina de Ecologia 48:98

Pereira TCJ, Pereira MLA et al (2013) Mesquite pod meal in diets for Santa Inês sheep: Ingestive behavior. Acta Scientiarum Anim Sci 35:201–206

Perera ANF, Pasiecznik NM (2005) Using invasive Prosopis to improve livelihoods in Sri Lanka. HDRA, Coventry

Perroni-Ventura Y, Montaña C, García-Oliva F (2010) Carbon–nitrogen interactions in fertility island soil from a tropical semi-arid ecosystem. Funct Ecol 24:233–242

Perveen A, Zeb S, Khan M, Qaiser M (2014) Seasonal fluctuations of airborne pollen grains count and its correlation with climatic factors from Khairpur; Sindh, Pakistan. Pak J Bot 46(1):299–306

Pickup AR (1999) Ascension Island Management Plan. Report from RSPB and Birdlife International

Pinto-Ruiz R, Gomez-Castro H, Guevara-Hernandez F, Hernandez-Sanchez D, Ruiz-Sesma B (2014) Preference and Ingestive behavior of sheep fed on tropical tree fruits. Revista Cientifica-Facultad De Ciencias Veterinarias 24:158–163

Polley HW, Johnson HB, Mayeux JH (1994) Increasing CO2: comparitive responces of the C4 grass schizachyrium and grassland invader Prosopis. Ecology 75:976–988

Potgieter LJ, Richardson DM, Wilson JRU (2013) Casuarina: biogeography and ecology of an important tree genus in a changing world. Biol Invasions 16:609–633

Poynton RJ (1990) The genus *Prosopis* in southern Africa. South Afr For J 152:62–66

Poynton RJ (2009) Tree planting in southern Africa, volume 3: other genera. Department of Agriculture, Forestry and Fisheries, Pretoria

Prado Silva RH, de Rezende ASC, da Silva Inácio DF (2016) Pectin-rich by-products in feeding horses – a review. Cogent Food Agricult 2:1193925. https://doi.org/10.1080/23311932.2016.1193925

Prieur-Richard AH, Lavorel S, Linhart YB, Dos Santos A (2002) Plant diversity, herbivory and resistance of a plant community to invasion in Mediterranean annual communities. Oecologia 130:96–104

Pyšek P, Jarošík V, Hulme PE, Pergl J, Hejda M, Schaffner U, Vilà M (2012) A global assessment of invasive plant impacts on resident species, communities and ecosystems: the interaction of impact measures, invading species' traits and environment. Glob Chang Biol 18(5):1725–1737

Qasem JR (2007) Chemical control of *Prosopis farcta* (Banks and Sol.) Macbride in the Jordan Valley. Crop Prot 26:572–575

Radauer C, Bublin M, Wagner S, Mari A, Breiteneder H. Allergens are distributed into few protein families and possess a restricted number of biochemical functions. J Allergy Clin Immunol 2008; 121: 847–852. e7. PMID: 18395549

Ræbild A, Diallo BO, Graudal L, Dao L, Sanou J (2003) Evaluation of a species and provenance trial of Prosopis at Gonsé, Nurkina Faso. FAO, Rome

Rashid M, Abbas SH, Rehman A (2014) The status of highly alien invasive plants in Pakistan and their impact on the ecosystem: a review. Innovare J Agricult Sci 2:1–4

Reddy CS, Rangaswamy M, Pattanaik C, Jah CS (2009) Invasion of alien species in wetland of Samaspur Bird Sanctuary, Uttar Pradesh, India. Asian J Water Environ Pollut 6:43–50

Reinoso H, Sosa L, Ramírez L, Luna V (2004) Salt-induced changes in the vegetative anatomy of Prosopis strombulifera (Leguminosae). Can J Bot 82(5):618–628

Rettberg S, Müller-Mahn D (2012) Human-environment interactions: the invasion of Prosopis juliflora in the Drylands of Northeast Ethiopia. In: Mol L, Sternberg T (eds) Changing deserts – integrating people and their environments. Whitehorse Press, Cambridge, pp 297–316

Richardson DM, Carruthers J, Hui C, Impson FAC, Miller JT, Robertson MP, Rouget M, Le Roux JJ, Wilson JRU (2011) Human-mediated introductions of Australian Acacias—a global experiment in biogeography. Divers Distrib 17:771–787

Riet-Correa F, de Andrade FR, Carvalho FK, Tabosa IM, Galiza GJ, Bernardino JN, Simões SV, Medeiros RM (2012) Use of *Prosopis juliflora* pods as food for sheep and goats. Pesqui Vet Bras 32:987–989

Ring J, Akdis C, Behrendt H, Lauener RP, Schäppi G, Akdis M, et al. Davos declaration: allergy as a global problem. Allergy 2012; 67: 141–143. PMID: 22235793

Robinson TP, van Klinken RD, Metternicht G (2008) Spatial and temporal rates and patterns of mesquite (*Prosopis* species) invasions in Western Australia. J Arid Environ 72:175–188

Ruiz TG, Zaragoza SR, Cerrato RF (2008) Fertility islands around *Prosopis laevigata* and *pachycereus hollianus* in the drylands of Zapotitlán Salinnas, Mexico. J Arid Environ 72:1202–1212

Ruskin FR (1980) Firewood crops. Shrub and tree species for energy production. National Academy of Sciences, Washington, D.C.

Sahli HF, Conner JK (2006) Characterizing ecological generalization in plant-pollinator systems. Oecologia 148:365–372

Sajad G, Sefidi K (2012) Comparison of sustainable forest management (SFM) trends at global and country levels: case study in Iran. J For Res 23(2):311–317

Sajjad A, Saeed S, Bashir MA (2012) Spatial variation in pollinator communities and reproductive performance of Prosopis juliflora (Fabaceae). J Pollinat Ecol 8(9):59–66

Salazar PC, Navarro-Cerrillo RM, Cruz G, Villar R (2018) Intraspecific leaf functional trait variability of eight Prosopis pallida tree populations along a climatic gradient of the dry forests of northern Peru. J Arid Environ

Samuel G (2010) Allelopathic effects of the invasive *Prosopis juliflora* on selected native plant species at Middle Awash, Southern Afar Rift of Ethiopia. MSc thesis, Addis Ababa University, Addis Ababa Ethiopia

Samuel M, Christopher M, Tibangayuka K, Moffat S, Mbaki M (2013) *Prosopis* L. Invasion in the South-Western Region of Botswana: the perceptions of rural communities and management options. Nat Resour 4:496–505

Satheesh Kumar S, Vittal BPR (1998) A preliminay survey of airborne pollen in Madras city. Aerobiologia 14:69–73

Sato T (2013) Beyond water-intensive agriculture: expansion of *Prosopis juliflora* and its growing economic use in Tamil Nadu, India. Land Use Policy 35:283–292
Sawal RK, Ratan R, Yadav SBS (2004) Mesquite *(Prosopis juliflora)* pods as a feed resource for livestock-A review. Asian Australas J Anim Sci 17:719–725
Schachtschneider K, February E (2013) Impact of *Prosopis* invasion on a keystone tree species in the Kalahari Desert. Plant Ecol 214:597–605
Schupp EW (1993) Quantity, quality and the effectiveness of seed dispersal by animals. Vegetatio 107:15–29
Seid MJ (2012) Household perception about *Prosopis juliflora* and its effect on pastoral livelihood diversification strategy: the case of Gewane District in Afar Regional State, Ethiopia. Intl J Agricult Sci Res 2:21–51
Senthilkumar P, Prince WS, Sivakumar S, Subbhuraam CV (2005) Prosopis juliflora—a green solution to decontaminate heavy metal (Cu and Cd) contaminated soils. Chemosphere 60:1493–1496
Shackleton RT, Le Maitre DC, Pasiecznik NM, Richardson DM (2014) *Prosopis*: a global assessment of the biogeography, benefits, impacts and management of one of the world's worst woody invasive plant taxa. AoB Plants 6:1–18
Shackleton RT, Le Maitre DC, Richardson DM (2015) Stakeholder perceptions and practices regarding *Prosopis* (mesquite) invasions and management in South Africa. Ambio 44:569–581
Shah RH, Baloch M, Khan AA, Ijaz M, Zubair M (2018) Bioherbicidal assessment of aqueous extracts of mesquite (Prosopis juliflora) on weeds control and growth, yield and quality of wheat. Planta Daninha 36
Shapaka TN, Cunningham PL, Joubert DF (2008) Invasive alien plants in the Daan Viljoen Game Park, vol 30. DINTERIA, Windhoek, pp 19–32
Sharma IK (1981) Ecological and economic importance of *Prosopis juliflora* in the Indian Thar Desert. J Taxon Bot 2:245–248
Sharma R, Dakshini KMM (1996) Ecological implications of seed characteristics of the native Prosopis cineraria and the alien P. juliflora. Vegetation 124:101–105
Sharma R, Dakshini KMM (1998) Intergration of plant and soil charecteristics and the ecological success of two *Prosopis* species. Plant Ecol 139:63–69
Sharmila S, Rebecca Jeyanthi L, Saduzzaman M (2013) Biodegradation of tannery effluent using Prosopis juliflora. Int J ChemTech Res 5:2186–2192
Shea KM, Truckner RT, Weber RW, Peden DB (2008) Climate change and allergic disease. J Allergy Clin Immunol 122:443–453
Shiferaw H, Teketay D, Nemomissa S, Assefa F (2004) Some biological characteristics that foster the invasion of *Prosopis juliflora* (Sw.) D.C.. at Middle Awash Rift Valley Area, north-eastern Ethiopia. J Arid Environ 58:135–154
Shimles A, Getachew M (2016) Contemporary Prosopis dilemma: perception of inhabitants of Sabian Kebele (Goro) towards invasive tree *Prosopis juliflora,* Eastern Ethiopia. J Agricult Biotechnol Sustain Develop 8(1):1–6
Shirke PA, Pathre UV (2004) Influence of leaf-to-air vapour pressure deficit (VPD) on the biochemistry and physiology of photosynthesis in Prosopis juliflora. J Exp Bot 55(405):2111–2120
Shirke PA, Pathre UV, Sane PV (2018) Adaptation strategies of two leaf cohorts of Prosopis juliflora produced in spring and monsoon. Photosynthetica:1–10
Shivpuri DN, Parkash D (1967) A study in allergy to *Prosopis juliflora* (Kabuli keekar). Ann Aller 25:643–648
Sigh A, Singh PK (2009) An ethnobotanical study of medicinal plants in Chandauli District of Uttar Pradesh, India. J Ethnopharmacol 121:324–329
Silva S (1986) Prosopis juliflora (SW) D.C. in Brazil. The current state of knowledge on Prosopis juliflora. II international conference on Propopis recife, 25–29 August, Brazil, pp 29–51
Silva S (1988) *Prosopis juliflora* (Sw) D.C. Brazil. In The Current State of Knowledge on *Prosopis juliflora.* (Eds.) M. A. Habit and J. C. Saavedra. FAO, Rome

Silva JHVD, Silva MBD, Silva ELD, Jordão Filho J, Ribeiro MLG, Costa FGP, Dutra Júnior WM (2003) Energia metabolizável de ingredientes determinada com codornas japonesas (Coturnix coturnix japonica). Rev Bras Zootec 32(6):1912–1918

Silva AMM, Silva AR, Pinheiro AM et al (2007) Alkaloids from Prosopis juliflora leaves induce glial activation, cytotoxicity

Silva VDA, Pitanga BPS, Nascimento RP et al (2013) Juliprosopine and Juliprosine from *Prosopis juliflora* leaves induce mitochondrial damage and cytoplasmic vacuolation on cocultured glial cells and neurons. Chem Res Toxicol 26:1810–1820

Silva TRM, Chung S, de Araújo TAT, de Azevedo KSP, dos Santos MC, Bicudo ÁDA (2015) Replacing corn by mesquite meal (*Prosopis juliflora*) in diets for juvenile Nile tilapia reared in low temperature. Revista Brasileira de Ciências Agrárias (Agrária) 10:460–465

Simpson BB (1977) Mesquite, its Biology in Two Desert Shrub Ecosystems. Dowden, Hutchinson and Ross, Stroudsberg, Pennsylvania, USA

Simpson BB, Neff JL, Moldenke AR (1977) Prosopis flowers as a resource. Mesquite: its biology in two desert scrub ecosystems. Simpson B. B. Ed. Dowden, Hutchinson and Ross, Stroudsburg, Pennsylvania

Singh S (2012) Phytochemical analysis of different parts of *Prosopis juliflora*. Intl J Curr Pharmaceut Res 4:59–61

Singh AB, Mathur C (2012) An aerobiological perspective in allergy and asthma. Asia Pac Allergy 2:210–222

Singh G, Shukla S (2012) Effect of *Prosopis julifora* (D.C..) tree on under canopy resources, dicersity and productivity of herbaceous vegetation in Indian desert. Arid Land Res Manag 26:151–165

Skolmen RG (Undated) *Prosopis pallida* (Humb. And Bonpl. Ex Willd.) H.B.K. Kiawe

Smit P (2004) Prosopis: a review of existing knowledge relevant to Namibia. J Scientific Soc 52:13–40

Smith KR, Woodward A, Campbell-Lendrum D, Chadee DD, Honda Y, Liu Q, et al. 2014. Human health: impacts, adaptation, and co-benefits. In: Field CB, Barros VR, Dokken DJ, Mach KJ, Mastrandrea MD, Bilir TE, et al. (eds) Climate Change 2014: impacts, adaptation, and vulnerability Part A: global and sectoral aspects contribution of Working Group II to the Fifth Assessment Report of the Intergovernmental Panel on Climate Change. Cambridge University Press, Cambridge, pp 709–754

Sohrabi S, Gherekhloo J, Mohassel MHR, Ghanbari A, Mahalati MN. 2011. Cardinal temperatures of three invasive weeds in Iran. 3rd International Symposium on Weeds and Invasive Plants, Acuna

Song U, Mun S, Ho CH, Lee EJ (2012) Responses of two invasive plants under various microclimate conditions in the Seoul metropolitan region. Environ Manag 49:1238–1246

Steele P, Breithaupt J, Labrada R (2008) Proceedings Expert Consultation (no. 4): increases food security control and management of *Prosopis*. FAO, Ethiopia

Steenkamp HE, Chown SL (1996) Influence of dense stands of an exotic tree Prosopis glandulosa Benson, on a savanna dung beetle (Coleoptera: Scarabeidae) assemblage in southern Africa. Biol Conserv 78:305–311

Steffan-Dewenter I, Tscharntke T (1999) Effects of habitat isolation on pollinator communities and seed set. Oecologia 121:432–440

Stromberg JC, Wilkins SD, Tress JA (1993) Vegetation-hydrology models: implications for management of *Prosopis velutina* (velvet mesquite) riparian ecosystems. Ecol Appl 2:307–314

Tabosa IM, Riet-Correa F, Barros SS, Summers BA, Simoes SVD, Medeiros RMT, Nobre VMT (2006) Neurohistologic and ultrastructural lesions in cattle experimentaly intoxicated with the plant *Prosopis juliflora*. Vet Pathol 43:695–701

Talavera S, Bastida F, Ortiz PL, Arista M (2001) Pollinator attendance and reproductive success in Cistus libanotis L. (Cistaceae). Int J Plant Sci 162:343–352

Teague WR, Ansley RJ, Kreuter UP, Pinchak WE, McGrann JM (2001) Economics of managing mesquite in northern Texas: a sensitivity analysis. J Range Manage 54:553–560

Teague WR, Grant WE, Kreuter UP, Diaz-Solis H, Dube S, Kothmann MM, Pinchak WE, Ansley RJ (2008) An ecological economic simulation model for assessing fire and grazing management effects on mesquite rangelands in Texas. Ecol Econ 64:611–624

Tegegn G (2008) Experiences on Prosopis management case of Afar region. FARM-Africa, London (September), ARM-Africa, London

Terink W, Immerzeel WW, Droogers P (2013) Climate change projections of precipitation and reference evapotranspiration for the Middle East and Northern Africa until 2050. Intl J Climatol 33(14):3055–3072

Tesoriere L, Butera D, Allegra M et al (2005) Distribution of betalain pigments in red blood cells after consumption of cactus pear fruits and increased resistance of the cells to ex vivo induced oxidative hemolysis in humans. J Agric Food Chem 53:1266–1270

Tessema YA (2012) Ecological and Economic Dimensions of the paradoxical invasive species – *Prosopis juliflora* and policy changes in Ethiopia. J Econ Sustain Develop 3:62–71

Thakur IS (1986) *Prosopis juliflora* pollen allergen induced hypersensitivity and anaphylaxis studies in guinea pigs. Biochem Int 13:915–912

Thakur IS (1989) Fractionation and analysis of allergenicity of allergens from Prosopis juliflora pollen. Int Arch Allergy Immunol 90(2):124–129

Thakur IS (1991) Purification and characterization of the glycoprotein allergen from Prosopis juliflora pollen. Biochem Int 23(3):449–459

Thakur IS, Sharma JD (1985) Isolation and characterization of allergens of Prosopis juliflora pollen grains. Biochem Int 11(6):903–912

Thomas J, El-Sheikh MA, Alfarhan AH, Alatar AA, Sivadasan M, Basahi M, Al-Obaid S, Rajakrishnan R (2016) Impact of alien invasive species on habitats and species richness in Saudi Arabia. J Arid Environ *127*:53–65

Thorp JR, Lynch RA (2001) The determination of weeds of national significance. National Weeds Strategy Executive Committee, Launceston, 234 pages

Throop HL, Archer SR (2008) Srub (*Prosopis velutina*) encroachment in a semidesert grassland: special-temporal changes in soil organic carbon and nitrogen pools. Glob Chang Biol 14:2420–2431

Tiedemann AR, Klemmedson JO (1973) Nutrient availability in desert grassland soils under Mesquite (Prosopis juliflora) trees and adjacent open areas. Soil Sci Soc Am J 37:107–111

Tilstone GH, Pasiecznik NM, PJC H, Wainwright SJ (1998) The growth of multipurpose tree species in the Almeria province of Spain and its relationship to native plant communities. Intl Tree Crops J 9(4):247–259

Timyan J (1996) BWA YO: important trees of Hati. South-East Consortium for International Development, Washington, DC

Torell LA, McDaniel KC (1986) Optimal timing of investements to control honey mesquite. J Range Manag 39:378–382

Tourenq C, Shuriqi MK (2010) Snapshot on introduced invasives in a desertic country, the United Arab Emirates. Aliens 30:49–51

Truman ST, Wallace HM (1999) Pollination and resource constraints on fruit set and fruit size of Persoonia rigida (Proteaceae). Ann Bot 83:145–155

Tschirley FH, Martin SC (1960) Germination and longevity of velvet mesquite seed in soil. J Range Manag 13:94–97

Valenta R, Steinberger P, Duchêne M, Kraft D (1996) Immunological and structural similarities among allergens: prerequisite for a specific and component-based therapy of allergy. *Immunol Cell Biol* 74:187–194

Vallejo VD, Arbeli Z, Tera'n W, Lorenz N, Dick RP, Roldan F (2012) Effect of land management and Prosopis juliflora (Sw.) D.C. trees on soil microbial community and enzymatic communities in intensive silvopastoral systems of Colombia. Agric Ecosyst Environ 150:139–148

Van den Berg EC (2010) Detection, quantification and monitoring Prosopis spp. in the Northern Cape Province of South Africa using Remote Sensing and GIS. MSc thesis, North-West University, Potchefstroom van der Linden PJ, Hanson CE (2007) Climate change 2007: impacts, adaptation and vulnerability. Contribution of Working Group II to the Fourth Assessment Report of the Intergovernmental Panel on Climate Change. Cambridge University Press, Cambridge, pp 7–22

van Klinken R (2012) Prosopis spp. – mesquite. In: Julien M, McFadyen R, Cullen J (eds) Biological control of weeds in Australia. CSIRO, Melbourne, Australia

van Klinken ED, Burwell CJ (2005) Evidence from a gelechidd leaf-tier on mesquite (Mimosaceae: Prosopis) that semi-concealed Lepidopteran biological control agents may not be at risk from parasitism in Australian rangelands. Biol Control 32:121–129

van Klinken RD, White AJ (2011) Overcoming seasonally fluctuating resources: Bruchid predation of mesquite (*Prosopis*) seed in dung. Biol Control 59:361–365

van Klinken RD, Graham J, Flack LK (2006) Population ecology of hybrid mesquite (*Prosopis* species) in Western Australia: how does it differ from native range invasions and what are the implications for impacts and management? Biol Invasions 8:727–741

van Klinken RD, Shepherd D, Parr R, Robinson TP, Anderson L (2007) Mapping mesquite (*Prosopis*) distribution and desnnity using visual areal surveys. Rangel Ecol Manag 60:408–416

van Wilgen BW, Forsyth GG, Le Maitre DC, Wannenburgh A, Kotze DF, van den Berg E, Henderson L (2012) An assessment of the effectiveness of a large, national-scale invasive alien plant control strategy in South Africa. Biol Conserv 148:28–38

Vaz AS, Kueffer C, Kull CA, Richardson DM, Vicente JR, Kühn I, Schröter M, Hauck J, Bonn A, Honrado JP (2017) Integrating ecosystem services and disservices: insights from plant invasions. Ecosyst Ser 23:94–107

Villagra PE, Boninsegna JA, Alvarez JA, Cony M, Cesca E, Villalba R (2005) Dendroecology of *Prosoois flexuosa* woodlands in the Monte desert: implications for their management. Dendrochronologia 22:209–213

Villagra PE, Vilela A, Giordano C, Alvarez JA (2010) Ecophysiology of Prosopis species from the arid lands of Argentina: what do we know about adaptation to stressful environments? In: Desert plants. Springer, Berlin/Heidelberg, pp 321–340

Vitousek PM (1986) Biological invasions and ecosystem properties: can species make a difference?. In Ecology of biological invasions of North America and Hawaii. Springer, New York, pp 163–176

Vitousek PM, D'Antonio CM, Loope LL, Westbrooks R (1996) Biological invasions as global environmental change. Am Scientist *84*(5):468–478

Vitousek PM, Dantonio CM, Loope LL, Rejmanek M, Westbrooks R (1997) Introduced species: a significant component of human-caused global change. N Z J Ecol 21:1–16

Wakie TT, Laituri M, Evangelista PH (2016) Assessing the distribution and impacts on *Prosopis juliflora* through participatory approaches. Appl Geogr 66:132–143

Walkie TT, Hoag D, Evangelista PH, Luizza M, Laituri M (2016) Is control through utilization a cost effective Prosopis juliflora management strategy? J Environ Manag 168:74–86

Walter K (2011) *Prosopis*, an alien amoung the sacred trees of south India. Dissertation. University of Helsinki, Helsinki

Ward CR, O'Brien CW, O'Brien LB, Foster DE, Huddleston EW (1977) Annotated checklist of New World Insects associated with Prosopis (Mesquite). Technical bulletin, vol 1557. United States Department of Agriculture, USA

Wardle DA, Bardgett RD, Callaway RM, Van der Putten WH (2011) Terrestrial ecosystem responses to species gains and losses. Science 332(6035):1273–1277

Warrag MOA (1995) Autotoxic potential of foliage on seed germination and early growth of mesquite (Prosopis juliflora). J Arid Environ 31(4):415–421

Warren A, Holechek J, Cardenas M (1996) Honey mesquite influence on Chihuahuan desert vegetation. J Range Manag 49:46–52

Wassel GM, Rizk AM, Abdel-Bary EF (1972) Phytochemical investigation of Prosopis juliflora D. C. I. Flavonoids and free sugars. Qual Plant Mater Veg 22:119–121

Wayne P, Forster S, Connelly J, Bazzaz FA, Epstein PR (2002) Production of allergenic pollen by ragweed (*Ambrosia artemisiifolia* L.) is increased in CO_2 enriched atmospheres. Ann Allergy Asthma Immunol 88:279–282

Weber JC, Larwanou M, Abasse TA, Kalinganire A (2008) Growth and survival of *Prosopis africana* provenances tested in Niger and related rainfall gradients in the West African Sahel. For Ecol Manag 256:585–592

White F (1962) Forest and Flora of Northern Rhodesia. Oxford University Press, Oxford

Wiggens IR, Proter M, Anderson EF (1971) Flora of the Galápagos Islands. Stanford University Press, Stanford

Wise RM, van Wilgen BW, Le Maitre DC (2012) Costs, benefits and management options for an invasive alien tree species: the case of mesquite in the Northern Cape, South Africa. J Arid Environ 84:80–90

Wojtusik T, Felker P, Russel EJ (1993) Cloning the erect, thornless, non-browsed nitrogen fixing trees of Haiti's principal fuelwood species (*Prosopis juliflora*). Agroferest Syst 21:293–300

Wronski T, Mosfer AN, Plath M (2012) Endemic Farasan Gazelle (Gazella gazella farasani) enhances the dispersal of invasive Prosopis juliflora on Farsan Islands, Saudi Arabia. Revue d'_ecologie – La Terre et la Vie 67:329–337

Yasin M, Animut G (2014) Replacing cottonseed meal with ground *Prosopis juliflora* pods; effect on intake, weight gain and carcass parameters of Afar sheep fed pasture hay basal diet. Trop Anim Health Prod 46:1079–1085

Zachariades C, Hofmann JH, Roberts A (2011) Biological control of mesquite (Prosopis species) (Fabaceae) in South Africa. Afr Entomol 19:402–415

Zaitoun S, Al-Ghzawi AA, Samarah N, Mullen R (2009) Pod production of Prosopis juliflora (Sw.) D.C.. As affected by supplementary and honeybee pollination under arid conditions. Acta Agric Scand 59:349–356

Zeila AA, Mwangi E, Swallew B (2004) Prosopis juliflora: boon or bane for dryland agroforestry. The Prunus Tribune, Jan–March

Ziello C, Sparks TH, Estrella N, Belmonte J, Bergmann KC, Bucher E, Brighetti MA, Damialis A, Detandt M, Galan C, Gehrig R, Grewling L, Gutierrez Bustillo AM, Hallsdottir M, Kockhans-Bieda MC, De Linares C, Myszkowska D, Paldy A, Sanchez A, Smith M, Thibaudon M, Travaglini A, Uruska A, Valencia-Barrera RM, Vokou D, Wachter R, de Weger LA, Menzel A (2012) Changes to airborne pollen counts across Europe. PLoS One 7:e34076

Zimmermann HG (1991) Biological control of Prosopis, *Prosopis* spp. (Fabaceae), in South Africa. Agric Ecosyst Environ 37:175–186

Zollner D (1986) Sand dune stabilization in central Somalia. Forest Ecol Manage 16:223–232

Chapter 8
Perennial Forage Grass Production on the Marginal Arabian Peninsula Land

M. Iftikhar Hussain, Nosheen Akhtar, Asad Sarwar Qureshi, and David Gallacher

Abstract Increasing water scarcity and soil salinization call for more diversified agricultural production systems that have the potential to sustain in marginal environments. Perennial forage grasses have been developed to replace conventional forages for the development of livestock sector in the Gulf and African regions. Perennial forage crops are known to reduce soil erosion, nitrogen leaching, and to increase water holding capacity, carbon sequestration and soil health. Here we review traditional and indigenous perennial forage grasses such as *Chloris gayana*, *Cenchrus divisus*, *Cenchrus ciliaris*, *Cenchrus setigerus* for saline and marginal Arabian Peninsula lands. The annual dry matter yields of *Chloris gayana* ranges between 35 and 60 t ha^{-1}, depending on the variety, climate and soil fertility conditions. It is persistent and produces drought-tolerant forage with a usual productive life of 3 years.

C. ciliaris is an appropriate fodder grass because of significant returns and nutritive values, drought resistance and high palatability. *C. ciliaris* is considered a 'wonder crop' for its ability to withstand strong wind, drought, salinity and heavy grazing, and rapidly responding to rain. *C. ciliaris* has high water-use efficiency and moderate-to-high tolerance to salinity. However, this plant species cannot survive under waterlogging condition. *C. divisus* is a wild relative of pearl millet with high growth potential under harsh environmental conditions. *C. divisus* has also

M. I. Hussain (✉)
Department of Plant Biology & Soil Science, Universidade de Vigo, Vigo, Spain

Research Institute of Science and Engineering, University of Sharjah, Sharjah, United Arab Emirates

N. Akhtar
Institute of Agricultural Sciences, University of the Punjab, Quaid-e-Azam Campus, Lahore, Pakistan

A. S. Qureshi
Natural Resources Management Division, International Center for Biosaline Agriculture (ICBA), Dubai, United Arab Emirates

D. Gallacher
Department of Interdisciplinary Studies, Zayed University, Dubai, United Arab Emirates

E. Lichtfouse (ed.), *Sustainable Agriculture Reviews 52*, Sustainable Agriculture Reviews 52, https://doi.org/10.1007/978-3-030-73245-5_8

been widely used for xerophytic landscaping and amenity horticulture under harsh conditions. Despite their advantages, these grasses need to fit into current cropping patterns and farming systems, and prove their utility in marginal lands affected by salinity and alkalinity. We conclude that producing perennial forage grasses on the marginal lands of Arabian Peninsula are feasible if appropriately managed to circumvent negative ecological and environmental effects. For large scale adoption, creating marketing opportunities is also vital.

Keywords Perennial forage grasses · Biomass · Marginal lands · Arabian Peninsula · Quality traits

8.1 Introduction

Water availability is central to agricultural development in arid regions such as the Arab Gulf States. Agricultural industries must have a strong partnership with water governance authorities and other industries that compete for water rights (Verhoeven 2015). Middle Eastern countries generally face degraded marginal lands, depleted aquifers, high population growth, and climate-change induced productivity decline. Gulf States have been increasing agricultural land area despite having very limited groundwater reserves and extremely high evapotranspiration rates. Gulf States are economically prosperous but are food insecure, being the greatest per-capita importers of food, pulses and cereals (Shahid and Al-Shankiti 2013).

Traditional Gulf agriculture consisted of land cultivation in or near wadis and oases, but today, large open-desert areas have become production zones through intensive inputs of water, fertilizers, and crop protection. Additionally, large areas have been afforested for non-commercial reasons, and amenity horticulture has flourished in urban landscapes. Enormous water resources have thus been devoted to these enterprises, but an increasing awareness of the costs associated with excessive water use is leading to closer scrutiny in the region. Excessive groundwater extraction causes a decline in both availability and quality of remaining water and risks the intrusion of saltwater into freshwater aquifers (Al-Zubari 1998; Al-Zubari et al. 2017; Environment Agency-Abu Dhabi (EAD)2016).

According to the United Nations World Water Assessment Programme, management of available water resources and the provision of access to drinking water and sanitation represent a global challenge for the twenty-first century (UNEP 2012). Gulf States (Bahrain, Kuwait, Oman, Qatar, Saudi Arabia, and the United Arab Emirates) have especially arid/hyper-arid climates (2% of global average precipitation) that make the problem of water scarcity more sever (World Bank 2017). The region is potentially vulnerable to natural land degradation process due to several anthropogenic factors like erosion, water scarcity, lack of organic matter, heat and thus threatening the sustainability of natural resources. This vast plateau is bounded by mountainous terrain and has been classified into 22 distinct agro-climatic zones

(De Pauw 2002). It covers a large landmass of about 3 million km^2 (Batanoumy 1986) and Saudi Arabia occupies by far the greater part. It is characterized as an ecologically fragile region which constitutes one of the largest contiguous arid zones in the world (De Pauw 2002).

The primary sources of freshwater are groundwater, most of which is non-renewable, and a limited amount of renewable surface water. The absence of administrative and legal regulations to control pumping encouraged unlawful drilling and resulted in inefficient irrigation practices that led to the loss of more than 50% of total applied water. The subsidies increased the irrigated areas as well as agricultural production but resulted in the mining of fossil aquifers. In many cases, aquifers have been either exhausted or highly polluted resulting in many farms to be abandoned in all countries. Water consumption by the United Arab Emirates (UAE), agricultural sector increased between 1990 and 2011 from 0.95 to 3.32 MT/year (Shahin and Salem 2014, 2015), mainly through an increase in the use of groundwater. Approximately 82% of the water requirement for artificial forests in Abu Dhabi emirate is met by groundwater, amounting to 0.58 MT/year, while the rest is met by non-conventional sources (Environment Agency-Abu Dhabi (EAD) 2016). Recharge of aquifers occurs at a very low rate across much of the Arabian Peninsula, such that it is essentially a non-renewable resource in locations not bordered by mountain ranges. Groundwater from deep, slowly recharging aquifers is thus often referred to as 'fossil' water.

8.2 Scarce Water Resources

Water scarcity in this arid region limits the agricultural development particularly and reduces the scope for its diversification. But inappropriate policies and technological solutions have placed emphasis on meeting self-sufficiency in food production leading to rapid growth of the agricultural sector in recent decades. As a consequence of this phenomenon, water requirements have increased beyond naturally sustainable levels, and depletion of groundwater aquifers is now common to most parts of this region (Odhiambo 2016). Groundwater in the Gulf region has one of the lowest absolute and per-capita available water. Groundwater reserves in the deep aquifers of the Arabian Shelf are estimated at 2.33 MT, while the average annual recharge rate is estimated at 2.7 MT (Al-Rashed and Sherif 2000). Increased population has caused a decline in the per-capita renewable water to the lowest in the world, from 1250 m^3 in 1950 to 100 m^3 in 2007 (World Resource Institute 2007) and 76.2 m^3 in 2014 (The World Bank 2008, 2017). Water resources are unequally distributed in space and time. In 2014 the per-capita renewable water varied from 0 m^3 in Kuwait (meaning no rain that year) to 311 m^3 in Oman (Food and Agriculture Organization 2017). Total water demand in all countries rose from 6.6 to 22.5 MT during the period of 1980–1990 and is expected to reach 36.7 MT by 2025 (Jaradat 2005). Since agriculture depends on irrigation and uses 80–90% of the water resources, the agricultural water demand is estimated to be 24.3 MT (Uitto and

Schneider 1997). Estimates for the Arabian Peninsula in 2010 were for 28.2 MT water requirement and 18.03 MT available groundwater.

Multi-layered aquifer systems extend over 1.5 million km^2 of the Arabian Peninsula (Food and Agriculture Organization 2010). Fossil groundwater is estimated to total 2.175 MT (Jaradat 2005). The main source for irrigation in the Arabian Peninsula countries is groundwater where the withdrawal reached 17,611 million m^3 while total recharge was only 6360 million m^3 in 1997 resulting in 63% annual recharge deficit (Abdulrazzak 1997). According to a report of Food and Agriculture Organization (FAO 2010); the maximum extraction of ground water is by Saudi Arabia (9% of the total) that is highest in the Middle East. Bulk of this storage is brackish with a big range of 200–20,000 ppm total solids dissolved in it. As extraction is increasing day by day, the salinity levels of these deeper fresh water aquifers are gradually becoming higher (Bazza 2005). Several aquifers in Bahrain, Oman, Qatar and UAE are already severely affected by this phenomenon. In Abu Dhabi Emirate, for example, groundwater extraction is 25 times faster than it can be naturally replenished (Environment Agency-Abu Dhabi (EAD) 2016). In 2016, Environmental Agency Abu Dhabi documented that groundwater extraction concept should have to be decreased substantially; otherwise water resources will be depleted rapidly that were deposited during the Ice age (10,000 years ago). With the depletion of the groundwater reserves and the increase in the demand for water, the proportion of water that comes from desalination will increase markedly. However, desalination is far from a sustainable solution, and presents significant economic and environmental challenges. With the depletion of the groundwater reserves and the increase in the water demand for agriculture, forestry, industry and domestic purposes. Therefore, demand will increase to secure water from other resources like, desalination that is not viable option and also have environmental and economic consequences and also not sustainable solution.

Currently the main fodder crops in the region are alfalfa (*Medicago sativa*) and Rhodes grass (*Chloris gayana*), both of which have a high water demand of around 48,000 m^3/ha/year (Peacock et al. 2003a). Use of marginal lands and lower quality water sources is becoming more common. Therefore there is a need to move to perennial forage grasses that are well adapted to the harsh environmental conditions.

8.3 Non-conventional Water and Perennial Forage Crops

Several non-conventional water sources are now used in the Middle East to address the imbalance between demand and conventional supply. Desalinated seawater and wastewater contributed around 7.2% and 1.7% of the total annual available water respectively (Al-Zubari 1998; Al-Zubari et al. 2017). In 2010 the production of desalinated water totaled 0.42 MT in Kuwait, 0.95 MT in UAE and 1.03 MT in Saudi Arabia (Food and Agriculture Organization 2010). UAE, Kuwait, Bahrain and Qatar all rely entirely on desalination for urban requirements. Treated wastewater has mainly been used for forestry and amenity horticulture, but also for forage production.

The sustainability of renewable water resources in Gulf Cooperation Council countries (GCC) is still debatable and required urgent attention from concerned authorities. According to Food and Agriculture Organization of United Nation (FAO), AQUASTAT report, among all the, UAE stands with highest water withdrawn as compared to the annual recharge than all other GCC member states. The Kuwait showed highest (2075%) water consumer, followed by UAE (1867%) and Saudi Arabia (936%). While fresh water withdrawal in Oman represents 9% of the renewable water resources. Arabian agriculture arable soils are very susceptible to salt-degradation, erosion and desertification. There are two ways for increasing food supply in the Arab Gulf; using more land and/or raising yields on existing farms. The major food crops require high quality soils and enough water supplies to achieve high yields. These factors are more likely not available in the Arab Gulf (Odhiambo 2016). In addition, introducing marginal land for food production is usually associated with more environmental problems, such as desertification, salinization, and exhaustion of aquifers that have been used for irrigation (Hussain et al. 2019). However, using the proper crop plants that could grow with minimum requirements and little disturbance for the soil could sustain the use of marginal lands. Here we provide examples of plants that have great potential to be grown in marginal lands of the Arab Gulf region. Previously, it was reported that crop rotation as well as tillage practices plays an important role in the changing of the physical properties of the soil layer (upper layers mostly) (Feng et al. 2011; Wuest et al. 2006). In Middle East and North Africa degraded marginal lands, introduction of new cropping systems and alternate stress tolerant crops might be a good option that will helps in rehabilitation of marginal soils and to use more efficiently fallow lands and to enhance carbon sequestration by employing the land for several years. This will also help in preventing the soil losses due to soil erosion. Moreover, cultivation of perennial crops will also minimize the tillage practices that will reduce the soil erosion. The perennial crops are beneficial to soil biology and help to improve soil stability and soil rhizosphere improvement. Because of extensive root system, production of perennial grasses showed high efficiency to use water and nutrient and are associated with erosion control, carbon sequestration, nutrient leaching reduction, and rehabilitation and restoration of degraded marginal lands. All these processes contribute and play a significant role to improve soil fertility, structure and increase the organic matter. Perennial crops also have advantages over annual crops in terms of biodiversity improvement, crop management, reduced herbicide and pesticide use and fertilizer inputs (Martinez-Mena et al. 2008).

The Arabian Peninsula is potentially an under-utilized resource for germplasm that can perform under challenging conditions of aridity and salinity (De Pauw 2002). A list of 152 species was compiled from field observations and consultations with Bedouin and international specialists, from which 27 were considered higher priority for forage production (Peacock et al. 2003a, b). Of these, the most promising included *Cenchrus ciliaris*, *Cenchrus setigerus*, *Coelachyrum brevifolium*, *Lasiurus scindicus*, *Panicum turgidum*, *Dichanthium foveolatum*, *Sporobolus ioclados* and *Stipagrostis plumosa*. The International Center for Biosaline Agriculture compared *C. ciliaris* and *Coelachyrum piercei* to the widely cultivated *Chloris gayana* in 2003 under irrigation with saline water (0–20 dS m^{-1}) and found *C. ciliaris* to be the most tolerant, producing 30 Kg ha^{-1} dry biomass (International Center for Biosaline Agriculture 2003). A later trial was conducted in Saudi Arabia to compare productivity and water use efficiency of five indigenous grasses (*C. ciliaris, P. turgidum, Coelachyrum piercei, Cenchrus divisus* and *Stipagrostis ciliata*) with *C. gayana*. In this study, *C. gayana* was the most affected by salinity (International Center for Agricultural Research in the Dry Areas (ICARDA) 2007).

El-Khatib and Hegazy (2001) conducted a study on three naturally growing populations of the perennial grasses *Panicum turgidum, Lasiurus scindicus* and *Cenchrus divisus* in Egypt, south west Saudi Arabia and Qatar. They reported that vegetative reproduction occurs by rhizome growth and sexual reproduction by tillering, where each tiller may end with a spike. The plants collected from Saudi Arabia and Egypt attained higher energy content than those collected from Qatar. The overall energy content of *P. turgidum* (over 6 kcal g^{-1} dry weight) is higher than that of *L. scindicus* and *P. divisium*. When the growth characteristics and energy content were taken as a measure of the grazing value, *P. turgidum* had better value as a forage plant than the other two species (El-Khatib and Hegazy 2001).

8.3.1 Chloris gayana

The livestock productivity in Arabian Peninsula is low due to poor genetic potential of indigenous breeds, feed shortage (both in quantity and quality), livestock diseases and parasites, lack of adequate livestock extension service, poor infrastructure and others. Natural pasture and crop residues are the main feed resources in these regions. However, feed produced from natural pasture is decreasing due to several factors that include overgrazing and land degradation. The nutritive value of crop residues and quality of natural pasture is usually low particularly in the dry seasons. The lower crude protein (CP) and metabolizable energy (ME) values of these pastures results in low livestock reproductive efficiency in these regions. Therefore, these feed resources require strategic supplementary feed stuffs such as agro-industrial by-products and cultivated improved forages. Rhodes grass is one possible perennial improved grass which can be grown on-farm and used by small-holder farmers.

Rhodes grass (*C. gayana*) is a stoloniferous multi-cut and multi-tillering perennial grass and has capacity to grow in different agro-climatic zones. Although it is native to Africa but now is widely grown in Asia and is a major source of animal feed in many Gulf countries. The forage demand in Saudi Arabia has increased significantly due to its growing livestock population, which has been estimated to be 3.5 million heads during 2010 (Bakhashwain 2010). Rhodes grass has also become increasingly popular in Africa, Australia, Japan, South America and the Middle East for its higher productivity, drought and salt tolerance (Ibrahim et al. 2006).

Rhodes grass is native to Africa where it emerges in woodlands, grasslands, alongside the roads, and river banks. It serves as one of the vital forage crops grown under crop rotational systems. Rhodes grass is a leafy plant with 1–2 m height and deep roots penetrating up to 4.5 m. However, it is not suitable for intercropping and relay cropping purposes. Rhodes grass is suitable in the mixed crop-livestock system. It can grow well with annual and perennial herbaceous legumes such as vetch, desmodium and clover. Due to its vigorous fibrous and deep rooting system and rapid establishment and surface covering ability, it is recognized as highly drought resistance and a valuable crop for soil conservation (Yossif and Ibrahim 2013). Rhodes grass is a fast growing plant, which can cover the surface within 3 months of sowing (Moore 2006). By doing so, it lowers soil temperature, improves soil stability, soil permeability and soil-water retention capacity (Valenzuela and Smith 2002). In many countries, Rhodes grass is used for soil stabilization. For example, in Australia, it is widely used for revegetating mine-disturbed soils (Moore 2006). In Hawaii, Rhodes grass trimmings are used as a mulch to protect soil erosion. This helps in the creation of valuable seed bed for vegetables because it is rich in organic matter and provides protection against wind and sun (Valenzuela and Smith 2002).

Rhodes grass is a suitable plant for different soils, from sands to fertile clays. However, it grows better in high altitude regions (1500–2500 masl) and areas where annual rainfall is between 500 and 750 mm (Arshad et al. 2014; Ecocrop 2014). It is not tolerant to seasonal waterlogging (FAO 2014). It is moderately tolerant to soil salinity [Electrical conductivity (ECe = 10 dS m^{-1})] and alkalinity (pH = 5.5–7.5) and water. Rhodes grass is best suited to well-drained moderate to high fertility soils. However, by adopting appropriate management strategies, it can also survive on infertile soils. Rhodes grass has some establishment problems on very acidic soils and is not tolerant to waterlogging. Therefore, over-irrigation should be avoided for especially in the initial growing season to ensure better germination. Rhodes grass is tolerant of Li but cannot survive in high Mn and Mg values (Cook et al. 2005). Rhodes grass does not grow well under shade (Ecocrop 2014; FAO 2014), however it can survive dry periods of up to 6 months and 15 days of water flooding (FAO 2014; Cook et al. 2005). For these reasons, Rhodes grass is widely grown in water stress regions such as India, Pakistan and Australia, where it is used as animal feed as it is a source of energy for animals (Arshad et al. 2014). Due to its perennial nature, farmers can benefit for several years after one planting.

Rhodes grass is suitable to grown in isolation or in combination with various other grasses such as *Paspalum dilatatum*, *Setaria sphacelata*, *Cenchrus ciliaris*, and *Panicum maximum* (Cook et al. 2005). In many other countries, it is also mixed with wheat, maize, sorghum or cotton crops. Yield is improved when grown with legume species such as alfalfa (*Medicago sativa*), stylo (*Stylosanthes guianensis*), soybean (*Neonotonia wightii*), centro (*Centrosema pubescens*), phasey bean (*Macroptilium lathyroides*), clover (*Trifolium* spp.), lotononis (*Lotononis bainesii*), *Desmodium uncinatum* and butterfly pea (*Clitoria ternatea*) (Cook et al. 2005).

Rhodes grass is a high quality forage when young (4 weeks of regrowth or less), with a crude protein value of over 5% (FAO 2014). However, its quality declines quickly as it matures with CP values decreasing 9–10% after 10 weeks of growth, and can drop to 8% after 15 weeks. At this CP value, it is not considered a healthy forage for ruminants (Cook et al. 2005). The decrease in CP values is higher before the first cut compared to subsequent cuts. The CP value of Rhodes grass is low in stems compared to the leaves (Mbwile and Udén 1997; Mero and Udén 1997).

Crude protein (CP) levels in Rhodes grass vary with the age and level of available nitrogen. The CP values may range from 17% on a dry matter (DM) basis in very young leaf to 3% in old leaves. The in vitro dry matter digestibility (IVDMD) varies from 40% to 80% (Cook et al. 2005). Young growth is very palatable, but it becomes less attractive after the plants produce seeds. Digestibility and CP values decrease as the grass matures and becomes hard. To avoid over-maturity regular cutting or grazing should be practiced and over-mature pasture should be slashed or burned. To avoid decline in the nutritive value after flowering, it is suggested keep plants in a leafy condition by steady defoliation. The nutritive value of Rhodes grass assessed by NDF, CP and IVDMD was found to be comparable to that of the tropical grasses *Cenchrus ciliaris*, *Bothriochloa insculpta* and *Panicum coloratum* (Mero and Udén 1997). Among temperate and tropical forages, NDF content in Rhodes grass was found to be comparable to that of *Cenchrus ciliaris*, but much higher than forage species such as fescue, ryegrass, white clover and alfalfa.

Rhodes grass can be used as green forage or hay. Rhodes grass sowing can be done from March to April and August to September in arid regions where summer comes early. One time sowing of Rhodes grass will give the production to the growers for about 3–4 years (Arshad 2012). Rhodes grass makes good hay if it is cut at the beginning of flowering. Old stands give low quality hay. It is not suitable for silage making. When preparing hay appropriate hay making procedures should be followed. Rhodes grass can be grazed 4–6 months after planting. Grazing may maintain Rhodes grass in a leafy and highly nutritive condition provided grazing is done for short periods (FAO 2014; Cook et al. 2005). Highest production is attained in the second year.

First cutting of the Rhodes grass should be done at least 6 months after sowing. However, during the second year and on-ward, harvesting can be done any time when it reaches the optimum harvesting stage. In areas where frost occurs it should be harvested before the onset of frost. Studies show that monthly cuttings are better than cutting in every 14-days interval in irrigated conditions (Cook et al. 2005; Yossif and Ibrahim 2013). If Rhodes grass is used for grazing, the pasture can be damaged by overgrazing. Therefore, it is better to use cut and carry approach when using Rhodes grass pasture.

On average, the annual productivity of Rhodes grass on farmers' field's ranges from 8.5 to 9.5 tons dry matter per ha under rain-fed conditions. The mean productivity of native pasture is 4.2 ton dry matter per ha in the central highlands of Ethiopia. Based on several studies, the annual dry matter yields of Rhodes grass ranges between 7 and 25 tha^{-1}, depending on the variety, climate and soil fertility conditions (Cook et al. 2005). Yields in the second year may be double than the establishment year depending on management and environmental conditions. Under favorable climatic, soil fertility and irrigation conditions, yields up to 35–60 t ha^{-1} dry matter can be achieved (Cook et al. 2005). Rhodes grass is persistent and drought tolerant when well grazed and fertilized but disappears after a few years if not well managed. The usual productive life of Rhodes grass is 3 years, which can be extended by providing optimum fertilization.

Rhodes grass is suitable to grow both under rainfed and irrigated conditions. For long-term sustainable production, good management and additional fertilizer (N) is recommended. The nutritive value of Rhodes grass can be enhanced through additional fertilizer and/or manure applications (Ecocrop 2014). Rhodes grass is receptive to N fertilizer for enhanced forage yields (Rahman 2007). However, excessive doses of N may cause buildup of nitrate in the forage which can be deleterious for cattle (Valenzuela and Smith 2002). The nitrogen rates up to 275–400 kg ha^{-1} in the presence of adequate phosphorus and potassium have shown linear increase in yield and CP content of the Rhodes grass. The split N applications after each cut or after grazing cycles have also been found healthier than one basic application (Skerman and Riveros 1990; Valenzuela and Smith 2002). Rhodes grass have shown to produce an annual dry matter yield of 35 t ha^{-1}with N application rate of 1200 kg ha^{-1} year^{-1} under center pivot irrigation in virgin desert lands of Saudi Arabia (Yossif and Ibrahim 2013). Higher fertilizer application rates require good irrigation amounts for better results. In irrigated areas, Rhodes grass can consume 20,000–35,000 m^3 ha^{-1} year^{-1} of water depending on the irrigation system used and climatic conditions. Under water deficit conditions, yields of Rhodes grass can drop to half of the irrigated areas (Arshad et al. 2000).

8.3.2 Cenchrus divisus

Cenchrus divisus is a perennial grass native to South Asia, the Middle East, and North Africa (Govaerts 2017). The species is widely known as *Pennisetum divisum* but was renamed after molecular phylogenetic studies showed overlap in *Cenchrus* and *Pennisetum* genera (Verloove et al. 2014). Global distribution of *Cenchrus divisus* was presented in Fig. 8.1 (Govaerts 2017). In the Cholistan desert of Pakistan, *C. divisus* is associated with better soils that have greater water retention (Akhter and Ahmad 2006). It produces green leaves even when soil moisture is low (Enright and Miller 2007) and is one of the most palatable native species of the Cholistan desert (Rafay 2012). A preference for soils high in calcium carbonate was observed in Bahrain (Abbas et al. 1991). Accessions from the northern United Arab Emirates (UAE) were associated with sandy soil types (Qureshi 2017) that were mid-range in pH and salinity (El-Keblawy et al. 2015). However, other reports indicated that production declines even at very low levels of salinity such as 60 mM NaCl (Khan and Ahmad 2007), with salt accumulating in the shoot tissues (Ashraf and Yasmin 1997). Seed germination is more sensitive to salinity than most other grass species, which are also usually sensitive. Germination was reduced to 20% by 100 mM NaCl (El-Keblawy et al. 2010) and salinity-induced dormancy is not alleviated by dormancy-regulating chemicals (El-Keblawy 2013).

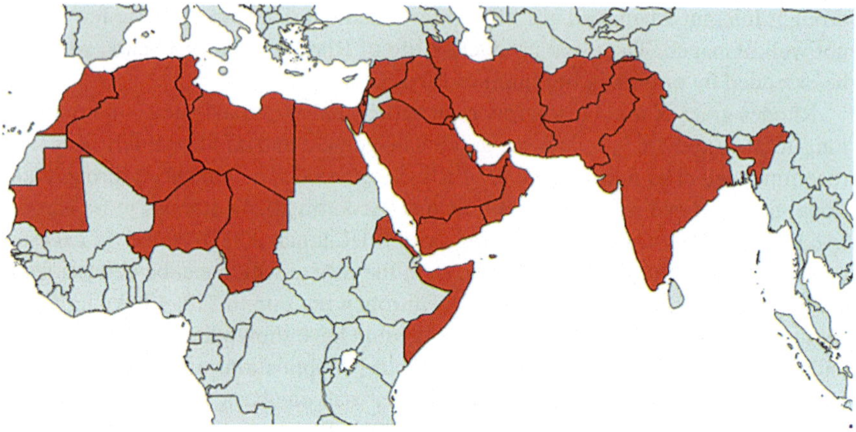

Fig. 8.1 Global distribution of *Cenchrus divisus*. (Source: Govaerts 2017)

Overgrazing is reported to have caused a decline of populations in Cholistan (Akhter and Ahmad 2006; Enright and Miller 2007) and the UAE (Gallacher 2010), and in a reduced seed bank in the UAE (Gallacher and Hill 2006). However several studies have indicated that the species can persist through heavy grazing; one from East Africa studying seed banks (Woldu and Saleem 2000), and two from the UAE studying plant communities (Fawzi and Ksiksi 2013; Gallacher and Hill 2006). There is regional variation in the nutritive values of naturally occurring populations (Alsherif 2018), but studies of native perennial grasses from the region have indicated that the species has less forage value than others, such as *Panicum turgidum* Forssk (El-Khatib and Hegazy 2001). Germplasm was collected from the UAE and Oman by ICARDA, but authors excluded the species from a shortlist of ten grasses prioritized for development, based on in situ observations and consultations in both countries (Peacock et al. 2003a, b). Later trials in Saudi Arabia showed that it performed less well than Rhodes grass (*Chloris gayana* Kunth) under limited irrigation (ICARDA 2007). Plot trials also indicated the species was less drought tolerant than *P. turgidum* and *C. pennisetiformis* Steud. (Ashraf and Yasmin 1995).

C. divisus is a wild relative of pearl millet (*C. americanus* (L.) Morrone) and is thus valuable as a genetic resource for improving the agricultural potential of the crop to harsh environmental conditions (Rao 2013). It has also been considered suitable for erophytic landscaping in India (Janakiram et al. 2018), Kuwait (Suleiman and Bhat 2004) and the UAE (Alam et al. 2017). A Qatari study rated it high for use in amenity horticulture under harsh conditions, but low on diversity of uses, particularly as it requires non-saline water and nitrogen fertilizer (Phondani et al. 2016). It is a preferred dietary species of spiny tailed lizard (*Uromastyx aegyptia microlepis*) in the UAE (Cunningham 2000) and is one of many species consumed by the Arabian sand gazelle (Cunningham 2013).

8.3.3 Cenchrus ciliaris

The genus name *Cenchrus* is derived from a 'Kenchros' which is the Greek name of a small millet (Wagner et al. 1990). In 1771, Carl Linnaaeus scientifically named the buffel grass as *Cenchrus ciliaris* L. This grass is an introduced as a perennial pasture grass, belongs to family poaceae, that is native to Africa and Western Asia. There are more than 30 Cenchrus species found in worldwide. It is an important perennial forage grass species in arid, semiarid and event desert regions of the world (McIvor et al. 2005). It has been planted widely in the United States, the Middle East, Mexico, India, Australia and South America (United states Department of Agriculture 2010). Other names by which this grass is known as dhaman grass, African foxtail grass, anjan grass kolukkatai. Buffel grass is a first choice to grazing animals, and is thought to increase the milk production in especially cows once they graze on this grass species. The *C. ciliaris* was widely distributed in different parts of the world and play significant role pastoralists, land rehabilitation and soil protection (Kharrat-Souissi et al. 2011). It was considered to be "wondered crop" for its ability to withstand strong wind, drought, salinity and heavy grazing and rapidly respond to rains (FAO 2010; Marshall et al. 2012). *C. ciliaris* has high water use efficiency; in irrigation area it gives the most effective results (Osman et al. 2008). This grass has a moderate tolerance to salinity but a lot of cultivars have maximum tolerance to salinity (FAO 2010; Cook et al. 2005). *C. ciliaris* could not survive under waterlogging condition as well as can tolerate the fire (Ecoport 2010). *C. ciliaris* occupies a diverse range of climate as describe in the Table 8.1.

Cenchrus ciliaris is a fast growing, short stoloniferous perennial that has a large, deep root system and able to out-compete native vegetation. It grows up to 1.7 meter tall depending upon the cultivar as well as growing conditions. Its stem is erect to prostrate with extensive branching. It has ability to respond quickly to the rain and flower early but at the same time as flowering it will continue to produce extra leaf and new shoots and they can produces a large amount of seeds. The flower or seed head resembles the typical "foxtail". The seeds of *C. ciliaris* are the strong competitor with high yield, fluffy and light weight and results in long distance spread through water and air (Hall 2001).

Cenchrus ciliaris is widely distributed in Arabian Peninsula. Irrigation with ground water has brought about a bringing down of the water table and also increased the salinity (Al-Senafy and Abraham 2004). Water is limited and the utilization is

Table 8.1 Temperature and rainfall ranges for production of *Cenchrus ciliaris*

Climatic factors	Ranges	References
Temperature	Opt. 30 °C Max. 50 °C Min. 5 °C	Christie (1975), De La Barrera and Catellanos (2007) and Cox et al. (1988)
Rainfall	From >250–2670 mm/annum	New South Wales Department of Primary Industries (2004) and Tix (2000)

high, which has put negative pressure on land resources, agricultural production and public health (Osman et al. 2008). Groundwater salinization has turned into a world-wide issue mainly in all countries of the Arabian Peninsula, where groundwater is the primary source of water. Plant's physiology can be affected due to the salinity for example delay in seed germination, high seedling mortality, poor harvest, stunted growth and decrease yield (Ahmad et al. 2010). The decrease in plant development subject to continuous increment in salinity is primarily identified with the increment in osmotic tension of soil solution which decreases retention of water by roots and additionally due to accumulation of various ions in toxic level (Tester and Davenport 2003).

Cenchrus ciliaris is considered as an appropriate for fodder grasses because of its significant returns and nutritive values, drought resistance and high palatability (Ruiz and Taleisnik 2013; Al-Dakheel et al. 2015). The composition of the grasses has an excellent influence on animal nutrition (Dewhurst et al. 2009). Chemical composition of fodder grasses species are consist of crude protein (CP), neutral detergent fiber (NDF), acid detergent fiber (ADF), lignin, and total digestible nutrients (TDN) (Guo et al. 2010). The chemical composition of the grasses can be affected by the phonological cycle of the grasses. In a few kinds of grasses the level of Crud protein (CP), Dry matter digestibility (DMD) and metabolic energy became diminished with increase of the harvesting period (Arzani et al. 2004). The impact of saline irrigation water and drought on biomass, growth, yield and quality attributes of *C. ciliaris* has been reported in Tables 8.2 and 8.3. *C. ciliaris* has likewise utilized as people drug for kidney pain, tumors, sores and injuries. It tends to be utilized as pain killer, and diuretic and as an emollient (Duke 1983).

In 2003 García-Dessommes with his colleagues revealed that five new genotypes of the *C. ciliaris* were produced without irrigation, which are less sensitive to harsh environmental condition as well as yielding progressively dry matter and crude protein (CP) content than common buffelgrass. Among the grasses, crude protein (CP) level are interrelated with numerous plant components such as digestibility, vitamins, Calcium (Ca) and Phosphorus (P). Be that as it may, all these decay to inadequate levels at about a similar time, also CP serves as a reliable measure of overall nutritional quality (Ganskoop and Bohnert 2001).

Different pathogens caused diseases in buffel grasses especially fungi and insects that limit the production. For example, *Pyricularia grisea* is a fungal pathogen that caused dieback disease in *C. ciliaris* (Rodríguez et al. 1999). Injury and death of buffel grass was occurred due to feeding of the seeds by the larvae of the moth (*Mampava rhodoneura* Turner) and spittle bugs (*Aeneolamia albofasciata*) (Cook et al. 2005).

Though, *C. ciliaris* is an economically important pasture species, it is also one of the most environmentally serious weed species. It is especially esteemed for its capacity to produce grazeable biomass rapidly in response to periodic or unpredictable rain on poor soil (Marshall et al. 2011). This grass is of most concern on sandy soils and is often also weedy on alluvial flats and riverine sites. Notwithstanding, a few examinations have revealed decreases in profitability of *C. ciliaris* pastures after some time, probably because of supplement constraints (Ibarra-Flores et al.

Table 8.2 Influence of salinity on quality and composition of *Cenchrus ciliaris*

Genotype	Salinity imposition	Quality indicator	Decrease(−)/ increase(+) over control	References
Cenchrus ciliaris cv. Biloela	10 dS.m^{-1}	Seeds per spikelet	0.01	Ruiz and Taleisnik (2013)
	1 5 dS.m^{-1}		5.79	//
	20 dS.m^{-1}		11.9	//
Cenchrus ciliaris cv. Texas	10 dS.m^{-1}		−15.3	//
	1 5 dS.m^{-1}		−16.7	//
	20 dS.m^{-1}		−20.4	//
Cenchrus ciliaris (Studied 40 genotype)	10 dS.m^{-1}	Fresh weight yield	−23.0	Al-Dakheel et al. (2015)
	15 dS.m^{-1}		−33.7	Al-Dakheel et al. (2015)
	10 dS.m^{-1}	Dry weight yield	−17.4	Al-Dakheel et al. (2015)
	15 dS.m^{-1}		−31.4	Al-Dakheel et al. (2015)
	10 dS.m^{-1}	Crud protein	17.53	Al-Dakheel et al. (2015)
	15 dS.m^{-1}		3.13	Al-Dakheel et al. (2015)
	10 dS.m^{-1}	Acid detergent fiber (ADF)	−0.29	Al-Dakheel et al. (2015)
	15 dS.m^{-1}		1.08	Al-Dakheel et al. (2015)
	10 dS.m^{-1}	Natural detergent fiber (NDF)	0.99	Al-Dakheel et al. (2015)
	15 dS.m^{-1}		0.62	Al-Dakheel et al. (2015)
	10 dS.m^{-1}	Ash	−1.7	Al-Dakheel et al. (2015)
	15 dS.m^{-1}		−3.03	Al-Dakheel et al. (2015)

(continued)

Table 8.2 (continued)

Genotype	Salinity imposition	Quality indicator	Decrease(−)/ increase(+) over control	References
Cenchrus ciliaris L.	200 mM NaCl	Fresh and dry weight	−50	Debouba et al. (2012)
	100 mM NaCl	Germination capacity (GC)	−25	Debouba et al. (2012)
	200 mM NaCl		−85	Debouba et al. (2012)
	300 mM NaCl		−90	Debouba et al. (2012)
	200 mM NaCl	Production of malonydialdehyde (MDA)	45	Debouba et al. (2012)
	300 mM NaCl		20	Debouba et al. (2012)
	300 mM NaCl	Total chlorophyll content	−35	Debouba et al. (2012)
		Carotenoïds contents	−45	Debouba et al. (2012)
RN 153	300 mM NaCl	Total fresh weight	−41.6	Quiroga et al. (2016)
RN 136			−52.9	Quiroga et al. (2016)
S. line			−8.53	Quiroga et al. (2016)
1–7-11			−66	Quiroga et al. (2016)
1–9-			−18	Quiroga et al. (2016)

(continued)

Table 8.2 (continued)

Genotype	Salinity imposition	Quality indicator	Decrease(−)/ increase(+) over control	References
RN 153	300 mM NaCl	Total sugar	−35.3	Quiroga et al. (2016)
RN 136			−22.77	Quiroga et al. (2016)
S. line			−42.86	Quiroga et al. (2016)
1–7-11			−64.03	Quiroga et al. (2016)
1–9-			−10.8	Quiroga et al. (2016)
RN 153		Total protein	131.08	Quiroga et al. (2016)
RN 136			−16.34	Quiroga et al. (2016)
S. line			131.99	Quiroga et al. (2016)
1-7-11			95.5	Quiroga et al. (2016)
1-9-			78.28	Quiroga et al. (2016)
Cenchrus ciliaris cv. Laredo	12,000 ppm	Average dry yield	−23	Abu-Alrub et al. (2018)
	18,000 ppm		−49	Abu-Alrub et al. (2018)
Cenchrus ciliaris cv. Pecos	12,000 ppm		−17.9	Abu-Alrub et al. (2018)
	18,000 ppm		−42	Abu-Alrub et al. (2018)
Cenchrus ciliaris cv. Laredo	6000 ppm	Acid-detergent fiber (ADF)	−7.71	Abu-Alrub et al. (2018)
	12,000 ppm		−14.62	Abu-Alrub et al. (2018)
	18,000 ppm		−22.3	Abu-Alrub et al. (2018)
Cenchrus ciliaris cv. Pecos	6000 ppm		−8.24	Abu-Alrub et al. (2018)
	12,000 ppm		−15.43	Abu-Alrub et al. (2018)
	18,000 ppm		−25.6	Abu-Alrub et al. (2018)

(continued)

Table 8.2 (continued)

Genotype	Salinity imposition	Quality indicator	Decrease(−)/ increase(+) over control	References
Cenchrus ciliaris cv. Laredo	6000 ppm	Crude protein (CP)	8.24	Abu-Alrub et al. (2018)
	12,000 ppm		18.55	Abu-Alrub et al. (2018)
	18,000 ppm		27.83	Abu-Alrub et al. (2018)
Cenchrus ciliaris cv. Pecos	6000 ppm		10.3	Abu-Alrub et al. (2018)
	12,000 ppm		19.6	Abu-Alrub et al. (2018)
	18,000 ppm		30.92	Abu-Alrub et al. (2018)
Cenchrus ciliaris	5 $dS.m^{-1}$	Average yield	−3.2	Qadir et al. (2008)
	10 $dS.m^{-1}$		−19.07	Qadir et al. (2008)
	20 $dS.m^{-1}$		−35.65	Qadir et al. (2008)

1999). For the long term, the economic advantages of this grass are uncertain. It has been utilized to restore productivity to degraded land and yielding economic advantages (Bisrat et al. 2004). In India, seeds of buffel grass are used as a human food in order to make bread and to be eaten as raw (Quattrocchi 2006). However there is small consumption of these seeds by human being. The *C. ciliaris* has a noteworthy negative impact on biodiversity. It has the ability to alter fire regimes and build up a positive feedback loop. It is also dangerous to the existing biological communities that have aesthetic and financial incentive in business sector other than farming, the executives of these rangelands for *C. ciliaris* control is prescribed as a high need.

8.3.4 Cenchrus setigerus

Cenchrus setigerus is an important nutritive perennial fodder grass, non-rhizomatous or shortly rhizomatous. Its false spike is dense, 1.5–9 cm long, with spikelets 3–4.5 mm long, each cluster containing one to three caryopses. *C. setigerus* grass grows from spring to summer. It matures in 8 weeks and seeds heavily. Yields range between 0.4 and 2.1 t/DM ha (Ecocrop 2014; FAO 2010). A hardy and drought-tolerant grass, *C. setigerus* is an interesting fodder in dry areas, though it is tussocky and lacks bulk. Once established, *C. setigerus* grass can withstand heavy grazing. In order to establish a thick stand, the grass seed should be broadcast in the old growth, after every 3 years. According to FAO (2010) reports, the grass is very suitable for

Table 8.3 Influence of drought stress on the grain yield of *Cenchrus ciliaris*

Cultivar	Drought stress (% soil water content)	Quality indicator	Change from control	References
E92011	20		−40	Mansoor et al. (2002)
	8			
E92014			−30	
E92017			−37	
E92020			−39	
E92022			−64	
E92027			−11	
E92029			−49	
E92032			−65	
E92036			−35	
E92039			−74	
E92041			−28	
E92047			−73	
E92049			−57	
Unspecified	75	Shoot fresh weight	−58	Nawazish and Hameed (2006)
	50		−67	
	75	Shoot dry weight	−66	
	50		−73	
RN1	30% at 45 °C	Aerial dry weight	−71	Tommasino et al. (2018)
RN49			−41	
RN51			−46	
J20			−40	
S6			−64	
Unspecified	15	Fresh weight	−32	Siddiqui et al. (2016)
		Dry weight	−48	
		Turgid weight	−23	
		Total free proline	+62	
		H_2O_2 content	+31	
		Superoxide dismutase (SOD)	+1200	
		Ascorbate peroxidase (APX)	+233	
		Catalase (CAT)	+191	

less productive and marginal lands where sufficient yield in not certain from other forage grasses. Meanwhile, it is a good palatable and ruminants like to graze.

Cenchrus setigerus grass is native of West Asia, North and East Africa and has been naturalized elsewhere in dry tropics and subtropics (Australia). It grows between 30°N and 30°S, at elevations between 500 and 800 m. Common in open dry bush and grassland, it is very tolerant of drought and heat and adapted to arid and dry lands. It can tolerate dry spells where annual rainfall as low as 200 mm. Optimal temperatures are probably 30–35 °C but it can survive frost. It responds very quickly to light rains but does not respond well to winter rains. *C. setigerus* grass is palatable but not very productive. Its feed value is highest during the pre-flowering stage (65% IVDMD), and much lower during the dry season (50% IVDMD) (Cook et al. 2005). Its drought resistance makes *C. setigerus* excellent for improving of low rainfall grazing lands. It is of minor value in erosion control but it has been proposed as a barrier against moving sand (FAO 2010). To increase the carrying capacity, it is a good option. In the Rajasthan arid zone, *C. setigerus* carries 1 sheep to 2.6 hectares in Jodhpur, and to 6.0 hectares in Pali (FAO 2010). In the same region, the values for stocking rate, DM production, number of lambs and wool production by hectare were higher in a sown pasture of *C. setigerus/Cenchrus ciliaris/Lasiurus setigerus* than in that of *C. setigerus/Cenchrus ciliaris* (Tran 2011).

8.4 Screening, Selection and Evaluation for Local Adaptation and Yield Potential

Buffelgrass (*Cenchrus ciliaris*) is a perennial forage reported to have sufficient potential for forage production and is tolerant toward harsh and marginal environmental conditions. It has the capability to withstand high temperature (45 °C), water scarcity and heat waves and can be grown on nutrient poor sandy marginal lands. In the present article, we overview its genetic diversity, production potential under different soil and water conditions and possible yield reduction, in different genotypes. According to long term studies conducted by Al-Dakheel and Hussain (2016); evaluated and screened genetic material of 160 Buffelgrass (*C. ciliaris*) at four saline irrigation waters (0, 10, 15, and 20 dS m^{-1}) in a pot culture trial and measured fresh and dry biomass. In general, average 4 harvests were achieved per year. Average annual dry biomass was in the range 122.5–148.9 g/pot in the control while it reduced a little bit and at medium salinity (10 dS m^{-1}) it produced the dry biomass (96.4–133.8 g/pot). The dry biomass was reported as 65.6–80.4 g/pot at 15 dS m^{-1}, and 55.4–65.6 g/pot at 20 dS m^{-1} (Al-Dakheel and Hussain 2016). Meanwhile, highest DW (148.9 g/pot) was found with accession 49 (PI 385321) that was originated from Tanzania while, lowest DW was obtained with accession 23 (PI 271206) that was originated from India. Other researchers also reported that it is very good pasture grass for arid regions of Africa, Australia and Asia (Arshadullah et al. 2011). The drought tolerance potential of *C. ciliaris* has also been documented and might be cultivated in hot and dry environments where the annual rainfall are lower than 100 mm. Buffelgrass is drought tolerant and usually present in regions receiving rainfall less than 100 mm. It is good forage during summer and winter season. Jorge et al. (2008), reported a wide geographical diversity among *C. ciliaris* population. Among the commercial cultivars tested, accession "Biloela" produced good higher biomass than other cultivars while Gayndah did not show stable yield at medium and higher salinity (Al-Dakheel and Hussain 2016). Based on the forage yield data, the top 40 accessions (out of 160) were identified and selected for further field experiments' during the year (2006–2013).

8.5 Agronomic Evaluation and Field Performance of Selected Accessions

The pots trials were very useful and results in the screening, identification and selection of 40 accessions for further investigations. However, further investigations were needed to study the performance under a range of biophysical field environments especially at marginal soil and using low quality saline water resources in order to save the fresh water for high purpose crops and to introduce the crop with good yield potential to the farmers. Hence field trials were conducted for 8 consecutive years under agro-ecological conditions of Dubai Emirate, UAE during 2006–2013 with 40 selected accessions. The irrigation was done by using three salinity levels (S1 = 5, S2 = 10 and S3 = 15 dS m^{-1}). The experimental plot, different growth stages, harvesting are illustrated in the Fig. 8.2. At lower salinity (5 dS m^{-1}), dry biomass was significantly higher (in range of 34.5–106 t ha^{-1}) and at 10 dS m^{-1} salinity, it ranged from 26.5 to 90.5 t ha^{-1}. However, at higher salinity, the dry biomass significantly decreased and was in the range of 24.5–73.0 t ha^{-1} (Al-Dakheel et al. 2015). From Ward's minimum-variance cluster analysis and accession group ranking, it was demonstrated that Grif 1639 (Pakistan) was the top performer and

Fig. 8.2 *Cenchrus ciliaris* at different stages in the field plots; (**a**) Seedling growth of *C. ciliaris* plantation; (**b**) Profound growth of *C. ciliaris*; (**c**) Front view of the *C. ciliaris* field plot; (**d**) View of the *C. ciliaris* plot after harvesting; (**e**) Preparation of bundles of harvested *C. ciliaris*; (**f**) *C. ciliaris* field harvested with whole produce

salt tolerant while MAK 9 (UAE) was most sensitive to salt stress. Other researchers also reported wide genotypic variation in salt tolerance among genotypes for biomass (Jorge et al. 2008). Buffelgrass accessions namely Grif 1639, PI161633, PI161637, PI279596, and PI365650 were the most salt tolerant and produced good yield and also were stable at different salinity levels. Most of the sensitive accessions (20, 21, 24, 25, 23) were originated from South Africa and showed poor performance against salinity with lower biomass.

The cultivation of perennial forage grasses on marginal lands will enhance forage supply for rural livestock, rural development, feedstock's for energy and soil sustainability (Ahmad 2010). That's why, there is significant interest in cultivation and adaptation of salt tolerant accessions among the farming communities for sustainable forage supply in UAE. Cenchrus spp. showed promising potential for animal feed due to the plant height, leaf size, branching, regeneration following harvest and capacity to survive under dry season. Meanwhile, buffelgrass also possess very useful characteristics that include good pasture grass and drought tolerant, and produce good yield and productivity. It also showed positive response to rains and resistant to grazing; thus it is a prominent forage in Australia (Buldgen and Francois 1998), North Africa (Kharrat-Souissi et al. 2011) and Ethiopia (Angassa and Baars 2000).

8.6 Stable and High Yielding Top Five Genotypes

The top 5 best-performing accessions showed significantly higher dry biomass than other accessions and produced an average 98.9, 82.5, and 66.8 t/ha at 5, 10, and 15 dS/m, respectively. The dry matter yield in lower 5 was documented as 41.7, 30.8 and 26.3 t/ha at above mentioned salinity levels respectively (Al-Dakheel and Hussain 2016). Within each salinity level, there was a good chance of selecting the genotypes, based on their relative performance (DW). In salt-degraded marginal lands of Dubai in general and UAE in particular, different buffelgrass accessions that had shown significant salt potential and possess good dry biomass yield are highly suitable for these areas. Meanwhile, Grif 1619 had proven most salt tolerant genotype with higher yield and productivity among all the accession. Other genotypes like 38 was high dry biomass producing at low (103.5 t/ha) and medium (76.5 t/ha) salinity but at high salinity, it was less stable and its dry biomass declined by more than 56% (45.5 t/ha). Another accession "17" was among the top ten genotypes at low salinity (5 dS/m) but its DW yield decreased at medium and high salinity displaying a classical pattern of a salinity sensitive genotype. Contrary to such pattern, genotype 12 was ranked 16 among the 40 genotypes at low salinity, while at medium and high salinity it ranked among the top 5 highest genotypes in dry biomass yield displaying better salt-tolerance. Among the 40 accessions (37, 38, 2, 12) were identified as salt-tolerant, high-yielding and stable genotypes at various salinity levels. These genotypes hold good salt tolerance potential and can be grown to enhance agriculture productivity in saline and marginal lands of Arabian Peninsula.

8.7 Nutritional and Quality Analysis

The quality analysis were conducted for crude protein (CP), neutral-detergent fiber (NDF), acid detergent fibre (ADF) and ash contents. Buffelgrass accessions showed different quality trait results. It was documented that *C. ciliaris* accessions with lower dry biomass had higher CP and other with higher yield indicated less CP (Al-Dakheel et al. 2015). In another study, Afzal and Ullah (2007), reported that to maintain animals meat quality, CP should have to be of sufficient amount in the diet of ruminants. The response of NDF in the accession were similar to those of attribute of CP. Accession 37 (Grif 1619) and 38 (MAF 74) had less than 47% NDF (38–47%), but accession Arshadullah et al. (2011), demonstrated that nutritional traits were significantly different among different genotypes of Buffelgrass.

8.8 Introducing Buffelgrass to Growers in Emirates and Arabian Gulf States

In 2016, under drip irrigation system, buffelgrass has almost banged exotic forages on a large number of farms in the UAE. In other six countries of the Peninsula, also its cultivations at pilot farms are scaling up. While mitigating the upscaling needs for quality seed, seed multiplication fields, as well as Seed Technology Units, have been established in Emirates, Oman, Qatar, Yemen and Jordan. This Buffel technology package is an output of a decade-long collaborative research for development program of ICBA with a parallel focus on capacity building and institutional strengthening of National Agricultural Research Services.

NARS support has asignificant impact on upscaling this technology package. The Oman government and Abu Dhabi Farmers' Services Center in UAE are supporting and educating growers to replace Rhodes grass the widely used forage species in the past with the more beneficial and less water consuming Buffelgrass. Nearly 10,500 farms in the Emirates of Abu Dhabi have stopped the cultivation of Rhodes grass as announced by ADFCA in May 2012 (Gulf News 2012) (Table 8.4).

Table 8.4 Species list

Family	Species	Common name
Agamidae	*Uromastyx aegyptia microlepis* Arnold, 1980	Spiny tailed lizard
Cercopidae	*Aeneolamia albofasciata* (Lallemand, 1939)	
Fabaceae	*Centrosema pubescens* Benth.	Centro
	Clitoria ternatea L.	Butterfly pea
	Desmodium uncinatum (Jacq.) DC.	
	Lotononis bainesii Baker	Lotononis
	Macroptilium lathyroides (L.) Urb.	Phasey bean
	Medicago sativa L.	Alfalfa/Lucerne
	Neonotonia wightii (Wight and Arn.) J.A.Lackey	Soybean
	Stylosanthes guianensis (Aubl.) Sw.	Stylo
	Trifolium repens L.	White clover
Poaceae	*Bothriochloa insculpta* (Hochst. ex A.Rich.) A. Camus	Creeping bluegrass
	Cenchrus americanus (L.) Morrone	Pearl millet
	Cenchrus pennisetiformis Steud.	Slender buffelgrass
	Cenchrus ciliaris L.	Buffelgrass
	Cenchrus divisus (J.F.Gmel.) Verloove, Govaerts and Buttler	Panicgrass
	Cenchrus setiger Vahl	Birdwood grass
	Chloris gayana Kunth	Rhodes grass
	Coelachyrum brevifolium Hochst. and Nees	
	Coelachyrum piercei (Benth.) Bor	
	Dichanthium foveolatum (Delile) Roberty	
	Festuca spp.	Fescue
	Lasiurus scindicus Henrard	
	Lolium spp.	Ryegrass
	Panicum coloratum L.	White buffalo grass
	Panicum maximum Jacq.	Guinea grass
	Panicum turgidum Forssk.	Panicgrass
	Paspalum dilatatum Poir.	Dallis Grass
	Pennisetum divisum – synonym for *Cenchrus divisus*	
	Pennisetum glaucum – synonym for *Cenchrus americanus*	
	Setaria sphacelata (Schumach.) Stapf and C.E.Hubb. ex Moss	African Bristle grass
	Sporobolus ioclados (Nees ex Trin.) Nees	
	Stipagrostis ciliata (Desf.) De Winter	Large Bushman Grass
	Stipagrostis plumosa (L.) Munro ex T.Anderson	
Pyriculariaceae	*Pyricularia grisea* Cooke ex Sacc. 1886	Rice Blast Fungus
Pyralidae	*Mampava rhodoneura* Turner, 1905	Buffel grass seed caterpillar

8.9 Impacts of Investments in Marginal Environments

There is a great opportunity to make heavy investment in the marginal lands to bring them under cultivation, to reduce erosion (soil, wind), to save fresh water resources. These are important pillars in the marginalized environment, to whom, plant breeder and agronomists should to concentrate to improve food security and to mitigate the climate change impact. There are several examples and successful stories from India and China where small and large investment from both public and private sector had a major impact on economic growth, human development, poverty reduction and it was comparable with investment in more favorable environments. Meanwhile impact of perennial crops on agro-ecosystem has been illustrated in Fig. 8.2. Adaptation of perennial cropping system on the marginal lands helps Climate-friendly agricultural practices by increasing the carbon stock in soil. Meanwhile, perennial crops also helps to reduce tillage, minimizing the need for chemical fertilizers, and managing livestock systems to reduce methane emissions. In this context low-emission farming systems include conservation agriculture, agroecology and organic farming.

By using the perennial crops over a long period of time might provide the following benefits to the soil biology;

1. Physical protection of soil carbon through minimum soil disturbance (tillage).
2. Enhancing the quality and density of plant and animal inputs to the soils
3. Improving soil biology through microbial diversity.
4. Maintaining continuous living plant cover on soils year-round (Table 8.5).

8.10 Conclusion

Using data for perennial forage grasses in marginal Arabian Peninsula, we identify salinity and drought as major abiotic stress factors that impact the plant growth and biomass yield. Meanwhile, it has been proposed that genetic diversity, salinity, drought and climate variability can lead to significant changes in growth response and ultimate yield. We discuss perspectives for different perennial forage grasses, their potential yield and nutritional profile and genetic diversity to improve grassland productivity in the marginal environment. Perennial forage grasses have a significant ability to accumulate green biomass over several seasons/years. This permits for organization of fodder reserves that can be used in times of scarcity and thus constitute a true "drought insurance" for ruminants. As fresh water resources are highly scare in the region, screening and selection of suitable forage grasses and particular genotypes will permit switching from traditional high water consuming forages (Rhodes grass) towards less water demanding forages with stable yields (*Cenchrus ciliaris*). It was concluded that, buffelgrass has potential to address the need for new genetics for this relatively new production system (irrigated forage production using low-quality water sources). However, higher biomass yields do

Table 8.5 Impact of perennial forage crops on agriculture biodiversity and plant-soil-water ecosystems

Buffering climate	Perennial provide protection cover to soils and moderate maximum and minimum soil temperature. Perennial absorbs intense, slow runoff, improve drainage and reduces risks of flooding. Above ground biomass of perennial forage grasses covers the surface while belowground biomass (deep extensive root network) stabilizes soil and enhances soil and agroecosystem resilience.
Carbon sequestration in the soil	Perennial can increase soil C pool from 0.1 to 3 mg ha^{-1} $year^{-1}$. Perennial can enhance soil C storage in deeper soil depths as grasses have greater root biomass than annual row crops. Marginally productive plants with low initial soil C rapidly accumulates soil C under perennial species.
Improving water quality	Perennial vegetation and residues protect the soil from water erosion. Perennial can improve water quality by reducing risks of nonpoint source pollution in runoff from agricultural fields. Perennials increase soil organic matter content, which can degrade pollutants in runoff.
Improving wildlife habitat	Perennial grasses from tropical and sub-tropical areas increase abundance of beneficial insects, birds that can help to maintain ecology and biodiversity. Perennials can provide cover, nesting and shelter for birds and wildlife even when moderately grazed or harvested at high quality cutting heights (10 cm).
Enhancing soil attributes	Perennials can provide soil biodiversity and microbial processes through their extensive and abundance root biomass, which enhances soil water holding capacity and decrease erosion. Growing perennials and incorporating residues help to improve soil physical, chemical and biological properties which directly enhances soil resilience. Perennials improve soil aggregation, soil macroporosity, water infilteration, water retention, capacity and soil fertility.

Source: Blanco-Canqui (2010), Vanloocke et al. (2017) and Werling et al. (2014)

not guarantee *C. ciliaris* success in the region and to be successful it must fit into current cropping patterns, farming systems and prove its worth in rigorous and properly designed on-farm trials in marginal lands affected by salinity and alkalinity and in the areas where the majority of the food crops could not be produced economically. For scaling up and large scale adoption by the farmers in the region, creating opportunities for marketing the produce are also vital. Consequently, more efforts will be needed to create awareness for its incorporation in the social, cultural and livestock feed to increase the market demand besides strengthening the efforts to improve forage yields combined with nutritional quality, production, harvesting and post-harvesting handling and storing for better utilization.

References

Abbas JA, Mohammed SA, Saleh MA (1991) Edaphic factors and plant species distribution in a protected area in the desert of Bahrain Island. Vegetatio 95:87–93

Abdulrazzak M (1997) The future of freshwater resources in the Arabian Peninsula. In: Uitto JI, Schneider J (eds) The fresh water resources in arid lands. The United Nations University, Tokyo

Abu-Alrub I, Marcum KB, Kabir N, Aran A, Hammadi MA (2018) Productivity and nutritional value of four forage cultivars compared to Rhodes grass irrigated with saline water. Aust J Crop Sci 12:203–209

Afzal J, Ullah MA (2007) Assessment of productive potential and utilization of rangelands and sown pastures in Pothowar Plateau. National Agriculture Research Centre, Islamabad, 10 pp
Ahmad F (2010) *Leptochloa fusca* cultivation for utilization of salt-affected soil and water resources in Cholistan desert. Rev Soc Nat 22:141–149. https://doi.org/10.1590/S1982-45132010000100010
Ahmad MSA, Ashraf M, Ali Q (2010) Soil salinity as a selection pressure is a key determinant for the evolution of salt tolerance in Blue Panic grass (*Panicum antidotale* Retz.). Flora 205:37–45
Akhter R, Arshad M (2006) Arid rangelands in the Cholistan desert (Pakistan). Science et Changement Planétaires/Sécheresse 17:210–217
Alam H, Khattak JZK, Ppoyil SBT, Kurup SS, Ksiksi TS (2017) Landscaping with native plants in the UAE: a review. Emirates J Food Agric 29:729–741
Al-Dakheel AJ, Hussain MI (2016) Genotypic variation for salinity tolerance in *Cenchrus ciliaris* L. Front Plant Sci 7:1090
Al-Dakheel AJ, Hussain MI, Rahman AQ (2015) Impact of irrigation water salinity on agronomical and quality attributes of *Cenchrus ciliaris* L. accessions. Agric Water Manag 159:148–154
Al-Rashed MF, Sherif MM (2000) Water resources in the GCC countries: an overview. Water Resour Manag 14:59–75
Al-Senafy M, Abraham J (2004) Vulnerability of groundwater resources from agricultural activities in southern Kuwait. Agric Water Manag 64:1–15
Alsherif E (2018) Exploration of unconventional fodder for arid land rehabilitation. Arid Land Res Manag 32:337–350
Al-Zubari W (1998) Towards the establishment of a total water cycle management and re-use program in the GCC countries. Desalination 120:3–14
Al-Zubari W, Al-Turbak A, Zahid W, Al-Ruwis K, Al-Tkhais A, Al-Mutaz I, Abdelwahab A, Murad A, Al-Harbi M, Al-Sulaymani Z (2017) An overview of the GCC unified water strategy (2016-2035). Desalin Water Treat 81:1–18
Angassa A, Baars RMT (2000) Ecological condition of encroached and non-encroached rangelands in Borana, Ethiopia. Afr J Ecol 38:321–328
Arshad I (2012) "Green fodder – a critical resource" published in SGS Pakistan Pvt. Ltd. News Letter, Pg. 03 – September 2012
Arshad M, Rao A, Akbar G, Akhtar M (2000) Eco-physiological adaptability of *Sporobolus iocladus*, from Cholistan, Pakistan. Pak J Biol Sci 2: 1604–1607. https://doi.org/10.3923/pjbs.1999.1604.1607
Arshad I, Medani KM, Khan ZA (2014) Effect of manual and artificial application of NPK fertilizers on the growth and yield of Rhodes Grass (*Chloris gayana* L. Kunth.) by using central pivot irrigation technology. Int J Res 7:236–239
Arshadullah M, Malik MA, Rasheed M, Jilani G, Zahoor F, Kaleem S (2011) Seasonal and genotypic variations influence the biomass and nutritional ingredients of *Cenchrus ciliaris* grass forage. Int J Agric Biol 13:120–124
Arzani H, Zohdi M, Fish E, Amiri GZ, Nikkhah A, Wester D (2004) Phenological effects on forage quality of five grass species. Rangel Ecol Manag 57:624–630
Ashraf M, Yasmin N (1995) Responses of four arid zone grass species from varying habitats to drought stress. Biol Plant 37:567
Ashraf M, Yasmin N (1997) Response of some arid zone grasses to brackish water. Der Tropenlandwirt-J Agric Trop 98:3–12
Bakhashwain AA (2010) Fodder yield and quality of Rhodes grass-alfalfa mixtures as affected by sowing rates in Makkah region. JKAU: Meteorol Environ Arid Land Agric Sci 21:19–33
Batanoumy KH (1986) Rangelands of the Arabian Peninsula with a special reference to the history of range management the Hemaan old Arabian Reserve System. Proceedings of the 2nd Int'l Rangeland Congress, Adelaide, Australia, pp 234–235
Bazza M (2005) Policies for water management and food security under water-scarcity conditions: the case of GCC countries. 7th Gulf water conference on Water Science and Technology Association, Kuwait, 19–23 November, FAO

Bisrat SA, Mullen BF, Grigg AH, Shelton HM (2004) Net primary productivity and rainfall use efficiency of pastures on reconstructed land following open-cut coal mining in central Queensland, Australia. Trop Grassl 38:47–55

Blanco-Canqui H (2010) Energy crops and their implications on soil and environment. Agron J 102:403–419

Buldgen A, Francois J (1998) Physiological reactions to imposed water deficit by *Andropogon gayanus* cv. Bisquamulatus and *Cenchrus ciliaris* cv. Biloela in a mixed fodder crop. J Agric Sci 131:31–38

Christie EK (1975) A note on the significance of *Eucalyptus populnea* for buffel grass production in infertile semi-arid rangelands. Trop Grassl 9:243–246

Cook BG et al (2005) Tropical forages. CSIRO, DPIandF(Qld), CIAT and ILRI, Brisbane

Cox JR, Martin-R MH, Ibarra-F FA, Fourie JH, Rethman JFG, Wilcox DG (1988) The influence of climate and soils on the distribution of four African grasses. J Range Manag 41:127–139

Cunningham PL (2000) Daily activity pattern and diet of a population of the Spiny tailed Lizard, *Uromastyx aegyptius microlepis*, during summer in the United Arab Emirates. Zool Middle East 21:37–46

Cunningham PL (2013) Plants included in the diet of Arabian sand gazelle (Reem) from Saudi Arabia. J King Saud Univ Sci 25:167–173

De La Barrera E, Castellanos AE (2007) High temperature effects on gas exchange for the invasive buffel grass. Weed Biol Manag 7:128–131

De Pauw E (2002) An agroecological exploration of the Arabian Peninsula. ICARDA, Aleppo, p 770

Debouba M, Kbaier N, Talbi S, Gouia H, Ferchichi A (2012) Seeds germination and shoot growth responses of a threatened poaceae (*Cenchrus ciliaris* L.) to increasing salt stress. Rev Écol 67:19

Dewhurst RJ, Delaby L, Moloney A, Boland T, Lewis E (2009) Nutritive value of forage legumes used for grazing and silage. Ir J Agric Food Res 48:167–187

Duke (1983) James A. Handbook of energy crops. Purdue University, Center for New Crops and Plant Products. Last updated 2 July 1998

Ecocrop (2014) Ecocrop database. FAO, Rome

Ecoport (2010) Ecoport database. Ecoport. http://www.ecoport.org

El-Keblawy A (2013) Impacts of dormancy-regulating chemicals on innate and salinity-induced dormancy of four forage grasses native to Arabian deserts. Grass Forage Sci 68:288–296

El-Keblawy A, Al-Ansari F, Al-Shamsi N (2010) Impact of dormancy regulating chemicals on salinity induced dormancy in *Lasiurus scindicus* and *Panicum turgidum*: two desert glycophytic grasses. Plant Growth Regul 62:163–170

El-Keblawy A, Abdelfattah MA, Khedr AHA (2015) Relationships between landforms, soil characteristics and dominant xerophytes in the hyper-arid northern United Arab Emirates. J Arid Environ 117:28–36

El-Khatib AA, Hegazy AK (2001) Growth and energy content of three forage grasses from the middle east rangelands. Acta Agron Hung 49:119–131

Environment Agency-Abu Dhabi (EAD) (2016) [Online] Available: https://www.ead.ae/

Enright NJ, Miller BP (2007) Livestock grazing impacts on desert vegetation, Khirthar National Park, Pakistan. Rangel Ecol Manag 60:680–684

FAO (2010) General summary Middle East region- Arabian Peninsula. AQUASTAT, FAOs' Information System on Water and Agriculture. http://www.fao.org/nr/water/aquastat/regions/meast/index.stm. Accessed Feb 2010

FAO (2014) Grassland index. A searchable catalogue of grass and forage legumes. FAO, Rome

FAO (2017) AQUASTAT Data. Retrieved from AQUASTAT: http://www.fao.org/nr/water/aquastat/data/query/index.html?lang=en. Accessed on 27 February, 2017

Fawzi MN, Ksiksi T (2013) Plant species diversity within an important United Arab Emirates ecosystem. Rev Ecol 68:25–36

Feng G, Sharratt B, Young F (2011) Soil properties governing soil erosion affected by cropping systems in the U.S. Pacific northwest. Soil Tillage Res 111:168–174

Gallacher DJ (2010) Arid rangeland degradation in an oil-rich Gulf state; inertia of perceived heritage and pro-agricultural policies. In: Veress B, Szigethy J (eds) Horizons in earth science research, vol 1. Nova Science Publishers, New York, pp 335–350

Gallacher DJ, Hill JP (2006) Effects of camel grazing on the ecology of small perennial plants in the Dubai (UAE) inland desert. J Arid Environ 66:738–750

Ganskoop D, Bohnert R (2001) Nutritional dynamics of 7 northern Great Basin grasses. J Range Manag 54:640–647

Govaerts R (2017) WCSP: World Checklist of Selected Plant Families. In: Roskov Y, Ower G, Orrell T, Nicolson D, Bailly N, Kirk P, Penev L (eds) Species 2000 and ITIS catalogue of life, 24th December 2018. Leiden, The Netherlands: species 2000: naturalis. Retrieved from http://www.catalogueoflife.org/col/details/database/id/24

Gulf News (2012) Farms stop cultivating Rhodes grass. Gulf News, May 20. Retrieved from http://gulfnews.com/news/uae/environment/farms-stop-cultivating-rhodes-grass-1.998624

Guo XL, Wilmshurst JF, Li Z (2010) Comparison of laboratory and field remote sensing methods to measure forage quality. Int J Environ Res Public Health 7:3513–3530

Hall TJ (2001) History and development of buffelgrass pasturelands in Queensland. In: Kyte J, Lees H, Loxtonand I, Cook B (eds) Buffelgrass symposium. Proceeding of buffelgrass workshop, 21–23 February. Department of Primary Industries, Bris-bane

Hussain MI, Muscolo A, Farooq M, Ahmad W (2019) Sustainable use and management of non-conventional water resources for rehabilitation of marginal lands in arid and semiarid environments. Agric Water Manag 221:462–476

Ibarra-Flores F, Cox JR, Martin-Rivera M, Crowl TA, Norton BE, Banner RE, Miller RW (1999) Soil physicochemical changes following buffelgrass establishment in Mexico. Arid Soil Res Rehabil 13:39–52

Ibrahim M, Rafiq M, Sultan A, Akram M, Goheer MA (2006) Green fodder yield and quality evaluation of maize and cowpea sown alone and in combination. J Agric Res 44:15–22

ICARDA-APRP (2007) Sustainable management of natural resources and improvement of the major production systems in the Arabian Peninsula. Final report, 2000–2005 ICARDA – Arabian Peninsula Regional Program. International Center for Agricultural Research in the Dry Areas (ICARDA), Aleppo, Syria

ICBA (2003) Biosalinity news. Newsletter for International Center for Biosaline Agriculture 14(1):3

Janakiram T, Sanikommu VRR, Maddukuri L, Durga (2018) Prospects of floriculture in arid and semi-arid regions of India. Presented at the national conference on arid horticulture for enhancing productivity and economic empowerment, Central Institute for Arid Horticulture, Bikaner

Jaradat AA (2005) Saline agriculture in the Arabian Peninsula: management of marginal lands and saline water resources. J Food Agric Environ 3:302–306

Jorge MAB, Van DEWouw M, Hanson J, Mohammed J (2008) Characterization of a collection of buffelgrass (*Cenchrus ciliaris*). Trop Grassl 42:27–39

Khan D, Ahmad R (2007) Effects of irrigation with amended dilutions of seawater on germination, growth and cations distribution in *Panicum turgidum* Forsk. A desert fodder graminoid. Int J Biol Biotechnol 4:149–157

Kharrat-Souissi A, Baumel A, Mseddi K, Torre F, Chaieb M (2011) Polymorphism of *Cenchrus ciliaris* L. a perennial grass of arid zone. Afr J Ecol 49:209–220

Mansoor U, Hameed M, Wahid A, RAO AR (2002) Ecotypic variability for drought resistance in *Cenchrus ciliaris* L. germplasm from Cholistan desert in Pakistan. Int J Agric Biol 3:392–397

Marshall NA, Friedel M, Klinken RD, van Grice AC (2011) Considering the social dimension of invasive species: the case of buffel grass. Environ Sci Pol 14:327–338

Marshall M, Lewis M, Ostendorf B (2012) Buffel grass (*Cenchrus ciliaris*) as an invader and threat to biodiversity in arid environments: a review. J Arid Environ 78:1–12

Martinez-Mena M, Lopez J, Almagro M, Boix-Fayos C, Albaladejo J (2008) Effect of water erosion and cultivation on the soil carbon stock in a semiarid area of south-east Spain. Soil Tillage Res 99:119–129

Mbwile RP, Udén P (1997) Effects of age and season on growth and nutritive value of Rhodes grass (*Chloris gayana* cv. Kunth). Anim Feed Sci Technol 65:87–98
McIvor KC, Cabernoch JL, Branch KD, Van Antwerp NM, Halili EC, Mastrototaro JJ (2005) inventors; Medtronic Minimed Inc, assignee. Glucose sensor package system. United States patent US 6,892,085, May 10
Mero RN, Udén P (1997) Promising tropical grasses and legumes as feed resources in Central Tanzania. I. Effect of different cutting patterns on production and nutritive value of six grasses and six legumes. Trop Grassl 31:549–555
Moore G (2006) Rhodes grass. Department of Agriculture and Food, Western Australia Bulletin 4690, Perth, Australia
Nawazish S, Hameed M (2006) Leaf anatomical adaptations of *Cenchrus ciliaris* L. from the salt range, Pakistan against drought stress. Pak J Bot 38:1723–1730
New South Wales (NSW) Department of Primary Industries (2004) Buffelgrass (online). Available at http://www.dpi.nsw.gov.au/agriculture/field/pastures-and-rangelands/species-varieties/a-z/buffel-grass. Accessed 05 Nov 2009
Odhiambo OG (2016) Water scarcity in the Arabian Peninsula and socio-economic implications. Appl Water Sci 7:1–14
Osman AE, Makawi M, Ahmed R (2008) Potential of the indigenous desert grasses of the Arabian Peninsula for forage production in a water-scarce region. Grass Forage Sci 63:495–503
Peacock JM, Ferguson ME, Al-Hadrami G, Saleh A, McCann IR, Al-Dakheel AJ (2003a) Desert forages of the Arabian Peninsula – the sustainable use of salt affected soils through conservation and evaluation. In: Prospects for saline agriculture. Kluwer Academic Publishers, Dordrecht, pp 43–56
Peacock JM, Ferguson ME, Alhadrami GA, McCann IR, Hajoj AA, Saleh A, Karnik R (2003b) Conservation through utilization: a case study of the indigenous forage grasses of the Arabian Peninsula. J Arid Environ 54:15–28
Phondani PC, Bhatt A, Elsarrag E, Alhorr YM, El-Keblawy A (2016) Criteria and indicator approach of global sustainability assessment system for sustainable landscaping using native plants in Qatar. Ecol Indic 69:381–389
Qadir I, Khan ZH, Majeed A, Yaqoob S, Khan RA, Anjum K (2008) Effect of salinity on forage production of range grasses. Pak J Sci 60:1–2
Quattrocchi U (2006) CRC world dictionary of grasses: common names, scientific names, eponyms, synonyms, and etymology. CRC Press/Taylor and Francis, Boca Raton
Quiroga M, Tommasino E, Griffa S, Ribotta A, Colomba EL, Carloni E, Grunberg K (2016) Genotypic variation in response to salinity in a new sexual germplasm of *Cenchrus ciliaris* L. Plant Physiol Biochem 102:53–61
Qureshi AS (2017) Sustainable use of marginal lands to improve food security in the United Arab Emirates. J Exp Biol Agric Sci 1:5. (Spl 1-SAFSAW)
Rafay M (2012) Studies on the productive potential and conservation strategy of major range grasses in the degrading range lands of Cholistan desert. PhD thesis, University of Agriculture Faisalabad Pakistan, Pakistan
Rahman AFI (2007) Effect seed rate and NPK fertilization on growth and yield of forage quality of Rhodes grass (*Chloris gayana* L. Kunth). MSc thesis, Faculty of Agriculture, University of Khartoum, Sudan
Rao NK (2013) Crop wild relatives from the Arabian Peninsula. Genet Resour Crop Evol 60:1709–1725
Rodríguez O, Gonzalez-Dominguez J, Krausz JP, Odvody GN, Wilson JP, Hanna WW, Levy M (1999) First report and epidemics of buffelgrass blight caused by *Pyricularia grisea* in South Texas. Plant Dis 83:398–398
Ruiz M, Taleisnik E (2013) Field hydroponics assessment of salt tolerance in *Cenchrus ciliaris* (L.): growth, yield, and maternal effect. Crop Pasture Sci 64:631–639
Shahid SA, Al-Shankiti A (2013) Sustainable food production in marginal lands – case of GDLA member countries. Int Soil Water Conserv Res 1:24–38

Shahin SM, Salem MA (2014) Food security in the United Arab Emirates (UAE): the great competition between the agricultural and forestry sector on irrigation resources. In: Proceedings of the International Research Conference on Science, Health and Medicine (IRCSHM), Dubai, UAE, October, pp 15–16

Shahin SM, Salem MA (2015) The challenges of water scarcity and the future of food security in the United Arab Emirates (UAE). Nat Res Conserv 3:1–6

Siddiqui ZS, Shahid H, Cho J, Park SH, Ryu TH, Park SC (2016) Physiological responses of two halophytic grass species under drought stress environment. Acta Bot Croat 75:31–38

Skerman PJ, Riveros F (1990) Tropical grasses. FAO plant production and protection series no. 23. FAO, Rome, pp 284–285

Suleiman MK, Bhat N (2004) Various attempts in greening the state of Kuwait. Arch Agron Soil Sci 50:353–360

Tester M, Davenport R (2003) Na^+ tolerance and Na^+ transport in higher plants. Ann Bot 91:503–527

The World Bank (2008) World development report: agriculture for development. The World Bank Group

The World Bank (2017) Renewable internal freshwater resources per capita (cubic meters), April 27. Retrieved from The World Bank http://data.worldbank.org/indicator/ER.H2O.INTR.PC?end=2014andstart=1962

Tix D (2000) *Cenchrus ciliaris* invasion and control in Southwestern U.S. Grasslands and Shrublands. University of Minnesota, Department of Horticultural Science. Retrieved from the University of Minnesota Digital Conservancy. https://hdl.handle.net/11299/59559

Tommasino E, López Colomba E, Carrizo M, Grunberg K, Quiroga M, Carloni E, Griffa S, Ribotta A, Luna C (2018) Individual and combined effects of drought and heat on antioxidant parameters and growth performance in Buffel grass (*Cenchrus ciliaris* L.) genotypes. S Afr J Bot 119:104–111

Tran G (2011) Birdwood grass (*Cenchrus setiger*). Feedipedia.org and Tables Régions Chaudes. A project by INRA, CIRAD and AFZ with the support of FAO. Last updated on March 11, 2011, 14:26

Uitto JI, Schneider J (1997) The fresh water resources in arid lands. The United Nations University, Tokyo

UNEP (2012) The UN-water status report on the application of integrated approaches to water resources management. Available from www.unwater.org/rio2012/report

United States Department of Agriculture (2010) *Pennisetum ciliare* (L.) Link Buffel- grass. Plants Profile. United States Department of Agriculture, Natural Resources Conservation Service

Valenzuela H, Smith J (2002) Rhodes grass. Univ. Hawaii, CTAHR Ext. Service, Sustainable Agriculture Cover Crops. SA-CC-3, Manoa, Hawaii

VanLoocke A, Twine TE, Kucharik CJ, Bernacchi CJ (2017) Assessing the potential to decrease the Gulf of Mexico hypoxic zone with Midwest US perennial cellulosic feedstock production. GCB Bioenergy 9:858–875

Verhoeven H (2015) The nexus as a political commodity: agricultural development, water policy and elite rivalry in Egypt. Int J Water Res Dev 31:360–374

Verloove F, Govaerts R, Buttler KP (2014) A new combination in Cenchrus (Poaceae: Paniceae), with lectotypification of *Panicum divisum*. Phytotaxa 181:59–60

Wagner WL, Herbst DR, Sohmer SH (1990) Manual of the flowering plants of Hawaii, 2 vols. University of Hawaii Press/Bishop Museum Press, Honolulu

Werling BP, Dickson TL, Isaacs R, Gaines H, Gratton C, Gross KL, Liere H, Malmstrom CM, Meehan TD, Ruan L, Robertson BA (2014) Perennial grasslands enhance biodiversity and multiple ecosystem services in bioenergy landscapes. Proc Natl Acad Sci 111:1652–1657

Woldu Z, Saleem MM (2000) Grazing induced biodiversity in the highland ecozone of East Africa. Agric Ecosyst Environ 79:43–52

WRI (World Resource Institute) (2007) Actual renewable water resources: per capita. Retrieved 2010, from Earth Trends. http://earthtrends.wri.org/text/water-resources/variable-694.html

Wuest SB, Williams JD, Gollany HT (2006) Tillage and perennial grass effects on ponded infiltration for seven semi-arid loess soils. J Soil Water Conserv 61:218–223

Yossif AM, Ibrahim YM (2013) Effect of fertilizers (urea, farmyard and chicken manure) on growth and yield of Rhodes grass (*Chloris gayana* L. Knuth.). Univ J Plant Sci 1:85–90

Chapter 9
Agricultural Sustainability and Food Security in Agroecological Zones of Tanzania

Msafiri Yusuph Mkonda

Abstract Agricultural sustainability is crucial for developing countries, including Tanzania whose economy and food security entirely depend on agriculture. Tanzania has seven agro-ecological zones with different potentials and challenges to attain agricultural sustainability. These agro-ecological zones are coastal, arid, semi-arid, plateau, southern and western highlands, northern highlands, and alluvial. To attain agricultural sustainability and food security in the country, it is essential to explore biophysical, economic and social dimensions. This chapter reviews the climatic situation, agricultural potentials and agronomic practices. Arid and semi-arid zones are more vulnerable to environmental stress, especially climate change, than plateau, alluvial, and northern and southern highlands. Efficient agricultural sustainability has increased peoples' income and food security in resilient agroecological zones by 50%, and by 10% in vulnerable zones. This has eventually improved the livelihoods of the people in resilient agro-ecological zones, and has allowed cultivation of few crop varieties such as sorghum and millet in vulnerable zones. Areas with the best agronomic practices such as animal manure fertilization have increased crop yields from 0.75 to 1.95 tons ha^{-1}. As a result, this yield increment has improved the livelihoods of about 70% of Tanzanian farmers who entirely depend in agriculture.

Keywords Agricultural sustainability · Agro-ecological zones · Climate change · Crop yields · Food security · Nutrient use efficiency · Organic fertilizations · Smallholder farmers · Soils fertility · Tanzania

---

M. Y. Mkonda (✉)
Department of Geography and Environmental Studies, Solomon Mahlangu College of Science and Education, Sokoine University of Agriculture, Morogoro, Tanzania

E. Lichtfouse (ed.), *Sustainable Agriculture Reviews 52*, Sustainable Agriculture Reviews 52, https://doi.org/10.1007/978-3-030-73245-5_9

9.1 Introduction

The essence of agricultural sustainability has increasingly attracted the attention of various stakeholders at both global and local levels (Pretty 2007). The definition of agricultural sustainability is mainly contextual in nature (Bationo et al. 2006; Solomon et al. 2007). However, the major concern for agricultural sustainability involves the need to develop technologies and practices that do not have adverse effects on environmental goods and services, are accessible to and effective for farmers, and lead to improvements in food productivity (Lal 1998; Pretty 2008; Mkonda and He 2018a). The increasing demand of food security and environmental conservation gives a prompt needs for agricultural sustainability at both global and local levels. This demand has been more extensive from the last two decades of twentieth century and the first decade of twenty-first century, and as well more pronounced in developing countries (Monfreda et al. 2008; Branca et al. 2013).

In most developing countries; food security is mainly obtained from agricultural production at local levels (Mkonda and He 2018b). This, in turn, necessitates the adoption of various technological aspects to meet the objectives of food security and environmental conservation (Lichtfouse et al. 2009). However, it has been difficult to meet these two objectives simultaneously. According to FAO (2008) and Monfreda et al. (2008) the per capita cereal production has decreased from 150 to 130 kg per person in most African countries while increasing in Asia and South America from 200 to 250 kg per person. Furthermore, FAO (2012) and Sieber et al. (2015) added that this situation has increased the demand of adopting best agriculture practices to attain food security and environmental conservation.

In this aspect, agricultural sustainability would increase agricultural yields to meet the demands of foods from global population increase (Vermeulen et al. 2012; IPCC 2014). The global food demand has significantly increased since the last decade of twentieth century and this is evidenced by the fact that, the growth of cereal grain has been at 1% while that of population is 3% (FAO 2006).

Globally, from 1980s to 2000s, the world population has grown from three billion to more than six billion, imposing an increasing impact of human footprint on the earth as consumption patterns change (Kitzes et al. 2008; Pretty 2007). The per capita agricultural production has outstripped population growth (Hazell and Wood 2008) in most countries. Specifically, for each person today, there has been an additional about 25% more food compared with that in 1960. Nevertheless, these cumulative figures, hide important regional differences especially between industrialized and developing countries. Although Europe and North American experience mega changes in this, Asia and Latin America also increased their per capita food production by 76% and 28%, respectively. However, in this aggregate figures; Africa, has managed badly, with production per person 10% lower today than in 1960.

For Africa having the poorest per capita food production and weak management systems of the environment, it is mandatory that AS should be given high attention by all stakeholders in order to simultaneously solve the problem of food shortage and environmental degradation (Mkonda and He 2018b). Now that, it is a role of scientists, climate practitioners, environmental experts and related experts to establish appropriate approaches towards succeeding this objective.

Agricultural sustainability is shown in Fig. 9.1 below. The figure stipulates the most important aspects of agricultural sustainability and its benefits to both the environment, crop production and community livelihoods. This figure applies to most countries and contexts at both global and local levels.

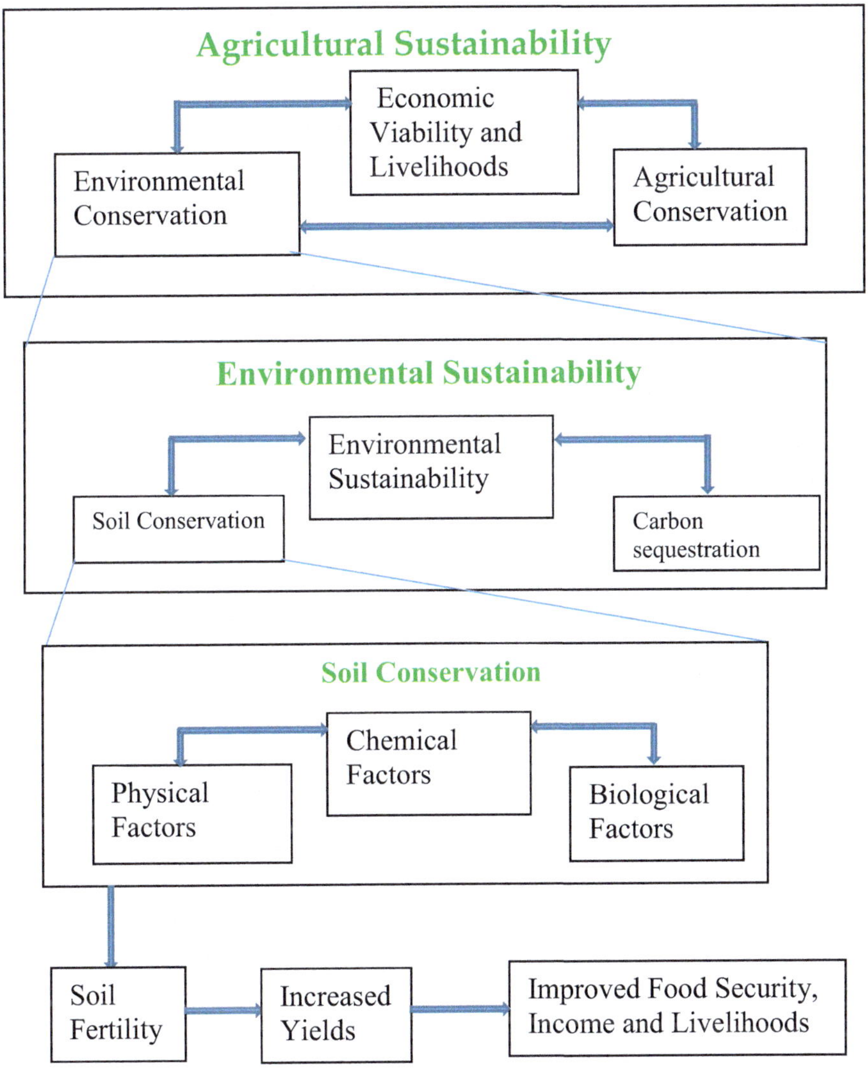

Fig. 9.1 Agricultural sustainability, environmental sustainability, and crop yields synergies. Sustainable agriculture creates favorable environmental conditions for crop production and environmental conservation. (Modified from Andrews 1998; Lal 1998; Pretty 2007)

In Tanzania agricultural sustainability is important aspect in attaining food security and environmental conservation (Andrews and Carroll 2001). It is also a tool for improving income in various communities (Ahmed et al. 2011). However, agricultural sustainability in the country is affected by the existence of diverse agroecological zones. These zones include coast, arid, semi-arid, plateau, southern and western highlands, northern highlands and alluvial (URT 2007). The alluvial zones seem to be more potential for irrigation agriculture due the influence of *Rufiji, Ruvu, Wami, Ruaha, Kilombero, Malagarasi* and *Pangani basins* that form important hydro-ecological zones (Altieri and Nicholls 2012; Lalika et al. 2017). These basins provide fruitful potentials for crop production.

For a precise view of the agroecological zones of Tanzania, Table 9.1 shows the sampled areas from region to village levels to explore the area. These regions, districts and villages are just representatives of the whole targeted area.

Although various studies have established that agricultural sustainability is important aspect to meet food security and environmental conservation, most of them have generalized it without giving a distinctive analysis of each agroecological zone. In doing so, these studies have largely conceded numerous shortcomings when a specific adaptation plan or strategy is needed within a specific agroecological zone. This has further adversely affected the implementations of programs and formulation of appropriate policies. Therefore, there is a need to establish a study that explores all agroecological zones of the country by treating each separately.

The main objective of this chapter is to assess the potentials and challenges influencing agricultural sustainability across different agroecological zones in Tanzania, and propose proper mechanism that can sustain agricultural sustainability. This, in turn, can improve environmental conservation, food security, and income. Explicitly, the results of this study can further synthesize the knowledge from diverse agroecological zones and thus, establishing the baseline for policy formulation. Finally, this can enhance agricultural sustainability in various agroecological zones of the country.

Table 9.1 Samples of the regions, districts and villages in the seven agro-ecological zones

Zone	Region	District	Village
Coast	Tanga	Pangani	Mkalamo
Arid	Mara	Serengeti	Ikoma
Semi-arid	Singida	Manyoni	Chikuyu
Plateau	Tabora	Uyui	Ufuluma
S&W Highlands	Iringa	Iringa rural	Sadani
N. Highlands	Kilimanjaro	Moshi rural	Makuyuni
Alluvial	Morogoro	Kilombero	Mang'ula

Source: Adopted from Mkonda et al. (2018) and Mkonda and He (2017d)

9.2 Study Site

The present chapter focuses on the seven agro-ecological zones of Tanzania, an Eastern African country with rich biodiversity. These zones include*; coastal, arid, semi-arid, plateau, southern and western highlands, northern highlands, plateau and alluvial plains* (Fig. 9.2). This classification bases on the altitude, precipitation pattern, dependable growing seasons and average water holding capacity of the soils and physiographic features (URT 2007). Agro-ecological zones refers to the geographical areas exhibiting similar climatic conditions that determine their ability to support rain-fed agriculture (Bockstaller et al. 1997; URT 2007). Therefore, the important aspects of different agro-ecological zones such as rainfall, soil fertility and agricultural systems determine sustainability of the area.

Coast zone covers different regions along the Indian Ocean. To the north; it covers Tanga except Lushoto, while to the coast it covers Coastal and Dares Salaam. Lastly, to the South it covers eastern Lindi and Mtwara (except Makonde Plateau).

The *arid* and *semi-arid zones* cover a number of regions and are grouped into two major parts. These are central and southern. The central region includes Dodoma, Singida, Northern Iringa, some parts of Arusha and Shinyanga (URT 2007). These regions are located at 1000–1500 m altitudes and receive unreliable unimodal rainfall ranging from 400 to 800 mm per annum. In general, the topography of the area is characterized by undulating plains with rocky hills and low scarps with a

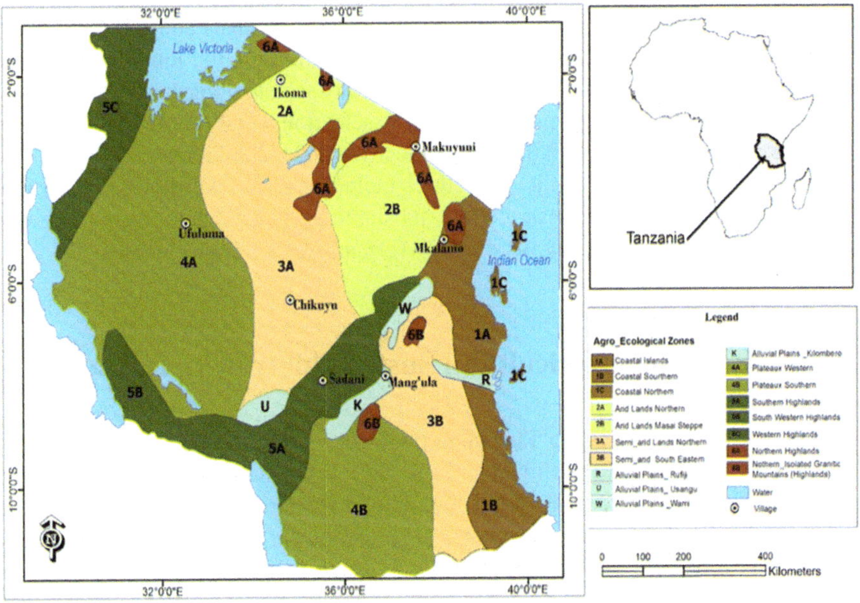

Fig. 9.2 Study area showing agro-ecological zones of Tanzania. (Adapted from Mkonda et al. 2018)

well-drained low fertile soils. It has an alluvial hardpan and saline soils in eastern Rift Valley and Lake Eyasi, and black cracking soils in Shinyanga (URT 2007).

The southern parts involves the regions of Morogoro (except Kiliombero and Wami Basins, and Uluguru Mountains), Lindi and southwest of Mtwara. These regions are located between 200 and 600 m altitudes and receive unreliable unimodal rainfall ranging from 500 to 800 mm per annum. Overall, the topography of the area is characterized by flat or undulating plains with rocky hills, moderate fertile loams and clays in south Morogoro, infertile sand soils in center (URT 2007; Mkonda and He 2017a). The growing season in both parts starts from December to March, however, it has been changing due change of onset and cessation of rainfall caused by global climate change.

As seen in Fig. 9.2 above; *Plateau zone* covers Tabora, Rukwa and Mbeya regions to the west. It also covers Kigoma and part of Mara regions to the north. Lastly, it covers Ruvuma and southern part of Morogoro regions to the south.

Southern and western highlands. This zone covers a broad ridge from northern Morogoro to Lake Nyasa. It also covers part of Iringa, Mbeya, Ufipa plateau in Sumbawanga and along the shore of Lake Tanganyika in Kigoma and Kagera regions.

Northern Highlands. This zone covers foot of Mt. Kilimanjaro and Mt. Meru, eastern Rift Valley to Eyasi, Mt. Uluguru in Morogoro, Pare Mountains in Kilimanjaro and Usambara Mountains in Tanga, Tarime highlands in Mara.

Lastly, the *Alluvial Plain*. This covers Kilombero (Morogoro), Rufiji (Coast), Usangu (Mbeya) and Wami (Morogoro).

To write this chapter, more than 80 publications, especially journal papers, were reviewed. The scientific papers published in authentic journals and mostly indexed in the web of science were given high priority. In addition, the journals with high impact factors and number of citations were equally selected. Furthermore, the most recent publications were given priority in this selection. Analyses and modification of some data were done to suit the study objectives and the guidelines of *Sustainable Agriculture Reviews*. All publication ethics including the seeking of permission to journal authors, where necessary, were substantially considered. The review was done to meet the standards of *Sustainable Agricultural Reviews* journal.

9.3 Principles of Agricultural Sustainability in Tanzania

The pressure from increasing Tanzanian population, weak agricultural technology, and increasing scenarios of climate change have compelled experts to recommend AS as the best strategy to limit global environmental degradations and food shortage cases (Branca et al. 2013; Ahmed et al. 2011; Rowhani et al. 2011a). There has been frequent occurrences of food shortage in Tanzania due to decline of agricultural productivity. This decline is evidenced by a number of scenarios such as frequent food insecurity, hunger and malnutrition cases (URT 2007, 2012; FAO 2012). This situation is more pronounced in rural areas.

In addition, UNEP (2011) and Poppy et al. (2014) asserted that there has been an acute decline of per capita grain per harvested areas in most developing countries. In fact, this is the most alarming indicator of agricultural unsustainability. In most sub-Saharan countries, including Tanzania, the per capita grain per harvested area has experience significant decline from 0.23 ha in 1980s to 0.12 ha in 2000s (FAO 2006; UNEP 2012; Branca et al. 2013). This decline has occurred severally even in recent years.

This challenge can be limited by adopting irrigation agriculture. However, irrigated land has declined from 0.047 ha in 1980s to 0.044 in 2000s (URT 2007; FAO 2013; Duru et al. 2015). This decline reveals that agricultural is considerably unsustainable in the area. This challenge is amplified by the presence of arid, semi-arid and tropical climates in the country (Giller et al. 2009; Branca et al. 2013; Chai et al. 2015; Pauline et al. 2016). According to URT (2007), 30% of Tanzanian land is under either arid or semi-arid climate and thus, experiencing the most consequences of climate stress and unsustainable agriculture (Doran and Zeiss 2000).

Therefore, the combination of several aspects such as arid and semi-arid tropics, long-term monoculture practices, and burning of grassland have significantly reduced the total carbon, nitrogen contents and other important minerals that can support sustainable agriculture (Bockstaller et al. 1997; Hartemink 1997; Medeiros et al. 1997; Sosovele et al. 1999; Monfreda et al. 2008; Msongaleli et al. 2015; Mkonda and He 2017b). It is obvious that the decline of important minerals in the soil hampers crop production and degrade the environment. This in turn, affects the livelihoods of over 70% of the Tanzanian smallholders who entirely depend on agriculture (Duru et al. 2015; URT 2014, 2012; Kangalawe et al. 2016). Again, this has direct effects to increased food shortage and poverty in the country (Paavola 2008; Lema and Majule 2009; Yanda 2015; Kangalawe 2016).

For the past two decades, the annual food deficit was approximated to 50% in most arid and semi-arid areas because; the little obtained yields were consumed within 3–6 months, thus, leaving the people under severe starvation for the rest of the year (URT 2007, 2012, 2014; Mkonda and He 2017c). More bad years still happen in the area and has skyrocketed food insecurity and abject poverty among the smallholder farmers (Ahmed et al. 2011; Rowhani et al. 2011b; Kangalawe and Lyimo 2013; URT 2014). From that point, the improvement of AS is very important to acquire both food security and income. According to the context, good agronomic practices is an immediate resolution (Andrews 1998; Lal 1998; Andrews and Carroll 2001).

9.4 Climate

The climate situation in the areas is considerably diverse almost in all agroecological zones of the country. Although these zones experience climate variability, some experience acute variations. As stipulated in the description of the study site section (Fig. 9.2), these zones have different climatic characteristics i.e. rains,

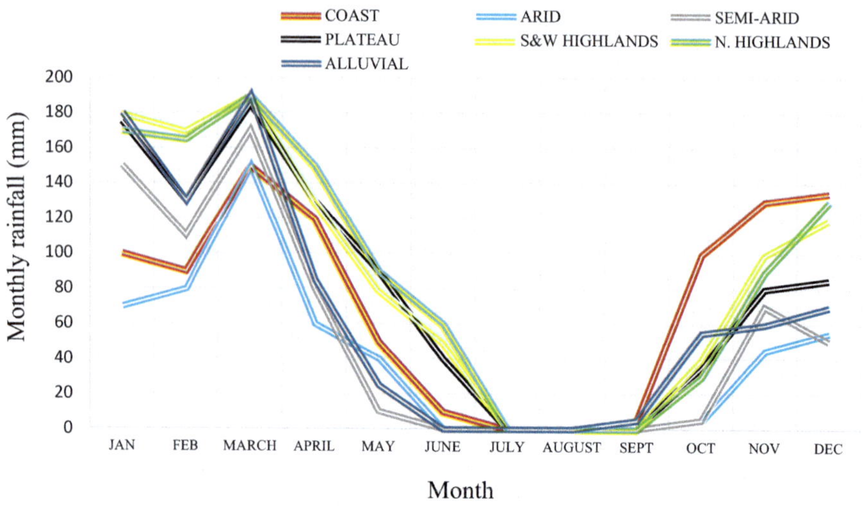

Fig. 9.3 The total monthly rainfall in different agro-ecological zones of Tanzania. (Adapted from Mkonda et al. 2018)

temperature etc. The temperature has also been varying over place and time though sometime is not so significant. Figure 9.3 above explicitly indicates the trends of the total monthly rainfall in all seven agroecological zones. Obviously, this climate variability has significantly affected the famers in their production process. So far, the vulnerability of the farmers varies from one agro-ecological zone to another.

This figure (i.e. 9.3) indicates that from May to October there is significant rain shortage in all agroecological zones. This is evidenced by the fact that about all agroecological zone experience a single rain regime in a year. Apparently, this exhibit that there is one growing season in a year (i.e. from November to April). Although there is almost 6 month of wet spells, the peak rain ranges from December to March. Actually this describes that the growing season is comprised of 4 months only (i.e. December to March). With this regards, it means that even the natural replenishment of the ecosystems is limited to those months (Bationo et al. 2006).

9.5 Biophysical Characteristics and Agricultural Situations in Agro-ecological Zones

9.5.1 Costal

This zone is mostly located to the coastal zone of Tanzania specifically along the Indian Ocean with an altitude about 3000 m above the sea level. The soil and topography in this zone is infertile sands on gently rolling uplands with alluvial soils in Rufiji (Doran and Parkin 1994). It also poses fertile clays on uplands and river flood plains (Birch-Thomsen et al. 2007). Likewise, there are infertile sand soils in some places. The northern part of the area is characterized by bimodal rainfall ranging from 750 to 1200 mm. The growing seasons in this part are October to December and March to June. The southern part experiences unimodal rainfall ranging from 800 to 1200 mm, and the growing season range from December to April.

In this zone, important crops like coconuts, cassava, rice and maize are grown under small scale farming system (Kalhapure et al. 2013). Alongside, mangoes are important fruits grown in the area. This is also accompanied by urban gardening. In addition, irrigation agriculture is predominantly done in the rea by using Pangani and Rufiji rivers.

9.5.2 *Arid*

This zone is mostly located in central and northern Tanzania (Fig. 9.2). It is located between the altitude 1300 and 1800 m above the see level (to the north) while the southern part is located between 500 and 1500 m above the sea level. The soil and topography in this zone is volcanic ash and sediments (Solomon et al. 2007). Soils vary in texture and are very susceptible to water erosion. In addition, the southern part is characterized by rolling plains of low fertility. The area is characterized by unimodal and unreliable rains ranging from 500 to 600 mm and 400 to 600 mm per annum in the northern and southern part, respectively. March to May form the growing season in the area.

Since annual mean rains is low; the drought resistant crops like millet, sorghum, sunflower and sesame are grown in these areas as major livelihoods of the people. Alongside, livestock keeping and organic soil management are important agricultural practices in the area. In some areas, Pangani River flood plain with saline and alkaline soil facilitates small scale irrigation in the area. However, the production per area is less than 0.5 tn/ha. This indicates that there is a need to improve agricultural system in the area in order to increase crop yields and food security.

9.5.3 *Semi-arid*

This zone is mostly located in central and southern Tanzania (Fig. 9.2). It is characterized by undulating plains with rocky hills and low scarps. It is located between the altitude 1000 and 1500 m above the see level (central part) while the south eastern part is located between 200 and 600 m above the sea level. The soil and topography in this zone is well drained soils with low fertility. It has alluvial hardpan and saline soils especially in the eastern Rift Valley and Lake Eyasi. In addition, it has black cracking soils in Shinyanga (URT 2007). It has a moderate fertile loams and clays in the south with infertile sand soils in the center. The area is characterized by unimodal and unreliable rains ranging from 500 to 800 mm and 600 to 800 mm per annum in the central and south eastern part, respectively. December to March form the growing season in the area.

In these areas, drought resistant crops like millet, sorghum and sesame are produced in these areas. Alongside, livestock keeping and organic soil management are important agricultural practices in the area. In some area; small scale irrigation is practiced to get vegetables and other horticultural products. However, the production per area is less than 0.8 tn/ha. This indicates that there is a need to improve agricultural system in the area to increase crop yields and food security.

9.5.4 Plateau

The area mostly located to the western part of Tanzania with an altitude between 800 and 1500 m above the sea level. In the western part; the area is characterized by wide sandy plains and Rift Valley scarps. Moreover, the flooded swamps of *Malagarasi* and *Ugalla* rivers have clay soils with high fertility upland plains with rock hills. The southern part is characterized by clay soils of low to moderate fertility. The western part of the area is characterized by unimodal rainfall ranging from 800 to 1000 mm. The growing seasons in this part are November to April.

In this zone, important crops like cashewnuts, cassava, beans and maize are mainly grown under small and medium farming system. Beside, mangoes and other citrus fruits grown in the area. In addition, irrigation agriculture is predominantly done in the rea by using *Malagarasi* and *Ugalla* rivers. Furthermore, in areas with organic fertilization, the yields have increased from 0.75 to 1.90 tn/ha (Mkonda and He 2018b).

9.5.5 Southern and Western Highlands

The area is mostly located to the western, northern and southern parts of Tanzania with an altitude ranging between 1200 and 1500 (south), 1400 and 2300 m (southwest), and 100 and 1800 m (west) above the sea level. These areas are characterized by unimodal, reliable, and local rains with 800–1400 mm (south), 800–1000 mm (southwest), and 1000–2000 mm (western). In the southern part; the area is characterized by undulating plains to dissected hills and mountains. It has moderate fertile clay soils with volcanic soils in Mbeya. In the south, the area has undulating plateau above Rift Valleys and sandy soils of low fertility. To the west, the area has north-south ridges separated by swampy valleys, loam and clay soils of low fertility in hills, with alluvium and ponded clays in the valleys. The growing seasons are this part are December–April (south), November–April (Southwest), while the western part has bimodal rains i.e. October–December and February–May.

In this area, numerous crops are produced based on the type of soils, rains and ecological gradients. In that respect, food crops such as maize, cassava, beans and maize are mainly grown under small, medium and large scale farming system. Beside, fruits like apples, mangoes, avocado, just to mention a few, are grown for household consumption and business. In addition, irrigation agriculture is predominantly done in the rea by using local swamps, seasonal rivers and wetlands.

9.5.6 Northern Highlands

The area mostly located to the northern part of Tanzania with an altitude between 1000 and 2500 m (northern) and 1000 and 2000 m (granitic) above the sea level. In the northern part; the area is characterized by volcanic uplands, volcanic soils from lavas and ash. The granite and mountainous areas are characterized by deep soils, arable and moderately fertile on upper slopes, shallow and stony on steep slopes. The northern part is characterized by bimodal rainfall ranging from 1000 to 2000 mm. The same rainfall characteristic exhibits in the granite areas. The growing seasons in the northern part are November to January and March to June, while that of granitic are October to December and March to June.

In this zone, important crops like coffee, banana, beans and maize are mainly grown under small and medium farming system. Agroforestry is also predominant agricultural system in the area where banana, beans, maize, and trees are intercropped in the areas (Kimaro et al. 2015). Kilimanjaro Region is a good example in this case.

9.5.7 Alluvial Plains

The area is mainly located in flood plains (URT 2007). It covers the regions of Morogoro (i.e. Kilombero and Wami), Coast (i.e. Rufiji) Mbeya (i.e. Usangu) Morogoro (i.e. Wami). Kilombero is characterized by clay plain with alluvial fans. It has unimodal and reliable rainfall ranging from 900 to 1300 mm per year. The growing season in the area ranges from November–April. Rufiji has mangrove swamp delta, alluvial soils, and sandy upstream, loamy down steam in floodplain. The area has unimodal, often inadequate rainfall ranging from 800 to 1200 mm. The growing season in the area ranges from December to April. Usangu has seasonally flooded clay soils in north, and alluvial fans in south. The area has unimodal rainfall ranging from 500 to 800 mm per annum. The growing season in the area ranges from December to March. Lastly, Wami has moderate alkaline black soils in the eastern part, and alluvial fans with well drained black loam in the western part. The area has unimodal rainfall ranging from 600 to 1800 mm. The growing season in the area exhibits the same characteristics with that of Usangu.

These biophysical potentials of *Alluvial plain* are very important to agricultural sector. Together, they have 44 million hectares of arable land that are potential for irrigation. If well harnessed, these hectares can significantly contribute to the development of agricultural industry in the country. Unfortunately, only 4% of these hectors have been harnessed (URT 2012, 2014).

9.6 Potentials of Agriculture Sustainability

For agricultural sustainability to be progressive and productive; there is a need to view the potentials that can facilitate its development (Pretty 2008). In developed countries, the development is viewed is more important than environmental conservation. This is a main reason for why the maximum potential photosynthetic yield in these countries is high i.e. 10^{-6} Mg/g (160 Mg/ha) compared to 3–5 Mg/ha of the developing countries. On that basis, the maximum yield potentials in developing countries, including Tanzania, is very low, and thus, becoming difficult to curb food shortage.

Agricultural sustainability is also significant at global level to accommodate the vast growing population as it is approximated that by 2050 there will be an increase in population for two billion and making over 8 billion people all dwelling on the Planet Earth. To feed all these population, we need to increase food production for 60% while conserving the environment at the same time (Poppy et al. 2014). Regrettably, about 80% of these two billion people will be living in developing countries particularly the sub-Saharan region (UNEP 2012).

Besides, as much as agriculture involves people and environment, there should be a balance among these aspects. People should meet their needs through utilization of environmental resources while conserving these resources (Lal 1998). In other perspective, the dimension of the quantity of output (Mg of yield/ha) should reflect the level of temporal and spatial conservation of environmental resources.

As described in section five of this chapter, each agroecological zone has its distinctive potentials to support agricultural sustainability (URT 2007; Lema and Majule 2009). Overall, the biophysical aspects like rivers, soil fertility, sustainable agricultural systems, and the use of agricultural technology are important aspects for AS (Herdt and Steiner 1995; Ahmed et al. 2011; Rowhani et al. 2011a).

In this regard, *alluvial plains* are considerably important for the development of agricultural sustainability due to their potentials. There are sufficient water resources and fertility soil in these areas that can support conservation agriculture while optimize yields, whereas, rice production is more predominant. Likewise, the presence of mangrove swampy delta and alluvial soils gives more options to sustain agriculture. However, this does not guarantee that there will be no degradation in these areas.

Furthermore, *coastal* areas are predominantly characterized by fertile loamy sand soils that can support numerous crops especially cassava, coconuts, beans and rice without environmental degradation. In addition, these are areas with numerous deltas and flood plains. The deltas are formed when permanent rivers such as Rufiji, Ruvu and Wami enter Indian Ocean. They are also formed due to seasonal rivers along the coast. Although these potentials support agricultural sustainability, the frequent floods have posed some serious challenges to meet adequate food security in these areas.

Likewise, *plateau* areas have a considerable potentials to support agricultural sustainability. This is evidenced by the presence of fertile clay soils with high fertility upland plains with rock hills. In addition, *Malagarasi* and *Ugalla* rivers are important aspects and potential for agricultural sustainability. These biophysical characteristics support the production of numerous types of crops. Retrospectively, these potentials have increased yields are reduced environmental degradation in various areas (Lichtfouse et al. 2009; UNEP 2011, 2012).

The southern, western and northern highlands exhibit moderate potentials of agricultural sustainability. This is due to their mixed characteristics ranging from rains to soils. Some areas are fertile especially the granite and mountain feet of Kilimanjaro and Meru. The availability of rivers and wetlands in the areas add significant potentials for agricultural sustainability. Though most agricultural systems are peasantry based, there has been numerous funded projects that emphasize on agricultural sustainability. For example; Southern Agricultural Growth Corridor project which involve *planting basins*, has brought significant achievement in maize and rice production. In addition, conservation agriculture in Arusha has raised farmers' yields while conserving the environment (Mkonda and He 2017a). In a nutshell, these zones prescribe the availability of real potentials that can support sustainability in Tanzania.

On the other hand, the *arid* and *semi-arid zones* possess limited potentials for agricultural sustainability. Most of these areas are characterized by infertile soils and excessive droughts. However, there are some few areas with fertile soils due to decomposition of volcanic sediments. Droughts and poor soils are mainly caused by very low rains (˂500 mm per year) and poor soil management, respectively. Overall, these potentials can slightly support agricultural sustainability in the area. Therefore, for to support agricultural sustainability, there should be proper soil organic management to conserve the environment and optimize yields.

9.7 Dominant Agricultural Systems

There are variety of agricultural practices conducted across the agro-ecological zones (Thierfelder and Wall 2009). This study intend to give an overview of the correlation between these agricultural practices and agricultural sustainability. Since agricultural sustainability assists to optimize yields and conserve the enrolment, it is discernible that the adoption of best agricultural practices would serve the two purposes. Numerous studies have indicated that, there has been an increased efforts to adopt agricultural practices that relate to actual situation (i.e. more especially climate and soils). This scenario has influenced numerous communities to adopt varied agricultural practices as seen in Table 9.2. This variation is considerably caused by the farmers' local priorities as stated above.

Table 9.2 Ranking of the farming systems in the study area

	Agro-ecological zones						
Farming type	CS	A	SA	P	S&WH	NH	All
Maize/legume system	4	4	4	3	1	1	3
Livestock/Sorghum-millet system	5	3	1	6	2	3	4
Pastoralism system	7	2	2	5	3	4	5
Agro-pastoralism system	6	1	3	4	4	6	6
Cassava/cashew/coconut system	1	5	5	1	5	5	7
Agroforestry	3	6	6	2	6	2	2
Wetland paddy	2	7	7	7	7	7	1

Source: Extracted from Mkonda et al. (2018) and Sosovele et al. (1999)
Abbreviations: *CS* Coast, *A* Arid, *SA* Semi-arid, *P* Plateau, *S&WH* Southern and Western Highlands, *NH* Northern Highlands, *All* Alluvial

In arid and semi-arid areas, small pastoralism and agro-pastoralism are main livelihoods. Table 9.2 above indicates the mentioned agricultural systems are highly ranked in these AEZs. This is supported by the findings by other studies (Lema and Majule 2009; Kangalawe and Lyimo 2013). A good example of the communities adopting pastoralism are the Maasai of Arusha, Singida, Shinyanga and some parts of Morogoro (URT 2007). Here, extensive semi-nomadic grazing is predominantly done. Beside, these agroecological zones are dominated by the cultivation of drought tolerant though in small scale (URT 2007).

As well, organic soil managements have been adopted in semi-arid areas to rise soil fertility, crop yields under climate change scenario (Vanlauwe 2004; Giller et al. 2009). Similarly, this practice serve as a sustainable measure for environmental conservation in the area (URT 2007; Branca et al. 2013). Among other things, they create favorable conditions that catalyze biological functions of mycorrhizas and other soil microorganisms (McDonagh et al. 2001). By doing so, it helps the biological and chemical interactions of important soil ingredients (Bationo et al. 2006; Birch-Thomsen et al. 2007; Wall et al. 2013; Kimaro et al. 2015).

In addition, maize and legume systems was highly ranked in the *southern, western* and *northern highlands*. Despite of the conducive biophysical condition; maize and beans are the main food crops in these agroecological zones. Possibly this can be among the main reasons for these crops to be high ranked in the area. In these areas, agroforestry and sorghum-millet systems are moderately adopted. These serve especially under drought incidences and climate change scenarios. However, in these areas, wetland paddy are lowly ranked agricultural practice due to the facts that the areas are dry and do not support this particular agricultural system.

Likewise, cassava/cashew/coconut system was highly ranked in *coast* and *plateau* agroecological zone systems. Essentially, the mentioned crops in this particular agricultural system are dominant food and cash crops. Cassava is the dominant food crop along the shore of Lake Tanganyika (i.e. more especially Kigoma and Rukwa regions). These areas are composed of deep and right soils that support tubers expansions. Similarly; coconuts, cashewnuts and palm trees are the dominant cash crops in these areas because climate is conducive to support the production of these crops.

Lastly, wetland paddy and some agroforestry are dominant agricultural systems in *alluvial plains*. The availability of Basin Rivers such as *Rufiji, Usangu, Wami* and *Kilombero* provide conducive environment irrigation agriculture and integration of crops (agroforestry). Apparently, Kilombero and Usangu are the reputable areas for paddy plantations (rice). They are source of food in most areas of the country especially Dar es Salaam; the main business hub of the country.

Generally, the level of adoptions of these agricultural systems is influenced by the biophysical characteristics of the area (water, soil, ecological gradients etc.) and to a certain extent is determined by the nature of the society.

9.8 Challenges of Agricultural Sustainability in the Study Area

The major challenge of agricultural sustainability in Tanzania is the existence of inappropriate balance between environmental conservation and food production. In this regards, Pretty (2007) asked that if environmental goods and services are to be protected or improved, what then happens to productivity? This shows the reality that in many incidences; it is difficult to meet the two targets simultaneously. Most agroecological zones of Tanzania especially the arid and semi-arid equally face this challenge due to their biophysical vulnerability (Glaser et al. 2001; Vanlauwe et al. 2014; Kimaro et al. 2015). Although there has been numerous measures to conserve the environment (especially in Kondoa District and Shinyanga Region), environmental degradation has consequently remained a major problem.

This is also evidenced even in industrialized countries (i.e. Europe) where intensive agricultural production has yielded more products at the expense of environmental degradation (Poppy et al. 2014). In Australia, organic farming in vineyard has conserved the environment though at the expense of lower yields (Rusinamhodzi et al. 2011). Therefore, under industrialized farming systems, intensive agriculture has been adopted to optimize yields despite the environmental consequences (more particularly soil degradation) that can happen.

According to Lal (1998), soil degradation is the loss of soils quality and fertility. It often prevails due to adoption of inappropriate agricultural systems such as monoculture and shifting cultivation. Similarly, the Tanzanian agroecological zones equally face this type of degradation. This occurs mainly due to unsustainable agricultural practices. This challenge is magnified by biophysical aspects such as excessive drought and soils types (Hartemink 1997; Glaser et al. 2001).

Inadequate agricultural technology is another serious challenge in various agroecological zones of Tanzania. Mainly this involves agricultural implements (especially tractors and chemicals). In Tanzania, agriculture is mainly under small scale farming thus, it mainly uses hand hoe to cultivate. This scenario limits the attainment of agricultural sustainability in various areas of the country. Here, both agricultural sustainability and food security are not obtained easily in different agroecological zones of the country (Andrews and Carroll 2001).

Another serious challenge is land use conflicts among different land users. Although there are plans on the land use; adherence to these plans has been a difficult process. There has been increased conflict between livestock keepers and farmers. Among these land users, each group claim to have right over these land resource. Specifically, livestock keepers have been grazing their herds on the farmers' farms, while the farmers have been killing livestock of the pastoral groups. This has been more pronounced in Morogoro Region where these groups under conflicts dwell. This mostly happens in arid and semi-arid agroecological zones and definitely, this scenario subsequently affects the sustainability of various areas. Besides, this land conflict happens between local communities and large scale investors whereas the latter have been grabbing the land from the local people. As response, these local communities have been encroaching into the investors land to acquire land for agriculture and other economic activities (Benjarninsen et al. 2009). This has increased degradation and unsustainable utilization of the land, thus, impeding AS in the areas.

Poverty among the farmers is another hindrances to agricultural sustainability in various agroecological zones. Although a recent report by World Bank (July 2020) indicates that Tanzania has raised from lower income to middle lower income countries; most people are still economically weak. In the country, most farmers (˃70%) entirely depend on agriculture as their main livelihoods. This situation has obviously lead to over utilization of land resources and thus, causing continuous degradation. Now that, this situation does not sustain AS in these areas.

9.9 Adaptations to Environmental Stress

As a response to environmental impacts; numerous adaptations have been in place to limit the associated impacts. However, these adaptations vary over agroecological zones. Table 9.3 below indicates those variations. These adaptations subsequently serve to optimize yields and environmental conservation in these areas.

Timing of farm operations and the adoptions of shorter cycle crops varieties are dominant adaptations across all agroecological zones. This is influenced by unreliable and erratic rains that has shortened the growing season. In addition, conservation agricultural practices such as mulching, agroforestry, crop rotation, and little tillage; are more dominant in *plateau, southern, western* and *northern* AEZs. This is influenced by the moderate rains across these agroecological zones which ranges from 800 to 1200 mm per annum. Besides, small scale irrigation are slightly practiced in *arid* and *semi-arid* agroecological zones due to prolonged drought (Osman-Elasha et al. 2006). Such environmental stress is significantly caused by unreliable rains which is less than 500 mmm per annum. These adaptations include a wide range of approaches designed to reduce the vulnerability and enhance the adaptive

Table 9.3 Comparison of farmers' adaptations to environmental stress presented in percentage

Adaptation activities	CS	A	SA	P	S&WH	NH	All
Timing of farm operations	80	75	75	50	60	50	65
Adopted shorter cycle crop varieties	50	35	45	60	50	70	60
Little tillage	40	50	55	65	50	60	40
Mulching	35	55	35	40	55	40	30
Agroforestry	20	45	25	45	35	50	55
Plating high yielding varieties	30	25	15	40	35	50	40
Practicing crop rotation	20	30	15	40	30	40	20
Small-scale irrigation	20	5	8	10	10	10	10

Source: Extracted from Mkonda et al. (2018) and Ayanlade et al. (2017)
Abbreviations: *CS* Coast, *A* Arid, *SA* Semi-arid, *P* Plateau, *S&WH* Southern and Western Highlands, *NH* Northern Highlands, *All* Alluvial

capacity of agricultural systems under climate change scenarios (Yanda 2015; Mkonda and He 2017c).

On the other hand, mitigation options involve activities that increase carbon stocks above and below ground, that reduce direct agricultural emissions (carbon dioxide, methane, nitrous oxides) anywhere in the lifecycle of agricultural production; and actions that prevent the deforestation and degradation (Bationo et al. 2006; Birch-Thomsen et al. 2007; Solomon et al. 2007). Among others, these mitigations includes; reduced or more efficient use of chemical fertilizers, management of water sources especially wetlands, reduced tillage, planting of biofuels and trees for fuel wood, use of improved feeding practices for livestock, and planting of fast-growing tree plantations (Glaser et al. 2001; Giller et al. 2009; FAO 2013).

9.10 Experience from Other Countries

Although agricultural sustainability is important all over the world, the Sub-Saharan African and other dryland parts of the world need it the most (Pretty et al. 2006). Most of these areas encounter food shortage and prolonged environmental degradation. These problems are caused by poor yields and unfriendly agricultural systems, respectively. Subsequently, this scenario has led to shrinking of farm sizes and inequitable land distribution patterns, depleted soils and limited use of fertilizer and soil amendments (either organic or inorganic). Besides, unreliable rainfall and lack of irrigation capacity, and limited access to improved varieties and seed distribution systems have equally magnified the extent of the problem (Hartemink et al. 2008; Okeyo et al. 2014).

Moreover, FAO (2013) pointed that most small-scale farms in Africa are less than 2 hectares and they are dependent on household members as a sole source of labour force. This situation embolden monoculture which is purely unsustainable for environmental conservation. To underpin this discussion, the study merely earmarked both sustainable and unsustainable agricultural systems that are practiced in East, West and Southern Africa.

According to various reports by FAO (2013) and IPPC (2014), east Africa is among the worst vulnerable regions in Africa, and this vulnerability is intensified by climate change impacts which have been hitting the region for a couple of years. For some couple years, the region has been implementing numerous climate based agricultural project to conserve the environment and improve yields (Mkonda and He 2017a). Mainly, the implementations of these projects are largely funded by FAO together with other local and international organs. However, the funding of most of these project have not been sustainable.

With this regard, most of these projects have not assured sustainable food security and environmental conservation in these areas due to a number of ecological factors (Solomon et al. 2007). Actually, high diversity in agroecological zones impedes the implementation of these projects and this is amplified by climate change impacts. For example, in Kenya, Rusinamhodzi et al. (2011) pointed out that the adoption of conservation agriculture under rain-fed maize production has increased yields in most lowland farms. However, this has been contrary in upland and dry areas (i.e. northeastern). In addition, Kimaro et al. (2015) asserted the same when proposing the optimization of yields along the Uluguru Mountain in Tanzania.

Furthermore, agroforestry systems (such as woodlots) has significant contributions to ecological improvements tenable for agricultural sustainability in various tropical areas (Christensen 1988; Nyadzi et al. 2006). The findings of these studies underpinned that the potential and actual optimization of yields had its base from adequate soil quality improvement in the area. This ecological situation can also play important functions in supporting agricultural sustainability in these areas. Lastly, these studies concluded by endorsing organic soil management against long-term chemical fertilization which appeared to affects the ecosystems especially in arid and semi-arid agroecological zones (Mkonda and He 2017b).

On the other hand, the Southern African countries such as Zimbabwe, Zambia, Malawi, Botswana, Mozambique and Angola; agricultural intensification has been in place to rise yields and environmental conservation (Wall et al. 2013). Among others, intensive agriculture in this region includes crop production, livestock rearing, forestry and fish farming (Nyong et al. 2007; Duru 2015). For example, Malawi attempts to improve fishing industry by applying different techniques like animal manure to feed the fish in the ponds (Thierfelder and Wall 2009).

Fortunately, this program has significantly increased yields especially *"tilapia"* that eventually raised income through selling. In this regard, the region is practicing both traditional and modern agricultural systems (Bationo et al. 2006). However, majority of the farming systems are traditionally practiced and they include extensive (i.e. shifting cultivation and nomadic herding) and intensive and specialized types of farming (such as compound farms and terrace farming). In this aspect, shifting cultivation is an extensive agricultural system (mainly *slash-and-burn*) is dominant in rural areas. However, the system degrades the environment as it involves serious deforestation (Pretty et al. 2006; Blythe 2013).

Likewise, the growing demand of organic products in the world market has raised the desire to adopt organic farming (Thierfelder and Wall 2009). Predominantly, this system gives little yields but of high value, and products are more marketable in developed countries than in the region (Wall et al. 2013). To a certain extent organic agriculture can conserve the environment by does not ensure food security in the region.

West Africa is another important region where agricultural sustainability need restoration to improve food security and environmental conservation (Nezomba et al. 2010). Despite the diversity of agricultural system in the region, the *fallow system* is predominantly practiced in rural areas. Here, important crops such as yams, sorghum, millet, maize and cassava are considerably cultivated in the region depending on the ecological zone (Nyong et al. 2007). In this aspect; *fallow system* involve the resting of the cultivated areas for regrowth of natural vegetation and rejuvenation of soil fertility (quality) through nutrient cycling, addition of litter and suppression of weeds. In most cases, the resting period can be 4–5 years however, ideally the longest period can range between 10 and 20 years. Considerably, these agricultural systems have not supported AS in the region.

For example, in Liberia, the traditional agriculture of the *Loma people* involves planting crops in fertile man-made soil known as *'anthropogenic dark earth'*. This man-made highly fertile soil, which is used for growing crops, forms in the same localized areas, building up over generations (Kareemulla et al. 2017). This soil is created from the deposits of charred and fresh organic matter, including manure, bones, ash, charcoal and ceramics. However, the sustainability of this farming systems is at "cross road" because it is limited by 'sacred' forests, which form around current settlements and cover areas of fertile man-made soil which used to be towns in the past. On top of that, customary laws prohibit these forests being cleared for farming, as some trees are believed to have mystical (*'medicinal'*) power, and also because of the presence of graves.

Mali is another study country from West Africa. The country is highly vulnerable to the threat of declining soil fertility and food security (Karlen et al. 2003; Kalra et al. 2013). Despite of numerous strategies to rise soil fertility (i.e. new green revolution) in the region, little has been achieved. With this respect, agricultural sustainability has not been considerably sustained in the area. Therefore, Mali is in need of long-term solutions for small-scale farmers to optimize crop production and environmental conservation.

Another example is picked from India, the Asian country. This is the second most populous country in the world, its priority has been to elevate agricultural yields, maintain food security and ensure the availability of industrial raw materials (Kalra et al. 2013). However, the country has great diversity in agro-climatic zones with as many as 127 zones under five agro-ecosystems such as rain-fed, arid, irrigated, coastal and hilly systems (Kareemulla et al. 2017). In that respect, there are spatial differences in agricultural systems responsible to meet food security and environmental conservation. Despite of that, these agricultural systems have to sustain agricultural sustainability in the region.

Another aspect that prompts agricultural differences in India is population density. The West Bengal, Bihar, Himachal Pradesh, Punjab, Bihar, Uttar Pradesh, Jharkhand and Kerala are among the major states with high population density of over 800 persons per square kilometer (Kalra et al. 2013; Kareemulla et al. 2017). Thus far, population necessitates the intensive agriculture rather that organic and extensive farming. Intensive agriculture can give more yields in a small geographical area but with serious environmental consequences. With this respect, India has still a long way to sustain agricultural sustainability.

China is another good example from Asia that can serve for the purpose of this study. The country is the most populous country in the Planet and has significant contributions to global agricultural sustainability (Tilman et al. 2002; Li et al. 2010). With diverse climatic region, China applies different farming systems to meet this spatial biophysical characteristics with respect of feeding the huge population and conserve the environment. In this respect, the intensive high-yield agriculture is dependent on addition of fertilizers, especially industrially produced ammonium (NH_4) and nitrate (NO_3). This is done to accrue high yields for food and industrial raw materials (Sharma and Minhas 2005).

Unfortunately, only 30–50% of applied nitrogen fertilizer 40%, 41% and ~45% of phosphorus fertilizer 42 is taken up by crops (Li et al. 2010). This means, a significant amount of the applied nitrogen and a smaller portion of the applied phosphorus is lost from agricultural fields and thus, polluting the environment.

While fertilization is highly emphasized in China, the agricultural systems in most dry areas is limited of irrigation (IPCC 2000; Sharma and Minhas 2005; Li et al. 2009). In arid land of northwest China, water consumption for agriculture accounts for approximately 90% of the total water uses but the average available water is less than 1635 × 108 m^3 per year, only 5.8% of the China average level (Li et al. 2010). Now that, this tells that AS is promising where there is no shocks or immediate demand of environmental services (Huang et al. 2012). However, in other areas where environmental resources are stressful, it has been difficult to optimize yields without degrading the environment. Therefore, agricultural sustainability in the country is not easily sustained.

Europe is another example of the region to be picked for this study. According to European Union (2012), Europe plays great roles in both practicing and funding numerous AS projects around the globe. The continent strongly believes that agriculture that is environmentally, economically and socially sustainable and can make a vital contribution in our response to the most urgent challenges especially reducing poverty and ensuring food security. This report further elaborates that, increasing demand of organic products at global level has raised organic agriculture in Europe. It is envisioned that apart from giving quality yields, this farming system ensures constant provision of environmental services.

For example, the southwest regions of Spain and southern Portugal the "*Dehesa*" is a very specific Mediterranean system of extensively grazed, wooded pasture that shows the multifunctional role of forests. Their intrinsic characteristics and management practices ensure the provision of a wide range of environmental services such as biodiversity, soil conservation, and carbon storage. In these areas, farmers rear *Iberian pig* species known as *'pata negra'*, which feed on corns of oak trees.

Besides, Europe has been a main partner and donor of the Global Rinderpest Eradication Campaign in collaboration with the World Organization for Animal Health (OIE) and FAO, contributing 390 million € over the last 50 years (www.oie.int/en/for-the-media/rinderpest/). The European Union is also supporting local communities in building capacities to restore and sustainably manage their dryland ecosystems, improve their marketing "activities" as well as support dialogues among stakeholders to share knowledge, ideas and priorities. A good example of the supported countries includes: Jordan, Mali, Botswana and Sudan which most of their areas are dryland. Generally, Europe has played a significant role in funding, donating and supporting the sustenance of AS in other parts of the world especially Africa, South America, and southeastern Asia.

9.11 Conclusion

This study assessed the influence of various agricultural systems on agricultural sustainability in various agroecological zones of Tanzania, and other parts of the world. It found that, most agricultural systems and practices can optimize yields although are not environmentally friends. For example, in most rural areas traditional agricultural systems like shifting cultivation, monoculture and fallows are the dominant. On the other hand, the study exhibited that funds from developed countries especially from Europe has helped to support agriculture that can increase yields and conserve the environment. However, this has not adequately helped the sustenance of agricultural sustainability in most agroecological zones. Therefore, agricultural sustainability can only be attained when integrative agricultural systems are adopted in the area. This will improve soil fertility that in turn increase yields. By doing so, the two major aspects (i.e. yields optimization and environmental conservation) will be attained.

Acknowledgement The author is thankful to the Editor-in-Chief of Sustainable Agricultural Reviews, Prof. Eric Lichtfouse for inviting him to write this book chapter. He is also appreciative to the reviewers for their constructive comments and insights.

References

Ahmed S, Deffenbaugh N, Hertel T, Lobell D, Ramankutty N, Rios A, Rowhani P (2011) Climate volatility and poverty vulnerability in Tanzania. Glob Environ Chang 21(2011):46–55. https://doi.org/10.1016/j.gloenvcha.2010.10.003

Altieri MA, Nicholls CI (2012) Agroecology scaling up for food sovereignty and resiliency. Sustain Agric Rev 11:1–30. https://doi.org/10.1007/978-94-007-5449-21

Andrews SS (1998) Sustainable agriculture alternatives: ecological and managerial implications of poultry litter management alternatives applied to agronomic soils. PhD dissertation, University of Georgia, Athens. https://doi.org/10.1155/2014/646502

Andrews SS, Carroll CR (2001) Designing a soil quality assessment tool for sustainable agroecosystem management. Ecol Appl 11(6):1573–1585. https://doi.org/10.2307/3061079

Ayanlade A, Radeny M, Morton JF (2017) Comparing smallholder farmers' perception of climate change with meteorological data: a case study from southwestern Nigeria. Weather Climate Extremes 15:24–33. https://doi.org/10.1016/j.wace.2016.12.001

Bationo A, Kihara J, Vanlauwe B, Waswa B, Kimetu J (2006) Soil organic carbon dynamics, functions and management in West African agro-ecosystems. Agric Syst 94(1):13–25. https://doi.org/10.1016/j.agsy.2005.08.011

Benjaminsen TA, Maganga F, Moshi JA (2009) The Kilosa killings: political ecology of a farmer–herder conflict in Tanzania. Dev Chang 40(3):423–445. https://doi.org/10.1111/j.1467-7660.2009.01558

Birch-Thomsen T, Elberling B, Bjarne F, Magid J (2007) Temporal and spatial trends in soil organic carbon stocks following maize cultivation in semi-arid Tanzania. Springer, East Africa. https://doi.org/10.1007/s10705-007-9116-4

Blythe JL (2013) Social-ecological analysis of integrated agriculture aquaculture systems in Dedza, Malawi. Environ Dev Sustain 15:1143–1155. https://doi.org/10.1007/s10668-012-9429-6

Bockstaller C, Girardin P, van der Werf HMG (1997) Use of agro-ecological indicators for the evaluation of farming systems. Eur J Agron 7:261–270

Branca G, Lipper L, McCarthy N, Jolejole M (2013) Food security, climate change, and sustainable land management. Agron Sustain Dev 33:635–650. https://doi.org/10.1007/s13593-013-0133-1

Chai Q, Gan Y, Zhao C, Xu H, Waskom R, Niu Y, Siddique K (2015) Regulated deficit irrigation for crop production under drought stress. Agron Sustain Dev 36(1):1–21. NOK Publishers, New York. https://doi.org/10.1007/s13593-015-0338-6

Christensen B (1988) Effects of animal manure and mineral fertilizer on the total carbon and nitrogen contents of soil size fractions. Biol Fertil Soils 5:304–307

Doran JW, Parkin TB (1994) Defining and assessing soil quality. In: Doran JW, Coleman DC, Bezdicek DF, Stewart BA (eds) Defining soil quality for a sustainable environment. SSSA, Inc, Madison. https://doi.org/10.2136/sssaspecpub35.cl

Doran J, Zeiss M (2000) Soil health and sustainability: managing the biotic component of soil quality. Appl Soil Ecol 15(1):3–11. https://doi.org/10.1016/S0929-1393(00)00067-6

Duru M (2015) How to implement biodiversity-based agriculture to enhance ecosystem services. Agron Sustain Dev 35:1259–1281. https://doi.org/10.1007/s13593-015-0306-1

Duru M, Therond O, Fares M (2015) Designing agroecological transitions. Agron sustain devaluation of farming systems. Eur J Agron 7:261–270

European Union (2012) Agriculture and rural development. http://ec.europa.eu/agriculture

FAO (2006) Conservation agriculture for SARD and food security in Southern and Eastern Africa, Terminal report, Kenya and Tanzania. FAO, Rome

FAO (2008) Investing in sustainable agricultural intensification: the role of conservation agriculture, A framework for action. FAO, Rome. http://agroecologie.cirad.f

FAO (2012) Conservation agriculture and sustainable crop intensification in Karatu District, Tanzania. Integrated crop management, vol 15. ISSN: 978-92-5-107173-1

FAO (2013) Climate-smart agriculture sourcebook. FAO, Rome. ISSN: 978-92-5-107721-4

Giller K, Witter E, Corbeels M, Tittonell P (2009) Conservation agriculture and smallholder farming in Africa: the heretics' view. Field Crop Res 114:23–34. https://doi.org/10.1016/j.fcr.2009.06.017

Glaser B, Lehmann J, Führböter M, Solomon D, Zech W (2001) Carbon and nitrogen mineralization in cultivated and natural savanna soils of Northern Tanzania. Biol Fertil Soils 33:301–309. https://doi.org/10.1007/s003740000324

Hartemink AE (1997) Soil fertility decline in some major soil groupings under permanent cropping in Tanga Region, Tanzania. Geoderma 75:245–229. https://doi.org/10.1016/S0016-7061(96)00087-0

Hartemink AE, Veldkamp T, Bai Z (2008) Land cover change and soil fertility decline in tropical regions. Turk J Agric For 32:195–213

Hazell P, Wood S (2008) Drivers of change in global agriculture. Philos Trans R Soc B 363:495–515. https://doi.org/10.1098/rstb.2007.2166

Herdt RW, Steiner RA (1995) Agricultural sustainability: concepts and conundrums. In: Barnett in East Africa: the impacts of resource variability on human security. Clim Change. https://doi.org/10.1007/s10584-010-9884-8

Huang Y, Li YP, Chen X, Ma YG (2012) Optimization of the irrigation water resources for agricultural sustainability in Tarim River Basin, China. Agric Water Manag 107:74–85. https://doi.org/10.1016/j.agwat.2012.01.012

IPCC (2000). Land use, land use change and forestry, a special report of the IPCC. Eds: Watson, R.T., Noble, I.R., Bolin, B., Ravondranath, N.H., Verardo, D.J., Dokken, D.J., Cambridge University Press, Cambridge

IPCC (2014) In: Field CB, Barros VR, Estrada RCG, Girma B, Kissel ES, Levy AN, MacCracken S, Mastrandrea PR, White LL (eds) Climate change 2014 impacts, adaptation, and vulnerability. Part A: global and sectoral aspects. Contribution of Working Group II to the Fifth Assessment Report of the Intergovernmental Panel on Climate Change. Cambridge University Press, Cambridge/New York

Kalhapure A, Shete B, Dhonde M (2013) Integrated nutrient management in maize (Zea Mays L.) for increasing production with sustainability. Int J Agric Food Sci Technol 4(3): 195–206. ISSN 2249-3050. http://www.ripublication.com/ijafst.htm. https://doi.org/10.18488/journal.87/2015.2.1/87.1.1.10

Kalra RKB, Tonts AM, Siddique KHM (2013) Self-help groups in Indian agriculture: a case study of farmer groups in Punjab, northern India. Agroecol Sustain Food Syst 37:509–530. https://doi.org/10.1080/10440046.2012.719853

Kangalawe R (2016) Climate change impacts on water resource management and community livelihoods in the southern highlands of Tanzania. Clim Dev. https://doi.org/10.1080/17565529.2016.1139487

Kangalawe R, Lyimo G (2013) Climate change, adaptive strategies and rural livelihoods in semi-arid Tanzania. Nat Resour 4(3):266–278. https://doi.org/10.4236/nr.2013.43034

Kangalawe R, Mung'ong'o C, Mwakaje A, Kalumanga E, Yanda P (2016) Climate change and variability impacts on agricultural production and livelihood systems in Western Tanzania. Clim Dev. https://doi.org/10.1080/17565529.2016.1146119

Kareemulla K, Venkattakumar R, Samuel MP (2017) An analysis on agricultural sustainability in India. Curr Sci 112(258):2

Karlen D, Ditzlerb C, Andrewsa S (2003) Soil quality: why and how? Geoderma 114:145–156. https://doi.org/10.1016/S0016-7061(03)00039-9

Kimaro AA, Mpanda M, Rioux J, Aynekulu E, Shaba S, Thiong'o M, Mutuo P, Abwanda S, Shepherd K, Neufeldt H, Rosenstock TS (2015) Is conservation agriculture 'climate-smart' for maize farmers in the highlands of Tanzania? Springer-Verlag, Berlin Heidelberg. Nutr Cycl Agroecosyst. https://doi.org/10.1007/s10705-015-9711-8

Kitzes J, Wackernagel M, Loh J, Peller A, Goldfinger S, Cheng D, Tea K (2008) Shrink and share: humanity's present and future ecological footprint. Philos Trans R Soc B 363:467–475. https://doi.org/10.1098/rstb.2007.2164

Lal R (1998) Basic concepts and global issues: soil quality and agricultural sustainability. In: Lal R (ed) Soil quality and agricultural sustainability. Ann Arbor Science, Chelsea, pp 3–12

Lalika MC, Meire P, Ngaga Y, Sanga G (2017) Willingness to pay for watershed conservation: are we applying the right paradigm? Echohydrol Hydrobiol 17:33–45. https://doi.org/10.1016/j.ecohyd.2016.12.004

Lema MA, Majule AE (2009) Impacts of climate change, variability and adaptation strategies on agriculture in semi-arid areas of Tanzania: the case of Manyoni District in Singida Region, Tanzania. J Environ Sci Technol 3:206–218. https://doi.org/10.5897/AJEST09.099

Li YP, Huang GH, Wang GQ, Huang YF (2009) FSWM: a hybrid fuzzy-stochastic water management model for agricultural sustainability under uncertainty. Agric Water Manag 96:1807–1818. https://doi.org/10.1016/j.agwat.2009.07.019

Li W, Li YP, Li CH, Huang GH (2010) An inexact two-stage water management model for planning cultural irrigation under uncertainty. Agric Water Manag 97:1905–1914. https://doi.org/10.1016/j.agwat.2010.07.005

Lichtfouse E, Navarrete M, Debaeke P (2009) Agronomy for sustainable agriculture. A review. Agron Sustain Dev 29:1. https://doi.org/10.1051/agro:2008054

McDonagh JF, Birch-Thomsen T, Magic J (2001) Soil organic matter decline and compositional change associated with cereal cropping in Southern Tanzania. Land Degrad Dev:13–26. https://doi.org/10.1002/ldr.419

Medeiros C, Serrano R, Martos J, Girón V (1997) Effect of various soil tillage systems on structure development in a Haploxeralf of central Spain. Soil Technol 11(2):197–204. https://doi.org/10.1016/S0933-3630(96)00131-6

Mkonda MY, He XH (2017a) Conservation agriculture in Tanzania. In: Lichtfouse (ed) Sustainable agriculture reviews 22. Springer, Cham, pp 309–324. https://doi.org/10.1007/978-3-319-48006-0_10

Mkonda MY, He XH (2017b) Long-term chemical fertilization in Tanzania. In: Lichtfouse (ed) Sustainable agriculture reviews 25. Springer, Cham, pp 261–276. https://doi.org/10.1007/978-3-319-58679-3_9

Mkonda MY, He XH (2017c) Yields of the major food crops: implications to food security and policy in Tanzania's semi-arid agro-ecological zone. Sustainability 9:1490. https://doi.org/10.3390/su9081490

Mkonda MY, He XH (2017d) Are rainfall and temperature really changing? Farmer's perceptions, meteorological data, and policy implications in the Tanzanian semi-arid zone. Sustainability 1:1. https://doi.org/10.3390/su9081412

Mkonda MY, He XH (2018a) Climate variability, crop yields and ecosystems synergies in Tanzania's semi-arid agro-ecological zone. Ecosyst Health Sustain 1:1. https://doi.org/10.1080/20964129.2018.1459868

Mkonda MY, He XH (2018b) Agricultural history nexus food security and policy framework in Tanzania. Agric Food Secur 7(1). https://doi.org/10.1186/s40066-018-0228-7

Mkonda MY, He XH, Festin ES (2018) Comparing smallholder farmers' perception of climate change with meteorological data: experiences from seven agro-ecological zones of Tanzania. Weather Clim Soc 2018. https://doi.org/10.1175/wcas-d-17-0036.1

Monfreda C, Ramankutty N, Foley JA (2008) Farming the planet. 2. Geographic distribution of crop areas, yields, physiological types, and net primary production in the year 2000. Glob Biogeochem Cycles 22:GB1022. https://doi.org/10.1029/2007GB002947

Msongaleli B, Rwehumbiza F, Tumbo S, Kihupi N (2015) Impacts of climate variability and change on rain-fed sorghum and maize: implications for food security policy in Tanzania. J Agric Sci (Canadian Center of Science and Education) 7(5). https://doi.org/10.5539/jas.v7n5p124

Nezomba H, Tauro TP, Mtambanengwe F, Mapfumo P (2010) Indigenous legume fallows (indifallows) as an alternative soil fertility resource in smallholder maize cropping systems. Field Crop Res 115:149–157. https://doi.org/10.1016/j.fcr.2009.10.015

Nyadzi G, Janssen B, Oenema O (2006) Analysis of the effects of rotational woodlots on the nutrition and yield of maize following trees in western Tanzania. Agric Ecosyst Environ 116:93–103. https://doi.org/10.1016/j.agee.2006.03.013

Nyong A, Francis A, Osman-Elasha B (2007) The value of indigenous knowledge in climate change mitigation and adaptation strategies in the African Sahel. Mitig Adapt Strateg Glob Chang 12:787–797. https://doi.org/10.1007/s11027-007-9099-0

Okeyo A, Muchera-Muna M, Mugwe J, Ngetich K, Mugendi D, Diels J, Shisanya C (2014) Defects of selected soil and water conservation technologies on nutrient losses and maize yields in the central highlands of Kenya. Agric Water Manag 137:52–58. https://doi.org/10.1016/j.agwat.2014.01.014

Osman-Elasha B, Goutbi N, Spanger-Siegfried E, Dougherty B, Hanafi H, Zakieldeen S, Sanjak A, Atti H, Elhassan H (2006) Adaptation strategies to increase human resilience against climate variability and change: lessons from the arid regions of Sudan. AIACC Working Paper No. 42

Paavola J (2008) Livelihoods, vulnerability and adaptation to climate change in Morogoro, Tanzania. Environ Sci Pol. https://doi.org/10.1016/j.envsci

Pauline N, Vogel C, Grab S, Liwenga E (2016) Smallholder farmers in the Great Ruaha River sub-Basin of Tanzania: coping or adapting to rainfall variability? Clim Dev. https://doi.org/10.1080/17565529.2016.1184607

Poppy GM, Chiotha S, Eigenbrod F, Harvey C, Honzák M, Hudson M, Dawson T (2014) Food security in a perfect storm: using the ecosystem services framework to increase understanding. Philos Trans R Soc Lond Ser B Biol Sci 369(1639):20120288. https://doi.org/10.1098/rstb.2012.0288

Pretty J (2007) The earth only endures. Earthscan, London. ISSN: 978-1-84407-432-7

Pretty J (2008) Agricultural sustainability: concepts, principles and evidence. Philos Trans R Soc B 363:447–465. https://doi.org/10.1098/rstb.2007.2163

Pretty J, Noble A, Dixon J, Hine J, Penning de Vries F, Morison J (2006) Resource-conserving agriculture increases yields in developing countries. Environ Sci Technol 40:1114–1119. https://doi.org/10.1021/es051670d

Rowhani P, Lobell DB, Linderman M, Ramankutty N (2011a) Climate variability and crop production in Tanzania. Agricultural and Forest Meteorology. Sci Direct 151(2011):449–460. https://doi.org/10.1016/j.agrformet.2010.12.002

Rowhani P, Degomme O, Guha-Sapir D, Lambin E (2011b) Malnutrition and conflict in East Africa: the impacts of resource variability on human security. Clim Chang. https://doi.org/10.1007/s10584-010-9884-8

Rusinamhodzi L, Corbeels M, van Wijk MT, Rufino MC, Nyamangara J, Giller KE (2011) A meta-analysis of long-term effects of conservation agriculture on maize grain yield under rain-fed conditions. Agron Sustain Dev 31:657–673. https://doi.org/10.1007/s13593-011-0040-2

Sharma BR, Minhas PS (2005) Strategies for managing saline/alkali waters for sustainable agricultural production in South Asia. Agric Water Manag 78:136–151. https://doi.org/10.1016/j.agwat.2005.04.019

Sieber S, Tscherning K, Graef F, Sergio U, Paloma G (2015) Food security in the context of climate change and bioenergy production in Tanzania: methods, tools and applications. Springer, Berlin/Heidelberg. https://doi.org/10.1007/s10113-015-0834-x

Solomon T, Snyman H, Smit G (2007) Cattle-rangeland management practices and perceptions of pastoralists towards rangeland degradation in the Borana zone of southern Ethiopia. J Environ Manag 82:481–494. https://doi.org/10.1016/j.jenvman.2006.01.008

Sosovele H, Schechambo F, Kisanga D (1999) Rethinking natural resource degradation in semi-arid Sub-Saharan Africa: the case of semi-arid Tanzania. ODI, University of Dar Es Salaam, Tanzania

Thierfelder C, Wall PC (2009) Effects of conservation agriculture techniques on infiltration and soil water content in Zambia and Zimbabwe. Soil Tillage Res 28:209–220. https://doi.org/10.1111/j.1475-2743.2012.00406.x

Tilman D, Cassman KG, Matson PA, Naylor R, Polasky S (2002) Agricultural sustainability and intensive production practices. Nature 418:671–677. https://doi.org/10.1038/nature01014

UNEP (2011) Food and ecological security: identifying synergy and trade-offs. UNEP, Nairobi. http://www.unep.org/ecosystemmanagement/Portals/7/Documents/unep_policy_series/FoodandEcologicalsolutionsJS.pdf

UNEP (2012). Avoiding future famines: ecological foundation sustainable food systems. Nairobi: United Nations Environment Program (UNEP). ISBN: 978-92-807-3261-0

URT (2007) United Republic of Tanzania, National Adaptation Program of Action (NAPA). Division of Environment, Vice President's Office, Dar es Salaam
URT (2012) [Tanzania] national sample census of agriculture 2007/2008 small holder agriculture: regional report – Singida Region (Volume Vm). United Republic of Tanzania, Ministry of Agriculture, Food Security and Cooperatives, Ministry of Livestock Development and Fisheries, Ministry of Water and Irrigation, Ministry of Agriculture, Livestock and Environment, Zanzibar
URT (2014) Review of food and agricultural policies in the United Republic of Tanzania, MAFAP country report series. FAO, Rome
Vanlauwe B (2004) Integrated soil fertility management research at TSBF: the framework, the principles, and their application. In: Bationo A (ed) Managing nutrient cycles to sustain soil fertility in Sub-Saharan Africa. Academy Science Publishers, Nairobi. ISBN: 978-92-9059-261-7
Vanlauwe B, Wendt J, Giller K et al (2014) A fourth principle is required to define conservation agriculture in sub-Saharan Africa: the appropriate use of fertilizer to enhance crop productivity. Field Crop Res 155:10–13. https://doi.org/10.1016/j.fcr.2013.10.002
Vermeulen S, Campbell B, Ingram J (2012) Climate change and food systems. Annu Rev Environ Resour 37(1):195–222. https://doi.org/10.1146/annurev-environ-020411-130608
Wall PC, Thierfelder C, Ngwira A, Govaerts B, Nyagumbo I, Baudron F (2013) Conservation agriculture in Eastern and Southern Africa. In: Jat RA, Graziano de Silva J (eds) Conservation agriculture: global prospects and challenges. CABI, Cambridge. ISBN-13: 9781780642598
Yanda P (2015) Climate change implications for management and use of agricultural biodiversity resources in Africa. Environ Ecol Res 3(2):35–43. http://www.hrpub.org. https://doi.org/10.13189/eer.2015.030201

Chapter 10
Soil Degradation, Resilience, Restoration and Sustainable Use

M. Iftikhar Hussain, Zainul Abideen, and Asad Sarwar Qureshi

Abstract Nearly 2 billion hectares of degraded land is diminishing ecosystem services and affect the living of 2.7 billion of the world population. This article presents strategies for the mitigation of land degradation and restoration of degraded lands. These strategies are operated either individually or in combination. Furthermore, the article describes physical, chemical and biological methodologies to mitigate degradation.

Keywords Organic pollutants · *Land degradation* · *Ecosystem services* · *Sustainability* · *Eco-restoration* · *Human well-being*

10.1 Introduction

Global land biosphere includes the terrestrial ecosystem that is supporting plan, soils, hydrological and microbial communities. Land itself play a critical role as life supporting system as well as providing basic structure for the system utility like provisioning for infrastructure development, provisioning for food fiber and fuel

M. I. Hussain (✉)
Department of Plant Biology & Soil Science, Universidade de Vigo, Vigo, Spain

Research Institute of Science and Engineering, University of Sharjah, Sharjah, United Arab Emirates

Z. Abideen
Dr Muhammad Ajmal Khan Institute of Sustainable Halophyte Utilization, University of Karachi, Karachi, Pakistan

A. S. Qureshi
Natural Resources Management Division, International Center for Biosaline Agriculture (ICBA), Dubai, United Arab Emirates

E. Lichtfouse (ed.), *Sustainable Agriculture Reviews 52*, Sustainable Agriculture Reviews 52, https://doi.org/10.1007/978-3-030-73245-5_10

production, climate regulation, biodiversity maintenance, carbon sequestration and many others. All these services, are directly or indirectly providing various benefits to the human beings. Whether it is a food, fiber or fuels production and others which are related directly to the humans or it is climate regulation, and biodiversity maintenance, providing better environment for the crop growth and the medicinally important plant species respectively. Apart from these services, the land plays an important role in filtering of water by removing various chemicals and potentially toxic elements, e.g. heavy metals, organic pollutants and other pollutants (Fleskens and Stringer 2014). Hence, the land dynamics intimately connected to human well-being, climate and biodiversity that make it a key resource supporting the ecosystem services (UNCCD 1994). Furthermore, the rapidly increasing seven billion populations is also facing the problem of food scarcity mostly due to enhanced competition for land, water and energy. Also it has been predicted that the global demand for food will continue to rise till 2050 (Godfray et al. 2010).

According to an estimate of FAO (2011), the world food production should have to triple to feed the growing population that will reach to 9.1 billion by 2050. This goal will be highly challenging because of other environmental catastrophe such as loss of fertile lands due to drought, salinity, flooding and urbanization (Hussain et al. 2016). This situation will be further exacerbated due to climate change scenarios (Gisladottir and Stocking 2005). Hence the focus has been made to enhance the agricultural productivities by various means of concurrent technologies like the developments of hybrids, transgenics, etc., via agro-biotechnological interventions and also by the means of biological implications. Further, the agricultural intensification has also been proposed to overcome the land exploitation, resource destruction and for the optimized production, still various issues are prevailing that hampers the environment.

These issues are basically the GHGs emission, land contamination and degradation, land invasion, loss of biodiversity from the unsuitable agricultural practices or from the mismanagement of the land resources. Already the changes made to increase food and water production during the recent decades has come at a cost of degradation of 60% of the ecosystem services (WHO 2005). Degradation of terrestrial production land area not only affects food production but also leads to deficiency of micronutrients, protein and calories of energy (Towhid 2014). Over 800 million people (mostly in the low income countries) suffer from deficiency protein, calories of energy and one or more micronutrients (like vitamin, zinc and iodine) (WHO 2005). Soil salinity is increasingly contributing to land degradation worldwide, with approximately 7% of the earth's land surface having salt-affected soils while sodium-affected soils are even more widespread (Flowers et al. 1997; Al Dakheel et al. 2018). The land degradation induced to salinity is continuous expansion in West Asia and Central Asian countries (Pakistan, India, Bangladesh, Kazakhstan, Kyrgyz Republic, Tajikistan, Turkmenistan, and Uzbekistan) (Qureshi et al. 2008; Qadir et al. 2009). Land degradation is mainly caused by anthropogenic activities, soil erosion, drought and salinity and erosion that will further exaberated by climate change (Hussain et al. 2016, 2018).

Land degradation leads to long term loss off ecosystem function and biological productivity or net primary productivity (Knowler 2004; WHO 2005). There is a clear link between the food security and land degradation but apart from that land degradation has other consequences as well. The effect of land degradation goes beyond food security as land regulates the ecosystem services like control losses of biodiversity, soil formation, nutrient cycling, water regulation, waste treatment, amelioration of xenobiotics, climate regulation, recreation and cultural activities (Foley et al. 2005; Bai et al. 2013). All these attributes have intrinsic value for which there could not be monetary valuation (Requier Desjardins et al. 2011).

The productive land ecosystems are not only source of basic nutrition preventing hunger but also a source of timber, fiber, and solid fuels (wood, crop stubble and animal dung). Thus it is important to prevent and limit the land degradation and sustainably manage the land resources by making adaptive changes as land degradation and climate) change will be one of the greatest challenges to sustainable development in future. Hence, the process has been mentioned in the priority goals of United Nations under the 15th rank out of 17 goals of sustainable development which depicts the global importance to combat deforestation, desertification, and reverse land degradation thus halting the biodiversity loss. In this context the present article addresses the challenges of land degradation, measures to control land degradation due to organic pollutants and heavy metals and action of international programs to control land degradation and sustainable land management.

10.2 Causes, Consequences and Restoration of Land Degradation

Land degradation has now arisen as a global problem that has been continuously addressed by the various researchers worldwide (Gisladottir and Stocking 2005; Bai et al. 2008; Campbell et al. 2008; Cai et al. 2011; Gibbs and Salmon 2015). The Global Assessment of Soil Degradation (GLASOD) is the pioneer commission of United Nations Environment Program (UNEP) that mapped the human-induced degradation around the world and is still utilized today for various interpretations like the studies related to bioenergy production, land-use changes (Lambin and Meyfroidt 2011; Nijsen et al. 2012). According to the study of commission, it has been estimated that degradation encompasses around 2 billion ha (22.5%) of agricultural and forest lands. It depicts that the most of the degradation occurred in the Asian and African sub-continents that figures around 453 and 321 million ha of land respectively. However, the least degradation was contributed by the Australia and Pacific regions that depicts around six million ha of land (Gibbs and Salmon 2015).

From another study that is on-going within the FAO's Global Assessment of Lands Degradation and Improvement project (GLADA) that aimed to enumerate more degradation events between 1981 and 2003 by utilizing the normalized difference vegetation index (NDVI), which is generally used to assess vegetation condition and the studies related to productivity (Bai et al. 2008; Gibbs and Salmon 2015). The GLADA project describes land degradation and estimates the decreasing trend in the net primary productivity (NPP) around 21% of the global land area that figures around 2.7 billion ha particularly in the tropical Africa, Southeast Asia, China, north central Australia, the Pampas, and swaths of the boreal forest in Siberia and North America (Gibbs and Salmon 2015).

Another global estimates have been highlighted by Cai et al. (2011), dealing with the bioenergy production on the marginal and degraded lands as nearly 1 billion ha of land is available for the production scenario. Further, it has also been estimated that around 24% of the global land area is affected by land degradation (Plieninger and Gaertner 2011) of which the soil erosion is contributing around 85% in the world such as in the Citarik, West Java, about 94–103 tons' ha^{-1} $year^{-1}$(Nakagoshi and Mabuhay 2014). Land degradation broadly affecting large number of people over a huge portion of the planet earth.

The main reason behind the scene is that mostly these areas are receiving significantly less rainfall and farmers are using more saline water in these degraded marginal lands to irrigate their crops (Malash et al. 2008). The land degradation situation is really very bad in central Asian states because more than 50% land has already been converted into unfertile marginal land because of water logging and continuous use of low quality highly saline-sodic water for irrigation (Kijne 2005; Qadir et al. 2009).

The global problem of land degradation is basically related to the desertification in the arid, semi-arid and dry sub-humid zones generally considered as drylands (Gisladottir and Stocking 2005). The process leading to the land degradation is basically due to the involvement of environmental interactions, socio-economic and historical attributes (Reynolds et al. 2007; Prince et al. 2009). These processes typically induced by cropping and grazing that has crossed the potential of the land during the adverse conditions such as droughts. Further the land degradation depicts the losses of soil organic carbon (SOC), nutrients, soil water storage and regulation and below ground diversity (Plieninger and Gaertner 2011). However, the other attributes such as solid waste dumping effluent discharge, land-use changes, unsuitable agro-practices, deforestation, etc. are frequently responsible for human-induced land degradation. Moreover, the degradation manifests itself in characteristics rather than productivity; for example, reduced biodiversity (Adeel et al. 2005; Prince et al. 2009). However, reduced net primary productivity (NPP) is chief indicator of the degradation comparative to the potential in the site. Particularly, this can be used as a key feature whether or not the effect on the productivity is the objective (Prince et al. 2009). Further, apart from these consequences, land degradation is also responsible for several other deleterious results that are seriously affecting human well-being.

10.3 Strategies for Revitalization

Revitalization is a process considered by different countries, whether developed or developing, and also by the different organizations and institutions worldwide. Revitalization has become a part of sustainable development goal in the context of zero net land degradation (ZNLD).

Though this goal might be ambitious, still it can be achieved as suggested as follows: i) First, even though the accurate spatial dimension of land degradation and desertification (LDD) is in debate (Wessels 2009), the fact that desertification, land degradation and drought (DLDD) prevails in many areas is undisputed (Safriel and Adeel 2005);ii) Remote sensing and time series data are being supported by ground trothing that have detected regions where non-degraded land has converted in the degraded lands.

In the present scenario, a significant association between different components of crop production chain (environment, agriculture and food production) exists that all together form the natural environment. This means that to taking care of the quality of nature and natural resources are not only a civilization requirement, but also a prerequisite for the operation of agricultural production and ultimately food security. Each agricultural activity requires firstly biophysical means such as suitable land, water and climate and secondly socioeconomic conditions such as credit, infrastructure, inputs and markets (Rapidel et al. 2015). Approximately 80% of African population lives in the rural sector and their state of food security depends directly on agricultural production or indirectly through providing for agricultural labor (Tomich et al. 1995). Whereas food security is affected by a myriad of factors including poverty, incomes and unemployment. Furthermore, climate change will have significant impacts on household food security through extreme weather events that will have direct and indirect negative effects on household food security. The productivity of marginal environment and its contribution to the economy, food security and poverty reduction depend on the services provided by well-functioning of degraded marginal ecosystems, including maintaining soil fertility, freshwater delivery, pollination and pest control. Access to food is more than ever a question of interest. Recent increases in world grain prices have added to the claims that we are facing a global food crisis. Alarming population growth, natural resources degradation, unfavorable climatic conditions and decrease in agricultural research and development have contributed to recent shortages in food crops and added weight to calls to increase the supply of agricultural commodities.

For the restoration of the degraded lands, different approaches have been adopted. Also there are several strategies for enhancing the productivity but there should be the integration of indigenous sustainable technologies for the efficient restoration practices. Though there are concerns whether adopting physical methods, chemical interventions or promoting biological methods are suitable or even combining all these practices simultaneously could give the better results. Therefore these methods have been separately discussed.

10.3.1 Physical Methods

Physical methods include various approaches that satisfactorily support the restoration of degraded lands in the different parts of the world. These approaches are associated with the suitable adoption of the indeginous practices for the sustainable land management (SLM). Approaches that are associated with SLM also enhance SOC and supplement to the soil organic matter in the degraded lands. In this regard, several approaches such as cover crops, compost, biochar, increasing soil organic matter, green manures, integrated crop-livestock interaction, and conservation tillage are most important that can help to increase soil fertility (Cowie et al. 2011).

The technical sink potential through these management practices is estimated to about 0.3–0.5 Pg C $year^{-1}$ for desertification control (Lal et al. 1999) and 0.4–1.0 Pg C $year^{-1}$ for restoring the salt affected soils (Lal 2010). It has been observed that the stubble management could be explicitly viable and beneficial for the microbial communitiesresiding in the soils of semi-arid regions of Western Australia.

Arginine ammonification was found around 105.7μg NH_4^+-N g^{-1} soil h^{-1} in the stubble treatments whereas it was 54.7μg NH_4^+-N g^{-1} soil h^{-1} in the stubble-burnt treatments. Further, the acid phosphatase was found to have activity of 243.4μg PNP g^{-1} soil h^{-1} in the former treatments as compared of the later one i.e., 161.0μg PNP g^{-1} soil h^{-1}. Similarly, the significant difference was also found for the β-glucosidase activity in the stubble and the stubble-burnt treatments as 158.9 and 122.8μg PNP g^{-1} soil h^{-1} respectively. Results also shown that above-ground organic matter retained during the 16-year history of the field experiment changed by around 11.7 t ha^{-1} with an average annual contribution of about 0.91 t ha^{-1}in the stubble treatments and 0.02 t ha^{-1} in the stubble-burnt treatments. Similarly, total soil nitrogen was also significant at $p < 0.007$(Hoyle et al. 2006).A study from the Indian Himalayas also suggest that the stubble retention under conservation tillage enhances the yied and the fodder production from the crops like wheat and rice. This also results in the carbon sequestration potential over the control by 0.61 Mg ha^{-1} $year^{-1}$.

Furthermore, intoducing cover crops can also restore carbon in the soil without incorporating major alterations to the agricultural system. These moderate alterations in practices broadly leads to sequest the carbon between 0.1 and 1 t C ha^{-1} $year^{-1}$ (Eagle and Olander 2012; Branca et al. 2013). In the annual ploughing systems, C and N losses from the topsoil can be reduced by incorporation of legume crops as compared to fallow and baren soils. It was further confirmed that this practice will help to maintain bacterial community and crop yield (Verzeaux et al. 2016). It has been estimated that around 0.12 Pg C $year^{-1}$ could be captured globally through this highly efficient management practicewhich would compensate 8% of direct annual GHG emissions from agriculture (Poeplau and Don 2015).

Use of green manure crops such as *Sesbania*sp.,sunnhemp, dhaincha, pillipesara, and clusterbeans form part of an efficient technology and economically viable strategy that can help in enhancing and conserving the soil orgnaic matter and nutrients especially nitrogen. The study reported that *Sesbania aculeata* helped to accumulatenitrogen around 133 kg ha^{-1} upto to 60 days of its growth with the dry matter production of around 23.2 t ha^{-1}. Similarly, Sunnhemp, cowpea and *Pillipesara* sp. accumulated 134, 74 and 102 kg N ha^{-1} with the dry matter production of around 30.6, 23.2 and 25.0 t ha^{-1} respectively at the 60 days of its growth (Kumar et al. 2013).

Conservation tillage (CT) is a practical technique for the utilization of crop residues to enhance the soil quality. The use of CT can affect carbon pool in the soil through its effect on carbon dynamics, aggregation and soil structure and interaction with cropping system. Conventional tillage usualy reduces SOC in all size fraction of the soil (sand, silt, fine silt and clay particles). However the CT has several merits of improving soil quality with major contribution in soil carbon sequestration. Estimated projections of carbon sequestration of the global soil can be made by estimating the area under CT and assuming a possible enhancement in carbon content. It has been estimated that the carbon sequestration potential could be enhanced by 4913 Tg of carbon if the land area under CT is enhanced by 417 million ha till 2020. This represents the global carbon sequestration rate 0.125 Pg year^{-1} that could result in the global increase of SOC content by 0.002% year^{-1}. Similarly, qulaity of soil and crop productivity and GHG emission from the soil can be managed through application of crop residue applied and biomass quantity and quality. It has also been observed that the crop residue management via conservaition tillage could contribute in the global carbon sequestration rate by 0.20 Pg year^{-1} with the global increase of SOC content by 0.001% year^{-1} (Lal 1997).

Marginal areas are therefore an integral component of the overall ecosystem of humans, plants and animals interacting with land, water and climate variables (Al-Dakheel et al. 2016). The continuum of natural resources from high potential regions to marginal zones is affected by the interaction among its components, which are constantly in a state of change. Some productive areas can be reduced into marginal resources because of poor management, such as the case of salinization of irrigated lands in Central Asia, Iraq and Pakistan (Bianchi et al. 2006). There are several agronomical, cultural and management practices such as, intercropping, contour cultivation, mulching'; vegetative like'tree planting, hedge barriers, grass strips'; structural practices that includes'graded banks or bunds, level bench terrace, dams'; or management measures such as'land use change, area closure, rotational grazing' that control land degradation and enhance productivity in the field (Schwilch et al. 2011). Others are examples of marginal regions being transformed into production zones through smart investments (such as tile drainage in the Egyptian delta), drip irrigation (throughout the Middle East and North Africa, treated urban wastewater reuse (UAE, Jordan, Tunisia), all of which involved the appropriate application of science and technology through substantial investments of public and private financial resources. These examples have expanded irrigation on large tracks of semi-arid zones and desert regions in South Asia, North Africa and Central Asia. The agro-industrial approach has focused on how to cost-effectively convert marginal environments to favorable ones through a combination of technology innovations and capital investments. Typically, better soil and water management techniques, together with innovations in technology applications, and innovative agronomic practices have brought about this transformation. The agro ecological approach by contrast explicitly incorporates natural resource management concerns in the analysis of marginal environments. It is in a better position to incorporate risks caused by climate change, unsustainable natural resource usage in assessing the sustainability of outcomes.

10.3.2 Chemical Methods

Chemically-driven methods are generally utilized to optimize the acidity or sodicity of the degraded lands and also contribute to overcomenutrient deficiencies. These methods are usually mediated by the application of fertiliser, lime or gypsum. Various studies suggest that the saline soils can be chemically treated for the reclamation purposes as the gypsum ($Ca_2SO_4.2H_2O$) is added to sodic and salt affected soils (Makoi and Verplancke 2010) to lower down the pH to the neutral. It has a potential to reverse the degradation and showing significant improvement i.e. $p \leq 0.05$. Gypsum application caused decline in saturated hydraulic conductivity (Ks) but couldn't reversed it completely because of the heavy storms leading to compaction of the land and thus the soil and or inequilibrium between Ca^{2+} and Mg^{2+} ions in the soil materials. Thus in this way the salt affected land i.e. 7% of total geographic land (Munns et al. 2002) can be treated to much extent.

Many other methods have been developed that relates with the chemical methods of restoration or rehabilitation of the degraded, contaminated or polluted lands (Wang et al. 2014; Swenson et al. 2015; Biswas et al. 2015). In this back drop, Swenson et al. 2015, suggested that several metabolites including those containing phosphates, dicarboxylates, both nitrogen and aromatic moieties, competitively sorb the ferrihydrite, an iron oxide mineral. They further investigated that the phosphates containing metabolites particularly the 2′-deoxyadenosine monophosphate, adenosine monophosphate, cytidine monophosphate, flavine adenine dinucleotide, glucose-6-phosphate, inosine monophosphate, NADH and uridine monophosphate showed highest (>90%) sorptivity of ferrihydrite with mineral concentration ranging between 4 and 32 mg/L. Even at the lower concentraions of mineral i.e., 0.5–4 mg/L, the phosphate containing metabolites were found with highest sorbtivity ranging from 40% to >90%.

Another study suggests that some low molecular weight organic acids (LMWOAs) like succinic acids, lactic acids, maleic acids and benzoic acids have critical role in the removal of polycyclic aromatic hydrocarbons (PAHs). Among all the organic acids studied, the citric acid was the amost dominant followed by the succinic acid. The LMWOAs, particularly the succinic and lactic acid and the deydrogenase activity in the soil sediments, enhanced the removal of mixed PAHs in the contaminated sediments. Dehydrogenase activity and concentrations of LMWOAs, was not only more tolerant to PAH contamination but also had a higher ability to remove mixed PAHs (Wang et al. 2014). Furthermore, other study also affirms that the citric acid generated the largest release of PAH (14.3 mg kg^{-1} to 73.5 mg kg^{-1}) from the soil and malic acid found to have lowest release (from around 35 to 45 mg kg^{-1}) amongst the three (citric, oxalic and malic acid) tested LMWOAs (Gao et al. 2015). Similarly, remediation of Cd-contaminated sites can be assisted by the addition of these three LMWOAs. Study showed that lower concentrations of these LMWOAs viz., citric acid (10 mmol kg^{-1}), malic acid (20 mmol kg^{-1}) and oxalic acid (10 mmol kg^{-1}) had a better potential of increasing Cd uptake (590.5%, 171.4% and 419.3% respectively from the control) and transport, mitigated the physiological toxicity of Cd in *Phytolacca americana* (pokeweed), and had great effect on the availability of cadmium in the soils (Liu et al. 2015a, b). It has also been reported that these LMWOAs (citric acid) secreted from the plant roots in the rhizosphere at the concentration between 150 and 314μg g^{-1} root dw can also assists *Phragmites australis* to overcome the high salinity stressed environment in the soil and hence it can help in enhancing the efficiency of phytoremediation (Rocha et al. 2015). A study postulating the beneficial interaction of nano-oxides (Maghemite and AMO) with LMWOAs for the remediation of contaminated soil has also been reported. The soil sample having different contaminant concentration namely Cu ($\simeq$21.2 mg kg^{-1}); Cd ($\simeq$6.39 mg kg^{-1}); Pb ($\simeq$1577 mg kg^{-1}); Zn ($\simeq$ 252 mg kg^{-1}); As ($\simeq$123 mg kg^{-1}) was collected and tested for sorption efficiency of the Maghemite and AMO ammended soil. In the study, amorphous Mn oxide (AMO) enhanced the stabilization of Pb, Zn, Cu, Cd and As in soil and hence showed an efficient stabilizing amendment for the studied contaminants at pH > 5. However, maghemite depicted poor sorption efficiency for Cu, Cd, Pb and Zn within 48 h of incubation (Vítková et al. 2015).

Moreover, heavy metal-immobilizing organoclay like Arquad® 2HT-75-bentonite treated with palmitic acid) (MIOC) has the potential to reduce metal toxicity and enhance PAHs (phenanthrene) degradation in a mixed-contaminated soil. The MIOC variably enhanced the microbial count from around 10 to about 43% as well as activities such as respiration by 3–44%; enzymatic activities up to 68% and simultaneously maintained phenanthrene in bioavailable form in a Cd-phenanthrene mixed-contaminated soil over a 21-day incubation period. Further, many heterogeneous metal systems like various minerals, bulk inorganic compounds, metal containing nanoparticles (NPs), zerovalent iron NPs and also the homogeneous and fenton-type metal systems such as permanganate, persulfate, catalyzed hydrogen peroxide propagation, tetrapyrrolic macrocycles and other electrokinetic methods are frequently used nowadays in the soil remediation process (Floris et al. 2016). Abiotic methods particularly based on the absorption mechanism are chiefly used under which many clay materials are exploited (Celis et al. 2000). A study reported an *in situ* stabilization of heavy metals by the utilization of various mineral amendments. Moreover, limestone reduced the bioavailability of Cd, Pb and Zn as observed by soil microbial activity after amendment (Lee et al. 2009). Moreover, soil containing the contaminants like Cd, Cu, Pb and Zn from a smelter site in Austria has also been stabilized by Apatite and Slovakite that resulted in the increase in the soil respiration and the enzymatic activity (Tica et al. 2011). Sodium montmorillonite could be utilized to study the competitive sorption of Cu, Pb and Cr estimating the possible inhibition by other less toxic metal ions (Zhu et al. 2011). Similarly, it has been observed that Cr (VI) was reduced by the action of citric acid while associating with Mn (II) and the two clay minerals viz., montmorillonite and kaolinite in the presence of dissolved organic carbon (DOC) (Sarkar et al. 2013). The birnessite (a Mn containing mineral) was found to be involved in the mineralization process of some phenolic compounds, particularly associating them to humic substances through oxidative radical mechanism (Li et al. 2012). Moreover, various other minerals like Mackinawite, Birnessite, Montmorillonite, Biotite, Zeolite, Vivianite are most commonly exploited as an adsorbent for various contaminants in the process of soil remediation.

Nanotechnology is used for the detoxification of pesticides (Table 10.1), slower release of pesticides by its encapsulation with nanoparticles (NPs), plant protection and fertilization, application of engineered NPs in the soil. (Aragay et al. 2012; Campos et al. 2015; Grillo et al. 2016; Maruyama et al. 2016; Gogos et al. 2012; Pan and Xing 2012). Polychlorocthanes and polychorobiphenyls were hydrodechlorinated by the utilization of bimetallic Pd/Fe nanoparticles (Lien and Zhang 2005; Chen et al. 2014). Further, Pd/Fe NPs associating with the activated carbon were also found to be involved in the dechlorination of polychlorobiphenyls (Choi et al. 2008). Initially, toxic nitrates were reduced by the action of Fe/Ni bimetallic nanoparticles (Kang et al. 2012), afterward the NP was associated with carboxymethyl-cellulose that acts as a stabilizer, todegrade lindane (Yang and Sun 2015). Interestingly, among metal NPs, the use of gold nanoflowers found to degrade efficiently the nitroaromatic pesticides inpresence of $NaBH_4$ (Mao et al. 2014). Further, a study also reported that Fe_2O_3 nanowires containing a Fe^0 core performed the aerobic oxidation of 4-nitrophenol (Ai et al. 2013). Moreover, the detoxification of a widespread azoic dye i.e., the hexavalent chromium and Orange 7 were achieved via composite nanotubes of Au particles surrounded by TiO_2 nanotubes, lighted with a xenon lamp (Luo et al. 2011). There are different combinations of metal or metal oxide NPs or even non-metal support have been developed to resolve the current environmental problems. FeSNPs stabilized on carboxymethyl-cellulose, were used for the immobilization of Hg in heavily contaminated lands, viacolumn and batch experiments (Gong et al. 2012). The results were promising, but still it is yet to perform the field experiments. SiO_2 nanoparticles embedded in a zwitterionic lipid derivative of choline were utilized in the bioremediation of PAHs (Wang et al. 2015). Similarly, various other NPs are also involved in the detoxification and remediation of polluted, contaminated and degraded lands (Table 10.1) that could further pave the way for developing human well-being.

However, considering the use of nanomaterials should be taken as a serious attention. Since the rapid growth in the use of NPs at a global level may have some demerits as well. Despite of the appealing properties of the novel NPs, there has been some attention that was drawn on environmental health and safety problems of nanomaterials in the recent years (Lee et al. 2010; Casals et al. 2012). Apprehension on the potential toxicity of the engineered NPs on the activity of soil microbes (Dinesh et al. 2012) and in the environment have been recently reviewed

Table 10.1 Role of nanoparticles (NP) in soil remediation

	Nanoparticles	Doses of NP	Target pollutants	Reduction (%)	Mechanism/Mode of action	References
1.	Bimetallic Pd/Fe	5.0 g	Polychloroethanes	87.0	Hydrodechlorination	Lien and Zhang (2005), Chen et al. (2014)
		1.0 g	Polychorobiphenyls	93.8	Hydrodechlorination	
2.	Fe	1.0 g	Polychorobiphenyls	95.0	Aerobic hydrodechlorination	Varanasi et al. (2007)
3.	Fe(III)oxide	1.0 g	Polychorobiphenyls	>90.0	Aerobic hydrodechlorination	Varanasi et al. (2007)
4.	V_2O_5/TiO_2	1.0 g	Polychorobiphenyls	>90.0	Aerobic hydrodechlorination	Varanasi et al. (2007)
5.	GAC/ZVI/Pd (granular activated carbon)	4.0 g	Polychorobiphenyls	90.0	Dechlorination	Choi et al. (2008)
6.	Bimetallic Fe/Ni	1.5 g	Toxic nitrate	91.0	Reduction	Kang et al. (2012)
7.	CMC-Fe/Ni (Carboxymethylcellulose)	0.1 g	γ-Hexachlorocyclohexane	100.0	Dehydrochlorination, Dichloroelimination, and Hydrogenolysis	Yang and Sun (2015)
8.	Au Nanoflower	100μL	Pendimethalin,	100.0	Oxidation	Mao et al. (2014)
			Trifluralin and	100.0	Oxidation	
			p-nitrophenol	100.0	Oxidation	
9.	Core–shell Fe@Fe_2O_3 nanowires	0.112	4-chlorophenol	50.8–72.2	Aerobic dechlorination	Ai et al. (2013)
10.	Au/TiO_2 nanotubes	–	Cr(VI)	>90.0	Photocatalytic reduction	Luo et al. (2011)
			Acid orange 7	98.7	Photocatalytic oxidation	
11.	CMC-FeS	1.0 g	Hg	96.0	Desorption	Gong et al. (2012)

(continued)

Table 10.1 (continued)

	Nanoparticles	Doses of NP	Target pollutants	Reduction (%)	Mechanism/Mode of action	References
12.	nZVI (nanoscale zerovalent iron)	0.1 g	Trichloroethene	82.0	Dechlorination	Zhang et al. (2010), Song and Carraway (2005), Yang and Sun (2015), Chen et al. (2014)
		0.1 g	*Chlorinated ethanes*	95.0	Dehydrochlorination	
		0.1 g	γ-	99.9	Dehydrochlorination	
		1.0 g	HexachlorocyclohexanePolychorobiphenyls	74.9	Hydrodechlorination	
		–	Trichloroethene	85.0	Reductive dechlorination	
13.	nZVI/C composite particles	–	Chlorinated hydrocarbons		Reductive dehalogenation	Sunkara et al. (2011)

[a]Zwitterionic lipid 1,2-dimyristoyl-*sn*-glycero-3-phosphocholine

(Maurer-Jones et al. 2013; Srivastava et al. 2015; Kookana et al. 2014). On the basis of trophic transfer studies, impact of NPs in soil and the unidentified consequences in the food chain, researchers suggested the accurately targeted and judicious utilization of these NPs (Gardea-Torresdey et al. 2014). Soil toxicity tests have been performed with NPs demonstrating the differences in the soil amended with CuO NPs and also its solution chemistry were examined (McShane et al. 2014). Further, it was also observed that the soil amendment having a combination of metal-based NPs resulted the inhibition of nodulation in *Medicago truncatula* plant and also inhibiting the nitrogen-fixing bacteria. These negative effects are larger with NPs than with bulk or solution materials of the same composition (Judy et al. 2015). Therefore, it is pertinent to monitor the doses of the NPs application in the soil for getting the positive response from the concerned environment and also maintaining its viability.

10.3.3 Biological Methods

The physicochemical strategies for remediation and restoration of contaminated and degraded sites like solidification/stabilization, soil vapor extraction, incineration, bioremediation, solvent extraction, chemical treatments are generally expensive, often performs incomplete remediation and restoration (Rayu et al. 2012). However, biological methods like bioremediation, offers an environmental friendly and economically feasible option for remediation and restoration of the contaminated land systems (Towhid 2014). Bioremediation is a process that uses microorganisms, green plants or their enzymes for remediation and restoration of contaminated and degraded sites (Megharaj et al. 2011). Bioremediation depending upon biological processes of plants and microbes not only removes the pollutants but also enhances the soil organic matter, biological activity and thus increasing the soil fertility (Tripathi et al. 2015).

Furthermore, according to an interesting study, Se-containing zeolitic tuff (modified Fe^{3+} ions used as an adsorbent for Se^{4+} and Se^{6+}) was utilized for the production of *Pleurotus ostreatus*. However, the fungus had potential to transform the inorganic Se into an organic compound (Jevtić et al. 2014). The finding is interesting as it is related with the agricultural soil, since the source of the Cd comes from the Cd-containing phosphate fertilizers that enters the food chain via the process of mobilization and hence causes adverse impacts on the human health. The potential of Fe-bearing phyllosilicates were also utilized to immobilize redoxsensitive pollutantsvia surface-mediated reduction which was used in the detoxification of Cr(VI). Hence, the integration of various minerals and the microorganisms could also be exploited for the bioremediation and the restoration of the contaminated, polluted and the degraded lands. For example, the novel use of either biotite or chlorite with *Geobacter sulfurreducens*, a microbe that has the potential to reduce Fe (III) under these minerals. Thestudy showed the reduction of Cr (VI) via biogenetic Fe (II) (Brookshaw et al. 2014). Also, a nano-bio integrated system, organized by Pd/Fe NPs and aerobic bacteria (*Sphingomonas* sp. strain) was found to have significant role in the efficient detoxification of the persistent pesticide lindane (1,2,3,4,5,6-hexachlorocyclohexane) (Singh et al. 2013).

With the upsurge in demand of energy, petroleum, nuclear and other resources are being widely exploited. Release of crude oil, refined petroleum products, nuclear wastes can contaminate soil leading to toxic effects on plants and soil microorganisms (Tyagi et al. 2010; Tripathi et al. 2015). Bioremediation is being widely accepted for clean-up of these contaminated soil systems. Arbuscular Mycorrhizal Fungi (AMF) are reported to immobilize Uranium (U) in soil and plant roots, reducing the root to shoot translocation of U. Intracellular acquisition of U by pyrophosphates, adsorption on negatively charged tissue of AMF and sequestration by glycoproteins are the main mechanism for the immobilization of U by AMF (De Boulois et al. 2008). Grasses like *Festuca arundinacea* Schreb and *Festuca pratensis* Huds. inoculated with endophytic fungi *Neotyphodium coenophialum* and *Neotyphodium uncinatum* helpedin the remediation of long term petroleum contaminated soil (Soleimania et al. 2010). A list of plant species having potential of phytoremediation of heavy metals and organic pollutants are demonstrated in Tables 10.2, 10.3 and 10.4. The removal of total petroleum hydrocarbons (TPH) was achieved around 68% in the rhizosphere of the plants contrasting to the controls which were found around 31% only. Also, the grasses inoculated with endophytes showed enhanced growth root and shoot biomass. There was also an increase in number of oil degrading microbes, soil dehydrogenase activity, and levels of water soluble phenols. Enhanced microbial activity increases the mineralization process and increase in phenolic compounds promotes cleavage of aromatic rings of organic pollutants (PAHs removal by 80–84% compared to the controls that were around only 56%) helping in hydrocarbon degradation (Soleimania et al. 2010).

Organochlorine compounds such as polychlorinated biphenyls (PCB), polycyclic aromatic hydrocarbons (PAH), and hexachlorocyclohexane (HCH), belonging to the category of persistent organic pollutants (POPs), were widely used during the 90s (Towhid 2014). These potentially hazardous compound are a serious threat to the environment thus strictly banned for application, however the leftover stockpile and already contaminated sites pose serious threat for contamination of the global soil resources in future (Towhid 2014). "Dig and dump" and "dig and incinerate" are the most common procedure frequently applied for remediation of POPs however, bioremediation has also come into the scene for remediation of POPs contaminated soils. Microorganisms dechlorinate the highly chlorinated organochlorine compounds under both aerobic and anaerobic conditions via various enzymatic pathways (Lal et al. 2010; Castro et al. 2013). Plant in combination of microbes are used for more effective remediation of the POPs (Castro et al. 2013; Towhid 2014) Dubey et al. 2016). Plant species transform, translocate, sequester, extract and detoxify the pollutants. Plant uptake of the pollutant depend upon various factors like octanol-water coefficient (log K_{ow}), acidity constant (PK_a), aqueous solubility (S_w), of the pollutant octanol solubility (S_o). POPs being lipophilic in nature bind to the lipid membrane of the roots of the plant (Tripathi et al. 2014). Similarly, the high lipophilic surfaces of the microorganisms like mycobacteria make it suitable

Table 10.2 Potential plant species for phytoremediation of heavy metals

Species	Contaminant	References
Allium schoenoprasum (Chive)	Ni, Co, Cd	Golan-Goldhirsh (2006)
Solanum Nigrum (Black Nightshade)	Cd	Wei et al. (2010)
Linum usitatissimum (Flax)	Cd	Bjelková et al. (2011)
Albizia amara (Bitter Albizia)	Cr	Shanker et al. (2005)
Casuarina equisetifolia (Coast Sheoak)		
Tectona grandis (Teak)		
Leucaena luecocephala (Leucaena)		Shanker et al. (2005)
Spirodela polyrhiza (Duckweed)		Appenroth (2010)
Allium fistulosum (Green Onion)	Pb	Cho et al. (2009)
Pteris cretica (Moonlight Fern)	Pb	Cho et al. (2009)
Pinus sylvestris (Pine)	Cd, Pb	Ostrowska and Skrzydlewska (2006)
Ricinus communis (Ricinus)	Cd, Pb	Zhi-Xin et al. (2007)
Grasses	Cd, Zn	Zhang et al. (2010)
Pennisetum americanum (Cattail Millet)		
Paspalum atratum (Atra Paspalum)		
Silphium perfoliatum (Cup Plant)		
Stylosanthes guianensis (Common Stylo)		
Brassica rapa (Field Mustard)	Cd, Cu, Zn	
Phragmites australis (Common Reed)	Cu, Hg, Pb	Weis and Weis (2004)
Spartina alterniflora (Smooth Cordgrass)		
Amorpha fruticosa (False Indigo-Bush)	Cu, Pb, Zn	Shi et al. (2011)
Vitex trifolia (Simpleleaf Chastetree)		
Glochidion puberum (Needlebush)		

(continued)

Table 10.2 (continued)

Species	Contaminant	References
Broussonetia papyrifera (Paper Mulberry)		
Styrax tonkinensis (Benzoin Tree)		
Species from brassica genus (Mustard Plant)	Heavy metals	Palmer et al. (2001)
Vetiveria zizanioides (Vetiver Grass)	Pb, Cu, Zn, Cd, Mn	Andra et al. (2009)
		Chen et al. (2004)
		Roongtanakiat and Chairoj (2001)
Eichhornia crassipes (Water Hyacinth)	Cd, Cu, Ni, Pb, Zn	Liao et al. (2004)
	Cd, Cr, Cu, Se	Zhu et al. (1999)
Brassica napus (Canola)	Cd, Cr, Cu, Ni, Pb, Zn	Marchiol et al. (2004)
		Saathoff et al. (2011)
Thlaspi caerulescens (Alpine Pennygrass)	Cr, Cd, Co, Cu, Mo, Ni, Pb, Zn, Mn	Robinson (1998)
Brassica juncea L. (Indian mustard)	Sharma	Belimov et al. (2005), Takeda and Isomura (2006), Turan and Esringu (2007), Singh and Fulekar (2012), Sharma (2016)
Brassica napus L. (canola)	Cd, Cu, Zn, Pb	Sheng and Xia (2006), Dell'Amico et al. (2008), Turan and Esringu (2007)
Cajanus Cajan (L.) Milsp. (pigeon pea)	As, Cd	Garg et al. (2015)
Cicer aeritinum L. (chickpea)	Cd, Pb, Cr, Cu	Wani et al. (2007), Kambhampati (2013), Dasgupta et al. (2011)
Cucumis sativus L. (cucumber)	Pb	Takeda and Isomura (2006)
Eichhornia crassipes L. (water hyacinth)	As, Cr, Zn, Cs, Co	Alvarado et al. (2008), Mishra and Tripathi (2009), Saleh (2012)
Jatropha curcas L. (purging nut)	Soil Fe, Al, Cu, Mn, Cr, As, Zn, Hg	Yadav et al. (2013), Marrugo-Negrete et al. (2015).
Lantana camara L. (lantana)	Pb	Alaribe and Agamuthu (2015)
Lens culinaris Medic. (lentil)	Pb	Wani et al. (2018)
Lepidium sativum L. (cress)	As, Cd, Fe, Pb, Hg	Gunduz et al. (2012)
Lactuca sativa L. (lettuce)	Cu, Fe, Mn, Zn, Ni, Cd,	
	Pb, Co, As	Achakzai et al. (2011), Rashid et al. (2014), Quainoo et al. (2015)

(continued)

Table 10.2 (continued)

Species	Contaminant	References
Medicago sativa L. (alfalfa)	Cd	Ghnaya et al. (2015)
Oryza sativa L. (rice)	Cu, Cd	Li et al. (2012)
Pistia stratiotes L. (water lettuce)	Cr, Cd, As	Akter et al. (2014), Farnese et al. (2014)
Pisum sativum L. (pea)	Pb, Cu, Zn, Fe, Cd, Ni, As, Cr, As	Malecka et al. (2008), Wani et al. (2017), Hegedűsová et al. (2009), Sharma et al. (2010);
Rapanus sativus L. (radish)	As, Cd, Fe, Pb, Cu	Hatano et al. (2016), Gunduz et al. (2012)
Spinacia oleracea L. (spinach) soil	Cd, Cu, Fe, Ni, Pb, Zn, Cr	Patel and Subramanian (2006), Salaskar et al. (2011), Pathak et al. (2013), Jahanbakhshi (2014)
Solanum nigrum L. (black nightshade)	Cd	Ji et al. (2011)
Sorghum bicolor L. (sorghum)	Cd, Cu, Zn, Fe	Pinto et al. (2004)
Zea mays L. (corn)	Cd, Pb, Zn, Cu	Mojiri (2011)

for uptake of soil adsorbed organic pollutants like PAH (Bogan et al. 2003). Bioavailability of the POPs is further increased by the biosurfactants producing microorganisms. Biosurfactants make POP-H_2O soluble aggregates helping in release of the POPs from the soil particles (Passatore et al. 2014). When these biosurfactants are produced by the microbes in the rhizosphere it further increases the bioavailability of the POPs for up taken by the plant roots. Inside the plants phytodegradation of the POPs occurs through transformation, conjugation and compartmentalization (Peuke and Rennenberg 2005). Among the various mechanisms oxidation reaction, hydrolysis and epoxide formation helps in phytodegradation of the POPs (Arslan et al. 2015). The microsomal enzymes of the plants like cytochrome P450, peroxidases and Flavin dependent monooxygenases helps in not only degradation of the POPs but other pollutants as well (Arslan et al. 2015).

Higher functional and metabolic diversity, ability to synthesize and degrade various compounds help microbes to degrade and protect themselves from the toxic effects of various recalcitrant compounds. Provided favourable conditions microorganisms help in remediation of organic pollutants as well as heavy metals. Like in a recent study Simpanen et al. (2014) found that nutrient amendments and circulation of β-cyclodextrin helped in bioremediation of creosote contaminated soil due to enhanced degradation capacity of soil microbes. Biostimulation with nutrient can perform better than bioaugmentation with certain microbes. In a comparative study of restoration of burned forest area it was found that biostimulation of soil with nutrient rich commercial formulation showed 70% hydrocarbon degradation and low molecular weight (LMW) PAHs after 60 days of time period however, bioaugmentation with *Trichoderma* sp. showed only 55% hydrocarbon degradation in the

Table 10.3 Potential plant species for phytoremediation of organic pollutants

Species	Contaminant	References
Brachiaria brizantha (Palisade grass)	Petroleum	Merkl et al. (2005)
Cyperus aggregatus (Inflated-scale Flatsedge)		
Gaillardia aristata (Blanket flower)	Petroleum	Liu et al. (2012)
Echinacea purpurea (Eastern Purple Cone Flower)		
Festuca arundinacea schreb (Fawn)		
Combined F. arundinacea (Fire Phoenix)		
Pisum sativum (Pea)	Diesel fuel	Palmroth et al. (2002)
Pinus sylvestris (Pine)		
Trifolium Repens (White Clover)		
S. alterniflora (Smooth Cordgrass)	Fuel	Lin et al. (2002)
Glycine max (Soybean)	Motor oil	Dominguez-Rosado et al. (2002)
Zea mays (maize)		
Trifolium pretense (Red Clover)		
Lolium arundinaceum (Fescue)	Anthracene, naphthalene, phenanthrene	White et al. (2006)
Cannabis sativa (Industrial Hemp)	Pyrene, chrysene	Campbell et al. (2002)
Trifolium Repens (White Clover)	Pyrene, Phenanthrene	Gao et al. (2010)
Festuca arundinacea (Tall Fescue)	Pyrene, Phenanthrene	Cheema et al. (2009)
Brassica napus (Rapeseed)	Pyrene, Phenanthrene	Sheng-Wang et al. (2008)
Phragmites communis (Common Reed)	Pyrene	Wang et al. (2008)
Typha orientalis (Bullrush)		
Vetiveria zizanioides (Vetiver grass)		
Rohdea japonica (Sacred Lily)		
Bolboschoenus planiculmis (Egorova)		
Morus (Mulberry)	PAHs	Olson and Fletcher (1999)
Cucurbita pepo (Zucchini)	p,p'-DDE	White (2009)

same time period (Andreolli et al. 2015). It would be better to apply microbial consortia for remediation and restoration than single microbe based bioaugmentation. A microbial consortium of *Pseudomonas* and *Aceinetobacter* sp. helped in 97% degradation of total petroleum hydrocarbon in the top (0.0–1.0 m) layer of soil (Suja et al. 2014).

Not only microbes but their products are also being utilized in bioremediation. In a recent study Liu et al. (2015a, b) showed that cell free enzymatic extracts of *Bacillus cereus* ZH-3 helps in degradation of the pyrethroid fenpropathrin in to simpler compounds like 3-phenoxybenzaldehyde, α-hydroxy-3-phenoxy-

Table 10.4 Potential plant species for phytoremediation of heavy metals and organic contaminants

Species	Contaminant	References
Avena sativa (Oat)	Zn	
	Phenanthrene	
Lolium perenne (Rye Grass)	Cu, Cd, As	
	Cu, Zn	
	Petroleum hydrocarbons (naphthalene, phenanthrene, anthracene)	White et al. (2006)
	Organic contaminants (creosote)	
Medicago sativa (Alfalfa)	Pyrene	
	Cd, Cr, Ni, Zn	
	Phenanthrene, pyrene	Sheng-Wang et al. (2008)
	Petroleum contamination	Liu et al. (2012)
Salix spp. (Willow)	Cd, organics(oil)	
	Zn, Cd, Ni, Cr, Pb, Cu	
	Cd	Robinson (1998)
Populus spp. (Poplar Trees)	BTEX	
	Cd	Robinson (1998)
	BTEX, Nutrient contamination	Schnoor et al. (1995)
Helianthus annuus (Sunflower)	Zn, Cu, Cd	Meers et al. (2010)
	Zn, Pb	Adesodun et al. (2010)
	Motor oil	Dominguez-Rosado et al. (2002)
Brassica juncea (Indian Mustard)	Cd, Cr, Cu, Ni, Pb, U, Zn	Blaylock et al. (1997)
		Liu et al. (2012)
		Singh and Sinha (2005)
	Motor oil	Dominguez-Rosado et al. (2002)

benzeneacetonitrile and phenol. These cell free enzyme extracts also have the potential to degrade other highly toxic pyrethroids like beta-cypermethrin, cyfluthrin, deltamethrin and cypermethrin. This study provides new insight for bioremediation of synthetic pyrethroid contaminated compounds. As we mentioned biological methods may improve the soil quality and the bioremediation is more sustainable than other remediation technologies still bioremediation is used for site clean-up for only 10% of the cases (Roelofsen et al. 2011). With increasing number of pollutant-degrading microorganisms, improved understanding of their genetics and biochemistry, bioremediation will gain popularity and acceptance for field application.

10.4 Conclusion

Since, the overall land degradation is being increased globally, therefore it has become a major challenge to address and mitigate for the optimal functioning of the ecosystem services and the human well-being. Further, there are various estimates being put forward by many researchers, organizations and nations. Hence, it is the need of the hour to immediately adapt the suitable and sustainable management practices for checking the degradation process and also restoring the already degraded lands that is prevailing worldwide. Though the various strategies have also been proposed by different researchers for the sustainable utilization of these partially viable and other degraded lands, but there are certain fallacies in the implementation of these practices either at the regional scale, national scale or even at the global level. Hence, we strongly support the concept and deliberations of the sustainable development goals for achieving ZNLD. For meeting the sustainable development goals that embraces the concept of a land degradation neutral world our policy measures need careful consideration. It is because carefully constructed policies including holistic thinking, approaches supporting sustainable land management practices could deliver benefits beyond control and restoration of land degradation. It requires to develop inclusion and partnership between local to global multiple stake holders for developing sustainable land management practices. Bioremediation is a technology that can help in solving the problem of sustainable remediation practices for remediation and restoration with offering multiple benefits as discussed earlier. However, we have to change our attitude towards the technology and acceptance of genetically modified organisms based remediation that can help in wide scale application of the technology.

References

Achakzai AK, Bazai ZA, Kayani SA (2011) Accumulation of heavy metals by lettuce (*Lactuca sativa* L.) irrigated with different levels of wastewater of Quetta city. Pak J Bot 43(6):2953. www.pakbs.org/pjbot/PDFs/43(6)/51

Adeel Z, Safriel U, Niemeijer D, White R (2005) Ecosystems and human well-being: desertification synthesis. World Resource Institute, Washington, DC. //hdl.handle.net/20.500.11822/8719

Adesodun JK, Atayese MO, Agbaje TA, Osadiaye BA, Mafe OF, Soretire AA (2010) Phytoremediation potentials of sunflowers (*Tithonia diversifolia* and *Helianthus annuus*) for metals in soils contaminated with zinc and lead nitrates. Water Air Soil Pollut 207(1–4):195–201. https://doi.org/10.1007/s11270-009-0128-3

Ai Z, Gao Z, Zhang L, He W, Yin JJ (2013) Core-shell structure dependent reactivity of Fe@Fe2O3 nanowires on aerobic degradation of 4-chlorophenol. Environ Sci Technol 47:5344–5352. https://doi.org/10.1021/es4005202

Akter S, Afrin R, Mia M, Hossen M (2014) Phytoremediation of chromium (cr) from tannery effluent by using water lettuce (*Pistia stratiotes*). ASA Univ Rev 8(2):149

Al Dakheel J, Tabet Aoul K, Hassan A (2018) Enhancing green building rating of a school under the hot climate of UAE; renewable energy application and system integration. Energies 11:2465. https://doi.org/10.3390/en11092465

Alaribe FO, Agamuthu P (2015) Assessment of phytoremediation potentials of *Lantana camara* in Pb impacted soil with organic waste additives. Ecol Eng 83:513–520. https://doi.org/10.1016/j.ecoleng.2015.07.001

Al-Dakheel AJ, Hussain MI, Qadir A, Harith A (2016) Impact of salinity on growth and development of eighteen date palm varieties from the Arabian Peninsula. 6th International Date Palm Award & Conference, Abu Dhabi, UAE

Alvarado S, Guedez M, Lue-Meru MP, Nelson G, Alvaro A, Jesus AC, Gyula Z (2008) Arsenic removal from water by bioremediation with the aquatic plants water hyacinth (*Eichhornia crassipes*) and lesser duckweed (*Lemna minor*). Bioresour Technol 99:8436–8440. https://doi.org/10.1016/j.biortech.2008.02.051

Andra SS, Datta R, Sarkar D, Saminathan SK, Mullens CP, Bach SB (2009) Analysis of phytochelatin complexes in the lead tolerant vetiver grass [*Vetiveria zizanioides* (L.)] using liquid chromatography and mass spectrometry. Environ Pollut 157(7):2173–2183. https://doi.org/10.1016/j.envpol.2009.02.014

Andreolli M, Lampis S, Brignoli P, Vallini G (2015) Bioaugmentation and biostimulation as strategies for the bioremediation of a burned woodland soil contaminated by toxic hydrocarbons: a comparative study. J Environ Manag 153:121–131. https://doi.org/10.1016/j.jenvman.2015.02.007

Appenroth KJ (2010) Definition of "heavy metals" and their role in biological systems. Soil heavy metals. Springer, Berlin, Heidelberg, pp 19–29

Aragay G, Pino F, Merkoçi A (2012) Nanomaterials for sensing and destroying pesticides. Chem Rev 112:5317–5338. https://doi.org/10.1021/cr300020c

Arslan M, Imran A, Khan QM, Afzal M (2015) Plant-bacteria partnerships for the remediation of persistent organic pollutants. Environ Sci Pollut Res 24:01–15. https://doi.org/10.1007/s11356-015-4935-3

Bai ZG, Dent DL, Olsson L, Schaepman, ME (2008) Global assessment of land degradation and improvement 1. Identification by remote sensing. Report 2008/01. *ISRIC* – World Soil Information, Wageningen

Bai Z, Dent D, Wu Y, de Jong R (2013) Land degradation and ecosystem services. In: Lal R, Lorenz K, Hüttl RF, Schneider BU, von Braun J (eds) Ecosystem services and carbon sequestration in the biosphere. Springer, Dordrecht, pp 357–381. https://doi.org/10.1007/978-94-007-6455-2

Belimov AA, Hontzeas N, Safronova VI, Demchinskaya SV, Piluzza G, Bullitta S, Glick BR (2005) Cadmium-tolerant plant growth-promoting bacteria associated with the roots of Indian mustard (*Brassica juncea* L. Czern.). Soil Biol Biochem 37(2):241–250. https://doi.org/10.1016/j.soilbio.2004.07.033

Bianchi FJ, Booij CJH, Tscharntke T (2006) Sustainable pest regulation in agricultural landscapes: a review on landscape composition, biodiversity and natural pest control. Proc R Soc B Biol Sci 273(1595):1715–1727

Biswas B, Sarkar B, Mandal A, Naidu R (2015) Heavy metal-immobilizing organoclay facilitates polycyclic aromatic hydrocarbon biodegradation in mixed-contaminated soil. J Hazard Mater 298:129–137. https://doi.org/10.1016/j.jhazmat.2015.05.009

Bjelková M, Genčurová V, Griga M (2011) Accumulation of cadmium by flax and linseed cultivars in field-simulated conditions: a potential for phytoremediation of Cd-contaminated soils. Ind Crop Prod 33(3):761–774. https://doi.org/10.1016/j.indcrop.2011.01.020

Blaylock JE, Salt DE, Dushenkov S, Zakharova O, Gussman C, Kapulnik Y, Ensley BD, Raskin I (1997) Enhanced accumulation of Pb in Indian mustard by soil-applied chelating agents. Environ Sci Technol 31:860–865. https://doi.org/10.1021/es960552a

Bogan BW, Lahner LM, Sullivan WR, Paterek JR (2003) Degradation of polycyclic aromatic and straight-chain aliphatic hydrocarbons by a strain of *Mycobacterium austroafricanum*. J Appl Microbiol 94:230–239. https://doi.org/10.1046/j.1365-2672.2003.01824.x

Branca G, Lipper L, McCarthy N, Jolejole MC (2013) Food security, climate change, and sustainable land management. A review. Agron Sustain Dev 33(4):635–650

Brookshaw DR, Coker VS, Lloyd JR, Vaughan DJ, Pattrick RAD (2014) Redox interactions between Cr(VI) and Fe(II) in bioreduced biotite and chlorite. Environ Sci Technol 48:11337–11342. https://doi.org/10.1021/es5031849

Cai X, Zhang X, Wang D (2011) Land availability for biofuel production. Environ Sci Technol 45:334–339. https://doi.org/10.1021/es103338e

Campbell S, Paquin D, Awaya JD, Li QX (2002) Remediation of benzo [a] pyrene and chrysene-contaminated soil with industrial hemp (*Cannabis sativa*). Int J Phytoremediation 4(2):157–168. https://doi.org/10.1080/15226510208500080

Campbell JE, Lobell DB, Genova RC, Field CB (2008) The global potential of bioenergy on abandoned agriculture lands. Environ Sci Technol 42:5791–5794. https://doi.org/10.1021/es800052w

Campos EV, de Oliveira JL, da Silva CM, Pascoli M, Pasquoto T, Lima R, Abhilash PC, Fraceto LF (2015) Polymeric and solid lipid nanoparticles for sustained release of carbendazim and tebuconazole in agricultural applications. Sci Rep 5:13809. https://doi.org/10.1038/srep13809

Casals E, González E, Puntes V (2012) Inorganic nanoparticles and the environment: balancing benefits and risks. In: Barceló D, Farré M (eds) Comprehensive analytical chemistry: analysis and risk of nanomaterials in environment and food samples, vol 59, pp 265–290. https://doi.org/10.1016/B978-0-444-56328-6.00008-6

Castro CB, Fernández AP, Kidd PS, Weyens N, Garrido BR, González MT, Acea MJ, Vangronsveld J (2013) Improving performance of *Cytisus striatus* on substrates contaminated with hexachloro-cyclohexane (HCH) isomers using bacterial inoculants: developing a phytoremediation strategy. Plant Soil 362:247–260. https://doi.org/10.1007/s11104-012-1276-6

Celis R, Hermosín HC, Cornejo J (2000) Heavy metal adsorption by functionalized clays. Environ Sci Technol 34:4593–4599. https://doi.org/10.1021/es000013c

Cheema SA, Khan MI, Tang X, Zhang C, Shen C, Malik Z, Ali S, Yang J, Shen K, Chen X, Chen Y (2009) Enhancement of phenanthrene and pyrene degradation in rhizosphere of tall fescue (*Festuca arundinacea*). J Hazard Mater 166(2–3):1226–1231. https://doi.org/10.1016/j.jhazmat.2008.12.027

Chen J, Wang W, Fang J, Varahramyan K (2004) Variable-focusing microlens with microfluidic chip. J Micromech Microeng 14(5):675. https://doi.org/10.1088/0960-1317/14/5/003)

Chen X, Yao X, Yu C, Su X, Shen C, Chen C, Huang R, Xu X (2014) Hydrodechlorination of polychlorinated biphenyls in contaminated soil from an e-waste recycling area, using nanoscale zerovalent iron and Pd/Fe bimetallic nanoparticles. Environ Sci Pollut Res 21:5201–5210. https://doi.org/10.1007/s11356-013-2089-8

Cho Y, Bolick JA, Butcher DJ (2009) Phytoremediation of lead with green onions (*Allium fistulosum*) and uptake of arsenic compounds by moonlight ferns (*Pteris cretica* cv Mayii). Microchem J 91(1):6–8. https://doi.org/10.1016/j.microc.2008.05.008

Choi H, Al-Abed SR, Agarwal S, Dionysiou DD (2008) Synthesis of reactive nano-Fe/Pd bimetallic system-impregnated activated carbon for the simultaneous adsorption and dechlorination of PCBs. Chem Mater 20:3649–3655. https://doi.org/10.1021/cm8003613

Cowie AL, Penman TD, Gorissen L et al (2011) Towards sustainable land management in the drylands: scientific connections in monitoring and assessing dryland degradation , climate change and biodiversity. Land Degrad Dev 260:248–260. https://doi.org/10.1002/ldr.1086

Dasgupta S, Satvat PS, Mahindrakar AB (2011) Ability of *Cicer arietinum* (L.) for bioremoval of lead and chromium from soil. Int J Eng Technol 24:338–341

De Boulois HD, Joner EJ, Leyval C, Jakobsen I, Chen BD, Roos P, Thiry Y, Rufyikiri G, Delvaux B, Declerck S (2008) Impact of arbuscular mycorrhizal fungi on uranium accumulation by plants. J Environ Radioact 99:775–784. https://doi.org/10.1016/j.jenvrad.2007.10.009

Dell'Amico E, Cavalca L, Andreoni V (2008) Improvement of Brassica napus growth under cadmium stress by cadmium-resistant rhizobacteria. Soil Biol Biochem 40:74–84. https://doi.org/10.1016/j.soilbio.2007.06.024

Dinesh R, Anandaraj M, Srinivasan V, Hamza S (2012) Engineered nanoparticles in the soil and their potential implications to microbial activity. Geoderma 173:19–27. https://doi.org/10.1016/j.geoderma.2011.12.018

Dominguez-Rosado E, Liggat JJ, Snape CE, Eling B, Pichtel J (2002) Thermal degradation of urethane modified polyisocyanurate foams based on aliphatic and aromatic polyester polyol. Polym Degrad Stab 78(1):1–5. https://doi.org/10.1016/S0141-3910(02)00086-1

Dubey RK, Tripathi V, Dubey PK, Singh HB, Abhilash PC (2016) Exploring rhizospheric interactions for agricultural sustainability: the need of integrative research on multi-trophic interactions. J Clean Prod 115:362–365. https://doi.org/10.1016/j.jclepro.2015.12.077

Eagle AJ, Olander LP (2012) Greenhouse gas mitigation with agricultural land management activities in the United States-a side-by-side comparison of biophysical potential. In: Advances in agronomy. Academic, New York, pp 79–179

FAO (2011) The state of food insecurity of the world: how does international price volatility affect domestic economies and food security? FAO, WFP, IFAD, Rome

Farnese FS, Oliveira JA, Gusman GS, Leão GA, Silveira NM, Silva PM, Ribeiro C, Cambraia J (2014) Effects of adding nitroprusside on arsenic stressed response of *Pistia stratiotes* L. under hydroponic conditions. Int J Phytoremediation 16(2):123–137. https://doi.org/10.1080/15226514.2012.759532

Fleskens L, Stringer LC (2014) Land management and policy responses to mitigate desertification and land degradation. Land Degrad Dev 25:1–4. https://doi.org/10.1002/ldr.2272

Floris B, Galloni P, Sabuzi F, Conte V (2016) Metal systems as tools for soil remediation. Inorg Chim Acta 455:429. https://doi.org/10.1016/j.ica.2016.04.003

Flowers TJ, Gracia A, Koyama M, Yeo AR (1997) Breeding for salt tolerance in crop plants. The role of molecular biology. Acta Physiol Plant 19:427–433

Foley JA, DeFries R, Asner GP, Barford C, Bonan G, Carpenter SR, Chapin FS, Coe MT, Daily GC, Gibbs HK, Helkowski JH, Holloway T, Howard EA, Kucharik CJ, Monfreda C, Patz JA, Prentice IC, Ramankutty N, Snyder PK (2005) Global consequences of land use. Science 309:570–574. https://doi.org/10.1126/science.1111772

Gao Y, Ren L, Ling W, Gong S, Sun B, Zhang Y (2010) Desorption of phenanthrene and pyrene in soils by root exudates. Bioresour Technol 101(4):1159–1165. https://doi.org/10.1016/j.biortech.2009.09.062

Gao Y, Yuan X, Lin X, Sun B, Zhao Z (2015) Low-molecular-weight organic acids enhance the release of bound PAH residues in soils. Soil Tillage Res 145:103–110. https://doi.org/10.1016/j.still.2014.09.008

Gardea-Torresdey JL, Rico CM, White JC (2014) Trophic transfer, transformation, and impact of engineered nanomaterials in terrestrial environments. Environ Sci Technol 48:2526–2540. https://doi.org/10.1021/es4050665

Garg N, Singla P, Bhandari P (2015) Metal uptake, oxidative metabolism, and mycorrhization in pigeonpeaand pea under arsenic and cadmium stress. Turk J Agric For 39(2):234–250. https://doi.org/10.3906/tar-1406-121

Ghnaya T, Mnassri M, Ghabriche R, Wali M, Poschenrieder C, Lutts S, Abdelly C (2015) Nodulation by Sinorhizobium meliloti originated from a mining soil alleviates Cd toxicity and increases Cd-phytoextraction in *Medicago sativa* L. Front Plant Sci 6:863. https://doi.org/10.3389/fpls.2015.00863

Gibbs HK, Salmon JM (2015) Mapping the world's degraded lands. Appl Geogr 57:12–21. https://doi.org/10.1016/j.apgeog.2014.11.024

Gisladottir G, Stocking M (2005) Land degradation control and its global environmental benefits. Land Degrad Dev 16:99–112. https://doi.org/10.1002/ldr.687

Godfray HCJ, Beddington JR, Crute IR, Haddad L, Lawrence D, Muir JF, Pretty J, Robinson S, Thomas SM, Toulmin C (2010) Food security: the challenge of feeding 9 billion people. Science 327:812–818. https://doi.org/10.1126/science.1185383

Gogos A, Knauer K, Bucheli TD (2012) Nanomaterials in plant protection and fertilization: current state, foreseen applications, and research priorities. J Agric Food Chem 60:9781–9792. https://doi.org/10.1021/jf302154y

Golan-Goldhirsh A (2006) Plant tolerance to heavy metals, a risk for food toxicity or a means for food fortification with essential metals: the Allium schoenoprasum model. In: Soil and water pollution monitoring, protection and remediation. Springer, Dordrecht, pp 479–486

Gong Y, Liu Y, Xiong Z, Kaback D, Zhao D (2012) Immobilization of mercury in field soil and sediment using carboxymethyl cellulose stabilized iron sulfide nanoparticles. Nanotechnology 23:294007–294020. https://doi.org/10.1088/0957-4484/23/29/294007

Grillo R, Abhilash PC, Fraceto LF (2016) Nanotechnology applied to bio-encapsulation of pesticides. J Nanosci Nanotechnol 1:1231–1234. https://doi.org/10.1166/jnn.2016.12332

Gunduz S, Uygur FN, Kahramanoglu I (2012) Heavy metal phytoremediation potentials of *Lepidum sativum* L, *Lactuca sativa* L, *Spinacia oleracea* L and *Raphanus sativus* L. Agric Food Sci Res 1:01–05. http://www.heraldjournals.org/hjafsr/archive.htm

Hatano KI, Kanazawa K, Tomura H, Yamatsu T, Tsunoda KI, Kubota K (2016) Molasses melanoidin promotes copper uptake for radish sprouts: the potential for an accelerator of phytoextraction. Environ Sci Pollut Res 23(17):17656–17663. https://doi.org/10.1007/s11356-016-6904-x

Hegedűsová A, Jakabová S, Vargová A, Hegedűs O, Pernyeszi T (2009) Use of phytoremediation techniques for elimination of lead from polluted soils. Nova Biotechnol 9(2):125. http://nbc-journal.fpv.ucm.sk/archive/revue nova biotechnologica

Hoyle FC, Murphy DV, Fillery IR (2006) Temperature and stubble management influence microbial CO2–C evolution and gross N transformation rates. Soil Biol Biochem 38(1):71–80. https://doi.org/10.1016/j.soilbio.2005.04.020

Hussain MI, Lyra DA, Farooq M, Nikoloudakis N, Khalid N (2016) Salt and drought stresses in safflower: a review. Agron Sustain Dev 36:4. https://doi.org/10.1007/s13593-015-0344-8

Hussain MI, Al-Dakheel AJ, Reigosa MJ (2018) Genotypic differences in agro-physiological, biochemical and isotopic responses to salinity stress in quinoa (*Chenopodium quinoa* Willd.) plants: prospects for salinity tolerance and yield stability. Plant Physiol Biochem 129:411. https://doi.org/10.1016/j.plaphy.2018.06.023

Jahanbakhshi S (2014) Optimization of phytoremediation in Cd-contaminated soil by using Taguchi method in *Spinacia oleracea*. Proc Int Acad Ecol Environ Sci 4(4):185

Jevtić S, Arčon I, Rečnik A, Babić B, Mazaj M, Pavlović J, Matijaševic D, Nikšić M, Rajić N (2014) The iron (III)-modified natural zeolitic tuff as an adsorbent and carrier for selenium oxyanions. Microporous Mesoporous Mater 197:92–100. https://doi.org/10.1016/j.micromeso.2014.06.008

Ji P, Sun T, Song Y, Ackland ML, Liu Y (2011) Strategies for enhancing the phytoremediation of cadmium-contaminated agricultural soils by *Solanum nigrum*. Environ Pollut 159(3):762–768. https://doi.org/10.1016/j.envpol.2010.11.029

Judy JD, McNear DH Jr, Chen C, Lewis RW, Tsyusko OV, Bertsch PM, Rao W, Stegemeier J, Lowry GV, McGrath SP, Durenkamp M (2015) Nanomaterials in biosolids inhibit nodulation, shift microbial community composition, and result in increased metal uptake relative to bulk/dissolved metals. Environ Sci Technol 49:8751–8758. https://doi.org/10.1021/acs.est.5b01208

Kambhampati MS (2013) EDTA enhanced phytoremediation of copper contaminated soils using chickpea (*Cicer aeritinum* L.). Bull Environ Contam Toxicol 91(3):310–313

Kang H, Xiu Z, Chen J, Cao W, Guo Y, Li T, Jin Z (2012) Reduction of nitrate by bimetallic Fe/Ni nanoparticles. Environ Technol 33:2185–2192. https://doi.org/10.1080/09593330.2012.665486

Kijne JW (2005) Towards a strategy for feasible investment in drainage for Aral Sea basin. Synthesis Report, PTRID FAO, June 2005
Knowler DJ (2004) The economics of soil productivity: local, national and global perspectives. Land Degrad Dev 15:543–561. https://doi.org/10.1002/ldr.635
Kookana RS, Boxall AB, Reeves PT, Ashauer R, Beulke S, Chaudhry Q, Cornelis G, Fernandes TF, Gan J, Kah M, Lynch I (2014) Nanopesticides: guiding principles for regulatory evaluation of environmental risks. J Agric Food Chem 62:4227–4240. https://doi.org/10.1021/jf500232f
Kumar A, Gupta DK, Kumar M (2013) Green manure crops: a boon for agricultural soil. Int J Agric Environ Biotechnol 6:193–198
Lal R (1997) Agroforestry: science, policy and practice. J Environ Qual 26:567
Lal R (2010) Carbon sequestration in saline soil. J Soil Salinity Water Qual 1:30–40
Lal R, Hassan HM, Dumanski JM (1999) Desertification control to sequester carbon and mitigate the greenhouse effect. In: Rosenberg RC, Izaurralde Malone EL (eds) Carbon sequestration in soils: science, monitoring and beyond. Battelle Press, Columbus, pp 83–107
Lal R, Pandey G, Sharma P, Kumari K, Malhotra S, Pandey R, Raina V, Hans-Peter E, Kohler HPE, Holliger C, Jackson C, Oakeshott JG (2010) Biochemistry of microbial degradation of hexachlorocyclohexane and prospects for bioremediation. Microbiol Mol Biol Rev 74:58–80. https://doi.org/10.1128/MMBR.00029-09
Lambin EF, Meyfroidt P (2011) Global land use change, economic globalization, and the looming land scarcity. Proc Natl Acad Sci U S A 108:3465–3472. https://doi.org/10.1073/pnas.1100480108
Lee SH, Lee JS, Choi YJ, Kim JG (2009) In situ stabilization of cadmium-, lead-, and zinc-contaminated soil using various amendments. Chemosphere 77:1069–1075. https://doi.org/10.1016/j.chemosphere.2009.08.056
Lee J, Mahendra S, Alvarez PJ (2010) Nanomaterials in the construction industry: a review of their applications and environmental health and safety considerations. ACS Nano 4:3580–3590. https://doi.org/10.1021/nn100866w
Li C, Zhang B, Ertunc T, Schaeffer A, Ji R (2012) Birnessite-induced binding of phenolic monomers to soil humic substances and nature of the bound residues. Environ Sci Technol 46(16):8843–8850
Liao SJ, Chang DY, Chen HJ, Tsou LS, Ho JR, Yau HT, Hsieh WH, Wang JT, Su YC (2004) Optimal process conditions of shrinkage and warpage of thin-wall parts. Polym Eng Sci 44(5):917–928. https://doi.org/10.1002/pen.20083
Lien HL, Zhang WX (2005) Hydrodechlorination of chlorinated ethanes by nanoscale Pd/Fe bimetallic particles. J Environ Eng 131:4–10. https://doi.org/10.1061/(ASCE)0733-9372(2005)131:1(4)
Lin Q, Mendelssohn IA, Suidan MT, Lee K, Venosa A (2002) The dose-response relationship between no. 2 fuel oil and the growth of the salt marsh grass, *Spartina alterniflora*. Mar Pollut Bull 44(9):897–902. https://doi.org/10.1016/S0025-326X(02)00118-2
Liu XQ, Huang W, Jiang YH, Zhu JC, Zhang CZ (2012) Preparation of a bio-based epoxy with comparable properties to those of petroleum-based counterparts. Express Polym Lett 6(4):293. https://doi.org/10.3144/expresspolymlett.2012.32
Liu J, Huang W, Han H, She C, Zhong G (2015a) Characterization of cell-free extracts from fenpropathrin-degrading strain *Bacillus cereus* ZH-3 and its potential for bioremediation of Pyrethroid-contaminated soils. Sci Total Environ 523:50–58. https://doi.org/10.1016/j.scitotenv.2015.03.124
Liu H, Liu YG, Zeng GM, Xie JL, Zheng BH, Tan XF, Wang DF, Sun ZC, Nie J, Jiang ZJ, Gan C (2015b) Mitigation mechanism of Cd-contaminated soils by different levels of exogenous low-molecular-weight organic acids and *Phytolacca americana*. RSC Adv 5:45502–45509. https://doi.org/10.1039/C5RA05700K
Luo S, Xiao Y, Yang L, Liu C, Su F, Li Y, Cai Q, Zeng G (2011) Simultaneous detoxification of hexavalent chromium and acid orange 7 by a novel Au/TiO2 heterojunction composite nano-tube arrays. Sep Purif Technol 79:85–91. https://doi.org/10.1016/j.seppur.2011.03.019

Makoi JH, Verplancke H (2010) Effect of gypsum placement on the physical chemical properties of a saline sandy loam soil. Aust J Crop Sci 4(7):556

Malash NM, Flowers TJ, Ragab R (2008) Effect of irrigation methods, management and salinity of irrigation water on tomato yield, soil moisture and salinity distribution. Irrig Sci 26(4):313–323. https://doi.org/10.1007/s00271-007-0095-7

Malecka A, Piechalak A, Morkunas I (2008) Accumulation of lead in root cells of *Pisum sativum*. Acta Physiol Plant 30(5):629–637. https://doi.org/10.1007/s11738-008-0159-1

Mao K, Chen Y, Wu Z, Zhou X, Shen A, Hu J (2014) Catalytic strategy for efficient degradation of nitroaromatic pesticides by using gold nanoflower. J Agric Food Chem 62:10638–10645. https://doi.org/10.1021/jf5034015

Marchiol L, Assolari S, Sacco P, Zerbi G (2004) Phytoextraction of heavy metals by canola (*Brassica napus*) and radish (*Raphanus sativus*) grown on multicontaminated soil. Environ Pollut 132(1):21–27. https://doi.org/10.1016/j.envpol.2004.04.001

Marrugo-Negrete J, Durango-Hernández J, Pinedo-Hernández J, Olivero-Verbel J, Díez S (2015) Phytoremediation of mercury-contaminated soils by *Jatropha curcas*. Chemosphere 127:58–63. https://doi.org/10.1016/j.chemosphere.2014.12.073

Maruyama CR, Guilger M, Pascoli M, Bileshy-José N, Abhilash PC, Fraceto LF, de Lima R (2016) Nanoparticles based on chitosan as carriers for the combined herbicides Imazapic and Imazapyr. Sci Rep 6:23854. https://doi.org/10.1038/srep23854

Maurer-Jones MA, Gunsolus IL, Murphy CJ, Haynes CL (2013) Toxicity of engineered nanoparticles in the environment. Anal Chem 85:3036–3049. https://doi.org/10.1021/ac303636s

McShane HVA, Sunahara GI, Whalen JK, Hendershot WH (2014) Differences in soil solution chemistry between soils amended with nanosized CuO or Cu reference materials: implications for nano-toxicity tests. Environ Sci Technol 48:8135–8142

Meers E, Van Slycken S, Adriaensen K, Ruttens A, Vangronsveld J, Du Laing G, Witters N, Thewys T, Tack FM (2010) The use of bio-energy crops (*Zea mays*) for 'phytoattenuation' of heavy metals on moderately contaminated soils: a field experiment. Chemosphere 78(1):35–41. https://doi.org/10.1016/j.chemosphere.2009.08.015

Megharaj M, Ramakrishnan B, Venkateswarlu K, Sethunathan N, Naidu R (2011) Bioremediation approaches for organic pollutants: a critical perspective. Environ Int 37:1362–1375. https://doi.org/10.1016/j.envint.2011.06.003

Merkl N, Schultze-Kraft R, Infante C (2005) Assessment of tropical grasses and legumes for phytoremediation of petroleum-contaminated soils. Water Air Soil Pollut 165(1–4):195–209

Mishra VK, Tripathi BD (2009) Accumulation of chromium and zinc from aqueous solutions using water hyacinth (Eichhornia crassipes). J Hazard Mater 164(2–3):1059–1063. https://doi.org/10.1016/j.jhazmat.2008.09.020

Mojiri A (2011) The potential of corn (Zea mays) for phytoremediation of soil contaminated with cadmium and lead. J Biol Environ Sci 5(13):17–22

Munns R, Husain S, Rivelli AR, James RA, Condon AT, Lindsay MP, Lagudah ES, Schachtman DP, Hare RA (2002) Avenues for increasing salt tolerance of crops, and the role of physiologically based selection traits. In: Progress in plant nutrition: plenary lectures of the XIV international plant nutrition colloquium. Springer, Dordrecht, pp 93–105. https://doi.org/10.1023/A:1021119414799

Nakagoshi N, Mabuhay JA (2014) Designing low carbon societies in landscapes. In: Ecological research monographs. Springer, Tokyo, Heidelberg, New York, Dordrecht, London. ISBN 978-4-431-54819-5 (eBook), pp 239–252. https://doi.org/10.1007/978-4-431-54819-5

Nijsen M, Smeets E, Stehfest E, Vuuren DP (2012) An evaluation of the global potential of bioenergy production on degraded lands. GCB Bioenergy 4:130–147. https://doi.org/10.1111/j.1757-1707.2011.01121.x

Olson PE, Fletcher JS (1999) Field evaluation of mulberry root structure with regard to phytoremediation. Biorem J 3(1):27–34. https://doi.org/10.1080/10889869991219172

Ostrowska J, Skrzydlewska E (2006) The comparison of effect of catechins and green tea extract on oxidative modification of LDL in vitro. Adv Med Sci 51:298. //pubmed.ncbi.nlm.nih.gov/17357329/

Palmer CE, Warwick S, Keller W (2001) Brassicaceae (Cruciferae) family, plant biotechnology, and phytoremediation. Int J Phytoremediation 3(3):245–287. https://doi.org/10.1080/15226510108500059

Palmroth MR, Pichtel J, Puhakka JA (2002) Phytoremediation of subarctic soil contaminated with diesel fuel. Bioresour Technol 84(3):221–228. https://doi.org/10.1016/S0960-8524(02)00055-X

Pan B, Xing B (2012) Applications and implications of manufactured nanoparticles in soils: a review. Eur J Soil Sci 63:437–456. https://doi.org/10.1111/j.1365-2389.2012.01475.x

Passatore L, Rossetti S, Juwarkar AA, Massacci A (2014) Phytoremediation and bioremediation of polychlorinated biphenyls (PCBs): state of knowledge and research perspectives. J Hazard Mater 278:189–202. https://doi.org/10.1016/j.jhazmat.2014.05.051

Patel M, Subramanian RB (2006) Effect of a chelating agent on lead uptake by *Spinacia olearea*. Pollut Res 25(1):77–79

Pathak C, Chopra AK, Srivastava S (2013) Accumulation of heavy metals in Spinacia oleracea irrigated with paper mill effluent and sewage. Environ Monit Assess 85(9):7343–7352. https://doi.org/10.1007/s10661-013-3104-8

Peuke AD, Rennenberg H (2005) Phytoremediation. EMBO Rep 6:497–501. https://doi.org/10.1038/sj.embor.7400445

Pinto AP, Mota AD, De Varennes A, Pinto FC (2004) Influence of organic matter on the uptake of cadmium, zinc, copper and iron by sorghum plants. Sci Total Environ 326(1–3):239–247

Plieninger T, Gaertner M (2011) Harnessing degraded lands for biodiversity conservation. J Nat Conserv 19:18–23. https://doi.org/10.1016/j.jnc.2010.04.001

Poeplau C, Don A (2015) Carbon sequestration in agricultural soils via cultivation of cover crops–a meta-analysis. Agric Ecosyst Environ 200:33–41. https://doi.org/10.1016/j.agee.2014.10.024

Prince SD, Becker-Reshef I, Rishmawi K (2009) Detection and mapping of long-term land degradation using local net production scaling: application to Zimbabwe. Remote Sens Environ 113:1046–1057. https://doi.org/10.1016/j.rse.2009.01.016

Qadir M, Noble AD, Qureshi AS, Gupta RK, Yuldashev T, Karimov A (2009) Salt-induced land and water degradation in the Aral Sea basin: a challenge to sustainable agriculture in Central Asia. In: Natural resources forum. Blackwell Publishing Ltd, Oxford, pp 134–149

Quainoo AK, Konadu A, Kumi M (2015) The potential of shea nut shells in phytoremediation of heavy metals in contaminated soil using lettuce (*Lactuca sativa*) as a test crop. J Bioremed Biodegr 6(1):1–7. https://doi.org/10.4172/2155-6199.1000268

Qureshi AS, McCornick PG, Qadir M, Aslam Z (2008) Managing salinity and waterlogging in the Indus Basin of Pakistan. Agric Water Manag 95:1–10. https://doi.org/10.1016/j.agwat.2007.09.014

Rapidel B, Ripoche A, Allinne C, Metay A, Deheuvels O, Lamanda N, Blazy JM, Valdés-Gómez H, Gary C (2015) Analysis of ecosystem services trade-offs to design agroecosystems with perennial crops. Agron Sustain Dev 35(4):1373–1390. https://doi.org/10.1007/s13593-015-0317

Rashid A, Mahmood T, Mehmood F, Khalid A, Saba B, Batool A, Riaz A (2014) Phytoaccumulation, competitive adsorption and evaluation of chelators-metal interaction in lettuce plant. Environ Eng Manag J 13(10):2583. www.eemj.icpm.tuiasi.ro/pdfs/vol13/no10/Full/18_52_Rashid_12

Rayu S, Karpouzas DG, Singh BK (2012) Emerging technologies in bioremediation: constraints and opportunities. Biodegradation 23:917–926. https://doi.org/10.1007/s10532-012-9576-3

Requier Desjardins M, Adhikari B, Sperlich S (2011) Some notes on the economic assessment of land degradation. Land Degrad Dev 22:285–298. https://doi.org/10.1002/ldr.1056

Reynolds JF, Smith DM, Lambin EF, Turner BL, Mortimore M, Batterbury SP, Downing TE, Dowlatabadi H, Fernández RJ, Herrick JE, Huber-Sannwald E (2007) Global desertification:

building a science for dryland development. Science 316:847–851. https://doi.org/10.1126/science.1131634

Robinson M (1998) 30 years of forest hydrology changes at Coalburn: water balance and extreme flows. Hydrol Earth Syst Sci 2(2/3):233 –238

Rocha ACS, Almeida CMR, Basto MCP, Vasconcelos MTSD (2015) Influence of season and salinity on the exudation of aliphatic low molecular weight organic acids (ALMWOAs) by *Phragmites australis* and *Halimione portulacoides* roots. J Sea Res 95:180–187. https://doi.org/10.1016/j.seares.2014.07.001

Roelofsen A, Boon WPC, Kloet RR, Broerse JEW (2011) Stakeholder interaction within research consortia on emerging technologies: learning how and what? Res Policy 40:341–354. https://doi.org/10.1016/j.respol.2010.10.015

Roongtanakiat N, Chairoj P (2001) Uptake potential of some heavy metals by vetiver grass. Agric Nat Resour 35(1):46–50. https://li01.tci-thaijo.org/index.php/anres/article/view/240319/163875

Saathoff AJ, Sarath G, Chow EK, Dien BS, Tobias CM (2011) Downregulation of cinnamyl-alcohol dehydrogenase in switchgrass by RNA silencing results in enhanced glucose release after cellulase treatment. PLoS One 6(1):16416. https://doi.org/10.1016/j.envpol.2004.04.001

Safriel UN, Adeel Z (2005) Dryland systems. In: Hassan R, Scholes R, Ash N (eds) Ecosystems and human well-being: current state and trends. Island Press, Washington, pp 623–662

Salaskar D, Shrivastava M, Kale SP (2011) Bioremediation potential of spinach (*Spinacia oleracea* L.) for decontamination of cadmium in soil. Curr Sci 101:1359–1363. www.jstor.org/stable/24079645

Saleh HM (2012) Water hyacinth for phytoremediation of radioactive waste simulate contaminated with cesium and cobalt radionuclides. Nucl Eng Des 242:425–432. https://doi.org/10.1016/j.nucengdes.2011.10.023

Sarkar B, Naidu R, Krishnamurti GSR, Megharaj M (2013) Manganese(II)-catalyzed and clay-minerals-mediated reduction of chromium(VI) by citrate. Environ Sci Technol 47:13629–13636. https://doi.org/10.1021/es401568k

Schnoor JL, Light LA, McCutcheon SC, Wolfe NL, Carreia LH (1995) Phytoremediation of organic and nutrient contaminants. Environ Sci Technol 29(7):318A–323A. https://doi.org/10.1021/es00007a002

Schwilch G, Bestelmeyer B, Bunning S, Critchley W, Herrick J, Kellner K, Liniger HP, Nachtergaele F, Ritsema CJ, Schuster B, Tabo R (2011) Experiences in monitoring and assessment of sustainable land management. Land Degrad Dev 22:214–225. https://doi.org/10.1002/ldr.1040

Shanker AK, Cervantes C, Loza-Tavera H, Avudainayagam S (2005) Chromium toxicity in plants. Environ Int 31(5):739–753. https://doi.org/10.1016/j.envint.2005.02.003

Sharma H (2016) Phytoremediation of lead using *Brasica juncea* and *Vetiveria zizanoides*. Int J Life Sci Res 4(1):91–96

Sharma A, Gontia I, Agarwal PK, Jha B (2010) Accumulation of heavy metals and its biochemical responses in *Salicornia brachiata*, an extreme halophyte. Mar Biol Res 6(5):511–518. https://doi.org/10.1080/17451000903434064

Sheng XF, Xia JJ (2006) Improvement of rape (*Brassica napus*) plant growth and cadmium uptake by cadmium-resistant bacteria. Chemosphere 64:1036–1042. https://doi.org/10.1016/j.chemosphere.2006.01.051

Sheng-wang PA, Shi-qiang WE, Xin YU, Sheng-xian CA (2008) The removal and remediation of phenanthrene and pyrene in soil by mixed cropping of alfalfa and rape. Agric Sci China 7(11):1355–1364. https://doi.org/10.1016/S1671-2927(08)60185-6

Shi X, Zhang X, Chen G, Chen Y, Wang L, Shan X (2011) Seedling growth and metal accumulation of selected woody species in copper and lead/zinc mine tailings. J Environ Sci 23(2):266–274. https://doi.org/10.1016/S1001-0742(10)60402-0

Simpanen S, Mäkelä R, Mikola J, Silvennoinen H, Romantschuk M (2014) Bioremediation of creosote contaminated soil in both laboratory and field scale: investigating the ability of methyl-β-cyclodextrin to enhance biostimulation. Int J Biodeterior Biodegrad 106:117–126. https://doi.org/10.1016/j.ibiod.2015.10.013

Singh A, Fulekar MH (2012) Phytoremediation of heavy metals by Brassica juncea in aquatic and terrestrial environment. In: The plant family Brassicaceae. Springer, Dordrecht, pp 153–169

Singh S, Sinha S (2005) Accumulation of metals and its effects in *Brassica juncea* (L.) Czern. (cv. Rohini) grown on various amendments of tannery waste. Ecotoxicol Environ Saf 62(1):118–127. https://doi.org/10.1016/j.ecoenv.2004.12.026

Singh R, Manickam N, Mudiam MK, Murthy RC, Misra V (2013) An integrated (nano-bio) technique for degradation of γ-HCH contaminated soil. J Hazard Mater 258:35–41. https://doi.org/10.1016/j.jhazmat.2013.04.016

Soleimania M, Afyuni M, Hajabbasi MA, Nourbakhsh F, Sabzalian MR, Christensen JH (2010) Phytoremediation of an aged petroleum contaminated soil using endophyte infected and non-infected grasses. Chemosphere 81:1084–1090. https://doi.org/10.1016/j.chemosphere.2010.09.034

Song H, Carraway ER (2005) Reduction of chlorinated ethanes by nanosized zero-valent iron: kinetics, pathways, and effects of reaction conditions. Environ Sci Technol 39(16):6237–6245. https://doi.org/10.1021/es048262e

Srivastava V, Gusain D, Sharma YC (2015) Critical review on the toxicity of some widely used engineered nanoparticles. Ind Eng Chem Res 54:6209–6233. https://doi.org/10.1021/acs.iecr.5b01610

Suja F, Rahim F, Taha MR, Hambali N, Razali MR, Khalid A, Hamzah A (2014) Effects of local microbial bioaugmentation and biostimulation on the bioremediation of total petroleum hydrocarbons (TPH) in crude oil contaminated soil based on laboratory and field observations. Int J Biodeterior Biodegrad 90:115–122. https://doi.org/10.1016/j.ibiod.2014.03.006

Sunkara B, Zhan J, Kolesnichenko I, Wang Y, He J, Holland JE, McPherson GL, John VT (2011) Modifying metal nanoparticle placement on carbon supports using an aerosol-based process, with application to the environmental remediation of chlorinated hydrocarbons. Langmuir 27(12):7854–7859. https://doi.org/10.1021/la200657m

Swenson TL, Bowen BP, Nico PS, Northen TR (2015) Competitive sorption of microbial metabolites on an iron oxide mineral. Soil Biol Biochem 90:34–41

Takeda K, Isomura M (2006) Method for forming ZnO film, method for forming ZnO semiconductor layer, method for fabricating semiconductor device, and semiconductor device. United States patent US 7,049,190. patents.google.com/patent/US7049190B2/en

Tica D, Udovic M, Lestan D (2011) Immobilization of potentially toxic metals using different soil amendments. Chemosphere 85:577–583

Tomich TP, Kilby P, Johnston BF (1995) Transforming agrarian economies: opportunities seized, opportunities missed. Cornell University Press, Ithaca

Towhid (2014) In: Osman KT (ed) Soil degradation, conservation, and remediation. Springer, Dordrecht, p 273

Tripathi V, Dubey RK, Edrisi SA, Narain K, Singh HB, Singh N, Abhilash PC (2014) Towards the ecological profiling of a pesticide contaminated soil site for remediation and management. Ecol Eng 71:318–325. https://doi.org/10.1016/j.ecoleng.2014.07.059

Tripathi V, Fraceto LF, Abhilash PC (2015) Sustainable clean-up technologies for soils contaminated with multiple pollutants: plant-microbe-pollutant and climate nexus. Ecol Eng 28:330–335. https://doi.org/10.1016/j.ecoleng.2015.05.027

Turan M, Esringu A (2007) Phytoremediation based on canola (Brassica napus L.) and Indian mustard (*Brassica juncea*) planted on spiked soil by aliquot amount of Cd, Cu, Pb, and Zn. Plant Soil Environ 53(1):7

Tyagi M, da Fonseca MR, de Carvalho CCCR (2010) Bioaugmentation and biostimulation strategies to improve the effectiveness of bioremediation processes. Biodegradation 22:231–241. https://doi.org/10.1007/s10532-010-9394-4

UNCCD (1994) General Assembly of United Nations Convention to Combat Desertification. http://www.unccd.int/Lists/SiteDocumentLibrary/conventionText/conv-eng.pdf

Varanasi P, Fullana A, Sidhu S (2007) Remediation of PCB contaminated soils using iron nanoparticles. Chemosphere 66(6):1031–1038. https://doi.org/10.1016/j.chemosphere.2006.07.036

Verzeaux J, Alahmad A, Habbib H, Nivelle E, Roger D, Lacoux J, Decocq G, Hirel B, Catterou M, Spicher F, Dubois F (2016) Cover crops prevent the deleterious effect of nitrogen fertilisation on bacterial diversity by maintaining the carbon content of ploughed soil. Geoderma 281:49–57

Vítková M, Komárek M, Tejnecky V, Sillerová H (2015) Interactions of nano-oxides with low-molecular-weight organic acids in a contaminated soil. J Hazard Mater 293:7–14. https://doi.org/10.1016/j.jhazmat.2015.03.033

Wang B, Lai Q, Cui Z, Tan T, Shao Z (2008) A pyrene-degrading consortium from deep-sea sediment of the West Pacific and its key member *Cycloclasticus* sp. P1. Environ Microbiol 10(8):1948–1963

Wang Y, Fang L, Lin L, Luan T, Tam NFY (2014) Effects of low molecular-weight organic acids and dehydrogenase activity in rhizosphere sediments of mangrove plants on phytoremediation of polycyclic aromatic hydrocarbons. Chemosphere 99:152–159. https://doi.org/10.1016/j.chemosphere.2013.10.054

Wang H, Kim B, Wunder SL (2015) Nanoparticle-supported lipid bilayers as an in situ remediation strategy for hydrophobic organic contaminants in soils. Environ Sci Technol 49:529–536. https://doi.org/10.1021/es504832n

Wani PA, Khan MS, Zaidi A (2007) Cadmium, chromium and copper in greengram plants. Agron Sustain Dev 27(2):145–153. https://doi.org/10.1051/agro:2007036

Wani RA, Ganai BA, Shah MA, Uqab B (2017) Heavy metal uptake potential of aquatic plants through phytoremediation technique-a review. J Bioremed Biodegr 8(404):2. https://doi.org/10.4172/2155-6199.1000404

Wani W, Masoodi KZ, Zaid A, Wani SH, Shah F, Meena VS, Wani SA, Mosa KA (2018) Engineering plants for heavy metal stress tolerance. Rendiconti Lincei. Sci Fis Natur 29(3):709–723. https://doi.org/10.1007/s12210-018-0702-y

Wei S, Li Y, Zhou Q, Srivastava M, Chiu S, Zhan J, Wu Z, Sun T (2010) Effect of fertilizer amendments on phytoremediation of Cd-contaminated soil by a newly discovered hyperaccumulator *Solanum nigrum* L. J Hazard Mater 176(1–3):269–273. https://doi.org/10.1016/j.jhazmat.2009.11.023

Weis JS, Weis P (2004) Metal uptake, transport and release by wetland plants: implications for phytoremediation and restoration. Environ Int 30(5):685–700. https://doi.org/10.1016/j.envint.2003.11.002

Wessels K (2009) Comments on proxy global assessment of land degradation. Soil Use Manag 5:91–92. https://doi.org/10.1111/j.1475-2743.2009.00195.x

White JC (2009) Optimizing planting density for p, p′-DDE phytoextraction by *Cucurbita pepo*. Environ Eng Sci 26(2):369–376. https://doi.org/10.1089/ees.2007.0347

White PM, Wolf DC, Thoma GJ, Reynolds CM (2006) Phytoremediation of alkylated polycyclic aromatic hydrocarbons in a crude oil-contaminated soil. Water Air Soil Pollut 169(1–4):207–220

WHO (2005) Ecosystem and human wellbeing, health synthesis. A report of the millennium ecosystem assessment. http://www.who.int/globalchange/ecosystems/ecosys.pdf

Yadav AK, Abbassi R, Gupta A, Dadashzadeh M (2013) Removal of fluoride from aqueous solution and groundwater by wheat straw, sawdust and activated bagasse carbon of sugarcane. Ecol Eng 52:211–218. https://doi.org/10.1016/j.ecoleng.2012.12.069

Yang J, Sun H (2015) Degradation of γ-hexachlorocyclohexane using carboxymethylcellulose-stabilized Fe/Ni nanoparticles. Water Air Soil Pollut 226:1–5. https://doi.org/10.1007/s11270-015-2553-9

Zhang X, Xia H, Li Z, Zhuang P, Gao B (2010) Potential of four forage grasses in remediation of Cd and Zn contaminated soils. Bioresour Technol 101(6):2063–2066. https://doi.org/10.1016/j.biortech.2009.11.065

Zhi-Xin N, Sun LN, Sun TH, Li YS, Hong WA (2007) Evaluation of phytoextracting cadmium and lead by sunflower, ricinus, alfalfa and mustard in hydroponic culture. J Environ Sci 19(8):961–967. https://doi.org/10.1016/S1001-0742(07)60158-2

Zhu YL, Pilon-Smits EA, Tarun AS, Weber SU, Jouanin L, Terry N (1999) Cadmium tolerance and accumulation in Indian mustard is enhanced by overexpressing γ-glutamylcysteine synthetase. Plant Physiol 121(4):1169–1177. https://doi.org/10.1104/pp.121.4.1169

Zhu Z, Li X, Zeng Y, Sun W, Zhu W, Huang X (2011) Application of cobalt oxide nanoflower for direct electrochemistry and electrocatalysis of hemoglobin with ionic liquid as enhancer. J Phys Chem 115:12547–12553

Chapter 11
Integrated Weed Management for Sustainable Agriculture

M. Iftikhar Hussain, Zainul Abideen, Subhan Danish, Muhammad Ahsan Asghar, and Khalid Iqbal

Abstract Increased world population will demand more food production, less water-consuming crops, better crop production techniques and better strategies for weed control. More production with rational use of available resources should enhance food security. Here, we review weed management approaches, which vary from crop to crop, with focus on organic weed management, non-chemical weed control, cover crops and allelopathy. Weeding practice in any crop depends on crop factors such as position in rotation, plant spacing and architecture; on field factors such as soil type and weed seed bank history; and on seasonal factors, e.g. rainfall. Crop losses can be reduced by integrated weed management with resistant crop varieties, rational use of agrochemicals, biocontrol agents, allelopathic crops, crop rotation and better cultural practices. Complementary weed control methods include grazing, herbicide application, land fallowing, biological control, cover crops, inclusion of competitive ability of crops, mechanical weeding, sowing time adjustment, irrigation methods, mulching and intercropping.

M. I. Hussain (✉)
Department of Plant Biology & Soil Science, Universidade de Vigo, Vigo, Spain

Research Institute of Science and Engineering, University of Sharjah, Sharjah, United Arab Emirates

Z. Abideen
Dr. Muhammad Ajmal Khan Institute of Sustainable Halophyte Utilization, University of Karachi, Karachi, Pakistan

S. Danish
Department of Soil Sciences, Faculty of Agricultural Sciences and Technology, Bahauddin Zakariya University, Multan, Punjab, Pakistan

M. A. Asghar
Department of Plant Molecular Biology, Agricultural Institute, ELKH Centre for Agricultural Research, Martonvásár, Hungary

K. Iqbal
Office of Research, Innovation & Commercialization (ORIC), University of Gujrat (UOG), Gujrat, Pakistan

E. Lichtfouse (ed.), *Sustainable Agriculture Reviews 52*, Sustainable Agriculture Reviews 52, https://doi.org/10.1007/978-3-030-73245-5_11

Keywords Allelopathy · Biological control · Weed control practices · Weeds · Cultural management · Cover crops · Mulching · Weeds

11.1 Introduction

Weeds are unwanted plants in the crop production sector that are seriously causing many problems for the farmers and interfere with the germination, growth and yield of crop plants (Bretagnolle and Gaba 2015). According to other researchers, "*Weeds are unwanted and undesirable plants which interfere with the utilization of land and water resources and thus adversely affect human welfare*" (Rao 2000). Therefore, weed management at an early growth stage is of paramount importance that needs immediate attention from farmers, plant scientists, breeders and extension workers (Bretagnolle and Gaba 2015). It is a critical aspect of the sustainable agricultural production to control weeds, in order to achieve the target yield and to maintain a good yield potential (Oerke 2006). In industrialized countries, large scale agriculture has been mechanized and is supported with inputs from mineral fertilizers and agro-chemical based plant protection products (Almberg et al. 2018). The large scale mechanized farming are totally dependent on the long-term use of herbicides or continuous tillage to control weeds that can have a significant negative impacts on the agriculture farm productivity and to the ecological environment. This will ultimately lead to the environmental pollution (Relyea 2005). Herbicide pollution may also damage soil biodiversity, natural vegetation, non-target crops and human health impact (Hussain et al. 2008, 2011a, b; Almberg et al. 2018). The agrochemical and deep tillage based weed control can increase soil erosion, disturb the soil biogeochemical cycle, function and contribute to climate change through greenhouse gases impact from the soil (Lal 2004; Gao et al. 2018). Another problem is the huge seed production by different weeds in field crops and thus causes big problems for the next crop (Manning 2004).

11.2 Harmful Aspects of Weeds

Perhaps one of the most essential advances in human history was the establishment of agriculture. Cultivation for food started to be one of the bases of any human society. With the development of agricultural practices, the rhythm of the life adapted to the rhythm of the crops. Along the years, this use led to higher transformations on soil for improving crop productivity. But it was in the first years of the last century when the agricultural practices started to be more and more technologically specialized and noxious to the environment. The development of new culture techniques got that only the concept of yield occupied the first place in the priorities of the farmers.

Pests and weeds were probably the primary biotic limitation that farmers found in the race for increasing the yield of their crops. Therefore, the research, development, and use of an industry of chemical products exclusively produced for the control of pests in the field started around 100 years ago. Their use is intensifying each year; for example, India experimented an increase of more than 350% in the use of herbicides in only 15 years (Alström 1990). These synthetic pesticides and herbicides (the most part of the pesticides are herbicides) were and are still extensively applied in the field and combined with new molecules for a more effective result.

Because of the severe loss caused by crop-weed interaction, it is imperative to consider the cost of weed management. This will help to understand which control strategies should be applied and at which cost. Although the most commonly known effects are those that either directly affect the crop through competition are the increase in production costs and/or decrease the quality of the crop, the lesser effects referring to animal or human health should also be considered (Ruchel et al. 2020). In agro-ecosystems, weeds caused severe damage to the crop plants that may rise upto 93% in the productivity loss. This loss in the productivity is because of competition for water, carbon dioxide, light, nutrients and space) (Druille et al. 2013). In a particular crop field, several weeds might exist but each exhibit different competitive ability. It also depends upon the crop species, for example; in soybean field, weed such as *Amaranthus viridis* L. is morphological and physiological similar and thus give tough time to the soybean and increase the competition for resources (Ruchel et al. 2020).

Weeds cause significant damage to crop plants and it can be divided into following categories.

- Plant competition: Weeds usually compete with the main crops for the soil, nutrient and water resources and impact on growth and yield of crops.
- Added protection costs: Weeds serve as a harbor for insects, pests and pathogens and hence increase the cost of production because farmers have to invest more economic resources to control all those pests to protect the field crop.
- Reduced quality of farm products: The presence of weed seeds in the cereals and grain crops increase the problem of weed infestation and hence growers again have to spent money to labor to clean the crop grains before selling them in the market. Cereals, grain crops and legumes are highly susceptible to weed seed infestation.

The weeds are also responsible to reduce the quality of animal feed and animals may die or abortion may occur while eating forages contaminated with noxious weeds branches, leaves. The weeds can increase the farm production costs due to engaging of farm labors for weed control practices.

11.3 Weed Management

11.3.1 Non-Chemical Weed Management

Non-chemical weed control includes all those crop production practices that can help in the management of weeds from vegetable and crop fields. Before the introduction of herbicide use in agriculture, mostly farmers were using seed cleaning, crop rotation and mulching. After the World War II, researchers and farmers put their focus on the synthetic herbicides. Farmers and researchers start using the herbicides because they think that other weed control practices are time consuming and labour intensive. Therefore, they ignored the non-chemical weed control options (Rask and Kristoffersen 2007). However, in the organic farming, farmers suffer a lot of economic losses because they avoid chemical weed control measures. However, to increase the crop production, yield, stability and sustainability in the food production systems and to avoid the environmental contamination, renewed the interest of all stake holders in the search of non-chemical weed control. Recent literature suggest that a lot of research has been conducted regarding the non-chemical weed management strategies and a wide range of options are developed and previous methods are improved.

The following methodologies can be used by farmers and weed control biologists in the crop production.

11.3.2 Cultural Weed Control

In the cultural weed management, several old and traditional approaches have to be changed. Instead of completely controlling the weeds, the emphasis should have to focus on reducing weed-crop competition and weed establishment at early growth stages (Blackshaw et al. 2007). The weed management through agronomical and cultural practices is widely accepted method as economic and environmental friendly. Among them, changing the sowing time, irrigation system, increasing resistance in crops against the weeds, mulching, cover crops and intercropping are the most important culture weed control practices (Rosset and Gulden 2020). Another study reported that cropping pattern, diversified tillage systems, and intercropping also play important role in the weed control. (Blackshaw et al. 2007). In this regard, selection of right crop variety, uniform and healthy seed and seed size, uniform germination, proper seedling depth, climate, and suitable planting pattern will definitely help in the effective weed management (Place et al. 2009; Korres et al. 2019; De Bruin and Pedersen 2008; Mierau et al. 2020).

The main cultural techniques used by farmers and weed scientists for improving weed control and management are listed below.

11.3.2.1 Crop Rotation

Several researchers reported that weeds tend to associate with certain crops which have similar growth and life cycle (Hunt et al. 2017). Therefore, rotating the crops with different growth cycle will led to the interruption of association of weed with that particular crop. This strategy can be achieved through choosing different planting time and other cultural practices and hence reducing the weed-crop competition and will highly disrupt the weed establishment (Mayerová et al. 2018).

11.3.2.2 Crop Competition and Cultivar Selection

Weeds always compete with the main crop through stealing the nutrients, water, and space and hence this weed-crop competition should have to be decreased. The cop capacity to compete with the obnoxious weed can be enhanced through ability of crop in suppression of weeds, decrease the weed growth and reduce the weed seed bank in the field plot soil rhizosphere (Zerner et al. 2016). Meanwhile, crop should have to maintain its yield and yield stability through rapid growth and its tolerance competition against the weed. There are several reports that genotype within a particular crop also showed different ability in the weed-crop competition episode. Therefore, knowing the life history of weeds (tall, dwarf), so preference should have to be given to a taller genotype with more competitive capacity against weeds to obtained more benefits (Datta et al. 2017).

11.3.2.3 Planting Pattern

The planting pattern also plays an important role in the crop-weed competition. For example, early plating of the crop will help to provide a competitive advantage over the weed. In this case, the main crop will germinate before the weed and can suppress the weed growth through interference in the sun light as required by the weed (Andrew and Storkey 2017; Mahajan et al. 2017, 2020). In this regard, increasing the planting density coupling with decreasing with row spacing will be a plus point for the crop and hence increase crop ability against the weeds (Bastiaans and Storkey 2017; Leclère et al. 2019).

11.3.2.4 Fertility Manipulation

Application of fertilizer can alter the nutrient status in the agro-ecosystem and hence are indirectly responsible to impact the weed growth and population dynamics (Kaur et al. 2018; Ekwangu et al. 2020). Nitrogen fertilizer are major factor to increase the weed growth as they are major consumer of nitrogen nutrient. Therefore, judicious use of N fertilizer, placement method, time and dosage will help to decrease the weed nterference with crop (Maqbool et al. 2020).

11.3.3 Mechanical Weed Control

Weed control has always been closely associated with farming. Mechanical weed control methods have a long history (Bond and Grundy 2001). It is very likely that the first method of weeding was by hand-pulling. This was followed by using a stick which became a hand-hoe. As agriculture became more mechanized, weed management in fields was successfully carried out with mechanical tools pulled first by animals and later by tractors (Zimdahl 2007). Although new technologies have been added in large-scale agriculture, old ones are still used effectively, especially in small-scale agriculture. The most effective mechanical method of weed management is complete burial of seedling weeds to 1 cm depth, or to cut them at or just below the soil surface (Bond and Grundy 2001). Mechanical weeders range from basic hand-held tools to sophisticated tractor-driven devices. For a successful mechanical weed management strategy it is necessary to have a good knowledge of crop-weed interaction, precise timing, frequency of application, and it is important to select the proper mechanical tool (Kurstjens and Perdok 2000). It is also essential to improve one's knowledge regarding mechanical methods of weed control if they are to become acceptable alternatives to chemical control.

11.3.3.1 Primary Tillage

Primary tillage is the first soil-working operation in conventional cropping systems which is carried out to prepare the soil for planting. Primary tillage is always aggressive and carried out at a considerable depth (Cloutier et al. 2007) in order to control annual and/or perennial weeds by burying a portion of germinable seeds and/or propagules at depths at which weed seeds are not able to emerge (Kouwenhoven 2000). The main tools used to perform primary tillage are mould-board ploughs, disc ploughs, diggers, and chisel ploughs (Leblanc and Cloutier 2001).

11.3.3.2 Secondary Tillage

With secondary tillage the soil is not worked aggressively or deeply (Cloutier et al. 2007). The aim of secondary tillage is to prepare the soil for planting or transplanting or it is used for carrying out the false seedbed technique (Leblanc et al. 2006). The equipment used to perform secondary tillage are cultivators, harrows (disc, spring tine, radial blade, and rolling) and power take-off machines (Cloutier et al. 2007). In conservation tillage this equipment could be used as a substitute for ploughs in primary tillage. Conservation tillage is useful for conserving or increasing the organic matter content in the soil and for saving time, fuel and money (Peruzzi and Sartori 1997). Although, reduced tillage techniques could cause some problems with weeds (Zimdahl 2007), farmers can optimally alternate primary and secondary tillage in order to optimize soil management by changing mechanical actions year after year and thus improving annual and perennial weed species control (Barberi 2002).

11.3.3.3 Cultivation Tillage

Cultivating tillage is carried out after crop planting in order to achieve a shallow tillage which loosens the soil and controls weeds (Leblanc and Cloutier 2001) For this purpose cultivators are used which can control weeds in different ways. The complete or partial burial of weeds and their seeds can be an important cause of mortality (Rasmussen 1996). Another mode of action is by uprooting and breakage of the weed root contact with the soil (Kurstjens and Perdok 2000; Kurstjens and Kropff 2001). It is preferable to carry out cultivation tillage when the soil is not too wet because it can damage the soil structure and favor the spread of perennial weeds (Leblanc and Cloutier 2001). Cultivators are generally classified according to their application in a crop (Cloutier et al. 2007): broadcast cultivators could be used both on and between the crop rows; inter-row cultivators are used only between crop rows; and intra-row cultivators which are used for removing weeds from the crop rows.

11.3.3.4 Hand Tools

Removing weeds or patch of weeds by hand is often the most effective way to prevent that weed from spreading and therefore from becoming a serious problem (Zimdahl 2007). Hand-weeding is more effective for annual rather than perennial weeds due to its capacity of vegetative reproduction. Hand hoes, push hoes and other traditional methods of hand-weeding are still used worldwide on horticultural crops. Hand-weeding is often used after mechanical inter-row weeding to deal with the weeds left in the crop row.

11.3.3.5 Cutting and Mowing

These methods are commonly used in turf, and can be used in vineyards, in orchards, in pastures and in forage crops if used in the appropriate way (Cloutier et al. 2007). Although, cutting and mowing techniques enable us to control the size of weeds and their seed production and to minimize the competition between weeds and crops (Donald 2007a, b). These techniques are seldom efficient enough to obtain a total weed control. Cutting and mowing weeds reduces their leaf area, slows their growth and decreases or prevents seed production (Zimdahl 2007).

11.3.4 Thermal Weed Control

Thermal weed means include use of fire, flaming, hot water, steam and freezing (Ascard et al. 2007), which provide rapid weed control without leaving chemical residues in the soil and water. Furthermore, thermal methods are selective towards the weeds, they do not disturb the soil therefore do not bring the buried seeds to the soil surface as in the case of cultivation methods. Although thermal weed methods do not leave chemical residues in the soil and water, this approach uses large amounts of fossil fuels per unit area. The effectiveness of thermal means on weeds can be influenced by several factors including temperature, exposure time and energy input (Zimdahl 2007; Hatcher and Melander 2003; Ascard et al. 2007). Since many of these methods only kill the shoots of target plants, they may regenerate and repeated treatments may be necessary.

11.3.4.1 Flaming

High-temperatures can damage many plant processes through coagulation and the denaturation of the proteins, the increase of membrane permeability and enzyme inactivation (Zimdahl 2007). The thermal dead point for most plant tissues is between 45 °C and 55 °C after prolonged exposure. Plant size at treatment influences flaming effectiveness much more than plant density. The most tolerant species cannot be controlled with one flaming regardless of the applications (Ascard 1994). By controlling the direction of the flamer it is possible to eliminate drift and to achieve some degree of insect and disease control. Although flaming is a successful type of weed control, it is not used much in crops due to its high cost and the effectiveness of other methods.

11.3.4.2 Steaming

Although steam was widely used for disinfecting soil in the past (Sonneveld 1979; Runia 1983), it has recently been found to be effective for weed control (Upadhyaya et al. 1993). The use of steam for weed control leads to a minor reduction in water quantity and provide better canopy penetration compared to hot water (Ascard et al. 2007). Weed damage is related to weed species, steam temperature, duration of exposure, and plant size. In the case of perennial weed species it is necessary to repeat exposure to steaming due to their ability to regenerate, while in the case of annual weed species the seed coat can offer some protection to stream (Ascard et al. 2007). Mobile soil steaming is commercially used to manage weeds and pathogens in the field (Bond and Grundy 2001; Pinel et al. 1999).

11.3.4.3 Solarization and Heat

Solarization is a process which uses the heat of the sun for controlling the weeds (Ascard et al. 2007). This technique consists in placing a cover (usually black or clear plastic) over the soil surface to trap solar radiation and cause an increase in soil temperatures to levels that kill plants, seeds, plant pathogens, and insects (Zimdahl 2007). In order to be effective for weed control there should be warm, moist soil and intense radiation throughout the day in order to raise the soil temperature enough to kill weed seeds and seedlings.

Cultural, mechanical and thermal control methods for weeds have developed independently and also have focused on weed control in different systems (Hatcher and Melander 2003). A common problem concerning non-chemical methods is that effective control needs more frequently repeated treatments than chemical weed management (Rask and Kristoffersen 2007), in fact non- chemical tools mainly affect the aboveground part of the plants, whereas systemic herbicides kill the entire plant and therefore only require one or two applications per year (Popay et al. 1992). Different factors could affect the frequency of the treatments such as weed species composition, weed cover, weed acceptance level, weed control methods, climate and type of soil surface. Although in reduced and low-input farming systems, the use of herbicides remains an integral part of the approach for managing the weeds, cultural and direct non-chemical weed management can make an important contribution to reducing chemical inputs within such systems (Bond and Grundy 2001). For this reason, the integration of cropping and weed management strategies is vital for the future success of a farming system that relies on non-chemical methods of weed management.

The development of ecological weed management depends on the collective ability of farmers and scientists to convert local weed information into an improved understanding of weed ecology. Special attention must be paid when developing principles of weed ecology that are applicable in improved farm planning and decision-making. Ecological weed management involves the use of different types of information and various control tactics for developing strategies for subjecting weeds to multiple, temporarily variable stresses. The selection pressure exerted by most ecological management tactics is less severe than the selection pressure from herbicides (Liebman and Davis 2000). Hence, "*flexible management using multiple ecological weed control tactics within a diverse cropping system may present sufficiently weak and contradictory selection pressures to avoid adaptation of weed species to management*" (Liebman et al. 2001). Herbicides are not excluded from the toolkit but are viewed as options rather than absolute requirements for crop production. They are used only when and where the application of other control tactics fails to reduce and maintain weeds at acceptable levels, and they are used in a manner that poses minimal risks to humans, other no-target organisms, and the environment.

Ecological weed management differs from traditional weed management in that the primary focus is on creating an environment un-favorable for weed establishment, growth and reproduction rather than on specific control tactics. Many of the components of an ecological management system are inextricably intertwined, thereby making it difficult to measure the individual contributions of specific elements of the systems. A better understanding of the underlying mechanisms that influence the success or failure of weeds in agro-ecosystems will further the development and adoption of ecological weed management systems for agricultural crops. The main purpose for developing ecological weed management strategies is to integrate the options and tools that are available to make the cropping system unfavorable for weeds and to minimize the impact of any weeds that survive. No single weed management tactic has proven to be the "*magic bullet*" for eliminating weed problems given the nature of weed communities. The best approach may be to integrate a cropping system plan and knowledge of ecological processes with all available weed control strategies into a comprehensive weed management system. The integration of ecological principles into weed management decision-making is a major challenge for weed science researchers and growers. Weed science must play a more important role in leading ecological research in agricultural systems. An expanded theory of applied ecology provides an excellent framework for expanded approaches to weed management because it allows for new and creative ways of meeting the challenge of managing weeds in ways that are environmentally and economically viable over the long term.

11.3.5 Integrated Weed Management

The increase of the world population will demand a higher food production, which can be achieved by increasing crop yields and applying a sustainable approach through a responsible use of land and water and enhanced food diversity. Relying solely on chemical weed management can be unsustainable both to the environmental impact of herbicides and their residues and when weed populations develop resistance to herbicides (Mace et al. 2007). No single management technique is perfect for all weed control situations, a multiple management actions is preferable for an effective control. One alternative is to apply current knowledge of agricultural practices on weed populations in order to design new cropping systems that would require small quantities of herbicides to manage weeds. Since the mid-1960s, integrated pest management has been promoted by the FAO worldwide as preferred strategy for pest control (Singh et al. 2006). Integrated pest management is "*a decision support system for the selection and use of pest control tactics singly or harmoniously coordinated into a management strategy, based on cost-benefit analyses that take into account the interests of and impacts on producers, society, and the environment*" (Norris et al. 2003). According to this definition it is necessary to evaluate all the available pest control techniques and the integration of appropriate measures that discourage the development of pest populations and keep pesticides and other interventions to levels that are economically justified and reduce or minimize risks to human health and the environment. Consequently, an integrated pest management aids the growth of a healthy crop with the least possible disruption to agro-ecosystems and encourages natural pest control mechanisms (FAO 2003). The strategy of using an integrated selection of management techniques has been developed for use in a variety of pest control situations, including plant pests, or weeds. Integrated weed management is what is known as integrated pest management with the focus on weeds. Integrated weed management relies on weed management principles that have proved to be suitable for long-term weed containment (Barberi 2002), by combining the use of cultural, mechanical, thermal, and chemical means based on ecological approach (Fig. 11.1).

According to an integrated weed management, a farmer needs to predict the likely outcome of different strategies, so that rational decisions can be made. Critically, weed management advice needs to be tailored to the individual field conditions, because weed infestations vary substantially, both within and between fields (Moss 2010). From the standpoint of crop protection, integrated weed management has three principal objectives (Liebman et al. 2001):

(a) **Weed density should be reduced to tolerable levels.** Several experimentations have demonstrated that the relationship between crop yield loss due to weed competition and weed plant density could be described by a hyperbola (Cousens 1985; Moffitt and Bhowmik 2006). However, this relationship is strongly affected by different factors such as weather and soil conditions (Lindquist et al. 1996). Despite this relationship, the total eradication of weeds may be excessively expensive depriving farmers of the ecological services that certain weeds provide.
(b) **The amount of damage that a given density of weeds inflicts on an associated crop should be reduced.** The crop yield damage caused by weeds could be reduced, not only by reducing weed density but also by minimizing the resource consumption, growth, and competitive ability of each surviving weed. These objectives can be reached by delaying weed emergence compared to crop emergence and increasing the proportion of available resources captured by crops.

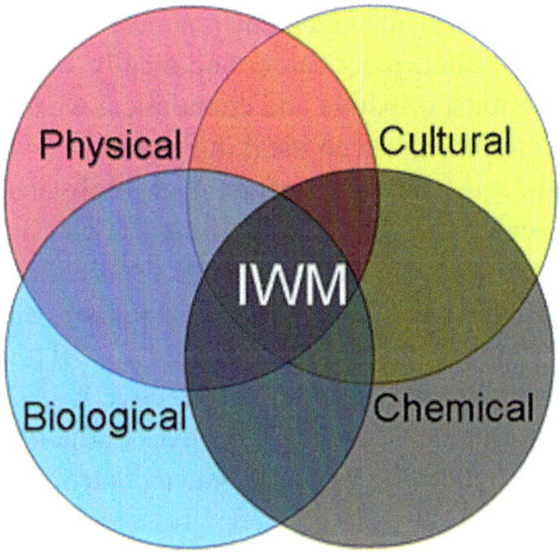

Fig. 11.1 Panorama of integrated weed management (IWM)

(c) **The composition of weed communities should be shifted toward less aggressive easier- to-manage species.** Weeds act differently in their relationship with the crops. As a consequence, it is desirable to balance the weed community composition within a agro- ecosystem toward a preponderance of species that crops and farmers can tolerate. This can be achieved by selectively and directly suppressing undesirable weed species, while manipulating environmental conditions to prevent their re-establishment.

An integrated weed management strategy minimizes the effect of weeds but certainly does not eliminate them all (Liebman et al. 2001). Weeds will be accepted as a normal and manageable part of the agricultural community. One goal of integrated weed management is to maintain weed populations below an economic threshold level by reducing emphasis on strategies of eradication and promoting a strategy of containment for potential increase in weed diversity (Cousens et al. 1987). The level of effectiveness depends on the balance between the characteristics of the weeds present and the management tools available to growers. The balance between weed communities and management tools could deteriorate if attention is not given to management of weeds at landscape level, and to preservation of herbicides and ecological control tactics in the face of evolutionary response of weeds. Management of weeds over large areas and long periods of time requires an expanded perspective on weed community dynamics and weed evolution (Cardina et al. 1999).

Integrated weed management attempts to address the underlying causes of a weed infestation, rather than just focusing on controlling visible weeds. This is done by targeting the different stages of the weed's lifecycle and undertaking measures that will prevent weed reproduction, reduce weed emergence, promote seed bank depletion and minimize weed competition with the desired crop (Streibig 1979; Ferron and Deguine 2005). This point of view provides an opportunity for producers to develop weed management systems that integrate prevention with control tactics. According to this approach, an integrated weed management may provide a more sustainable technique for crop production by reducing the reliance on external inputs that characterizes conventional agriculture.

In order to reach the desired level of weed suppression it is necessary to use specific weed management practices. An integrated weed management plan must be individual, practical, economically sound and flexible. It is important to note that the most effective and economical weed control plan always requires several types of approach. In an ideal integrated weed management strategy, it is essential to consider the cultural, mechanical, biological and chemical methods contained in the weed management toolbox and each component contributes to the overall level of weed control like several "*little hammers*" (Liebman and Gallandt 1997). All successful weed control programs analyze the various tactics available to control the existing weed problem (Fig. 11.2) and this is followed by making a plan to integrate the various tools into an effective weed control system according to the crop, the environment, and the objectives of the farm. Without this knowledge it is impossible to evaluate the impact of weed control tactics on a given weed population. When the strategy is well defined, growers must consider "*what can be done*" (scientific

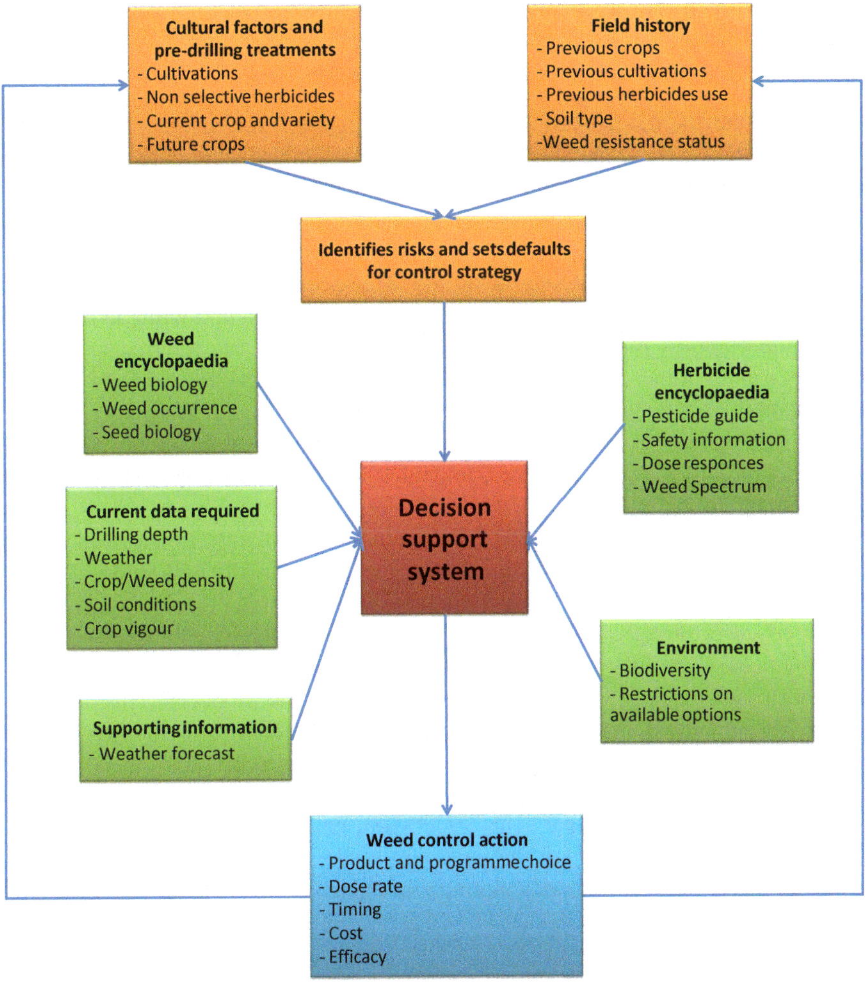

Fig. 11.2 Framework for developing decision systems to improve weed management. (Modified from Clarke 2002)

question) and "*what should be done*" (moral question). Indeed, some methods that could be scientifically possible may not be socially, culturally, politically, or environmentally desirable (Zimdahl 2007). A final important step is to evaluate the yearly success of a weed management program to verify and identify tactics to consider in the future. The long term aims of a good crop production practices is to decrease the problems cause by weeds and if these weed problems continue or increase it will be necessary to modify the original management plan (Fig. 11.2).

11.4 Allelopathy in Ecological Weed Management

In the evolutionary process, plants have developed several routes of biosynthesis of secondary metabolites as result of the selective pressure suffered along the history. At the beginning, these metabolites appeared as compounds without vital functions in plant metabolism, but nowadays we know that interaction between organisms through chemical substances sustains natural and biological ecosystems (Hunter and Menges 2002; Inderjit and Mallik 2002; Florentine 2003). This kind of chemical interference between plants is known as allelopathy. Pliny, the Elder (Caius Plinius Secundus, 23 A.D.), a roman scholar and naturalist, first wrote in his books about the complex phenomenon of chemical interaction. He referred about how chickpea and barley scorch up cornland, and noted also the toxic effects of black walnut on the surrounding. Authors have also appointed the much stronger negative effects among species from different ecosystems than among those native from the same community (Steenhagen and Zimdahl 1979; Tang et al. 1995; Callaway and Aschehoug 2000).

The intensity of the allelopathic effect in the field will depend, between others, on the different transformations that the organic compounds will suffer after the release to the environment. These toxic metabolites can be stored in the vacuole, polymerised or directly liberated, but anyway, they will be finally released to the environment where they can act as allelopathic agents on the metabolism of neighbouring plants, and giving usually an advantage to the producer (Hussain et al. 2008; Hussain and Reigosa 2011). Allelochemicals are released to the environment in different ways (by volatilization, leaching, exudation or decomposition) and can act in a direct or an indirect way on the receptor plants. Their chemical nature is complex and diverse (organic acids, aldehydes, coumarins, quinones, flavonoids, alkaloids, terpenoids, etc), but the most part of them comes from three principal biosynthetic routes, the route of shikimic acid (benzoic and cinnamic acids and their derivatives, coumarins, glycosides, alkaloids, etc), and the routes of acetic and mevalonic acids (terpenoids, steroids, complex quinones, etc) (Hussain and Reigosa 2017). Allelopathic compounds are present in almost all plants and they can be found in many parts of the plant like in roots, seeds, leave, fruits, stems (Hussain and Reigosa 2014).

Since the beginning, in the research of allelopathy we have data showing that allelopathic activity is closely related to environmental conditions. Several studies verified that an environmental stress situation (light, water, temperature, etc.) produces an increase in the production of allelochemicals as a physiological response (Mwaja et al. 1995; Zobel and Lynch 1997). Noguchi found that light stressed plants produced much more DIBOA than control plants, enhancing also the inhibitory capacity of maize plants on other species (Kato-Noguchi 1999). A stressed plant will produce, in general, more allelochemicals than a plant in optimal conditions, which suggests that allelopathy is used by organisms as a response to stress situations. The induction of stress that increases the concentration of the allelochemicals influences also the intensity and duration of allelopathy (Einhellig 1995). Therefore, in the study of allelopathic phenomenon a compromise between the most precise lab physiological studies and the approach to the "natural" field conditions in the experimental design must be found, avoiding confusions between the allelopathic mode of action and the plant response to some other stress.

But the interaction of allelochemicals and environmental stresses has another aspect in the allelopathic occurrence, because the ability of the receptor plant to respond against these compounds is also very different depending on the stresses taking place in the medium. The allelopathic activity investigation has three main components on which scientists base the research to demonstrate the existence of chemical interference (Inderjit and Weston 2000). An ecological component (demonstration of occurrence in nature of chemical interference), a chemical component (search and characterization of chemical compounds responsible of this interference) and a physiological component (identification of the mechanism of action at physiological, biochemical, genetic level in the plant). And on this last component, the search of specific mode or modes of action, will be focused the main part of the present work.

Allelopathy, which has long been an important part of the field of ecology, began to be understood and applied into sustainable agriculture. Without knowing what allelochemicals are, the phenomenon of allelopathy was first recorded 2000 years ago. Agricultural applications such as crop rotation, mulching, cover crop and intercropping, which aim at reducing dependence on external energy, are all related to allelopathy. Current research on allelopathy usage mostly focuses on Integrated Pest Management including weed, insect and pathogen control. Moreover, the development of technology will lead to our better understanding of allelopathy, which will be fully applied to transforming conventional agriculture into sustainable agriculture.

11.4.1 Allelopathy in Traditional and Sustainable Agriculture

The recent tendencies in weed management defend the damage and/or displacement of weeds in the agroecosystem but not the total eradication of them (as usually happens with commercial herbicides). At this point, the role of allelopathic compounds can be determinant because of their high potential as possible natural herbicides (Dayan et al. 1999; Duke et al. 2000). Weed management must protect environmental quality and human health, and allelochemicals released from live crops and crop residues can be used this way to damage weeds and improve crop performance (Liebman 2001).

11.4.1.1 Crop Rotation

Allelopathic plants conduct self-poisoning (Wang et al. 2011). In traditional farming, farmers found decrease crop yield on second and further cultivation. The decrease of crop yield is caused by accumulated self-toxic chemicals released by one kind of crop that is planted continuously. For example, many crops such as rice, wheat, corn, sugar cane, soy beans, broad beans and tea plantations have the obvious self-poisoning phenomenon (Wang et al. 2011). According to Wang et al., rice stubble and straw can produce some toxic substances in the process of their decomposition, which will inhibit the growth of rice seedlings. Rice and wheat cropping system experiences great weed problems (Farooq et al. 2011).

Crop rotation is used to increase crop yield by avoiding the accumulation of self-toxic allelochemicals in continuous mono-cropping. Other benefit of crop rotation includes weed suppression. Sorghum releases weed suppressing allelochemicals into the soil. Crops following sorghum take advantage from this natural herbicide left by sorghum and have less weed problems (Farooq et al. 2011). According to Farooq et al., growing pearl millet, maize and sorghum after harvesting wheat and before growing rice can control the weed growth when growing rice for 45 days and minimize the input of herbicide when growing rice.

11.4.1.2 Cover Crop

Apart from weed problems, agriculture also faces soil erosion, insect, nematodes and disease pathogen problems. Cover crop can overcome these problems, reduce or eliminate input of chemical fertilizer, pesticide, nematicide and insecticide, and provide dodder. In cover crop applications, specific allelopathic cover crops are used aiming at specific crop and weed. For example, legume cover crops are used in maize to reduce the population of barnyard-grass; barley is grown as a cover crop for weed control in soybean-suppressed weeds such as crabgrass and barnyard-grass; velvet bean, jack bean and hyacinth beans are used to effectively control weed in rubber plantations (Farooq et al. 2011). The application of cover crop substitutes the use of a chemical herbicide so as to reduce the development of herbicide resistance in weed species and chemical pollutions into the environment.

11.5 Case Studies on Allelopathy and Feasible Applications

11.5.1 Intercropping

Among all the allelopathic interactions, most of them are typically negative in character, while any positive effect rarely presents. The negative effect of inhibiting or delaying plant growth and/or seed germination is usually defined as phytotoxic potential. A common example of phytotoxic potential is the weed problem in agriculture. Many weeds can produce and release allelochemicals to interact with other crop competitors (Kong 2010).

In addition to maximizing crop yields, intercropping with allelopathic plants or crops is a natural method of weed control. Intercropping, which is the agricultural application of planting two or more crops together, can substitute the use chemical herbicide thus reduce energy consumption, environmental contamination and herbicide resistance of crops.

Successful weed control approaches of intercropping plants include mint, winter savory and the genus *Ocimum*. Intercropping sweet potato with mint, winter savory or genus *Ocimum* has great effect of suppressing even toxic weeds (Farooq et al. 2011).

Some intercropping practices increase agricultural production but some affect crop yield in the opposite way. The existence of allelopathy within paired crops affects crop yields. For example, sorghum has some inhibitory effect on weeds, crop that is inter-planted with sorghum can effectively control weeds and improve crop yields (Farooq et al. 2011); intercropping of corn and soybeans as well as wheat and pea intercropping also increase production. However, allelochemicals released by tomato root and plant significantly inhibit the growth of cucumber. Thus, they should not be planted together. Moreover, root exudates of apple inhibit the growth of wheat. All of these examples are evidence that choosing the right match of intercrops can improve agricultural productivity effectively.

11.5.2 The Applications of Allelopathic Water Extracts

Allelopathic water extracts are water-soluble allelochemicals extracted from plants (Farooq et al. 2011). Although using allelopathic water extracts is a more environmentally friendly method of weed management than using synthetic chemicals as herbicides, the rate and frequency of certain applied allelochemicals should be carefully considered. Farooq et al. reported that, in allelopathic water extract applications, using a combination of two or more allelopathic compounds extends the effect of weed control more than using only one kind of allelopathic compound. This is due to the enhancement of the allelopathy caused by allelopathic interactions in the mixture. According to Farooq et al., experimental results showed that applying sunflower and sorghum extracts (each at 12 L $ha^{-}1$) can reduce little seed canary biomass by 36–55% and wild oat biomass by 42–60% in wheat; when apply a half amount of the same combined extracts, wheat yield increased by 89% in the first year.

Combination of herbicide and Allelopathic water extracts reduces the rate of synthetic herbicide, keeps the full effect of weed control and increases crop yield (Farooq et al. 2011). When applying the combination of herbicide and allelopathic water extracts, the input of herbicide is reduced by a rate of 50–60%. And the ratio of herbicide can be reduced by more than 60% when applying the combination of herbicide and mixture of allelopathic water extracts.

11.6 Allelopathy to Build a Sustainable Agroecosystem

Current research achievement about allelopathy has shown that allelopathy will play a crucial role in sustainable agriculture because the interspecies chemical interaction (allelopathy) is a common phenomenon in nature. And plants' allelopathic function is gaining more and more attention due to the urgent needs of reducing the dependence on synthetic chemicals. Herbicide, pesticide and fertilizer which are widely used in conventional agriculture are synthetic chemicals that pollute the environment and require great amount of energy input.

Weed problems have been reducing crop yield, causing economic concerns. Instead of using herbicides to control weeds, it is wise to use allelopathic weeds to manage crop pests. *Ageratum conyziodes*, *Ambrosia trifida*, and *Lantana camara* are invasive weeds that contain different kinds of plant-growth-inhibitors. However, these growth-inhibitors can also have a pest control effect on crops. Studies of allelopathins released by targeted weeds and in-field practices have shown that these three allelopathic weeds have effects on microbes, nematodes and water hyacinth (Kong 2010). Although weed-crop intercropping may have great effect on pest management, the crop yield in weed-crop intercropping may not be able to reach the crop yield in crop-crop intercropping with regard to land use efficiency.

Neem seed oil contains allelochemicals such as azadirachtin, nimbin and salannin that inhibits the appetite of insects like nymphs, aphids, green leafhopper, whitefly, and pine weevil. Also, allelopathic water extracts of fine grain, mustard and sunflower can be used to control pests such as aphids and sucking insects. Farooq et al. concluded that fine grain water extract effectively killed 62.5% of aphids, which was the most effective natural pesticide. When ethanol extracts from the leaves of the California pepper tree is diluted with a concentration of 4.3 and 4.7%, the water extract kills 97% of elm leaf beetles (Farooq et al. 2011).

Some organs of neem tree release allelochemicals that inhibit the development of root-knot nematodes. The inhibitive effect can last for as long as 16 weeks. A practicable application is to apply neem leaves and neem cakes to the soil. Allelochemicals from rice can not only control the growth of weed but also inhibit fungal pathogens by inhibiting the germination of spores (Farooq et al. 2011).

11.7 Limitations of Allelopathic Applications

There are limitations of using allelopathy in weed management due to the complex allelopathic mechanism. Allelopathic plants which contain more than one kind of allelochemicals release specific types of allelochemicals under specific environmental stress that can have abiotic or biotic effects. The released allelochemicals can also be interference or simulation, which become new environmental abiotic and biotic influences that will exert to the host plant itself and the acceptor. For example, when using crop rotation as a method of weed management, inhibition of crop growth is observed in sorghum-wheat crop rotation system. Because the allelochemicals sorghum releases not only suppresses weed but also inhibits the development of wheat (Farooq et al. 2011). As a result, it is difficult to control the type and amount of allelochemicals that the host plants release. The reactions between released allelochemicals and the abiotic and biotic factors in soil are also hard to estimate. In addition, isolating and synthesizing allelochemicals are expensive and energy consuming. Moreover, allelopathic potentiality is affected by the amount of nutrient in plants (Bhadoria 2011).

On the other hand, one allelopathic weed species that Chui studied is native to the Americas, and is a poisonous weed that is dangerous to humans and animals. This weed species, *Lantana camara*, having been introduced into agriculture in southern China, has become an aggressively invading species that is seriously damaging resources in certain ecosystems such as forests, tea plantations and orchards. With strong allelopathy, *Lantana camara* can inhibit the growth of surrounding plants. Wang et al. suggested that using native enemy species in its origin country is the proper way to control weeds. Because when an exotic species is introduced to a non-native region, it lacks regulation from other natural enemies, which results in an uncontrolled increase in distribution and abundance (Wang et al. 2011). Before considering the introduction of exotic species as a method of weed management, its invasion mechanism should be carefully studied.

There are other problems that limit the use of allelopathy in sustainable agriculture. Take allelopathic nematodes control as an example. The cost of the allelopathic rotation crop is greater than its benefit compared to that of using nematicide; and allelopathic crops may not adapt to the physical environment (Halbrendt 1996) (Table 11.1).

Table 11.1 Allelopathic compounds isolated from plants, that showed inhibitory impact on germination and seedling growth of weeds/plants

Compounds	Botanical source	Sensitive weeds/plant species	References
Phenolic compounds	*Acacia melanoxylon* R. Br. (flowers and phyllodes aqueous extract).	Cocksfoot (*Dactylis glomerata*), perennial ryegrass (*Lolium perenne*), common sorrel (*Rumex acetosa*), lettuce (*Lactuca sativa*)	Hussain et al. (2011a, b)
Flavonoids		Cocksfoot (*Dactylis glomerata*), perennial ryegrass (*Lolium perenne*), common sorrel (*Rumex acetosa*), lettuce (*Lactuca sativa*)	Hussain et al. (2011a, b)
Glucosinolates,	Mustard (Brassica sp.)	Spiny sowthistle *(Sonchus asper L. Hill),* scentless	Petersen et al. (2001), Haramoto and Gallandt (2004)
Isothiocyanates	Garden radish (*Raphanus sativus*)	*mayweed (Matricaria inodora L.),* smooth pigweed *(Amaranthus hybridus L.),* barnyardgrass *(Echinochloa cruss-galli L. Beauv.),* slender meadow foxtail or blackgrass *(Alopecurus myosuroides Huds.), Alhagi spp., Cachia maritime,* Shepherd's-purse*(Capsella bursapastoris L.),* morning glory *(Convolvulus arvensis L.), dodders (Cuscuta spp.),* wild carrot or bird's nest *(Daucus carota L.),* shortpod mustard, *buchanweed or* hoary mustard *(Hirschfeldia incana L.), Ochtodium aegyptiacum (L.),* shortfruit hedgemustard *(Sisymbrium polyceratium L.)*	Haramoto and Gallandt (2004), Vig et al. (2009), Rice et al. (2007)i
Sorgoleone	Sorghum (*Sorghum bicolor* L. Moench)	Littleseed canarygrass *(Phalaris minor Retz.),* lesser swinecress *(Coronopus didymus L.),* purple nutsedge *(Cyperus rotundus L.),* black nightshade *(Solanum nigrum L.),* redroot pigweed *(Amaranthus retroflexus L.),* common ragweed *(Ambrosia atrtemisiflora L.),* sicklepod *(Cassia obtusifolia L.)*	Głąb et al. (2017), Dayan et al. (2010), Sarr et al. (2020), Tibugari et al. (2019), Besançon et al. (2020), Mareya et al. (2020), Tibugari et al. (2020a, b)
Momilactone	Rice (*Oryza sativa* L.), moss (Hypnum plumaeform)	Barnyardgrass, *(Echinochloa colonum L.),* livid amaranth*(Amaranthus lividus L.),* hairy crabgrass *(Digitaria sanguinalis L.),* annual meadow grass, annual bluegrass *(Poa annua L.)*	Quan et al. (2019a, b), Bajsa-Hirschel et al. (2020), Gu et al. (2019)

(continued)

Table 11.1 (continued)

Compounds	Botanical source	Sensitive weeds/plant species	References
Artemisinin	Annual wormwood (*Artemisia annua* L.)	Redroot pigweed, pitted morning-glory *(Ipomoea lacunose L.),* common purslane *(Portulaca oleracea L.),* annual wormwood, duckweed *(Lemna minor L.),* algae *(Pseudokirchneriella subcapitata)*	Bachheti et al. (2020), Li et al. (2019)
Leptospermone	Bottle brush (*Callistemon citrinus*), manuka (*Leptospermum scoparium* J.R., G. Forst)	Barnyard grass, hairy crabgrass, yellow foxtail *(Setaria glauca L.),* California red oat *(Avena sativa L.),* Indian mustard *(Brassica juncea L.),* curly dock *(Rumex crispus L.)*	Mallet et al. (2019), Shirgapure and Ghosh (2020), Travlos et al. (2020), Mushtaq et al. (2020), Mehdizadeh and Mushtaq (2020)
Essential oils	Eucalyptus (*Eucalyptus* sp.)	Barnyard grass, *Cassia occidentalis,* annual ryegrass *(Lolium rigidum)*	Mushtaq et al. (2020), Mehdizadeh and Mushtaq (2020), Ibáñez and Blázquez (2019), Krumsri et al. (2019), Khare et al. (2019)
Sarmentine, Garcienone	Pepper (Piper sp.)	Barnyard grass, redroot pigweed, crabgrass, *Sprangletop (Leptochloa filiformis Lam.),* dandelion *(Taraxacum sp.),* lambsquarter or wild spinach *(Chenopodium album L.),* annual bluegrass, bindweed, wild mustard, curly dock	Suwitchayanon et al. (2019), Feng et al. (2019), Rob et al. (2019), Jaramillo-Colorado et al. (2019)

Acknowledgement This work was supported by the European Commission, grant number 771367-2, for the project **Increasing the efficiency and competitiveness of organic crop breeding** "ECOBREED".

References

Almberg KS, Turyk ME, Jones RM, Rankin K, Freels S, Stayner LT (2018) Atrazine contamination of drinking water and adverse birth outcomes in community water systems with elevated atrazine in Ohio, 2006–2008. Int J Environ Res Public Health 15:1889. https://doi.org/10.3390/ijerph15091889

Alström S (1990) Fundamentals of weed management in hot climate peasant agriculture. Crop production science, vol 11. Swedish University of Agricultural Sciences, Uppsala, p 271

Andrew IK, Storkey J (2017) Using simulation models to investigate the cumulative effects of sowing rate, sowing date and cultivar choice on weed competition. Crop Prot 95:109–115. https://doi.org/10.1016/j.cropro.2016.05.002

Ascard J (1994) Dose response models for flame weeding in relation to plant size and density. Weed Res 34:377–385. https://doi.org/10.1111/j.1365-3180.1994.tb02007

Ascard J, Hatcher PE, Melander B, Upadhyaya MK (2007) Thermal weed control. In: Upadhyaya MK, Blackshaw RE (eds) Non-chemical weed management. Principles, concepts, and technology. CABI, London

Bachheti A, Sharma A, Bachheti RK, Husen A, Pandey DP (2020) Plant allelochemicals and their various applications. In: Mérillon JM, Ramawat K (eds) Co-evolution of secondary metabolites, Reference series in phytochemistry. Springer, Cham. https://doi.org/10.1007/978-3-319-96397-6_14

Bajsa-Hirschel J, Pan Z, Duke SO (2020) Rice momilactone gene cluster: transcriptional response to barnyard grass (*Echinochloa crus-galli*). Mol Biol Rep 47:1507–1512. https://doi.org/10.1007/s11033-019-05205-8

Barberi P (2002) Weed management in organic agriculture: are we addressing the right issues? Weed Res 42(3):177–193. https://doi.org/10.1046/j.1365-3180.2002.00277.x

Bastiaans L, Storkey J (2017) Descriptive and mechanistic models of crop-weed competition. In: Hatcher PE, Froud-Williams RJ (eds) Weed research: expanding horizons. Wiley, Hoboken, pp 33–60

Besançon TE, Dayan FE, Gannon TW, Everman WJ (2020) Conservation and divergence in sorgoleone production of sorghum species. J Environ Qual 49(2):368–377. https://doi.org/10.1002/jeq2.20038

Bhadoria PBS (2011) Allelopathy: a natural way towards weed management. J Exp Agric Int 1(1):7–20. https://doi.org/10.9734/AJEA/2011/002

Blackshaw R, Anderson R, Lemerle D (2007) Cultural weed management. In: Upadhyaya M, Blackshaw R (eds) Non-chemical weed management: principles, concepts and technology. CAB International, Wallingford, pp 35–47

Bond W, Grundy AC (2001) Non-chemical weed management in organic farming systems. Weed Res 41(5):383–405. https://doi.org/10.1046/j.1365-3180.2001.00246.x

Bretagnolle V, Gaba S (2015) Weeds for bees? A review. Agron Sustain Dev 35:891–909. https://doi.org/10.1007/s13593-015-0302-5

Callaway RM, Aschehoug ET (2000) Invasive plants versus their new and old neighbors: a mechanism for exotic invasion. Science 290:521–523. https://doi.org/10.1126/science.290.5491.521

Cardina J, Webster TM, Herms CP, Regnier EE (1999) Development of weed IPM: levels of integration for weed management. J Crop Prod 2:239–267. https://doi.org/10.1201/9781003075202-13

Clarke J (2002) Developing decision-support systems to improve weed management. In: Naylor REL (ed) Weed management handbook, 9th edn. Blackwell Science Ltd, Oxford, pp 311–322. https://doi.org/10.1002/9780470751039

Cloutier DC, Van der Weide RY, Peruzzi A, Leblanc ML (2007) Mechanical weed management. In: Upadhyaya MK, Blackshaw RE (eds) Non-chemical weed management. Principles, concepts, and technology. CABI, London

Cousens R (1985) An empirical model relating crop yield to weed and crop density and a statistical comparison with other models. J Agric Sci 105(3):513–521. https://doi.org/10.1017/S0021859600059396

Cousens R, Brain P, O'donovan JT, O'sullivan A (1987) The use of biologically realistic equations to describe the effects of weed density and relative time of emergence on crop yield. Weed Sci 35(5):720–725. https://doi.org/10.1017/S0043174500060872

Datta A, Ullah H, Tursun N, Pornprom T, Knezevic SZ, Chauhan BS (2017) Managing weeds using crop competition in soybean [*Glycine max* (L.) Merr.]. Crop Prot 95:60–68. https://doi.org/10.1016/j.cropro.2016.09.005

Dayan FE, Romagni JG, Tellez M, Rimando A, Duke S (1999) Managing weeds with natural products. Pestic Outlook 10:185–188

Dayan FE, Rimando AM, Pan Z, Baerson SR, Gimsing AL, Duke SO (2010) Sorgoleone. Phytochemistry 71(10):1032–1039. https://doi.org/10.1016/j.phytochem.2010.03.011

De Bruin JL, Pedersen P (2008) Effect of row spacing and seeding rate on soybean yield. Agron J 100:704–710. https://doi.org/10.2134/agronj2007.0106

Donald WW (2007a) Control of both winter annual and summer annual weeds in no-till corn with between-row mowing systems. Weed Technol 21(3):591–601. https://doi.org/10.1614/WT-05-141.1

Donald WW (2007b) Between-row mowing systems control summer annual weeds in no-till grain sorghum. Weed Technol 21(2):511–517. https://doi.org/10.1614/WT-05-128.1

Druille M, Omacini M, Golluscio RA, Cabello MN (2013) Arbuscular mycorrhizal fungi are directly and indirectly affected by glyphosate application. Appl Soil Ecol 72:143–149. https://doi.org/10.1016/j.apsoil.2013.06.011

Duke SO, Dayan FE, Romagni JG, Rimando AM (2000) Natural products as sources of herbicides: current status and future trends. Weed Res 40:99–111. https://doi.org/10.1046/j.1365-3180.2000.00161.x

Einhellig FA (1995) Allelopathy: current status and future goals. In: Inderjit (ed) Allelopathy: organisms, processes and applications, ACS symposium series 582. American Chemical Society, Washington, DC

Ekwangu J, Anguria P, Andiku C, Tenywa JS, Bisikwa J, Wanyera N, Ugen MA (2020) Fertilizer micro-dosing and timing of weeding for enhancing finger-millet production in eastern Uganda. J Agric Sci 12(11):290–298. https://doi.org/10.5539/jas.v12n11p290

FAO (2003) World agriculture: towards 2015/2030. Interim Report. Rome

Farooq M, Jabran K, Cheema ZA, Wahid A, Siddique KH (2011) The role of allelopathy in agricultural pest management. Pest Manag Sci 67(5):493–506. https://doi.org/10.1002/ps.2091

Feng G, Chen M, Ye HC, Zhang ZK, Li H, Chen LL, Chen XL, Yan C, Zhang J (2019) Herbicidal activities of compounds isolated from the medicinal plant *Piper sarmentosum*. Ind Crop Prod 132:41–47. https://doi.org/10.1016/j.indcrop.2019.02.020

Ferron P, Deguine JP (2005) Crop protection, biology control, habitat management and integrated farming. A review. Agron Sustain Dev 25:17–22. https://doi.org/10.1051/AGRO:2004050

Florentine SK (2003) Allelopathic effects of *Eucalyptus victrix* L. on grasses. Allelopath J 11(1):77–84

Gao B, Huang T, Ju X, Gu B, Huang W, Xu L, Rees RM, Powlson DS, Smith P, Cui S (2018) Chinese cropping systems are a net source of greenhouse gases despite soil carbon sequestration. Glob Chang Biol 24(12):5590–5606. https://doi.org/10.1111/gcb.14425

Głąb L, Sowiński J, Bough R, Dayan FE (2017) Allelopathic potential of sorghum (*Sorghum bicolor* (L.) Moench) in weed control: a comprehensive review. Adv Agron 145:43–95. https://doi.org/10.1016/bs.agron.2017.05.001

Gu CZ, Xia XM, Lv J, Tan JW, Baerson SR, Pan ZQ, Song YY, Zeng RS (2019) Diterpenoids with herbicidal and antifungal activities from hulls of rice (Oryza sativa). Fitoterapia 136:104183. https://doi.org/10.1016/j.fitote.2019.104183

Halbrendt JM (1996) Allelopathy in the management of plant-parasitic nematodes. J Nematol 28(1):8–14

Haramoto ER, Gallandt ER (2004) Brassica cover cropping for weed management: a review. Renewable Agric Food Syst 19(4):187–198. https://doi.org/10.1079/RAFS200490

Hatcher P, Melander B (2003) Combining physical, cultural, and biological methods: prospects for integrated non-chemical weed management strategies. Weed Res 43:303–322. https://doi.org/10.1046/j.1365-3180.2003.00352.x

Hunt ND, Hill JD, Liebman M (2017) Reducing freshwater toxicity while maintaining weed control, profits, and productivity: effects of increased crop rotation diversity and reduced herbicide usage. Environ Sci Technol 51(3):1707–1717. https://doi.org/10.1021/acs.est.6b04086

Hunter ME, Menges ES (2002) Allelopathic effects and root distribution of *Ceratiola ericoides* (Empetraceae) on seven rosemary scrub species. Am J Bot 89(7):1113–1118. https://doi.org/10.3732/ajb.89.7.1113

Hussain MI, Reigosa MJ (2011) Allelochemical stress inhibits growth, leaf water relations, PSII photochemistry, non-photochemical fluorescence quenching and heat energy dissipation in three C_3 perennial species. J Exp Bot 62:4533–4545. https://doi.org/10.1093/jxb/err161

Hussain MI, Reigosa MJ (2014) Higher peroxidase activity, leaf nutrient contents and carbon isotope composition changes in *Arabidopsis thaliana* are related to rutin stress. J Plant Physiol 171(15):1325–1333. https://doi.org/10.1016/j.jplph.2014.05.009
Hussain MI, Reigosa MJ (2017) Evaluation of photosynthetic performance and carbon isotope discrimination in perennial ryegrass (*Lolium perenne* L.) under allelochemicals stress. Ecotoxicology 26(5):613–624. https://doi.org/10.1007/s10646-017-1794-3
Hussain MI, González L, Reigosa MJ (2008) Germination and growth response of four plant species towards different allelochemicals and herbicides. Allelopath J 22:101–110
Hussain MI, Gonzalez L, Reigosa MJ (2011a) Allelopathic potential of *Acacia melanoxylon* on the germination and root growth of native species. Weed Biol Manag 11(1):18–28. https://doi.org/10.1111/j.1445-6664.2011.00401.x
Hussain MI, González L, Souto C, Reigosa MJ (2011b) Ecophysiological responses of native plants to phytotoxic effect of *Acacia melanoxylon* R. Br. Agrofor Syst 83:149–166. https://doi.org/10.1007/s10457-011-9433-0
Ibáñez MD, Blázquez MA (2019) Phytotoxic effects of commercial *Eucalyptus citriodora, Lavandula angustifolia*, and *Pinus sylvestris* essential oils on weeds, crops, and invasive species. Molecules 24(15):2847. https://doi.org/10.3390/molecules24152847
Inderjit, Mallik AU (2002) Can *Kalmia angustifolia* interference to black spruce (*Picea mariana*) be explained by allelopathy? Forest Ecol Manag 160(1):75–84. https://doi.org/10.1016/S0378-1127(01)00463-7
Inderjit, Weston LA (2000) Are laboratory bioassays for allelopathy suitable for prediction of field responses? J Chem Ecol 26:2111–2118. https://doi.org/10.1023/A:1005516431969
Jaramillo-Colorado BE, Pino-Benitez N, González-Coloma A (2019) Volatile composition and biocidal (antifeedant and phytotoxic) activity of the essential oils of four Piperaceae species from Choco-Colombia. Ind Crop Prod 138(5):111463. https://doi.org/10.1016/j.indcrop.2019.06.026
Kato-Noguchi H (1999) Effect of light-irradiation on allelopathic potential of germinating maize. Phytochemistry 52:1023–1027. https://doi.org/10.1016/S0031-9422(99)00365-9
Kaur S, Kaur R, Chauhan BS (2018) Understanding crop-weed-fertilizer-water interactions and their implications for weed management in agricultural systems. Crop Prot 103:65–72. https://doi.org/10.1016/j.cropro.2017.09.011
Khare P, Srivastavac S, Nigam N, Singh AK, Singh S (2019) Impact of essential oils of *E. citriodora, O. basilicum* and *M. arvensis* on three different weeds and soil microbial activities. Environ Technol Innov 14:100343. https://doi.org/10.1016/j.eti.2019.100343
Kong CH (2010) Ecological pest management and control by using allelopathic weeds (*Ageratum conyzoides, Ambrosia trifida,* and *Lantana camara*) and their allelochemicals in China. Weed Biol Manag 10(2):73–80. https://doi.org/10.1111/j.1445-6664.2010.00373.x
Korres NE, Norsworthy JK, Mauromoustakos A (2019) Effects of palmer amaranth (*Amaranthus palmeri*) establishment time and distance from the crop row on biological and phenological characteristics of the weed: implications on soybean yield. Weed Sci 67:126–135. https://doi.org/10.1017/wsc.2018.84
Kouwenhoven JK (2000) Mouldboard ploughing for weed control. In: Cloutier D (ed) 4th EWRS Workshop on physical and cultural weed control, Elspeet, The Netherlands, pp 19–22. KROMP, B., 1999. Carabid beetles in sustainable agriculture: a review on pest control efficacy, cultivation impacts and enhancement. Agric Ecosyst Environ 74:187–228
Krumsri R, Boonmee S, Hisashi KN (2019) Evaluation of the allelopathic potential of leaf extracts from *Dischidia imbricata* (blume) steud. on the seedling growth of six test plants. Not Bot Horti Agrobot 47(4):1019–1024. https://doi.org/10.15835/nbha47411598
Kurstjens DAG, Kropff MJ (2001) The impact of uprooting and soil-covering on the effectiveness of weed harrowing. Weed Res 41:211–228. https://doi.org/10.1046/j.1365-3180.2001.00233.x
Kurstjens DAG, Perdok UD (2000) The selective soil covering mechanism of weed harrows on sandy soil. Soil Tillage Res 55(3–4):193–206. https://doi.org/10.1016/S0167-1987(00)00128-8

Lal R (2004) Carbon emission from farm operations. Environ Int 30(7):981–990. https://doi.org/10.1016/j.envint.2004.03.005

Leblanc ML, Cloutier DC (2001) Susceptibility of row-planted soybean (*Glycine max*) to the rotary hoe. J Sustain Agric 18(4):53–62. https://doi.org/10.1300/J064v18n04_07

Leblanc ML, Cloutier D, Stewart KA (2006) Rotary hoe cultivation in sweet corn. HortTechnology 10(16):583–589. https://doi.org/10.21273/horttech.16.4.0583

Leclère M, Jeuffroy MH, Butier A, Chatain C, Loyce C (2019) Controlling weeds in camelina with innovative herbicide-free crop management routes across various environments. Ind Crop Prod 140:111605. https://doi.org/10.1016/j.indcrop.2019.111605

Li ZR, Amist N, Bai LY (2019) Allelopathy in sustainable weeds management. Allelopath J 48(2):109–138

Liebman M (2001) Weed management: a need for ecological approaches. In: Liebman M, Mohler CL, Staver CP (eds) Ecological management of agricultural weeds. Cambridge University Press, Cambridge, p 544. https://doi.org/10.1017/CBO9780511541810

Liebman M, Davis AS (2000) Integration of soil, crop and weed management in low-external-input farming systems. Weed Res 40:27–47. https://doi.org/10.1046/j.1365-3180.2000.00164.x

Liebman M, Gallandt ER (1997) Many little hammers: ecological management of crop-weed interactions. In: Jackson LE (ed) Ecology in agriculture. Academic, Oralndo. https://doi.org/10.1016/B978-012378260-1/50010-5

Liebman M, Mohler C, Staver CP (2001) Ecological management of agricultural weeds. Cambridge University Press, Cambridge. https://doi.org/10.1017/CBO9780511541810

Lindquist JL, Mortensen DA, Clay SA, Shemik R, Kells JJ, Howatt K, Westra P (1996) Stability of corn (*Zea mays*)-velvetleaf (*Abutilon theophrasti*) interference relationships. Weed Sci 44(2):309–313. https://doi.org/10.1017/S0043174500093930

Mace K, Morlon P, Munier-Jolain NM, Quere L (2007) Time scales as a factor in decision- making by French farmers on weed management in annual crops. Agric Syst 93(1–3):115–142. https://doi.org/10.1016/j.agsy.2006.04.007

Mahajan G, Kaur G, Chauhan BS (2017) Seeding rate and genotype effects on weeds and yield of dry-seeded rice. Crop Prot 96:68–76. https://doi.org/10.1016/j.cropro.2017.01.008

Mahajan G, Hickey L, Chauhan BS (2020) Response of barley genotypes to weed interference in Australia. Agronomy 10(1):99. https://doi.org/10.3390/agronomy10010099

Mallet C, Romdhane S, Loiseau C, Béguet J, Martin-Laurent F, Calvayrac C, Barthelmebs L (2019) Impact of leptospermone, a natural β-triketone herbicide, on the fungal composition and diversity of two arable soils. Front Microbiol 10:1024. https://doi.org/10.3389/fmicb.2019.01024

Manning R (2004) Against the grain: how agriculture has hijacked civilization. North Point Press, New York

Maqbool MM, Naz S, Ahmad T, Nisar MS, Mehmood H, Alwahibi MS, Alkahtani J (2020) The impact of seed burial depths and post-emergence herbicides on seedling emergence and biomass production of wild oat (*Avena fatua* L.): implications for management. PLoS One 15(10):1–15. https://doi.org/10.1371/journal.pone.0240944

Mareya CR, Tugizimana F, Steenkamp P, Piater L, Dubery IA (2020) Lipopolysaccharides trigger synthesis of the allelochemical sorgoleone in cell cultures of *Sorghum bicolor*. Plant Signal Behav 15(10):1796340. https://doi.org/10.1080/15592324.2020.1796340

Mayerová M, Madaras M, Soukup J (2018) Effect of chemical weed control on crop yields in different crop rotations in a long-term field trial. Crop Prot 114:215–222. https://doi.org/10.1016/j.cropro.2018.08.001

Mehdizadeh M, Mushtaq W (2020) Biological control of weeds by allelopathic compounds from different plants: a bioherbicide approach. In: Natural remedies for pest, disease and weed control. Academic, London, pp 107–117. https://doi.org/10.1016/B978-0-12-819304-4.00009-9

Mierau A, Johnson EN, Gulden RH, Weber JD, May WE, Willenborg CJ (2020) Minimizing competition between glyphosate resistant volunteer canola and glyphosate resistant soybean: impact of soybean planting date and rate. Weed Technol 34(2):220–228. 1–9. https://doi.org/10.1017/wet.2019.97

Moffitt LJ, Bhowmik PC (2006) Design for rectangular hyperbolic crop-weed competition. Ann Appl Biol 149:87–89. https://doi.org/10.1111/j.1744-7348.2006.00063.x
Moss SR (2010) Non-chemical methods of weed control: benefits and limitations. In 17th Australasian weed conference. CAWs, Christchurch, pp 14 –19
Mushtaq W, Siddiqui MB, Hakeem KR (2020) Allelopathy: potential for green agriculture. Springer Nature, Cham
Mwaja VN, Masiunas JB, Weston LA (1995) Effects of fertility on biomass, phytotoxicity and allelochemical content of cereal rye. J Chem Ecol 21:81–97. https://doi.org/10.1007/BF02033664
Norris RF, Caswell-Chen EP, Kogan M (2003) Concepts in integrated pest management. Prentice Hall, Upper Saddle River
Oerke EC (2006) Crop losses to pests. J Agric Sci 44:31–43. https://doi.org/10.1017/S0021859605005708
Peruzzi A, Sartori L (1997) Guide to the choice and the proper use of tillage operative machines. Edagricole, Bologna. 255 pp
Petersen J, Belz R, Walker F, Hurle K (2001) Weed suppression by release of isothiocyanates from turnip rape mulch. Agron J 93(1):37–43. https://doi.org/10.2134/agronj2001.93137x
Pinel MPC, Bond W, White JG, De Courcy WM (1999) Field vegetables: assessment of the potential for mobile soil steaming machinery to control diseases, weeds and mites of field salad and related crops, Final report on HDC project FV229. Horticultural Development Council, East Malling
Place GT, Reberg-Horton SC, Dunphy JE, Smith AN (2009) Seeding rate effects on weed control and yield for organic soybean production. Weed Technol 23:497–502. https://doi.org/10.1614/WT-08-134.1
Popay I, Hoskins G, Lewthwaite R (1992) Weed control in urban environments in New Zealand. In: Popay AJ (ed) Proceedings of the New Zealand plant protection conference. New Zealand Plant Protection Society, Rotorna, pp 231–234. https://doi.org/10.30843/nzpp.1992.45.11267
Quan NV, Tran HD, Xuan TD, Ahmad A, Dat TD, Khanh TD, Teschke R (2019a) Momilactones A and B are α-amylase and α-glucosidase inhibitors. Molecules 24(3):482. https://doi.org/10.3390/molecules24030482
Quan NV, Xuan TD, Tran HD, Dieu NT (2019b) Inhibitory activities of momilactones A, B, E, and 7-ketostigmasterol isolated from rice husk on paddy and invasive weeds. Plants 8(6):159. https://doi.org/10.3390/plants8060159
Rao VS (2000) Principles of weed science, 2nd edn. Science Publishers, Enfield
Rask AM, Kristoffersen P (2007) A review of non-chemical weed control on hard surfaces. Weed Res 47(5):370–380. https://doi.org/10.1111/j.1365-3180.2007.00579.x
Rasmussen J (1996) Mechanical weed management. In: Proceedings second International Weed Control Congress, Copenhagen, Denmark, pp 943–948
Relyea RA (2005) The lethal impacts of roundup and predatory stress on six species of North American tadpoles. Arch Environ Contam Toxicol 48:351–357. https://doi.org/10.1007/s00244-004-0086-0
Rice AR, Johnson-Maynard JL, Thill DC, Morra MJ (2007) Vegetable crop emergence and weed control following amendment with different Brassicaceae seed meals. Renewable Agric Food Syst 22(3):204–212. https://doi.org/10.1017/S1742170507001743
Rob M, Iwasaki A, Suzuki R, Suenaga K, Kato-Noguchi H (2019) Garcienone, a novel compound involved in allelopathic activity of *Garcinia Xanthochymus* hook. Plan Theory 8(9):301. https://doi.org/10.3390/plants8090301
Rosset JD, Gulden RH (2020) Cultural weed management practices shorten the critical weed-free period for soybean grown in the northern Great Plains. Weed Sci 68(1):79–91. https://doi.org/10.1017/wsc.2019.60
Ruchel Q, Zandoná RR, Fraga DS, Agostinetto D, Langaro AC (2020) Effect of high temperature and recovery from stress on crop–weed interaction. Bragantia 79(4):457–466. https://doi.org/10.1590/1678-4499.20200138

Runia WT (1983) A recent development in steam sterilization. Acta Hortic 152:195–200

Sarr PS, Ando Y, Nakamura S, Deshpande S, Subbarao GV (2020) Sorgoleone release from sorghum roots shapes the composition of nitrifying populations, total bacteria, and archaea and determines the level of nitrification. Biol Fertil Soils 56(2):145–166. https://doi.org/10.1007/s00374-019-01405-3

Shirgapure KH, Ghosh P (2020) Allelopathy a tool for sustainable weed management. Arch Curr Res Int 20(3):17–25. https://doi.org/10.9734/acri/2020/v20i330180

Singh HP, Batish DR, Kohli RK (2006) Handbook of sustainable weed management. Food Products Press, New York. https://doi.org/10.1201/9781482293593

Sonneveld C (1979) Changes in chemical properties of soil caused by steam sterilization. In: Mulder D (ed) Developments in agricultural and managed-forest ecology 6: soil disinfestation. Elsevier, New York

Steenhagen DA, Zimdahl RL (1979) Allelopathy of leafy spurge (*Euphorbia esula*). Weed Sci 27(1):1–3. https://doi.org/10.1017/S0043174500043356

Streibig JC (1979) Numerical methods illustrating the phytosociology of crops in relation to weed flora. J Appl Ecol 16(2):577–587. https://doi.org/10.2307/2402532

Suwitchayanon P, Ohno O, Suenaga K, Kato-Noguchi H (2019) Phytotoxic property of Piper retrofractum fruit extracts and compounds against the germination and seedling growth of weeds. Acta Physiol Plant 41(3):33. https://doi.org/10.1007/s11738-019-2824-y

Tang CS, Cai WF, Kohl K, Nishimoto RK (1995) Plant stress and allelopathy. In: Allelopathy, organisms, processes and applications. American Chemical Society, Washington DC, pp 142–157

Tibugari H, Chiduza C, Mashingaidze AB, Mabasa S (2019) Quantification of sorgoleone in sorghum accessions from eight southern African countries. S Afr J Plant Soil 36(1):41–50. https://doi.org/10.1080/02571862.2018.1469794

Tibugari H, Chiduza C, Mashingaidze AB (2020a) Farmer knowledge, attitude and practices on sorghum allelopathy in five sorghum producing districts of Zimbabwe. S Afr J Plant Soil 37(2):152–159. https://doi.org/10.1080/02571862.2019.1706003

Tibugari H, Chiduza C, Mashingaidze AB, Mabasa S (2020b) High sorgoleone autotoxicity in sorghum (*Sorghum bicolor* (L.) Moench) varieties that produce high sorgoleone content. S Afr J Plant Soil 37(2):160–167. https://doi.org/10.1080/02571862.2020.1711539

Travlos I, Rapti E, Gazoulis I, Kanatas P, Tataridas A, Kakabouki I, Papastylianou P (2020) The herbicidal potential of different pelargonic acid products and essential oils against several important weed species. Agronomy 10(11):1687. https://doi.org/10.3390/agronomy10111687

Upadhyaya MK, Polster DF, Klassen MJ (1993) Weed control by superheated steam. Weed Sci Soc Am Abstr 33:115

Vig AP, Rampal G, Thind TS, Arora S (2009) Bio-protective effects of glucosinolates–a review. LWT-Food Sci Technol 42(10):1561–1572. https://doi.org/10.1016/j.lwt.2009.05.023

Wang C, Zhu M, Chen X, Qu B (2011) Review on allelopathy of exotic invasive plants. Proc Eng 18:240–246. https://doi.org/10.1016/j.proeng.2011.11.038

Zerner MC, Rebetzke GJ, Gill GS (2016) Genotypic stability of weed competitive ability for bread wheat (Triticum aestivum) genotypes in multiple environments. Crop Pasture Sci 67(7):695–702. https://doi.org/10.1016/j.cropro.2018.08.001

Zimdahl RL (2007) Fundamental of weed science. Academic, Burlington

Zobel AM, Lynch JM (1997) Extrusion of UV-absorbing phenolics in Acer spp. in response to UV and freezing temperature. I. UV A-absorbing compounds on the surface of Acer saccharum and Acer platanoides autumn leaves. Allelopath J 4:269–276

Chapter 12
Electrical Impedance Spectroscopy in Plant Biology

M. Iftikhar Hussain, Ali El-Keblawy, Nosheen Akhtar, and Ahmed S. Elwakil

Abstract Electrical impedance spectroscopy (EIS) is a non-invasive approach to characterize the electrical properties such as resistance and capacitance of many materials including biological tissues. For measurement, the plasma membrane acts as an electrical insulator that controls the movement of ions and electrolytes across the cells. Under application of a voltage to a tissue, an electric current flows through cell walls, from cell to cell and in fluids, giving rise to an electrical impedance. The magnitude of the impedance and its phase angle can be measured at multiple frequencies using an analyzer. Here we review EIS theory, instrumentation, application, model validation and data assimilation for studying of physiological and biochemical changes in biological tissues. EIS thus allows to understand cellular and synthetic membranes, cell biophysics and ionic systems. Ions are the main current carriers inducing the impedance inside tissues. The symplastic and apoplastic resistances form a parallel impedance circuitry at a given frequency. EIS allows to assess physiological attributes, leaf water content, root size and fruit quality traits. Impedance dispersions are described using Nyquist graphs. Best fitting circuit parameters can be optimized.

Keywords Impedance analysis · Cole-Cole electrical model · Non-invasive measurements · Growth · Plant organs · Phytotoxicity

---

M. I. Hussain (✉)
Department of Plant Biology & Soil Science, Universidade de Vigo, Vigo, Spain

Research Institute of Science and Engineering, University of Sharjah, Sharjah, United Arab Emirates
e-mail: iftikhar@uvigo.es

A. El-Keblawy
Department of Applied Biology, University of Sharjah, Sharjah, UAE

N. Akhtar
Institute of Agricultural Sciences, University of the Punjab, Quaid-e-Azam Campus, Lahore, Pakistan

A. S. Elwakil
Department of Electrical and Computer Engineering, University of Sharjah, Sharjah, UAE

E. Lichtfouse (ed.), *Sustainable Agriculture Reviews 52*, Sustainable Agriculture Reviews 52, https://doi.org/10.1007/978-3-030-73245-5_12

Abbreviations

C	capacitance
DCE	distributed circuit element
EIS	electrical impedance spectroscopy
R	resistance
R_1	Extracellular resistance
R_2	intracellular resistance
C_1	extracellular capacitance
C_2	intracellular capacitance

12.1 Introduction

Bio-electrical measurement are non-destructive methods are gaining significant attention among the scientific community due to their global application at field and laboratory scale and capacity to provide pertinent information on plant health, biochemistry and plant physiology via measuring the electrical resistance and capacitance. Advances in electrochemical measurement and large-scale deployments allow these studies with high spatial and temporal resolution at both laboratory and field scales (Singha et al. 2014). EIS have been used to measure the physiological status of biological tissues as well as in medical sciences (Miklavčič et al. 2006). Several collaborators demonstrated EIS application in body mass index calculation (Meeuwsen et al. 2010), and nutritional and hydration status evaluation (Rösler et al. 2010). Electrical impedance has been studied to determine plant physiological status as done by Zhang and Willison (1991) who analyzed potato tubers and carrot roots using a double-shell (Cole-Cole) model. This model was applied to analyze the electrical impedance response of Scots pine needles (Zhang et al. 1995) too. Several researchers demonstrated EIS application in plant stress physiology like, damaged tissue detection in bruised apples (Jackson and Harker 2000), fruit biochemical properties, pH, sugar content, ripening (Liu 2006), willow root system (Cao et al. 2011), water status in tomato (He et al. 2011), phosphorus and potassium deficiencies detection in *Trifolium subterraneum* (Greenham et al. 1982). Moreover, there are few studies that focused to use EIS for evaluating heavy metal stress in plants.

Electrical impedance spectroscopy (EIS), measurements, are non-invasive, fast and less sensitive to environmental variables and can therefore be used to monitor a variety of physiological and biochemical processes at cell level. Since the applications of electrical impedance technique on intracellular biochemical responses have not been fully researched, the present article will demonstrate the impedance measurements on evaluating aluminium stress on electrical impedance measurements in desert shrub, *Calotropis procera* and propose new approaches for plant stress physiology determinations. If the process of plant growth and development can be monitored in a predictable manner, then the intracellular properties can be demonstrated using impedance technology. Data assimilation, selection of suitable equivalent electrical model and using appropriate impedance software to obtain the best electrical data fitting is performed in order to exploit and highlight the entire process of electrical impedance spectroscopy.

In this review, we present the latest advancements of electrical impedance estimation of biological tissue, with focus on:

- Plant cell anatomy
- Theory, instrumentation, measurements and modeling
- Application, experimentation and case studies
- Plant biotic and abiotic/biotic stress and electrical impedance spectroscopy

We discuss the status of biological tissues under abioticand biotic stress that is measured using impedance-based techniques.

12.2 Plant Cell Anatomy

The plant cell is usually consisting of cell wall (outer layer), an inner cell membrane (also known as plasmalemma). The cell membrane comprised of phospholipids that are embedded with protein layers. It controls the movement and entry of different kind of solutes and act as a semipermeable membrane, except water that can pass freely across it. The out layer (cell wall) consist of rigid cell layers and is absent in animal cells. Inside the cell, there are different cell organelles which include nucleus, Golgi bodies, ribosomes, endoplasmic reticulum, vacuole and chloroplast. The cytoplasm is the viscous fluid between the plasma membrane and cell organelles except the nucleus (Hall et al. 1974). The vacuole is surrounded by a thinner membrane known as tonoplast (Fig. 12.1).

12.3 Electrical Impedance Spectroscopy Theory

The lipid bilayer of plasma membrane behaves as a conductor or capacitor during the impedance measurements. The membrane contributes to stop the passage of AC current at low frequencies. The complete organization composed of the extracellular medium, the lipid bilayer membrane, and the intracellular medium forms the conductor–dielectric–conductor structure also known in electrical engineering as a distributed resistance-capacitance (R-C) structure. From a physiological view point; when an alternating voltage difference is applied to a plant tissue, it causes polarization and relaxation, leading to changes in the amplitude and phase of the resultant AC current signal (Repo et al. 2000) that flows in the tissue. Based on these changes the impedance of the sample, which is the ratio of the applied AC voltage to the resultant AC current, can be determined. EIS is usually applied throughout a range of frequencies to produce a spectrum of measurements. That is because the polarization of a particular component depends on the sample's tissue properties as well

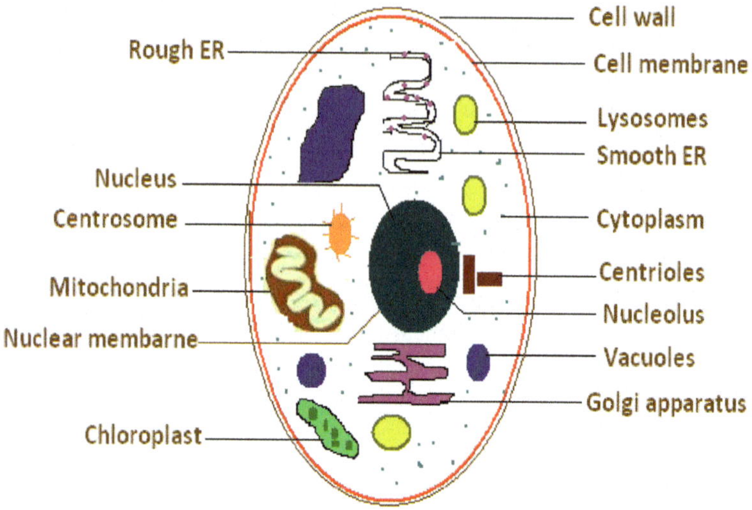

Fig. 12.1 A typical plant cell anatomy. ER: endoplasmic reticulum

as on the applied frequency. Therefore, it is usually not enough to measure the impedance at a particular fixed frequency but a complete spectrum that spans a wide range of frequencies is necessary. In biological samples, the proportion of current passing through the apoplastic and symplastic spaces in a tissue depends on AC frequency.

Biological tissues consist of cells and extracellular matrix, an intricate network of macro and small molecules, ions, and water. Both extracellular and intracellular medium can be considered as liquid electrolytes whose viscosity varies with the temperature, decreasing or increasing ions mobility (and consequently the resistance). Inside the cell, beside ions and charged molecules (protein, organic acids, etc.), there are numerous membrane structures with a completely different electrical response. Thus, the impedance of the intracellular medium must be a mixture of conductive and capacitive properties (Ivorra 2003). Cell membranes separate intracellular space from the extracellular ones contributing to make a barrier for the passage of ions and large molecules. They are formed by a lipid bilayer to which is associated proteins, transport organelles, ionic channels, and ionic pumps, the basic elements of the membrane active role. However, the electrochemical behaviour of a plant tissue which changes due to their physical structure, internal chemical properties or combination of both (Azzarello et al. 2012) reflects in it the electrical impedance of the plant. Generally, a cell represent as a capacitor because it is regarded as a dielectric, whereas intracellular and extracellular fluids can simulate a resistor (Ferreira et al. 2013). Electrical energy passing through the plant cell through ions can be dissipated by those resistors, and stored by capacitors.

Electrical impedance determines how an applied alternating-current flow generated by an external electrical field is "impeded" by biological tissues (Ferreira et al. 2013). A combination of resistive and capacitive effects contributes to alternating-current flow opposition, which is related to the impedance magnitude (mainly, due to resistive effects) and the impedance phase (mainly, due to capacitive effects) of electrical impedance. At low frequency, the current flows in the apoplastic space of the tissues where ions are the main current carriers, which determine the total impedance. At higher frequencies, the current can flow inside the cell organelles. The symplastic space becomes conductive at high frequencies and the symplastic and apoplastic resistance form a parallel impedance circuitry (Repo et al. 2000). The plant cell may be simulated by an equivalent electrical circuits that comprised of capacitors and resistors that are arranged in series or parallel. Thus, a tissue block can be presumed to be a network of cells, which contain numerous arrays of mini-circuits (Varlan and Sansen 1996).

As a general phenomenon, during the EIS measurements, an electrical voltage is applied to a plant tissue. This caused an immediate change in the cell that showed the ionic polarization and relaxation occurs after removing the electric field, according to the time constant of the polarized system. The change in relaxation time with frequency from one value to another is called dispersion, and is due to ionic movements. The principal mechanism of dispersion is the accumulation of charges on both sides of a dielectric (i.e. cell membranes). Electrical impedance can also be estimated by measuring the voltage signal developed across that sample by injecting an AC current signal to it. In this case, the impedance (Z) is obtained by dividing the voltage signal measured by the current signal applied. Mathematically, Z is complex quantity and it will have a particular phase angle that depends on the tissue properties. When the real and imaginary parts of the impedance are measured at different frequencies, an impedance spectrum is obtained (Repo et al. 2000). Typically, the complex impedance graph of a biological sample comprised of real versus imaginary part of the impedance and might show single or a double dispersion (depending upon the type and complexity of plant tissues, e.g. stems, leaves, roots, fruits etc).

12.4 Methodology, Instrumentation and Impedance Measurements

EIS studies the impedance response of an object under test by measuring its electrical impedance (Z) and its phase angle (θ) at several frequency points (ωi: i = 1, 2, 3, ...) from the voltage-current data at the object surface. In EIS, a constant amplitude sinusoidal voltage or current signal is injected to the object surface at different frequencies and the boundary current or voltage are measured at each frequency to estimate the impedance using an array of electrodes attached to the object surface. A general set-up for EIS measurements is shown in Fig. 12.2.

If the injected signal is a voltage, the setup is known as a potentiostatic setup. Here, impedance Z is the quotient of the applied voltage $v(t)$ and the resultant current $i(t)$:

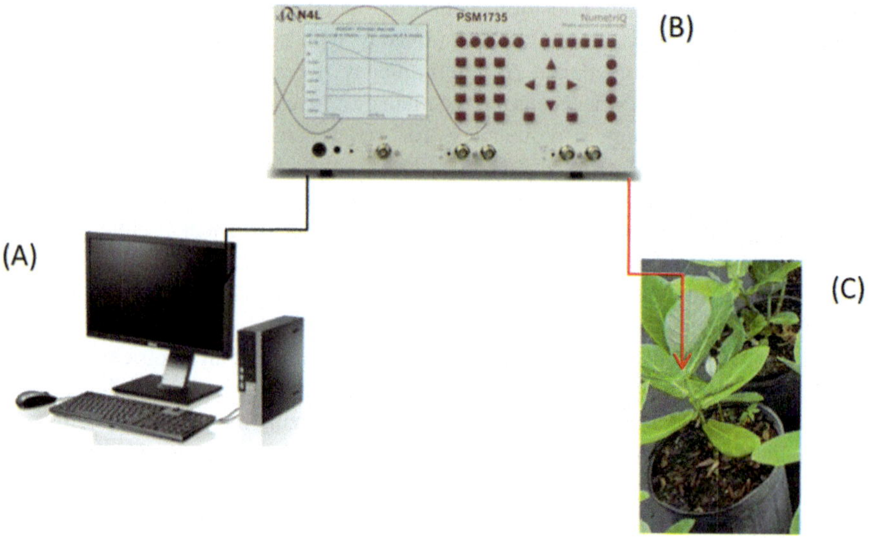

Fig. 12.2 Electrical impedance measurements setup (**a**) desktop computer, (**b**) impedance analyzer, (**c**) *Calotropis procera* plants

$$Z(t) = \frac{v(t)}{i(t)} = |Z| \angle \theta$$

where θ is the phase difference between the voltage and the current (Pänke et al. 2008).

In Cartesian coordinates, impedance becomes a complex number, constituted of two components in the following equations;

$$Z(\omega) = Z_r(\omega) + jZ_i(\omega),$$

where

$$Z_r(\omega) = |Z| \cos(\theta)$$

is the real component,

$$Z_i(\omega) = |Z| sen(\theta)$$

is the imaginary component or reactance,

$$\text{and}\,\theta = \tan^{-1} \frac{Zi}{Zr}$$

is again the impedance phase angle. The phase angle can vary from 0 to 90°. When the angle is 0 the circuit is pure resistive and at 90° it is pure capacitive, while at 45°

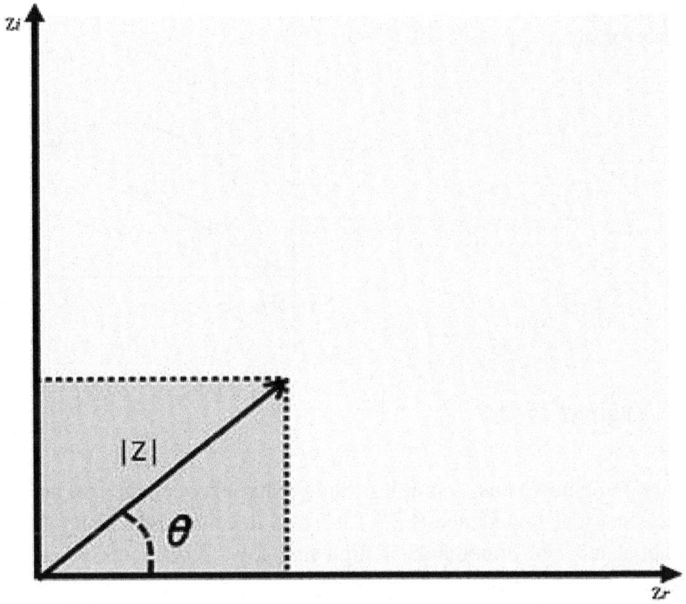

Fig. 12.3 Vector diagram showing relationship between resistance (Zr), reactance (Zi), and impedance

the circuit has an equal amount of capacitive reactance and resistance (Liedtke 1997) and this is known as the Warburg impedance.

Finally, the relation between impedance and its individual component (resistance and reactance) can be represented as a vector (see Fig. 12.3), whose magnitude is:

$$|Z(\omega)| = \sqrt{Z_r^2 + Z_i^2}.$$

The real and imaginary parts of Z describe the resistance and reactance, respectively. If the real part is plotted on the X-axis and the imaginary part on the Y-axis, an impedance spectrum, using the frequency as the parametric variable, is obtained (Fig. 12.2).

12.5 Data Collection and Processing

Once the raw data from the experiment has been obtained, it is important to extract characteristic parameters to analyze the system properties. There are different ways the data can be plotted, such as the Nyquist and Bode plane plots.

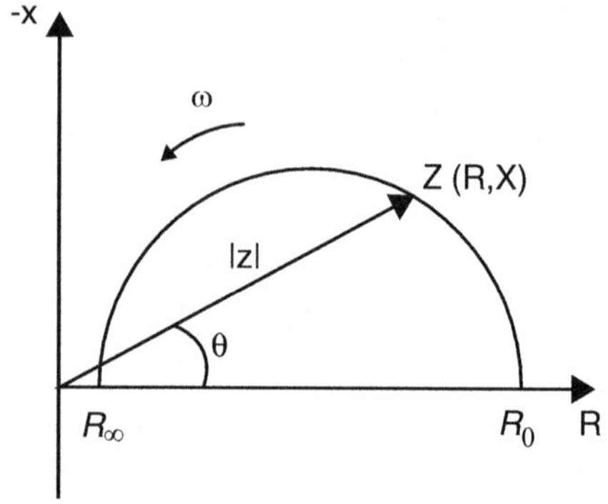

Fig. 12.4 Nyquist plot with impedance vector

12.5.1 Nyquist Plot

The Nyquist Plot showed us a complex relationship between the real and imaginary part of the electrical impedance (Fig. 12.4) and developed by Harry Nyquist. EIS measurements may be conducted at high and low frequencies where electrode-electrolyte interface represents a capacitive behavior. Consequently, the imaginary part of the impedance is always negative. In complex Nyquist graphs, the sign convention has been adopted in order to plot the data in the first quadrant (Gabrielli 1993). Generally, there is no intercepts of real axis at extremely low and high frequencies, owing to the existing equipment not being able to measure such frequencies. Therefore, the values of R_0 and R∞ are calculated by extrapolation and interpolation, respectively from the values obtained over a limited frequency range (Cornish et al. 1993). The disadvantage of this data presentation is that it cannot directly indicate the frequency at which each point on the arc was measured.

12.6 Bode Plot

The bode plot comprised of two subplots, Bode magnitude and Bode phase as developed by Hendrik Bode. The Bode magnitude showed graph of magnitude logarithm (Log |Z|) against the frequency while Bode phase represents graph of phase against the frequency (Fig. 12.5).

The Nyquist plot is a relationship between the real and imaginary parts of the total impedance at all measured frequencies where the total impedance at each individual frequency is given by;

$$Z_{tot} = Z_{\mathrm{Re}} + jZ_{Img} \tag{12.1}$$

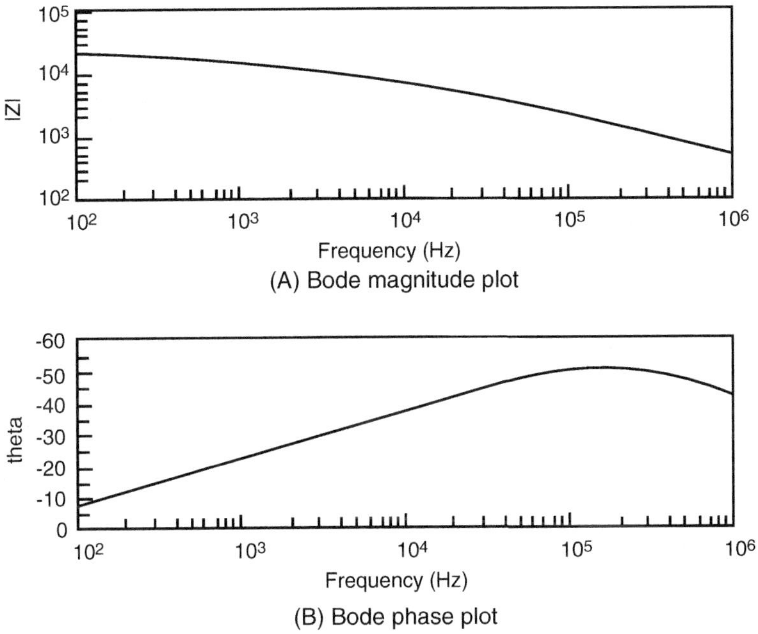

Fig. 12.5 Bode plot

The magnitude and phase angle of this impedance can be obtained as;

$$\left|Z_{tot}\right| = \sqrt{Z_{\mathrm{Re}}^2 + Z_{Img}^2}\,\Omega \tag{12.2}$$

and

$$\measuredangle Z_{tot} = \tan^{-1}\left(\frac{Z_{Img}}{Z_{\mathrm{Re}}}\right) \tag{12.3}$$

Bio-impedance analysis and modeling usually relies on the Nyquist plot which relates Z_R to $-Z_I$ both measured in units of Ω. The fact that biological tissues show capacitive (rather than inductive) behavior means that Z_I is always negative and for that reason $-Z_I$ is plotted on the Y- axis. At every individual frequency, Z_{tot} is formed of two elements connected in series: a pure resistance with value R = ZR and a pure capacitance with value in Farads given by $C = 1/(2\pi f \times Z_I)$ where "f" is the frequency in Hertz. This simple model can offer information but only at a single applied frequency, which is usually not enough and therefore EIS measurements are conducted at multiple frequencies resulting in complex impedance spectra.

Typically, the spectrum of plant tissues is composed of one or two arcs in the complex plane, depending on the sample under study and the range of frequencies used (Zhang and Willison 1992). However, it is imperative to mention here that there are several recent attempts to propose cheap and portable impedance measurement devices such as those reported in Uvanesh et al. (2015), Hoja and Lentka (2013), and Chabowski et al. (2015). However, all "direct" impedance measurement techniques suffer from a major draw-back which is the necessity of having data post-processing using a suitable software optimization (data fitting) technique. The parameters of the best fitting circuit are then estimated using an optimization technique. The freely available EIS Spectrum Analyzer software (Bondarenko and Ragoisha 2005) is an example of software that can do this.

12.6.1 Selection of an Equivalent Circuit

To analyze and interpret experimental data, it is best to have an equivalent circuit model that provides a representation of the electrical properties and represent a realistic picturc of the impedance data in the electrical makeup (Srinivas et al. 2003). In order to verify the validity of the chosen circuit, the quality of the fit to the experimental curve must be evaluated. Whenever possible, the model should be verified before being used. Srinivas et al. (2003) outlined three principles for equivalent circuit selection (i) intuition as to what kinds of impedance are expected to be present in the sample and whether they are connected in series or in parallel, (ii) examination of the experimental data to see whether the response is consistent with the proposed circuit, (iii) inspection of the resistance and capacitance values that are obtained in order to check that they are realistic for the investigated continuum. The choice of the equivalent model is crucial to understand the characteristics of the materials under investigation. The choice depends on the characteristic of the system under study and on the intuition of the researcher. There can be potential problems caused by the fact that equivalent circuits are rarely unique and several circuit models can have identical or very similar impedances (Lackermeier et al. 1999).

The selection of an equivalent model is crucial to understand the characteristics of the sample under investigation. The Cole electric models have been significantly used to monitor electrical properties of plant samples during EIS measurements (Freeborn 2013). The impedance spectra can be modeled by an equivalent circuit through well-known double- dispersion Cole model (Cole and Cole 1941); as given by Freeborn et al. (2014);

$$Z_{tot} = R_{\infty} + \frac{R_1}{1 + s^{\alpha_1} R_1 C_1} + \frac{R_2}{1 + s^{\alpha_2} R_2 C_2} \quad (12.4)$$

This model is composed of:

- $R\infty$ is the very high frequency resistance, usually very small in value.
- R_1, C_1 and $\alpha 1$ are the parameters for the first dispersion while R_2, C_2 and $\alpha 2$ are the parameters for the second dispersion.

It is important to note that $\alpha 1,2$ are unit-less and are known as the dispersion coefficients which measure the distribution of relaxation times in the material and hence its closeness to ideal capacitive behavior ($\alpha = 1$). It is also important to note that $C_{1,2}$ are pseudo capacitances.

12.7 Case Study I: Apple Quality Evaluation Via Electrical Impedance Spectroscopy

Bioimpedance has been commonly used to monitor the fruit quality through indirect measurement of physiological changes in fruits and vegetables. Several attributes of Cole model (R_0, R_1, C, and α), as an equivalent circuit are widely used to highlight the impedance properties of biological tissues. In this study, the variability of the Cole parameters extracted from magnitude-only measurements (from 200 Hz to 1 MHz) of apples in a two electrode setup was examined to understand the impact of electrode placement on the parameters evaluated using this technique (Fig. 12.6). EIS was measured from the latitudinal line of four different apple varieties (Granny Smith, Fuji, Red Delicious, and Spartan) and Cole impedance traits were extracted in MATLAB from the collected measurements using a nonlinear least squares fitting method. These extractions indicated that the parameters R0 and R1 had the highest variability based on the electrode location, whereas the dispersion coefficient (α) had the lowest variability. Further research is also needed to determine whether these parameters can be monitored as an indicator of aging or ripening, physical damage, quality assessment, or any other type of assessment that is employed in the food processing industry.

12.8 Case Study II: Monitoring of Aluminium Stress in Plant Tissue via Impedance

Al^{3+} toxicity has been described in different vegetations, trees and crops and possible adaptive mechanisms (exclusion, detoxification and even survival in Al-rich environment through the release of organic compounds and phenolic acids) was previously reported by several researchers (Haridasan 2008; Tolrà et al. 2005). Hussain et al. (2018) evaluated the Al stress response in *C. procera* through electrical impedance spectroscopy. According to their study, the complex impedance spectra (Nyquist plots) showed that the impedance of leaves and roots of *C. procera* had different electrical responses under various Al concentrations after 10 and 20 days and had different arcs (Fig. 12.7). For each Al concentration, a complete arc was obtained at the higher frequencies (above 2 kHz and up to 1 MHz), but a semi-circle was detected at lower frequencies (lower than 1 kHz down to 1 Hz). After 10 days, the arcs of the five Al concentrations of leaves at the higher frequencies were completely overlapped, but those of roots were separated; higher impedance values were obtained in the controls, with the lowest being observed at 60 mg/l (Fig. 12.7). A

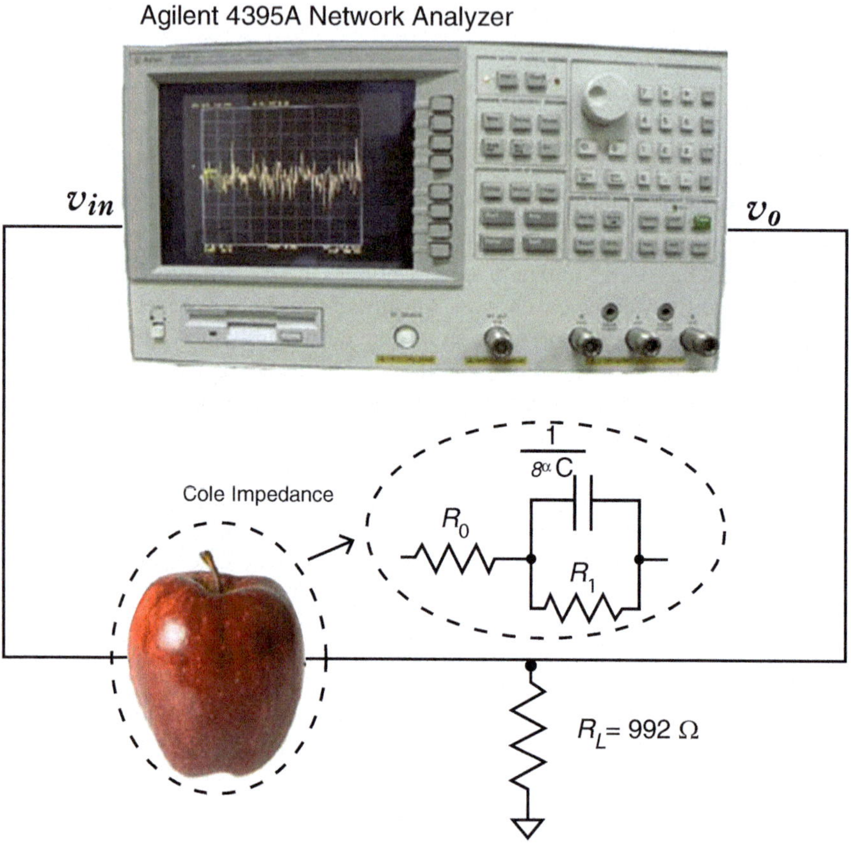

Fig. 12.6 Experimental test setup to collect magnitude response using Agilent 4395A network analyzer when an apple is used as a component in a simple filter circuit

more or less similar trend was observed at the lower frequencies. Generally, the impedance (imaginary and real) values were higher in control leaves in comparison with control roots. At the lowest frequency (1 Hz), the impedance values after 10 days were higher in controls of both leaves and roots than those treated with different Al concentrations. The impedance values after 10 days were comparable to those after 20 days, but with different trends. After 20 days, the arcs of the different concentrations in leaves were completely separated, but those of roots were overlapped (Hussain et al. 2018).

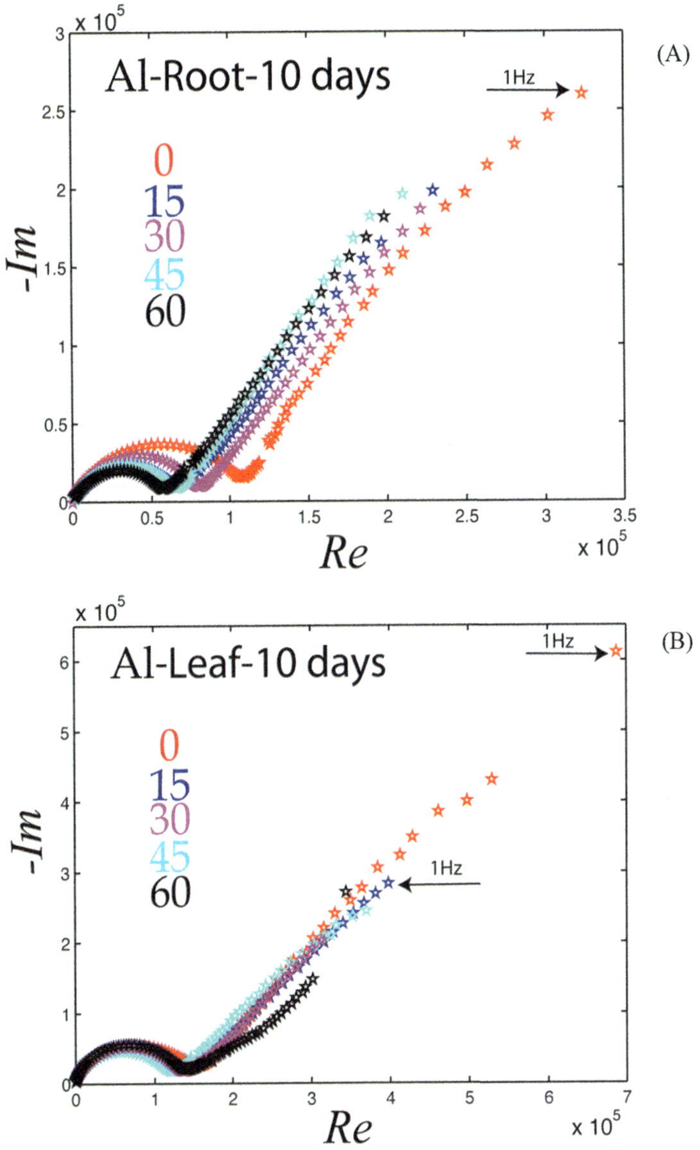

Fig. 12.7 Nyquist plots related to the incubation of *Calotropis procera* with different aluminium concentrations (0, 15, 30, 45 and 60 ppm) for 10 days and showed the impact of Al on (**a**) roots, and (**b**) leaves organ. The term "Re" refers to real part (Z) and "-Im" to imaginary part of impedance (Z'). In each curve, frequency increased from right (1 Hz) to left (1 MHz) (Source: Hussain et al. 2018)

12.9 Plant Stress and Electrical Impedance Spectroscopy

12.9.1 Plant Phenotypic Plasticity

Cao et al. (2011), used electrical impedance spectroscopy (EIS) for hydroponically raised willows (*Salix schwerinii*) to estimate the root system size and area. The lumped models parameters were correlated with the contact area of the roots and/or stems raised in the hydroponic solution. While, Dietrich et al. (2012) found that the capacitance of barley (*Hordeum vulgare*) appeared to be determined, not by the mass of their root system, but by the cross-sectional area of roots at the solution surface. They observed that capacitance was not linearly related to the mass of roots in solution when root systems were partly submerged and (ii) that excising the root below the solution surface had a negligible effect on the capacitance measured. Several researchers have reported the correlations between EIS and root mass (van Preston et al. 2004; Ozier-Lafontaine and Bajazet 2005; McBride et al. 2008; Tsukahara et al. 2009). A form of single frequency (128 Hz) measurement, the earth impedance method, was introduced to estimate the absorbing root surface area in the field (Aubrecht et al. 2006; Čermák et al. 2006). The impedance was related to the basal area over a large range of stem diameters, which was further assumed to be related to the absorbing root surface area (Čermák et al. 2006). Based on the single-frequency measurements, equivalent models were formulated using resistors in the case of willow cuttings with their root system raised in hydroponic systems (Cao et al. 2011). In that particular study it was found that the resistance decreased in relation to an increase in the contact surface area of roots using the solution. The single-frequency alternating current was used to assess the capacitance or resistance of the root system, since those attributes were assumed to provide a measurement of the active root surface area (Aubrecht et al. 2006; Čermák et al. 2006).

12.9.2 Plant Water Relations

Electrical impedance parameters for water uptake in *Schefflera arboricola* were used to measure the electromagnetic properties. Results showed that the capacitance of leaf and stem was periodically decreased by 51 and 0.8 PF/h during night and increased by 62.3 and 18 PF/h during the day respectively and its resistance increased by 3.6 KΩ/h at night and decreased by 92.3 KΩ/h during the day (Sinha and Tabib-Azar 2016). Greenham (1966) observed that electrical impedance in crown tissue of alfala was decreased during freezing. In earlier, the impedance analytical approaches were used to study the physiological status of plant cell (Zhang and Willison 1992). de León et al. (2010) exhibited that scheduled irrigation based on the trunk shrinkage signal intensity was not always possible due to the temporal changes in the reference values that occurred as trees aged.

12.9.3 Evaluation of Metal and Salt Toxicity through Impedance

In 2009 Jocsak with his colleagues used electrical impedance technique to study the effect of flood and cadmium stress on Pea roots. They concluded that EIS can be used for detecting structural changes in plants caused by environmental stresses. Moreover, sudden flooding encouraged the growth of vacuoles, which act as an electrical conductor in the cell and decreased the resistance and capacitance values (Niki and Gladish 2001).

Tavakkoli et al. (2012) worked on electro-physical response of barley to salinity under laboratory and field conditions. The results showed significant difference in physiological response because field conditions were more stressful as compare to laboratory due to the soil heterogeneity and the occurrence of different abiotic stresses. Other physical factors such as the availability of water and nutrients, air and soil temperatures and moisture content could also account for the observed differences. Hamed et al. (2016) stated a negative correlation was observed between the resistance and the hydroponic culture condition ($R = -0.9$), while it was positive between the latter and the capacitance ($R = 0.96$). The opposite correlations were observed between sand culture condition and both electrical impedance parameters. Under controlled greenhouse conditions, the electrical resistance was positively correlated with the treatment of plant with 100 mM NaCl ($R = 0.85$), while its correlation with 400 mM NaCl treatment was negative ($R = -0.53$). The differences in the electrical impedance between hydroponic (liquid) and sand (solid) culture conditions could be explained by the influence of osmotic pressure which is higher in the liquid solution. Sand-grown plants were likely to have more time to adapt to the salt concentration than plants in hydroponic systems.

12.9.4 Environmental Stresses and Impedance Measurements

Electrical impedance can be used to measure the different environmental stress. Protocols for such evaluations have been developed. Similarly, several protocols for estimating the resultant injury such as differential thermal analysis (Fiorino and Mancuso 2000), visible injury (Cannel and Sheppard 1982), chlorophyll fluorescence (Havaux 1987), electrolyte leakage (Steponkus 1984) and electrical resistance (Mancuso 2000) are currently used.

Many explorations have been directed in the field towards impedance measurements and environmental stress (Repo et al. 1995; Stout 1988; Repo et al. 1997; Mancuso and Azzarello 2002; Mancuso and Rinaldelli 1996). Repo and collaborators (2000) revealed the relationship between frost hardiness (FH) of stems and needles with equivalent circuit EIS parameters. Repo et al. in 1994 demonstrated that distributed circuit element (DCE) model was used to calculate the extracellular resistance according to the impedance spectra. However, the intracellular resistance in plants has been increased with the frost hardening due to the increased concentration of the intracellular sap and impaired intracellular ions mobility (Pauly and Schwan 1966). Equivalent circuit parameter of impedance analysis is a useful tool for studying the extracellular and intracellular resistance as well as plant physiology (Repo et al. 2004).

Another researcher Mancuso with his colleague in 2002, worked on stem impedance analysis and electrolyte leakage of leaf and shoots of ten cultivars of olive to determine heat injury. They found a significant decrease in *Z ratio*, together with an increase in electrolyte leakage, in all genotypes after heat treatment. In addition to this Jócsák et al. (2010) revealed that resistance (Rs) of roots of hydroponically grown seedlings decreased continuously to one fifth of the starting value over the period of our experiment. In a recent study of Bazihizina et al. (2015) showed the work on intra and extracellular resistance of plant against the different concentration of Ni^{2+}. They reported an increase in the intracellular resistance at higher Ni^{2+} concentration (1000 and 3000μM).

12.9.5 Plant Water Relation

Numerous researches have demonstrated the plat water relation by using electrical impedence and proved that changes occurred rapidly at root level, through rapid systemic long-distance signals; mediate stomatal behavior (Schachtman and Goodger 2008). On the contrary, several studies have shown that heavy metal caused blockage of water transport from roots to above parts of the plant (Przedpelska-Wasowicz and Wierzbicka 2011; Kholodova et al. 2011). Similarly, many researchers have studied that surplus amount of Ni^{2+} impedes photosynthesis through various mechanisms, e.g., the disruption of the photosynthetic electron transport, negative effects on the water-splitting site of PSII (Velikova et al. 2011) and the inhibition of photosynthetic pigment biosynthesis (Küpper et al. 2002). A summary of application of impedance spectroscopy (EIS) in various biological systems under different abiotic and biotic stresses was discussed in Table 12.1.

12.9.6 Fruit Injury and Ripening Characteristics

Electrical impedance spectroscopy (EIS) measurements have been widely researched to monitor physiological changes in fruits and vegetables in a nondestructive manner. Freeborn et al. (2017), reported that Cole bioimpedance model ($R0$, $R1$, C, and α) parameters were extracted using EIS techniques from four different varieties (Granny Smith, Fuji, Red Delicious, and Spartan). The Cole impedance parameters were extracted in MATLAB from the collected measurements using a nonlinear least squares fitting method. These extractions indicated that the

Table 12.1 Application of impedance spectroscopy in various biological systems under different abiotic and biotic stresses

Reference	Plant/part	Experiment	Results
Väinölä and Repo (2000)	*Rhododendron* leaves	Real and imaginary levels of impedance at 42 frequencies between 80 Hz and 1 MHz.	Electrical impedance spectroscopy (EIS) is an adequate method for studying frost hardiness, both after controlled freeze tests and without artificial freezing.
Bazihizina et al. (2015)	*P. guajava*, root	The physiological characterization in response to high nickel (Ni) concentrations	The results showed that the physiological changes indicate a good Ni^{2+} tolerance up to 300μM $NiSO_4$ suggesting a potential role for the phytostabilization of polluted soils.
Chloupek et al. (2010)	*Hordeum vulgare* L, root	Electrical capacitance on drought tolerance	They observed small root system size (RSS) barley with low grain yield and malt quality in dry environments which is genetically diverse varieties
McBride et al. (2008)	Zea mays, root	Root mass determination through root electrical capacitance (C_{root}) method.	They concluded that different maize genotypes have different root dry mass vs. C_{root} relationships. The maize plant roots at the bottom of smaller have a negative effect on the strength of the root mass-C_{root} relationship.
Preston et al. (2004)	*Populus deltoides x P. nigra*)	Root mass assessment in young hybrid polar trees (*Populus deltoides x P. nigra*) using the electrical capacitance.	EIS was used to estimates root mass for young trees in a variety of applications, such as afforestation, agroforestry.
Pitre et al. (2010)	*Salix viminalis* × *Salix schwerinii*	Measurement of leaf biomass, stem height and root biomass	EC also showed good correlations with stem and leaf biomass, as well as with stem height. EC was used to estimate the below-ground biomass in willow and may become useful in screening varieties for differences in root biomass traits.
Tavakkoli et al. (2012)	Barley	Evaluation of electro- physiological responses of plants to salinity under hydroponic and filed conditions	They showed significant difference in physiological response of plant to salinity under laboratory and field condition. Field conditions were more stressful because of the soil heterogeneity and the occurrence of different abiotic stresses.

(continued)

Table 12.1 (continued)

Reference	Plant/part	Experiment	Results
Ellouzi et al. (2011)	*C. maritima*	Adaptive mechanisms to tolerate high salinity.	They observed transient changes in electric resistance after few hours of salt stress and induced by the osmotic shock.
Mancuso and Rinaldelli (1996),	*Oleaeuropea*L. Leaves	Electrical impedance parameters	Results exhibited an evident reduction in extra-and intracellular resistance values for non-mycorrhizal plants with increased NaCl concentration
Repo et al. (2004)	*Betulapendula* clones	Effects of elevated concentrations of ozone and carbon dioxide	They determined that elevated O3 reduced both relaxation time and the extracellular resistance, indicating cell membrane damage, while elevated CO2 increased intracellular resistance, indicating changes in symplastic composition.
Pottosin et al. (2014)	Pea roots	Effect Ca2+ effluxes in plant due to stress	They determined that Ca2+ effluxes were linked to a coupled Ni2+ effect on PM H+ and Ca2+ -ATPase and caused Ca2+ extrusion.

parameters $R0$ and $R1$ had the highest variability based on the electrode location, whereas the dispersion coefficient (α) had the lowest variability.

12.10 Conclusion

EIS is a non-invasive technique that has been developed and applied to study extra-cellular and intracellular resistance and capacitance in plant tissues. A new strategy was developed based on research team expertise in EIS technique and the present review helps to understand the application of EIS to study electrochemical cross-talk's within the biological tissues. Nyquist graphs showed EIS of a particular biological tissue in terms of arcs, complete circles or semicircle at a particular range of frequencies. An improvement in this technique was achieved through using the distributed model that fits the biological tissue impedance data well compared with the double dispersion Cole-Cole model due to the complexity of plant sample. The methodology, instruments, modeling, data assimilation and their applications for plant tissues has been elaborated. EIS is a multi-frequency impedance procedure, which can give more extravagant data than single frequency impedance estimation. Along these lines, EIS can be used to evaluate single frequency for detection and recognition, and afterward, single or multiple frequencies may be employed for concluding high-speed investigations. Electrical impedance spectroscopy studies in combination with other molecular techniques (use of PCR, chlorophyll fluorescence imaging, X- ray fluorescence), allowed us to reconcile the plant stress physiology phenomena robustly.

References

Aubrecht L, Stanek Z, Koller J (2006) Electrical measurement of the absorption surface of tree roots by the earth impedance method: 1. Theory. Tree Physiol 26:1105–1112. https://doi.org/10.1093/treephys/26.9.1105

Azzarello E, Masi E, Mancuso S (2012) Electrochemical impedance spectroscopy. In: Volkov AG (ed) Plant Electrophysiology. Springer, Heidelberg, pp 205–223. https://doi.org/10.1007/978-3-642-29119-7_9

Bazihizina N, Redwan M, Taiti C, Giordano C, Monetti E, Masi E, Azzarello E, Mancuso S (2015) Root based responses account for Psidiumguajava survival at high nickel concentration. J Plant Physiol 174:137–146. https://doi.org/10.1016/j.jplph.2014.10.011

Bondarenko A, Ragoisha G (2005) Progress in Chemometrics research. Nova Science Publishers, New York. ISBN 1-59454-257-0

Cannel MGR, Sheppard LJ (1982) Seasonal changes in the frost hardiness of provenances of Piceasitchensisin Scotland. Forestry 55:137–153. https://doi.org/10.1093/forestry/55.2.137

Cao Y, Repo T, Silvennoinen R, Lehto T, Pelkonen P (2011) Analysis of the willow root system by electrical impedance spectroscopy. J Exp Bot 62:351–358. https://doi.org/10.1093/jxb/erq276

Čermák J, Ulrich R, Staněk Z, Koller J, Aubrecht L (2006) Electrical measurement of tree root absorbing surfaces by the earth impedance method: 2. Verification based on allometric relationships and root severing experiments. Tree Physiol 26(9):1113–1121. https://doi.org/10.1093/treephys/26.9.1113

Chabowski K, Piasecki T, Dzierka A, Nitsch K (2015) Simple wide frequency range impedance meter based on AD5933 integrated circuit. Metrol Measur Syst 22(1):13–24. https://doi.org/10.1515/mms-2015-0006

Chloupek O, Dostál V, Středa T, Psota V, Dvořáčková O (2010) Drought tolerance of barley varieties in relation to their rootsystem size. Plant Breed 129:630–636. https://doi.org/10.1111/j.1439-0523.2010.01801.x

Cole KS, Cole RH (1941) Dispersion and adsorption in dielectrics. Alternating current characteristics. J Chem Phy 9:341–352. https://doi.org/10.1063/1.1750906

Cornish BH, Thomas BJ, Ward LC (1993) Improved prediction of extracellular and Total body water using impedance loci generated by multiple frequency bioelectrical impedance analysis. Phys Med Biol 38:337–346. https://doi.org/10.1088/0031-9155/38/3/001

de León LBP, Ferrer P, Sánchez JRC, Molina DSI (2010) Soil capacitance sensors and stem dendrometers: useful tools for irrigation scheduling of commercial orchards? Spani J Agri Res 2:52–65. https://doi.org/10.5424/sjar/201008S2-1348

Dietrich RC, Bengough AG, Jones HG, White PJ (2012) A new physical interpretation of plant root capacitance. J Exp Bot 63(17):6149–6159. https://doi.org/10.1093/jxb/ers264

Ellouzi H, Ben Hamed K, Cela J, Munné-Bosch S, Abdelly C (2011) Early effects of salt stress on the physiological and oxidative status of Cakilemaritima (halophyte) and Arabidopsis thaliana (glycophyte). Physiol Plant 142(2):128–143. https://doi.org/10.1111/j.1399-3054.2011.01450.x

Ferreira E,Sequeira M, Cortez A, Pereira H, Pereira T, Almeida V, Cardoso J, Correia C, Vasconcelos T, Duarte I, Nazaré N (2013) Assessment of physiological states of plants in situ-an innovative approach to the use of electrical impedance spectroscopy. In: Proceeding of fifth international conference on Bio informBio comp Syst Biotech (BIOTECHNO), Lisbon, Portugal, pp 1–8. ISBN: 978-1-61208-260-8

Fiorino P, Mancuso S (2000) Differential thermal analysis, deep supercooling and cell viability in organs of Oleaeuropaeaat subzero temperatures. Adv Hort Sci 14:23–27. https://www.jstor.org/stable/42882052

Freeborn T (2013) A survey of fractional-order circuit models for biology and biomedicine. IEEE J Emerg Sel Top Circuits Syst USA 3(3):416–424. https://doi.org/10.1109/JETCAS.2013.2265797

Freeborn TJ, Maundy B, Elwakil AS (2014) Extracting the parameters of the double-dispersion Cole bioimpedance model from magnitude response measurements. Med Biol Eng Comput 52(9):749–758. https://doi.org/10.1007/s11517-014-1175-5

Freeborn TJ, Elwakil AS, Maundy B (2017) Variability of Cole-model bioimpedance parameters using magnitude-only measurements of apples from a two-electrode configuration. Int J Food Prop 20(sup1):S507–S519. https://doi.org/10.1080/10942912.2017.1300810

Gabrielli C (1993) Use and applications of electrochemical impedance techniques. Schlumberger Technical Report, England

Greenham CG (1966) Bruise and pressure injury in apple fruits. J Exp Bot 17:404–409. https://doi.org/10.1093/jxb/17.2.404

Greenham CG, Randell PJ, Muller WJ (1982) Studies of phosphorus and potassium deficiencies in *Trifolium subterraneum* based on electrical measurements. Can J Bot 60:634–644. https://doi.org/10.1139/b82-084

Hall JL, Flowers TJ, Robert RM (1974) Plant cell structure and metabolism, 1st edn. Longman Group Limited, London. https://doi.org/10.1111/j.1365-2621.1984.tb13243.x

Hamed KB, Zorrig W, Hamzaoui AH (2016) Electrical impedance spectroscopy: A tool to investigate the responses of one halophyte to different growth and stress conditions. Comput Electron Agric 123:376–383. https://doi.org/10.1016/j.compag.2016.03.006

Haridasan M (2008) Adaptações de plantas nativas do bioma Cerrado em solos ácidos. Braz J Plant Physiol 20:183–195. https://doi.org/10.1590/S1677-04202008000300003

Havaux M (1987) Effects of chilling on the redox state of the primary electron acceptor QA of photosytem II in chilling-sensitive and resistant plant species. Plant Physiol Biochem 25:735–743. ISSN: 0981-9428

He JX, Wang ZY, Shi YL, Qin Y, Zhao DJ, Huang L (2011) A prototype portable system for bioelectrical impedance spectroscopy. Sens Lett 9:1151–1156. https://doi.org/10.1166/sl.2011.1384

Hoja J, Lentka G (2013) A family of new generation miniaturized impedance analyzers for technical object diagnostics. Metrol Measur Syst 20(1):43–52. https://doi.org/10.2478/mms-2013-0004

Hussain MI, El-Keblawy A, Elwakil AS (2018) Aluminum influence on *Calotropis procera* seedling growth, nutrient accumulation and electrochemical attributes. Flora 248:34–42. https://doi.org/10.1016/j.flora.2018.08.012

Ivorra A (2003) Bioimpedance monitoring for physicians: an overview. Centre Nacional de Microelectronica, Biomed App Grouppp 1–35

Jackson PJ, Harker FR (2000) Apple bruise detection by electrical impedance measurement. Hort Sci 35:104–107. https://doi.org/10.21273/HORTSCI.35.1.104

Jócsák I, Droppa M, Horváth G, Bóka K, Vozáry E (2010) Cadmium-and flood-induced anoxia stress in pea roots measured by electrical impedance. Zeit Natur C 65(1–2):95–102. https://doi.org/10.1515/znc-2010-1-216

Kholodova V, Volkov K, Abdeyeva A, Kuznetsov V (2011) Water status in *Mesembryanthemum crystallinum* under heavy metal stress. Env Exp Bot 71(3):382–389. https://doi.org/10.1016/j.envexpbot.2011.02.007

Küpper H, Šetlík I, Spiller M, Küpper FC, Prášil O (2002) Heavy metal-induced inhibition of photosynthesis: targets of in vivo heavy metal chlorophyll formation. J Phycol 38(3):429–441. https://doi.org/10.1046/j.1529-8817.2002.01148.x

Lackermeier AH, McAdams ET, Moss GP, Woolfson AD (1999) In vivo ac impedance spectroscopy of human skin: theory and problems in monitoring of passive percutaneous drug delivery. Ann N Y Acad Sci 873:197–213. https://doi.org/10.1111/j.1749-6632.1999.tb09468.x

Liedtke RJ (1997) Principles of bioelectrical impedance analysis. RJL Systems Internet site. Detroit, MI
Liu X (2006) Electrical impedance spectroscopy applied in plant physiology studies. M.Sc. thesis, RMIT University, Melbourne, Australia. https://doi.org/10.3390/ijerph17030759
Mancuso S (2000) Electrical resistance changes during exposure to low temperature and freezing measure chilling tolerance in olive tree (*Olea europaea* L.) plants. Plant Cell Env 23:291–299. https://doi.org/10.1046/j.1365-3040.2000.00540.x
Mancuso S, Azzarello E (2002) Heat tolerance in olive. Advan Hort Sci 16:125–130. https://doi.org/10.1400/14109
Mancuso S, Rinaldelli E (1996) Response of young mycorrhizal and non-mycorrhizal plants of Olive tree (*Olea europea* L.) to saline conditions. II. Dynamics of electrical impedance parameters of shoots and leaves. Advan Hort Sci 10:135–145. https://www.jstor.org/stable/42881418
McBride R, Candido M, Ferguson J (2008) Estimating root mass in maize genotypes using the electrical capacitance method. Arch Agron Soil Sci 54:215–226. https://doi.org/10.1080/03650340701790658
Meeuwsen S, Horgan GW, Elia M (2010) The relationship between BMI and percent body fat, measured by bioelectrical impedance, in a large adult sample is curvilinear and influenced by age and sex. Clin Nutr 29:560–566. https://doi.org/10.1016/j.clnu.2009.12.011
Miklavčič D, Pavšelj N, Hart FX (2006) Electric properties of tissues. In: Wiley encyclopedia of biomedical engineering. Wiley, Hoboken. https://doi.org/10.1002/9780471740360.ebs0403
Niki T, Gladish DK (2001) Changes in growth and structure of pea primary roots (*Pisum sativum* L. cv. Alaska) as a result of sudden flooding. Plant Cell Physiol 42:694–702. https://doi.org/10.1093/pcp/pce086
Ozier-Lafontaine H, Bajazet T (2005) Analysis of root growth by impedance spectroscopy (EIS). Plant Soil 277:299–313. https://doi.org/10.1007/s11104-005-7531-3
Pänke O, Balkenhohl T, Kafka J, Schäfer D, Lisdat F (2008) Impedance spectroscopy and biosensing. Adv Biochem Eng Biotechnol 109:195–237. https://doi.org/10.1007/10_2007_081
Pauly H, Schwan H (1966) Dielectric properties and ion mobility in erythrocytes. Biophys J 6:621–639. https://doi.org/10.1016/S0006-3495(66)86682-1
Pitre FE, Brereton NJB, Audoire S, Richter GM, Shield I, Karp A (2010) Estimating root biomass in Salix viminalis × Salix schwerinii cultivar "Olof" using the electrical capacitance method. Plant Biosyst 144(2):479–483. https://doi.org/10.1080/11263501003732092
Pottosin I, Velarde-Buendia AM, Bose J, Fuglsang AT, Shabala S (2014) Polyamines cause plasma membrane depolarization, activate Ca2+−, and modulate H+-ATPase pump activity in pea roots. J Exp Bot 65(9):2463–2472. https://doi.org/10.1093/jxb/eru133
Preston GM, McBride RA, Bryan J, Candido M (2004) Estimating root mass in young hybrid poplar trees using the electrical capacitance method. Agrofor Syst 60:305–309. https://doi.org/10.1023/B:AGFO.0000024439.41932.e2
Przedpelska-Wasowicz EM, Wierzbicka M (2011) Gating of aquaporins by heavy metals in *Allium cepa* L. epidermal cells. Protoplasma 248(4):663–671. https://doi.org/10.1007/s00709-010-0222-9
Repo T, Zhang MIN, Ryyppö A, Vapaavuori E, Sutinen S (1994) Effects of freeze-thaw injury on parameters of distributed electrical circuits of stems and needles of Scots pine seedlings at different stages of acclimation. J Exp Bot 45:823–833. https://doi.org/10.1093/jxb/45.6.823
Repo T, Leinonen M, Pääkkönen T (1995) The application of electrical impedance spectroscopy in assessing the frost hardiness of Scots pine. In: Gersing E, Schafer M (eds) Proceedings of 9th international conference electrical bio-impedance. ICPRBI, Heidelberg, pp 370–373
Repo T, Hiekkala P, Hietala T, Tahavanainen L (1997) Intracellular resistance correlates with initial stage of frost hardening in willow (Salix viminalis). Physiol Plant 101:627–634. https://doi.org/10.1111/j.1399-3054.1997.tb01047.x
Repo T, Zhang G, Ryyppö A, Rikala R (2000) The electrical impedance spectroscopy of Scots pine (*Pinus sylvestris* L.) shoots in relation to cold acclimation. J Exp Bot 51:2095–2107. https://doi.org/10.1093/jexbot/51.353.2095

Repo T, Oksanen E, Vapaavuori E (2004) Effects of elevated concentration of ozone and carbon dioxide on the electrical impedance of leaves of silver birch (*Betula pendula*) clones. Tree Physiol 24:833–843. https://doi.org/10.1093/treephys/24.7.833

Rösler A, Lehmann F, Krause T, Wirth R, von Renteln-Kruse W (2010) Nutritional and hydration status in elderly subjects: clinical rating versus bioimpedance analysis. Arch Gerontol Geriatr 50:e81–e85. https://doi.org/10.1016/j.archger.2009.06.007

Schachtman DP, Goodger JQ (2008) Chemical root to shoot signaling under drought. Tren Plant Sci 13(6):281–287. https://doi.org/10.1016/j.tplants.2008.04.003

Singha K, Day-Lewis F, Johnson T, Slater L (2014) Advances in interpretation of subsurface processes with timelapse electrical imaging. Hydrol Process 29:1549–1576. https://doi.org/10.1002/hyp.10280

Sinha K, Tabib-Azar M (2016) Effect of light and water on Schefflera plant electrical properties. J Sci Res Rep:1–11. https://doi.org/10.9734/JSRR/2016/19174

Srinivas K, Sarah P, Suryanarayana SV (2003) Impedance spectroscopy study of polycrystalline $Bi_6Fe_2Ti_3O_{18}$. Bull Mat Sci 26:247–253. https://doi.org/10.1007/BF02707799

Steponkus PL (1984) Role of the plasma membrane in freezing injury and cold acclimation. Ann Rev Plant Phyiol Plant Mol Biol 35:543–584. https://doi.org/10.1146/annurev.pp.35.060184.002551

Stout DG (1988) Effect of cold acclimation on bulk tissue electrical impedance I. measurements with Birdsfoot trefoil at subfreezing temperatures. Plant Physiol 86:275–282. https://doi.org/10.1104/pp.86.1.283

Tavakkoli E, Fatehi F, Rengasamy P, McDonald GK (2012) A comparison of hydroponic and soil-based screening methods to identify salt tolerance in the field in barley. J Exp Bot 63(10):3853–3867. https://doi.org/10.1093/jxb/ers085

Tolrà RP, Poschenrieder C, Luppi B, Barceló J (2005) Aluminium induced changes in the profiles of both organic acids and phenolic substances underlie Al tolerance in Rumex acetosa L. Env Exp Bot 54:231–238. https://doi.org/10.1016/j.envexpbot.2004.07.006

Tsukahara K, Yamane K, Yamaki Y, Honjo H (2009) A nondestructive method for estimating the root mass of young peach trees after root pruning using electrical capacitance measurements. J Agri Meteorol 65:209–213. https://doi.org/10.2480/agrmet.65.2.6

Uvanesh K, Das S, Champaty B, Singh V, Das K, Ghosh P, Kar J, Anis A, Al-Zahrani S, Pal K (2015) Development of a continuous impedance monitoring system: a study to monitor d-glucose concentration. Int J Electrochem Sci 10:112–120

Väinölä A, Repo T (2000) Impedance spectroscopy in frost hardiness evaluation of Rhododendron leaves. Ann Bot 86:799–805. https://doi.org/10.1006/anbo.2000.1237

Varlan AR, Sansen W (1996) Nondestructive electrical impedance analysis in fruit: normal ripening and injures characterization. Electro Magneto Biol 15:213–227. https://doi.org/10.3109/15368379609012878

Velikova V, Tsonev T, Loreto F, and Centritto M (2011) Changes in photosynthesis, mesophyll conductance to CO_2, and isoprenoid emissions in *Populus nigra* plants exposed to excess nickel. Environmental Pollution 159(5): 1058–1066. https://doi.org/10.1016/j.envpol.2010.10.032

Zhang MIN, Willison JHM (1991) Electrical impedance analysis in plant tissues 11. J Exp Bot 42:1465–1475. https://doi.org/10.1093/jxb/42.11.1465

Zhang MIN, Willison JHM (1992) Electrical impedance analysis in plant tissues: the effect of freeze-thaw injury on the electrical properties of potato tuber and carrot root tissues. Can J Plant Sci 72:545–553. https://doi.org/10.4141/cjps92-068

Zhang MIN, Repo T, Willison JHM, Sutinen S (1995) Electrical impedance analysis in plant tissues: on the biological meaning of Cole-Cole α in scots pine needles. Eur Biophys J 24:99–106. https://doi.org/10.1007/BF00211405

Chapter 13
Non-symbiotic Bacteria for Soil Nitrogen Fortification

Yash Flora, Preeti Rabha, Anushka Shinde, Pamela Jha, and Renitta Jobby

Abstract Although dinitrogen N_2 is the most abundant compound of atmosphere, plants are unable to feed directly on dinitrogen, and thus require N-fixing microbes to transform N_2 into ammonia NH_3. On one hand, some plants such as legumes have established a symbiotic relation with soil microbes to harness nitrogen. On the other hand, non-leguminous plants associate with non-symbiotic nitrogen fixers and utilize nitrates present in the soil. Experiments reveal their capability to fix 0.012–0.02 g of nitrogen per g of carbon source consumed in vitro. This nitrogen-fixing cycle provides environmentally benefits such as reducing nitrogen-based greenhouse gas emissions and N leaching to groundwater. Here we review how non-symbiotic nitrogen fixers fix soil nitrogen, with focus on catalysis by the molybdenum-iron protein and involved genes. Plant growth-promoting effects of major nitrogen fixers such as *Azospirillum, Azotobacter, Cyanobacteria, and Beijerinckia* are presented. The role of vitamins, enzymes and, indole acetic acid and gibberellic acid is described.

Keywords Biological nitrogen fixation · Non-symbiotic nitrogen fixers · *Azotobacter* · *Azospirillum* · *Cyanobacteria* · *Beijerinckia* · Nitrogenase · Dinitrogen · Nitrogen cycle · Diazotrophs · PGPR

Yash Flora, Preeti Rabha and Anushka Shinde contributed equally with all other contributors.

Y. Flora · P. Rabha · A. Shinde · P. Jha · R. Jobby (✉)
Amity Institute of Biotechnology (AIB), Amity University Maharashtra, Navi Mumbai, Maharashtra, India
e-mail: rjobby@mum.amity.edu

E. Lichtfouse (ed.), *Sustainable Agriculture Reviews 52*, Sustainable Agriculture Reviews 52, https://doi.org/10.1007/978-3-030-73245-5_13

Abbreviations

mg	milligram
kg	kilogram
BNF	Biological Nitrogen Fixation
ATP	Adenosine triphosphate
ADP	Adenosine diphosphate
kDa	kilodalton
g	gram
PGPR	Plant Growth Promoting Rhizobacteria
y^{-1}	per year
µm	micron
N per ha	Nitrogen per hectare
MoFe protein	Molybdenum-Iron protein
Tg	Teragram
PSB	Phosphate Solubilising Bacteria

13.1 Introduction

Nitrogen is an essential element for the synthesis of proteins, nucleic acids and other organic compounds for all organisms. Daniel Rutherford and Scheele and Cavendish discovered nitrogen independently in 1772. The origins of the name "nitrogen" are the Greek words "nitron genes" meaning "nitre" and "forming". Nitrogen (N) is a Group 15 element in the Periodic table. Its atomic number is 7, and atomic weight is 14.00674. There are two naturally occurring isotopes: ^{14}N (99.632 atom %) and ^{15}N (0.368 atom %) (Cartigny and Busigny 2017).

Nitrogen is the most limiting element for the growth of plant and productivity in both aquatic and terrestrial ecosystems (Wagner 2011). Although dinitrogen is a bountiful element in the atmosphere, it is biochemically unavailable for plants and most microbes as they utilize only reduced or oxidized forms of nitrogen. The two atoms in dinitrogen are triple-bonded, requiring a high level of energy to dissociate and reduce to ammonia (Figg et al. 2012). Several biological systems are able to convert dinitrogen into other useful forms of reactive nitrogen, primarily, nitrite, and nitrate, and to incorporate ammonia into organic compounds, mainly, amino acids (Galloway 1998; Dahal 2016). Almost all living beings such as the plants, animals and many prokaryotes depend on the supply of fixed nitrogen for their nitrogen requirement. For all living organisms N is an essential nutrient and hence nitrogen fixation is an important process in the nitrogen cycle (Sessitsch et al. 2002). About 15,000 mg kg^{-1}, average concentration of N is required for the adequate growth of the plant shoot, which is the highest as compared to the other nutrients (Marschner 1995).

Nitrogen fixation is not a biological process exclusively. There are two natural processes for the fixation of atmospheric nitrogen present in the biosphere: lightning and biological nitrogen fixation. Nitrogen fixation by abiotic means can occur either by lightning or fires that results in oxidation of N_2 to nitrate (NO_3^-). The NO_3^- thus formed are deposited in the terrestrial ecosystem as they are washed out from the atmosphere by precipitation. Lightning makes about 1% ammonia of the net nitrogen fixed per year (Igarashi and Seefeldt 2003). Biological nitrogen fixation is carried out by a group of microorganisms with sporadic distribution among the bacterial and archaeal domains (Vitousek et al. 1997) and fixes about 50% of the total nitrogen per year (Igarashi and Seefeldt 2003). This chapter talks in brief on nitrogen cycle and the process of biological nitrogen fixation. The focus of the chapter is on the non-symbiotic nitrogen fixing bacteria. It includes a detailed study of major nitrogen fixing bacteriasuch as *Azotobacter, Azospirillum, Cyanobacteria, Beijerinckia* and *Drexia*. In addition the mechanism and the genetic mechanism involved by the bacteria in order to fix the atmospheric nitrogen is also explained in this chapter. Lastly, the application of these nitrogen fixers and their effects seen on plants is described.

13.2 Nitrogen Cycle and Biological Nitrogen Fixation

The process of conversion of dinitrogen (N_2) to non-gaseous nitrogen ammonium compounds (NH_4^+) is referred to as biological fixation of nitrogen. In order to fix nitrogen an enzymatic reduction of N_2 to ammonia (NH_3), mostly as ammonium (NH_4^+) in solution is required. Ammonia or ammonium is a very crucial molecule for the biosynthesis of amino acids and other nitrogen containing biomolecules. With ongoing research till date the current data suggests that the only organisms able to perform the biological nitrogen fixation are the prokaryotes. And hence, N-fixing prokaryotes are the source for the nitrogen supply to all the eukaryotes.

There are various processes that involve the removal of nitrogen from the biosphere and especially from the soil. A few examples include the combustion and the biological denitrification process, the conversion of NO_3^- into gaseous forms of nitrogen such as N_2O, NO and N_2, resulting in the return of nitrogen to the atmosphere. The flow of nitrogen into the hydrosphere either in a soluble form or a particulate form is because of leaching, erosion or sedimentation that results in the transportation of nitrogen to inaccessible areas. The importance of biological nitrogen fixation in a specific ecosystem depends on the nutrient status of the systems, which in turn is linked to the development stage of the ecosystem (Bürgmann et al. 2003; Giller and Day 1985).

Phototrophic diazotrophs are basically bacteria and archaea that have the ability to fix nitrogen gas into a more useful form such as ammonia and obtain their energy from sunlight to synthesize these organic compounds (Puri et al. 2015). These can be important colonizers of oligotrophic soil surfaces in early succession stages of soil development. Heterotrophs that are free-living resulting in biological nitrogen fixation could help in high accumulation of carbon, e.g. forest soils with high litter inputs and decaying plant debris (Giller and Day 1985). The biological nitrogen fixation won't have a remarkable contribution to the nitrogen budget in places where the system is enriched with high mineral nitrogen concentrations or other unfavorable conditions (Bürgmann et al. 2003).

Fixed nitrogen availability in an ecosystem does not only depend on nitrogen inputs and outputs, but also on nitrogen bound mineralization (ammonization) in biomass and organic material. In fact, internal fluxes are greater than inputs and outputs in order of magnitude (Sandaa et al. 1998). Plant uptake is a major sink for mineral nitrogen in most terrestrial systems (Miller et al. 1999; Sandaa et al. 1998). Plants take up both NH_4^+ and NO_3^-, but most of the plants usually uptake one of these atop the other (Pesaro et al. 2003). The cation NH_4^+ in most soils is immobile and thus can be associated with sites of cation exchange or can be irreversibly fixed between the interlayer sites of clay minerals (Weaver and Graham 1994).

Under aerobic conditions, nitrifying microorganisms (ammonium oxidizers and nitrite oxidizers) will quickly transform mineral nitrogen into NO_3^-. NO_3^- as compared to NH_4^+ is much more mobile in soil and can percolate into groundwater (NO_3^- leaching) if plants or microorganisms do not use it. Groundwater used as drinking water is contaminated because of the nitrate accumulation resulting in a health risk. The nitrogen lost because of denitrification is restored and compensated because of nitrogen fixation which is a crucial step in the global nitrogen cycle (Dixon and Kahn 2004). The element nitrogen is an integral part of proteins and nucleic acids that are present in all living organisms. The conversion of NO_3 to N_2 results in the completion of nitrogen cycle by returning the nitrogen to the atmosphere, this is an anaerobic process described as denitrification, the detailed step of which are depicted in Fig. 13.1 (Weaver and Graham 1994).

Nitrogen exists in either the reduced or oxidized forms in the global nitrogen cycle. The conversion of ammonia into oxides of nitrogen is called nitrification, which is undertaken by nitrifying bacteria like *Nitrosomonas* and *Nitrobacter* which is a reversible reaction. The reductive conversion of nitrogen oxides back to dinitrogen is called denitrification and is undertaken by denitrifying bacteria like *Thiobacillus denitrificans*, and *Micrococcus denitrificans*.

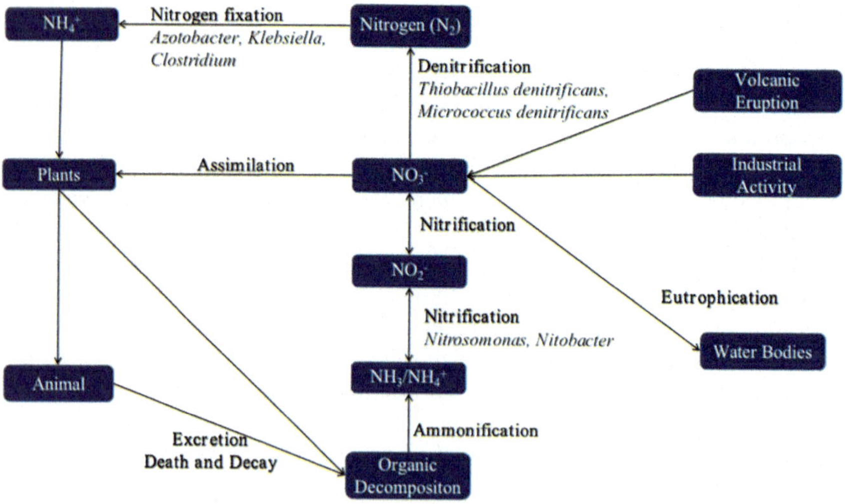

Fig. 13.1 Biological nitrogen fixation and the various biological and non-biological contributors

13.3 Non-symbiotic Nitrogen Fixation

Berthelot in 1888 clearly indicated the potential of bacteria in the soil to fix atmospheric nitrogen (N_2) when he found that unsterilized soil had acquired nitrogen while sterile soil had not. Berthelot reasoned on the basis of logical speculation that some free-living soil microorganisms were responsible (Keneddy and Islam 2001). The organisms that live an independent life without any association with other living organisms but are also able to fix atmospheric or soil nitrogen are called non-symbiotic organisms (Kumar 2012). Some of the free-living nitrogen-fixing organisms derive energy directly from sunlight, while others derive energy from organic soil (Hillel 2008).

Fixing the atmospheric nitrogen gas by heterotrophic diazotrophs such as bacteria and archaea results in obtaining a usable substance, ammonia, but depends on organic substances for their energy source such as on carbon source, e.g. straw and the most common soil isolates include (*Azotobacter, Desulfovibrio* and *Desulfotomaculum Azomonas, Beijerinckia* and *Derxia*, Enterobacter and *Klebsiella, Clostridium* and *Bacillus, Azospirillum*) (Havelka et al. 1982; Roger and Watanabe 1986). Whereas the organisms that derive their energy using the process of photosynthesis are termed as autotrophic bacteria (Havelka et al. 1982; Roper and Ladha 1995).

Generally, bacteria and cyanobacteria are considered non-symbiotic nitrogen fixers. Based on their nutrition they are classified as aerobic (*Azotobacter*), anaerobic (*Clostridium*), facultative anaerobic (*Enterobacter*), photosynthetic (*Rhodospirillum*) (Kumar 2012).

13.4 Mechanism of Nitrogen Fixation

Biological fixation of nitrogen consists of numerous regulatory and functional products of genes (Zehr et al. 1996). The breakdown of N_2 is catalyzed by MoFe protein (molybdenum-iron protein, NifDK) or the nitrogenase, and Fe protein (iron protein, NifH) or the nitrogenase reductase. The MoFe protein clusters of molybdenum-iron-sulfur-homocitrate are the actual sites where the N_2 substrate and other substrates such as protons, acetylene are binding and reduced (Sessitsch et al. 2002). The Fe protein is responsible for shutting electrons with at least two MgATPs per electron to the MoFe protein (Hamelin et al. 2002) (Fig. 13.2). The complete process can be written as follows:

$$N_2 + 10H^+ + 8e^- + nMgATP -$$
$$2NH_4^+ + H_2 + nMgADP + nPi\left(\text{Chakraborty and Samaddar 1995}\right)$$

The energy required in N_2-fixation is very high, consuming at least 8 mol ATP per mol NH_4^+ produced. Under natural conditions, the ratio is probably significantly

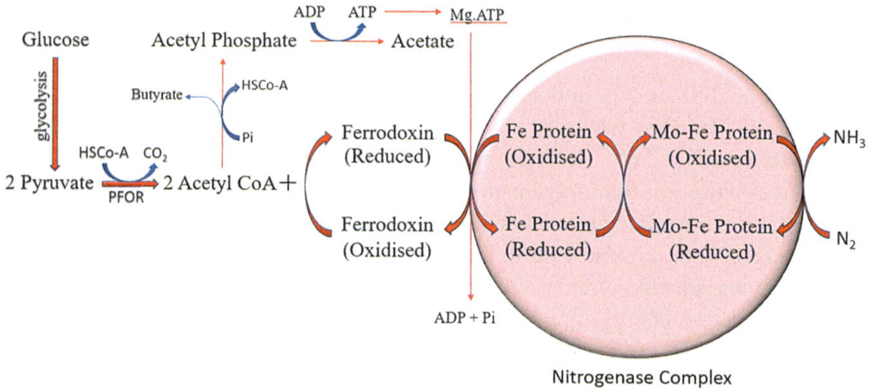

Fig. 13.2 Mechanism of nitrogen fixation, where: *PFOR* pyruvate ferrodoxin oxido-reductase, *HSCo-A* coenzyme A (no acyl group), *ADP* adenosine diphosphate, *ATP* adenosine triphosphate, *Pi* inorganic phosphate

higher, as indicated by in situ studies (Hubner et al. 1993). Proton reduction is an obligate part of the process, with a minimum production of 1 mol H_2 per mol N_2 reduced. This is again an optimum ratio that will often be higher in situ, thus further increasing the ATP and reducing equivalent demand for the reaction (Hubner et al. 1993).

The number of genes involved in biological nitrogen fixation varies from organisms to organisms. NifDK-encoded MoFe protein has a tetrameric structure ($\alpha 2\beta 2$) whereas *nifH* gene product of Fe protein has a homo-dimer structure ($\alpha 2$) (Halbleib and Ludden 2000). The molecular weight of Mo-Fe protein is approximately 250 kDa and the molecular weight of Fe protein is about 65 kDa (Christiansen et al. 2001).

These genes comprise the nif regulation, together with accessory genes and regulatory genes which codes for enzymes required in metal cluster synthesis and electron transfer (Chakraborty and Samaddar 1995). The *nifD* gene and *nifK* gene are usually part of the same operon. These genes are arranged as *nifDK* and *nifH* is also included frequently in the same operon as *nifHDK*. The nitrogenase metal clusters (FeMo-co) biosynthesis requires the *nifE* and *nifN* genes (Doolittle 1999).

13.5 Genetic Mechanism of Nitrogen Fixation

The earliest studies of nitrogenase were done on an enzyme isolated from *Clostridium pasteurianum* instead of the nif genes present in *Klebsiella pneumoniae* has shown better efficiency of nitrogen fixation. The nitrogenase assembly requires the products of nif genes which are involved in the development of Fe protein, the synthesis of FeMo cofactor (FeMo-co) and FeS clusters formation. For the synthesis of Fe-Mo cofactor several genes are required including *nifQ, nifB, nifV, nifS, nifE, nifN, nifU, nifX* and *nifY.* The assemblage of Fe-S clusters is done by gene nifS and nifU (Hu and Fay 2007). For the maturation of Fe protein *nifM, nifS, nifU* and *nifH* genes and their products are required. The products of *nifM* and *nifS* gene are involved in the synthesis of *nifH* gene product (Roberts *et al.* 1978). *nifE* and *nifN* gene products are required for the biosynthesis of FeMo cofactor. The product of *nifB* gene acts as a precursor of FeMo-co containing iron and sulfur and the product of *nifQ* gene acts as a precursor for FeMo-co containing molybdenum-sulfur. Homocitrate synthase is the product of *nifV* genes and is needed for the FeMo-co synthesis. *nifW* gene is responsible for stabilizing dinitrogenase and preventing oxygen inactivation of proteins (Cheng 2008). Nitrogen fixation involves electron transfer from Mo-Fe protein to Fe protein and hydrolysis of MgATP to MgADP by component II is required and the *nifJ* and *nifF* genes are involved in the transfer of electrons from pyruvate to flavodoxin and flavodoxin to nitrogenase respectively (Hill and Kavanagh 1980). The product of *nifF* is flavodoxin and transports an electron to nitrogenase and the product of *nifJ* is pyruvate oxidoreductase and transports an electron from pyruvate to flavodoxin. The *nifA* gene requires a positive regulatory protein for the activation of transcription of the other genes while the nifL gene is responsible for inactivation of nitrogenase (Beringer and Hirsch 1984).

In *K. Pneumoniae,* the *nif* mutant genes are found between the genes required for Shikimic acid uptake (shi A) and histidine (his) synthesis. The size of the *nif* gene of *K. Pneumoniae* is 24 kb which contains 20 genes. Mo-Fe protein consists of 'a' and 'b' subunits which are the product of *nifD* and *nifK* gene respectively (Beringer and Hirsch 1984). Fe protein is a dimer protein consisting of *nifH* gene (Roberts et al. 1978).

The *nif* genes and *his* genes of *K. pneumoniae* are transferred to *E. coli* which requires histidine for nitrogen fixation using a gene-mobilizing plasmid. A conjugative plasmid, pRDI is selected for transfer of *nif* and *his* genes to other bacterial genome (Dixon et al. 1976). The constitutive expression of *nifA* and *nifC* (nitrogen regulator) gene product is carried out by removing the *nifL* gene (Beringer and Hirsch 1984). It has been observed that in *E.coli* carrying the *nif* plasmid pRDI had shown the nitrogenase activity. The mutants of *Azotobacter vinelandii* had shown the nitrogenase activity when pRDI was inserted to them (Dixon et al. 1976).

13.6 Microbial Contributors of Nitrogen Fixation

13.6.1 Azotobacter

Dutch microbiologist and botanist Beijerinck discovered the genus *Azotobacter* in 1901 (Beijerinck 1901). The spp. of *Azotobacter* are free-living gram negative (Gandora et al. 1998) oval or spherical bacteria forming thick-walled cysts (Salhia 2013; Jnawali et al. 2015). They are aerobic and heterotrophic non-symbiotic bacteria that can fix an average of 20 kg N/ha/year (Kizilkaya 2008).

Azotobacter spp. comprises of 6 species, namely, *Azotobacter chroococcum*, *Azotobacter vinelandii*, *Azotobacter beijerinckii*, *Azotobacter nigricans*, *Azotobacter armeniacus* and *Azotobacter paspali* (Sivasakthi et al. 2017). Out of these, *Azotobacter chroococcum* is an aerobic free-living nitrogen fixer to be found first and is the most prevalent species (Kizilkaya 2008; Jnawali et al. 2015).

Azotobacter spp are found in various plant soil and rhizosphere, depending on their physico-chemical and microbiological properties (Kizilkaya 2008). *Azotobacter* requires pH ranging from 5.5 to 8.5 for their growth but the optimal pH is between 7.0 and 7.5 (Channal et al. 1989; Kaushik and Sethi 2005; Akhter et al. 2012). The optimum temperature for their growth is 28–32 °C and the maximum temperature is around 38 °C and minimum is 22 °C. Overall, the *Azotobacter spp.* is susceptible to acidic pH, high concentration of salts and temperature (Tchan and New 1989).

13.6.1.1 Plant Growth Promoting Effect of *Azotobacter*

Beneficial effects have been noticed by *Azotobacter* that includes enhanced growth of plant and increase in the yield, synthesis of biological material, trigger the rhizospheric microbes, and production of phytopathogenic inhibitors (Chen 2006; Lenart 2012). They are also reported to produce growth-promoting material like phosphate solubilization, PGR production of auxins, gibberellins, cytokinins (Nagananda et al. 2010; Damir et al. 2011) and are shown to have an antagonistic activity against pathogens (Kizilkaya 2008).

Azotobacter manufactures and releases substantial quantities of biologically active materials such as biotin, vitamin B, heteroauxins, pantothenic acid, and gibberellins, nicotinic acid, etc. (Narula and Guptha 1986). A large number of chemical fertilizers and other substances such as pesticides can be replaced using sp. *Azotobacter*. Evidence persists that *Azotobacter* has the ability to increase the height, root length of the plant and also increase the production of dry matter with an additional bonus of controlling the diseases in plants (Sivasakthi et al. 2017; Kloepper et al. 2004).

13.6.2 Azospirillum

Azospirilla are characterized as free-living gram-negative nitrogen-fixing rhizosphere bacteria. (Hartmann and Zimmer 1994). *Azospirillum* genus bacteria,since ages are known as rhizobacteria-promoting plant growth (Okon 1994).

At present, 17 species of *Azospirillum* have been identified. They include *A. lipoferum*, *A. brasilense*, *A. amazonense*, *A. halopraeferens*, *A. irakense*, *A. largimobile*, *A. doebereinerae*, *A. oryzae*, *A. melinis*, *A. canadense*, *A. zeae*, *A. rugosum*, *A. picis*, *A. thiophilum*, *A. formosense*, *A. fermentarium*, *A. humicireducens* and *A. himalayense*. Out of these, *A. brasilense* and *A. lipoferum* are the two

Azospirillum species that are widely studied and described in depth (Rodrigues et al. 2015).

13.6.2.1 Plant Growth Promoting Effect of *Azospirillum*

Azospirillum species are capable of producing and directly benefiting plants through associative nitrogen fixation, phytohormone synthesis, mainly auxins, gibberellins, cytokinins and nitric oxide. These help to promote and regulate plant growth and hormonal balance (Spaepen et al. 2008; Bashan and De-Bashan 2010).

Indole-3-acetic acid (IAA) is a vital auxin produced by *Azospirillum* (Glick 2014; Mehnaz 2015). *Azospirillum* produces IAA at all stages of growth (Malhotra and Srivastava 2009) and four IAA biosynthesis pathways exist (Duca et al. 2014). Out of the four, three are tryptophan-dependent pathways and one tryptophan-independent pathway (Fig. 13.3). The most momentous pathway is the IPA pathway in *Azospirillum* as 90% of IAA is synthesized by them through this pathway (Glick 2014).

The major visual impact after *Azospirillum* inoculation include changes in morphology of the roots resulting in increase in length, number of lateral and adventitious roots and branching of root hair (Bashan and Levanony 1985; Cohen et al. 2015). One such example is of the rice plant that resulted in an increase in different plant criterion such as height, tiller number, dry matter yield and N uptake after inoculating with *Azospirillum lipoferum* (Nayak et al. 1986; Murty and Ladha 1988). *Azospirillum sp.* inoculation improved shot growth, straw yield and N uptake for wetland rice under acidic conditions (Govindan and Bagyaraj 1995). Hossain

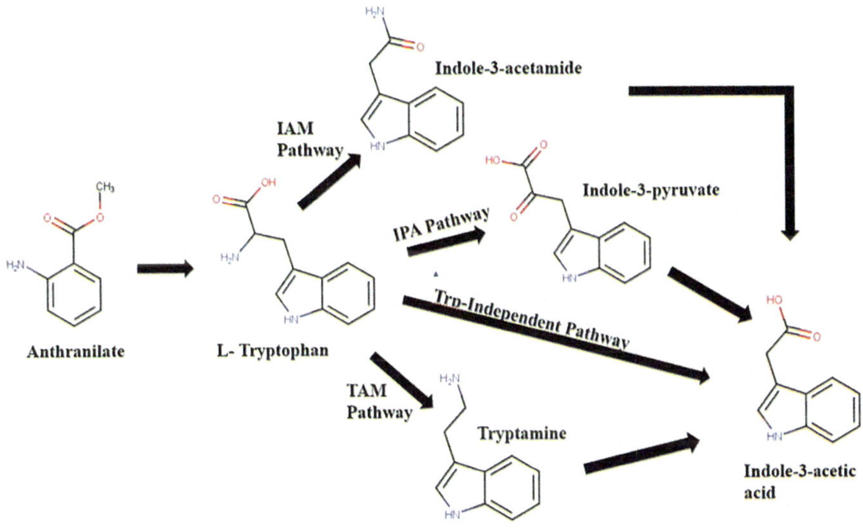

Fig. 13.3 Indole-3-acetic acid pathways identified in *Azospirillum* species

et al. demonstrated that inoculation of *Azospirillum* resulted in increase of plant height; increase in the number of leaves and also the length and breadth of leaf;and fresh and dry weight of rice plant (Hossain et al. 2015).

Antimicrobial activity against different bacterial and fungal isolates were seen in plants when iron was denied. This is because of the production of iron-hungry siderophores by *A. lipoferum* (Shah et al. 1992).

13.6.3 Cyanobacteria

One of the oldest life forms are the *cyanobacteria,* a blue-green algae, which are a group of prokaryotes that have the ability of photosynthesis (Boone and Castenholz 2001). The *cyanobacteria* are a diverse biochemical and morphological class of bacteria (Latysheva et al. 2012). *Cyanobacteria* in the N_2 cycle are key biocatalysts (Vitousek et al. 2002). A remarkable property of *cyanobacteria* is that being a photosynthetic microorganisms it has the ability to fix N_2 in the free-living state and as well as in symbiosis (Bothe 1982) with a extensive range of temperatures, salinity, water potential, pH and irradiance. Extreme environments like hyper-saline waters, hot springs, freezing environments and arid deserts can be easily survived by *Cyanobacteria* (Singh 2014). *Cyanobacteria* are able to survive temperatures ranging from 45 to 70 °C (Castenholz 1978) and pH below 4–5 (Pfennig 1969) with an optimal pH range of 7.5–10 (Fogg 1956). As they are widely distributed throughout the area they reflect a wide range of physiological properties and environmental stress tolerance (Singh et al. 2013). About 42% of global nitrogen fixation of 240 Tg N_2 y^{-1} is carried out by *Trichodesmium,* a cyanobacterial genus (Berman-Frank et al. 2003).

They have a considerable and ever-increasing impact on social, economic and environmental aspects of human life and society. They have both harmful and beneficial effects on human life. Among the harmful effects, these blue greens produce toxic compounds in humans and livestock through water supplies. These toxins also make swimming and recreation in lakes and pools hazardous and anaesthetic. Poisoning can cause not only the death of animals but humans too. Nuisance and undesirable conditions are also developed in water bodies due to the excess algal blooms, which overwhelm the water surface. On the other hand, there are many beneficial effects too, most important of them are the fixation of atmospheric nitrogen (Kumar et al. 2014).

Conserving certain organic matter, nitrogen, phosphate and moisture within the soil helps to improve the quality of the soil and this is seen when the soil is inoculated with *cyanobacteria* (Maqubela et al. 2009). The crust of the biological soil, which is the most nutritious and important part, sometimes undergo degradation due to natural or anthropogenic abnormalities. However, the *cyanobacterial* inoculation greatly accelerates the recovery process and restores the area's ecosystem (Wang et al. 2009).

13.6.3.1 Plant Growth Promoting Effect of *Cyanobacteria*

Naturally, *cyanobacteria* are located abundantly in paddy soils and result in improving the fertility and texture of the soil at a zero cost that benefits most of the rice fields (Prasanna et al. 2013). Cyanobacterial soil inoculation enhances the establishment and nitrogen fixation capability of bacterial populations under the soil surface (Prasanna et al. 2012). The cost incurred for preparing the inoculum of the *cyanobacteria* is found to be approximately one third of that of the chemical fertilizers. Moreover a 23–28% availability of fixed nitrogen is observed when inoculated with cyanobacteria for the first crop that is grown in the soil and then the percentage increases in the following years (Prasanna et al. 2013). As promising biofertilizers, the microflora that has the ability to produce oil that includes the *cyanobacteria* play a critical role in the production of the crops (Ashraf et al. 2013). *Cyanobacteria* are able to manufacture a range of substances that include amino acids, cytokinins, gibberellins, auxins (Sood et al. 2011). It is observed that inoculation with *cyanobacteria* has a positive effect on the growth of paddy crop as they help in seed germination, elongation of the shoot and root (Singh et al. 2016). The ability to synthesize IAA by free-living as well as symbiotic cyanobacteria species include *Anabaena, Nostoc, Chlorogloeopsis, Calothrix, Gloeothece, Cylindrospermum, Anabaenopsis* and *Plectonema* that further help in the enhancement of plant growth (Natarajan et al. 2012).

Cyanobacteria have the ability to manufacture extracellular polymeric substances (EPS) that improve soil water holding capacity and prevent erosion, thus enhancing soil microbial biomass, nitrogen, carbon, and humus content thus results in retention of the moisture and helps in the formation of quality soil that indirectly results in healthy crop growth (Prasanna et al. 2013).

13.6.4 Beijerinckia

Beijerinckia are gram-negative rods with a diameter of 0.5–1.5μm and are commonly found in tropical soils. These organisms are motile as they consist of peritrichous flagella. They are aerobic organisms and are catalase positive. The optimum temperature for their growth is 20–30 °C. They have large quantities of nitrogenase enzymes and hence are able to reduce nitrogen. According to the studies the mineral composition of the laterite soil has been found to have some significant correlation with that of the *Beijerinckia's* mineral requirements (Becking 1961), and hence a conclusion could be drawn that the adaptation of *Beijerinckia* is seen due to these limiting conditions. One example is when the rhizosphere of Paspalum grass was inoculated with *beijerinckia,* it was observed that the plant had greater amount of fixed nitrogen when compared to the non-inoculated plant that resulted in positive development of the grass (Ruschel and Britto 1966). Improvement in the growth as well as the yield of the plant was reported when *Beijerinckia indica* (Pati and Chandra 1981) and *Beijerinckia* sp. (Nandi and Sen 1981) that were isolated from the phyllosphere and other sources were sprayed on plants. *Beijerinckia derxii* has been shown to have a positive influence on the growth of non-diazotrophic bacteria and are able to survive as well as grow in nitrogen-free media when cultivated with them (Barbosa et al. 2000). Even though a considerable amount of information about the features of this genus is known, till date we are yet to find the biology and the contribution of *Beijerinckia* to the habitat structure and function in which they are found.

13.6.5 Derxia

Jennsen et al in 1960 were the first one to propose that the genus *Derxia*, belonged to the family Alcaligenaceae of the order Burkholderiales in the class Betaproteobacteria (Garrity et al. 2005). They are gram-negative rods, obligately aerobic, consisting of a thick capsule, that are motile with the help of a short polar flagellum, and catalase-negative. Under both aerobic conditions and reduced oxygen pressure, cells are able to fix molecular nitrogen (Jennsen et al. 1960). *Derxia* are able to grow at temperature 15–40 °C with an optimum temperature of 25–35 °C, at pH 6–7 but preferring more of a weak acid pH of about 6.0 and with 0–4% NaCl, optimum being 0–1%.

13.7 Applications of Non-symbiotic Nitrogen Fixers as Biofertilizers

Crop plants require higher concentrations of fixed nitrogen than what is naturally available to them. To meet this requirement, nitrogenous fertilizers were introduced. Although, upon realising the detrimental effects of the prior, usage of symbiotic nitrogen fixers was preferred; however, biological nitrogen fixation by these symbionts is restricted to the legume family. To fully employ the process of biological nitrogen fixation, numerous experiments have been performed to understand the mechanism of non-symbiotic nitrogen-fixing bacteria and their effects on plant growth as biofertilizers.

Reynders and Vlassak (1982) carried out a study to check the efficiency of *Azospirillum brasilense* as a biofertilizer for intensive cropping of wheat. The experiment was conducted in two parts; once in the winter and then in spring. In winter, 10 wheat cultivars were sown in clay soil in three different locations, having different nitrogen doses. Both strains of *Azospirillum brasilense* were inoculated by overhead spraying. In spring, 4 spring-wheat cultivars were sown in four different locations with nitrogen doses in loamy soil using the same method of inoculation of strains as used during winter. Upon immersing the same seeds in centrifuged bacterium culture and drying, they were also sown into clay soil with varying nitrogen doses. The winter wheat cultivars showed significant increase in grain yield and nitrogen content with only 80 kg of N per ha, grain yield increased to 8000 for which the controls required 156.68 kg of N per ha.

An experiment carried out at a greenhouse in Egypt at the University of Cairo in 1989 by M. Fayez to screen for unconventional bacteria that might fix nitrogen and could then be used as biofertilizers. They found 27 pure isolates which mostly belonged to the following families, *Rhizobiaceae, Pseudomonadaceae, Achromobacteriaceae, Enterobacteriaceae, Micrococcaceae, Bacillaceae and Streptomycetaceae*, as well as some yeasts (Fayez 1990). Their effects on the growth of wheat and barley were examined in a greenhouse. Out of these only *Pseudomonas* and *Bacillus* had a nitrogen fixing effect in case of wheat and barley respectively (Fayez 1990). However, upon multi-strain inoculation along with either *Azospirillum* and/or *Azotobacter* a very high value of nitrogenase activity was recorded (Fayez 1990) and there was significant increase in growth of both wheat and barley in each case.

An experiment conducted by Baldani and group, studied effects of certain endophytic diazotrophs on rice plants. *Herbaspirillum seropedicae* and *Burkholderia* spp. were the selected strains. Being diazotrophs they colonize the roots, stems and leaves of the cereals endophytically, and hence suffer much less competition from other microorganisms and possibly excrete some of the fixed nitrogen directly into the plant (Baldani et al. 2000). The results showed that N_2-fixing ability of *H. seropedicae* strains ranged from 54% to 31%. *Burkholderia* spp. showed similar effects as the ones observed for *H. seropedicae.* When the rice plants were grown along with the most potent nitrogen fixing strains of *H. seropedicae* and *Burkholderia* endophytes as inoculum, grains with higher nitrogen content were produced (Baldani et al. 2000).

The latest study performed by (Zaidi et al. 2017) aimed at evaluating the nitrogen-fixing ability of phosphate solubilising bacteria (PSB), *Azotobacter* spp., a germinator and PSB + germinator along with full and half dosage of phosphorus and nitrogen by monitoring growth of radish. The outcome suggested that the germinator negatively impacted the growth and yield attributes of the plant. Marketable yields of radish are strongly correlated with root fresh weight. Maximal root diameter was observed upon administration of either PSB or *Azotobacter* spp. along with the recommended dose of fertilizers; and maximum root length was achieved in plants that were given half the recommended dosage of phosphorus with *Azotobacter* spp.

13.8 Conclusion

For a higher yield, crop plants need to be provided with all the crucial nutrients that it requires, the most essential ones being nitrogen, phosphorus and potassium. In today's cropping techniques, mass availability of nitrogen, especially in a form the plant can uptake easily is arduous. To overcome this major issue, modern agriculture employs the use of synthetic fertilizers to increase nitrogen availability as well as yield of the crop. However, the use of these synthetic fertilizers poses a serious harm to the environment as well as the biodiversity of the area.

The aforementioned problem makes bio-fertilizers even more promising as a means to replenish soil nitrogen. They are organic products containing either free-living or symbiotic nitrogen fixers. Free-living nitrogen fixers live in the rhizosphere and convert atmospheric nitrogen to ammonia (biological nitrogen fixation) for its easy uptake by the plants. Some microbes even synthesize and release phytohormones into the rhizosphere which enhance plant growth upto being absorbed by the plant. There also exist some partially symbiotic microbes (diazotrophs) that associate with non-nodulating plants either endophytically or by associating with root surfaces. This is termed as Associative Nitrogen Fixation (ANF) and is similar to free-living nitrogen fixation, excluding the association between the plant and microbe and occurs irrespective of the plant's nitrogen demand.

Genes involved in nitrogen fixation are very well explored and can prove to be a great tool if genetic manipulations are to be made to the microbes to enhance their nitrogen fixing abilities or to increase phytohormone production. Increasing the efficiency of the microbes is crucial in order to get a maximal crop yield and eventually eliminating the use of synthetic fertilizers.

References

Akhter M, Hossain S, Hossain S, Datta R (2012) Isolation and characterization of salinity tolerant Azotobacter sp. Greener J Biol Sci 2(3):43–51. https://doi.org/10.15580/GJBS.2012.3.110312207

Ashraf MA, Asif M, Zaheer A, Malik A, Qasim Ali Q, Rasool M (2013) Plant growth promoting rhizobacteria and sustainable agriculture: a review. Afr J Microbiol Res 7(9):704–709. https://doi.org/10.5897/AJMR12.936

Barbosa HR, Thuler DS, Shirakawa MA, Miyasaka NRS (2000) Beijerinckia derxii stimulates the viability of non-N2-fixing bacteria in nitrogen-free media. Braz J Microbiol 31:168–173. https://doi.org/10.1590/S1517-83822000000300004

Bashan Y, De-Bashan LE (2010) How the plant growth-promoting bacterium *Azospirillum* promotes plant growth—a critical assessment. Adv Agron 108:77–136. https://doi.org/10.1016/S0065-2113(10)08002-8

Bashan Y, Levanony H (1985) An improved selection technique and medium for the isolation and enumeration of *Azospirillum brasilense*. Can J Microbiol 31:947–952. https://doi.org/10.1139/m85-178

Becking JH (1961) Studies on nitrogen-fixing bacteria of the genus Beijerinckia. II Mineral nutrition and resistance to high levels of certain elements in relation to soil type. Plant Soil XIV:297–322. https://doi.org/10.1007/BF01666290

Beijerinck MW (1901) Ueber Oligonitophile Mikroben, Zentralblatt für Bakteriologie, Parasitenkunde, Infektionskrankheiten and Hygiene. Abteilung II 7:561–582

Beringer JE, Hirsch PR (1984) Genetic Engineering and nitrogen fixation. Biotechnol Gen Engin Rev 1:65–88. https://doi.org/10.1080/02648725.1984.10647781

Berman-Frank I, Lundgren P, Falkowski P (2003) Nitrogen fixation and photosynthetic oxygen evolution in cyanobacteria. Res Microbiol 154:157–164. https://doi.org/10.1016/S0923-2508(03)00029-9

Boone DR, Castenholz RW (2001) The Archaea and the deeply branching and phototrophic bacteria. In: Garrity GM (ed) Bergey's manual of systematic bacteriology, 2nd edn. Springer, New York, pp 33–38. https://doi.org/10.1007/978-0-387-21609-6

Bothe H (1982) Nitrogen fixation. In: Carr NG, Whitton BA (eds) The biology of cyanobacteria: botanical monographs 19. University of California Press, Los Angeles, pp 87–105

Bürgmann H, Widmer F, Sigler WV, Zeyer J (2003) mRNA extraction and RT-PCR protocol for detection of *nifH* gene expression of *Azotobacter vinelandii* in soil. Appl Environ Microbiol 69:1928–1935. https://doi.org/10.1128/aem.69.4.1928-1935.2003

Cartigny P, Busigny V (2017) Nitrogen isotopes. In: Encyclopedia of Earth sciences series, pp 1–13. https://doi.org/10.1007/978-3-319-39193-9_197-1

Castenholz RW (1978) The biogeography of hot spring algal through enrichment cultures. Mitt Int Ver Limnol 21:296–315. https://doi.org/10.1080/05384680.1978.11903973

Chakraborty B, Samaddar KR (1995) Evidence for the occurrence of an alternative nitrogenase system in *Azospirillum brasilense*. FEMS Microbiol Lett 127:127–131. https://doi.org/10.1111/j.1574-6968.1995.tb07461.x

Channal T, Ahagawadi A, Gowder T, Udupde S, Patil P, Mannikeri I (1989) *Azotobacter* population is influenced by soil properties in some soils of northern Karnataka. Curr Sci 58(2):70–71

Chen J (2006) The combined use of chemical and organic fertilizers and/or biofertilizer for crop growth and soil fertility. In: International workshop on sustained management of the soil–rhizosphere system for efficient crop production and fertilizer use, Thailand, pp 1–10. http://www.fftc.agnet.org/htmlarea_file/activities/20110719102200/7.pdf. https://doi.org/10.30058/SE.200706.0001

Cheng Q (2008) Perspectives in biological nitrogen fixation research. J Integr Plant Biol 50(7):784–796. https://doi.org/10.1111/j.1744-7909.2008.00700.x

Christiansen J, Dean DR, Seefeldt LC (2001) Mechanistic features of the Mo-containing nitrogenase. Annu Rev Plant Physiol Plant Mol Biol 52:269–295. https://doi.org/10.1146/annurev.arplant.52.1.269

Cohen AC, Bottini R, Pontin M, Berli FJ, Moreno D, Boccanlandro H, Travaglia CN, Piccoli PN (2015) *Azospirillum brasilense* ameliorates the response of *Arabidopsis thaliana* to drought. Physiol Plant 152(1):79–90. https://doi.org/10.1111/ppl.12221

Dahal B (2016) Diversity of free-living nitrogen fixing bacteria in the badlands of South Dakota .Theses and dissertations, p 688. http://openprairie.sdstate.edu/etd/688

Damir O, Mladen PI, Bozidar S, Sran N (2011) Cultivation of bacterium *Azotobacter chroococcum* for preparation of biofertilizers. Afr J Biotechnol 10(16):3104–3111. https://doi.org/10.5897/AJB10.1086

Divan Baldani V, Baldani J, Döbereiner J (2000) Inoculation of rice plants with the endophytic diazotrophs Herbaspirillum seropedicae and Burkholderia spp. Biol Fertil Soils 30(5-6):485–491. https://doi.org/10.1007/s003740050027

Dixon R, Kahn D (2004) Genetic regulation of biological nitrogen fixation. Nat Rev Microbiol 2(8):621–631. https://doi.org/10.1038/nrmicro954

Dixon RA, Cannon F, Kondorosi A (1976) Construction of P plasmid carrying nitrogen fixation genes from Klebsiella pneumoniae. Nature 260:268–271. https://doi.org/10.1038/260268a0

Doolittle WF (1999) Phylogenetic classification and the universal tree. Science 284:2124–2128. https://doi.org/10.1126/science.284.5423.2124

Duca D, Lorv J, Patten CL, Rose D, Glick BR (2014) Indole-3-acetic acid in plant-microbe interactions. Antonie Van Leeuwenhoek 106:85–125. https://doi.org/10.1007/s10482-013-0095-y

Fayez M (1990) Untraditional N2-fixing bacteria as biofertilizers for wheat and barley. Folia Microbiol 35(3):218–226. https://doi.org/10.1007/BF02820488

Figg TM, Holland PL, Cundari TR (2012) Cooperativity between low-valent iron and potassium promoters in dinitrogen fixation. Inorg Chem 51:7546–7550. https://doi.org/10.1021/ic300150u

Fogg GE (1956) The comparative physiology and biochemistry of the blue-green algae. Bacteriol Rev 20:148–165

Galloway JN (1998) The global nitrogen cycle: changes and consequences. Environ Pollut 102:15–24. https://doi.org/10.1016/S0269-7491(98)80010-9

Gandora V, Gupta RD, Bhardwaj KKR (1998) Abundance of *Azotobacter* in great soil groups of North–West Himalayas. J Indian Soc Soil Sci 46(3):379–383

Garrity GM, Bell JA, Lilburn T (2005) Order I. Burkholderiales ord. nov. In: Brenner DJ, Krieg NR, Staley JT, Garrity GM (eds) Bergey's manual of systematic bacteriology, vol 2, 2nd edn. Springer, New York, pp 575–763

Giller KE, Day JM (1985) Nitrogen-fixation in the rhizosphere: significance in natural and agricultural systems. In: Fitter AH (ed) Ecological interactions in soil. Plants microbes and animals. Blackwell Scientific Publications, Oxford. ISBN: 0632013869

Glick B (2014) Bacteria with ACC deaminase can promote plant growth and help to feed the world. Microbiol Res 169(1):30–39. https://doi.org/10.1016/j.micres.2013.09.009

Govindan M, Bagyaraj DJ (1995) Field response of wetland rice to Azospirillum inoculation. J Soil Biol Ecol 15:17–22

Halbleib CM, Ludden PW (2000) Regulation of biological nitrogen fixation. J Nutr 130:1081–1084. https://doi.org/10.1093/jn/130.5.1081

Hamelin J, Fromin N, Tarnawski S, Teyssier-Cuvelle S, Aragno M (2002) nifH gene diversity in the bacterial community associated with the rhizosphere of Molinia coerulea, an oligonitrophilic perennial grass. Environ Microbiol 4:477–481. https://doi.org/10.1046/j.1462-2920.2002.00319.x

Hartmann A, Zimmer W (1994) Physiology of Azospirillum. In: Okon Y (ed) Azospirillum/plant associations. CRC Press, Boca Raton, pp 15–39

Havelka UD, Boyle MG, Hardy RWF (1982) Biological nitrogen fixation. In: Nitrogen in agricultural soils. Agronomy monograph No. 22, pp 365–422. https://doi.org/10.2134/agronmonogr22.c10

Hill S, Kavanagh EP (1980) Roles of *nifF* and *nifl* gene products in electron transport to nitrogenase in *Klebsiella pneumoniae*. J Bacteriol 141:470–475

Hillel D (2008) Soil in the environment. Elsevier/Academic, New York. ISBN: 9780080554969

Hossain M, Jahan I, Akter S, Rahman N, Rahman SM (2015) Effects of Azospirillum isolates from paddy fields on the growth of rice plants. Res Biotechnol 6(2):15–22

Hubner P, Masepohl B, Klipp W, Bickle TA (1993) nif gene expression studies in Rhodobacter Capsulatus: ntrC- independent repression by high ammonium concentrations. Mol Microbiol 10:123–132. https://doi.org/10.1111/j.1365-2958.1993.tb00909.x

Hu Y, Fay AW (2007) P-cluster maturation on nitrogenase MoFe protein. Proc Natl Acad Sci USA 104(25):10424–10429. https://doi.org/10.1073/pnas.0704297104

Igarashi R, Seefeldt L (2003) Nitrogen fixation: the mechanism of the mo-dependent nitrogenase. Crit Rev Biochem Mol Biol 38(4):351–384. https://doi.org/10.1080/10409230391036766

Jennsen HL, Petersen EJ, De PK, Bhattacharya R (1960) A new nitrogen-fixing bacterium: Derxia gummosa nov. gen. nov.spec. Arch Mikrobiol 36:182–195. https://doi.org/10.1007/BF00412286

Jnawali A, Ojha R, Marahatta S (2015) Role of azotobacter in soil fertility and sustainability – a review. Adv Plants Agric Res 2(6):250–253. https://doi.org/10.15406/apar.2015.02.00069

Kaushik A, Sethi V (2005) Salinity effects on nitrifying and free diazotrophic bacterial populations in the rhizosphere of rice. Bull Natl Inst Ecol 15:139–144

Keneddy I, Islam N (2001) The current and potential contribution of asymbiotic nitrogen fixation to nitrogen requirements on farms: a review. Aust J Exp Agric 41:447–457. https://doi.org/10.1071/EA00081

Kizilkaya R (2008) Nitrogen fixation capacity of Azotobacter spp. strains isolated from soils in different ecosystems and the relationship between them and the microbiological properties of soils. J Environ Biol 30(1):73–82

Kloepper JW, Reddy SM, Rodreguez-Kabana R, Kenney DS, Kokalis-Burelle N, Ochoa NM (2004) Application for rhizobacteria in transplant production and yield enhancement. Acta Hortic 631:217–229. https://doi.org/10.17660/ActaHortic.2004.631.28

Kumar J, Patel J, Kumar R, Khan S (2014) Chronic response of three different cyanobacterial species on growth, pigment, and metabolic variations to the high molecular weight polycyclic aromatic hydrocarbon – pyrene. Polycycl Aromat Compd 34:143–160. https://doi.org/10.1080/10406638.2013.867514

Kumar N. (2012) Biotechnology in plant science. Himalaya Publishing House. https://doi.org/10.1155/2012/831460

Latysheva N, Junker V, Palmer W, Codd G, Barker D (2012) The evolution of nitrogen fixation in cyanobacteria. Bioinformatics 28(5):603–606. https://doi.org/10.1093/bioinformatics/bts008

Lenart A (2012) Occurrence characteristics and genetic diversity of *Azotobacter chroococcum* in various soils of southern Poland. Pol J Environ Stud 21(2):415–424

Malhotra M, Srivastava S (2009) Stress-responsive indole-3-acetic acid biosynthesis by Azospirillum brasilense SM and its ability to modulate plant growth. Eur J Soil Biol 45(1):73–80. https://doi.org/10.1016/j.ejsobi.2008.05.006

Maqubela MP, Mnkeni PNS, Issa MO, Pardo MT, D'Acqui LP (2009) *Nostoc* cyanobacterial inoculation in South African agricultural soils enhances soil structure, fertility and maize growth. Plant Soil 315:79–92. https://doi.org/10.1007/s11104-008-9734-x

Marschner H (1995) Mineral nutrition of higher plants, 2nd edn. Academic, London, p 889

Mehnaz S (2015) Azospirillum: a biofertilizer for every crop. In: Arora NK (ed) Plant microbes symbiosis: applied facets. Springer, New Delhi

Miller DN, Bryant JE, Madsen EL, Ghiiorse WC (1999) Evaluation and optimization of DNA extraction and purification procedures for soil and sediment samples. Appl Environ Microbiol 65:4715–4724

Murty MG, Ladha JK (1988) Influence of Azospirillum inoculation on the mineral uptake and growth of rice under hydroponic conditions. Plant Soil 108:281–285. https://doi.org/10.1007/BF02375660

Nagananda GS, Das A, Bhattacharya S, Kalpana T (2010) In vitro studies on effect of biofertilizers (*Azotobacter* and *Rhizobium*) on seed germination and development of *Trigonella foenum-graecum L.* using a novel glass marble containing liquid medium. Int J Bot 6:394–403. https://doi.org/10.3923/ijb.2010.394.403

Nandi AS, Sen SP (1981) Utility of some nitrogen-fixing microorganisms in the phyllosphere of crop plants. Plant Soil 63:465–476. https://doi.org/10.1007/BF02370046

Narula N, Guptha KG (1986) Ammonia excretion by *Azotobacter* in liquid culture and soil in the presence of manganese and clay minerals. Plant Soil 93:205–209. https://doi.org/10.1007/BF02374222

Natarajan C, Prasanna R, Gupta V, Dureja P, Nain L (2012) Characterization of the fungicidal activity of *Calothrix elenkinii* using chemical methods and microscopy. Appl Biochem Microbiol 48:51–57. https://doi.org/10.1134/S0003683812010115

Nayak DN, Ladha JK, Watanabe I (1986) The fate of marker Azospirillum lipoferum inoculated into rice and its effect on growth, yield and N fixation of plants studied by acetylene reduction, N feeding and N dilution techniques. Biol Fertil Soils 2:7–14. https://doi.org/10.1007/BF00638955

Okon Y (1994) Azospirillum/plant associations. CRC Press, Boca Raton, p 175

Pati BR, Chandra AK (1981) Effect of spraying nitrogen-fixing phyllospheric bacterial isolates on wheat plants. Plant Soil 61:419–427. https://doi.org/10.1007/BF02182022

Pesaro M, Bürgmann H, Widmer F, Zeyer J (2003) Soil DNA content – a novel quantitative bulk parameter for description of microbiological soil characteristics. In Dilly O (ed) Proceedings of the COST Action831 Workshop on "Evaluating Soil Quality", Kiel, 2001, Budapest, Hungary

Pfennig N (1969) Rhodopseudomonas acidophila, sp. n., a new species of the budding purple nonsulfur bacteria. J Bacteriol 99:597–602

Prasanna R, Chaudhary V, Gupta V, Babu S, Kumar A, Singh R, Shivay YS, Nain L (2013) Cyanobacteria mediated plant growth promotion and bioprotection against Fusarium wilt in tomato. Eur J Plant Pathol 136(2). https://doi.org/10.1007/s10658-013-0167-x

Prasanna R, Joshi M, Rana A, Shivay YS, Nain L (2012) Influence of co-inoculation of bacteria cyanobacteria on crop yield and C-N sequestration in soil under rice crop. World J Microbiol Biotechnol 28:1223–1235. https://doi.org/10.1007/s11274-011-0926-9

Puri A, Padda KP, Puri CP (2015) Can a diazotrophic endophyte originally isolated from lodgepole pine colonize an agricultural crop (corn) and promote its growth? Soil Biol Biochem 89:210–216. https://doi.org/10.1016/j.soilbio.2015.07.012

Reynders L, Vlassak K (1982) Use ofAzospirillum brasilense as biofertilizer in intensive wheat cropping. Plant Soil 66(2):217–223. https://doi.org/10.1007/BF02183980

Roberts GP, Macneil T, Macneil D, Brill WJ (1978) Regulation and characterization of protein products coded by the nif genes of Klebsiella pneumoniae. J Bacteriol 136:267–279

Rodrigues A, Bonifacio A, de Araujo F, Junior M, Vale Barreto Figueiredo M (2015) Azospirillum sp. as a challenge for agriculture. In: Bacterial metabolites in sustainable agroecosystem, pp 29–51. https://doi.org/10.1007/978-3-319-24654-3_2

Roger PA, Watanabe (1986) Technologies for utilizing biological nitrogen fixation in wetland rice: potentialities, current usage, and limiting factors. Fertil Res 9:39–77. https://doi.org/10.1007/BF01048695

Roper M, Ladha J (1995) Biological N_2 fixation by heterotrophic and phototrophic bacteria in association with straw. Plant Soil 174(1–2):211–224. https://doi.org/10.1007/BF00032248

Ruschel A, Britto D (1966) Fixação assimbiótica de nitrogênio atmosférico em algumas gramíneas e na tiririca pelas bactérias do gênero *Beijerinckia* Derx. Pesq Agropec Bras 1:65–69

Salhia B (2013) The effect of Azotobacter chrococcumas nitrogen biofertilizer on the growth and yield of Cucumis sativus. Deanery of Higher Education Faculty of Science, Master of Biological Sciences, Botany, The Islamic University Gaza

Sandaa RA, Enger O, Torsvik V (1998) Rapid method for fluorometric quantification of DNA in soil. Soil Biol Biochem 30:265–268. https://doi.org/10.1016/S0038-0717(97)00110-7

Sessitsch A, Gyamfi S, Stralis-Pavese N, Weilharter A, Pfeifer U (2002) RNA isolation from soil for bacterial community and functional analysis: evaluation of different extraction and soil conservation protocols. J Microbiol Methods 51:171–179. https://doi.org/10.1016/s0167-7012(02)00065-9

Shah S, Karkhanis V, Desai A (1992) Isolation and characterization of siderophore, with antimicrobial activity, from Azospirillum lipoferum M. Curr Microbiol 25:347–351. https://doi.org/10.1007/BF01577233

Singh H, Ahluwalia AS, Khattar JIS (2013) Induction of sporulation by different nitrogen sources in *Anabaena naviculoides*, a diazotrophic strain capable of colonizing paddy field soil of Punjab (India). Vegetos 2 6(1):283–292. https://doi.org/10.5958/j.2229-4473.26.1.041

Singh JS (2014) Cyanobacteria: a vital bio-agent in eco-restoration of degraded lands and sustainable agriculture. Clim Change Environ Sustain 2:133–137

Singh J, Kumar A, Rai A, Singh D (2016) Cyanobacteria: a precious bio-resource in agriculture, ecosystem, and environmental sustainability. Front Microbiol 7:529. https://doi.org/10.3389/fmicb.2016.00529

Sivasakthi S, Saranraj P, Sivasakthivelan P (2017) Biological nitrogen fixation by Azotobacter sp. – a review. Indo-Asian J Multidiscip Res 3(5):1274–1284. https://doi.org/10.22192/iajmr.2017.3.5.6

Sood A, Singh PK, Kumar A, Singh R, Prasanna R (2011) Growth and biochemical characterization of associations between cyanobionts and wheat seedlings in co-culturing experiments. Biologia 66(1):104–110. https://doi.org/10.2478/s11756-010-0133-4

Spaepen S, Dobbelaere S, Croonenborghs A, Vanderleyden J (2008) Effects of *Azospirillum brasilense* indole-3-acetic acid production on inoculated wheat plants. Plant Soil 312:1–23. https://doi.org/10.1007/s11104-008-9560-1

Tchan YT, New PB (1989). *Azotobacteraceae*. In: Holt JG, Williams et al (eds) Bergey's manual of systematic bacteriology volume 1. Baltimore, pp 220–229

Vitousek P, Aber J, Howarth R, Likens G, Matson P, Schindler D et al (1997) Human alteration of the global nitrogen cycle: sourced and consequences. Ecol Appl 7(3):737–750. https://doi.org/10.1890/1051-0761(1997)007[0737:haotgn]2.0.co;2

Vitousek P, Cassman K, Cleveland C, Crews T, Field C, Grimm N, Howarth R, Marino R, Martinelli L, Rastetter E, Sprent J (2002) Towards an ecological understanding of biological nitrogen fixation. Biogeochemistry 57:1–45. https://doi.org/10.1023/A:1015798428743

Wagner SC (2011) Biological nitrogen fixation. Nat Educ Knowl 2:14

Wang W, Liu Y, Li D, Hu C, Rao B (2009) Feasibility of cyanobacterial inoculation for biological soil crusts formation in desert area. Soil Biol Biochern 41:926–929. https://doi.org/10.1016/j.soilbio.2008.07.001

Weaver RW, Graham PH (1994) Legume nodule symbionts. In: Weaver RW, Angle S, Bottomley P, Bezdicek D, Smith S, Tabatabai A, Wollum A (eds) Methods of soil analysis. Part 2. Microbiological and biochemical properties. Soil Science Society of America, Madison, pp 199–222

Zaidi A, Khan M, Saif S, Rizvi A, Ahmed B, Shahid M (2017) Role of nitrogen-fixing plant growth-promoting rhizobacteria in sustainable production of vegetables: current perspective. In: Microbial strategies for vegetable production, pp 49–79. https://doi.org/10.1007/978-3-319-54401-4_3

Zehr JP, Braun S, Chen YB, Mellon M (1996) Nitrogen-fixation in the marine environment: relating genetic potential to nitrogenase activity. J Exp Mar Biol Ecol 203:61–73. https://doi.org/10.1016/0022-0981(96)02570-1

Chapter 14
Nutritional and Medical Benefits of Moringa Oleifera, the Miracle Tree

Iqra Aslam and Ali Asghar

Abstract The plant *Moringa oleifera* occurs in sub-Himalayan areas of Asia and in Africa. Named the Miracle Tree, moringa contains many micro and macro-nutrients of great value for human nutrition, in particular for the malnourished population. Moringa provides vitamins, minerals and essential amino acids that are helpful in preventing protein malnutrition. *M. oleifera* is also a powerful antioxidant, anticancer, antidiabetic and anti-inflammatory agent. Leaves, pods, roots, seeds and flowers of *M. oleifera* contain bioactive compounds. After oil extraction from seeds, the left-over seed cake can be used as a protein-rich soil fertilizer. Moringa roots also reduce soil degradation.

Keywords Moringa · Leaves · Essential amono acids · Nutrition · Protein

14.1 Introduction

Moringa oleifera is grown in tropics and sub-tropics zones of the world. Common names of moringa are "horseradish tree" and "drumstick tree". *M. oleifera* belongs to the *Moringaceae* family and is the only species among its 13 species that is widely grown and used all over the world due to its exceptionally nutritious benefits. Studies showed that in different regions of world there are different names of *M. oleifera* like benzolive, marango, drumstick tree, mlonge, horseradish tree, kelor, saijihan, sajna and mulangay (Fahey 2005). Trees of moringa produces triangular shaped dry fruits and the seeds are easily dispersed by the wind. In world *M. oleifera* has been found in many areas like Sri Lanka, Bangladesh, Afghanistan, West Asia, Africa, America, Paraguay, Caribbean Islands, from Mexico to Peru and Brazil (Oliveira et al. 1999). Moringa can grow vigorously under the tropical insular

I. Aslam · A. Asghar (✉)
National Institute of Food Science and Technology, University of Agriculture, Faisalabad, Pakistan
e-mail: ali.asghar@uaf.edu.pk

E. Lichtfouse (ed.), *Sustainable Agriculture Reviews 52*, Sustainable Agriculture Reviews 52, https://doi.org/10.1007/978-3-030-73245-5_14

climate with a temperature of 25–30 °C. Moringa is cultivated in soil that is loamy or sandy and pH from slightly acidic to slightly alkaline. 250–3000 mm is the required rainfall for it (Thurber and Fahey 2010). Moringa can thrives best in the hot dry land or humid tropics. Moringa can also grow in less fertile soils, and the amazing thing about moringa is that drought has a little effect on it. Moringa is expected that after seeding moringa seeds are germinated within 5–12 days and it is cultivated at the depth of 2 cm in soil (Aslam et al. 2005). *M. oleifera* has wide applications in the food and medicinal industry and is also used for different commercial and industrial purposes. Every part of this plant is a garner of different essential nutrients.

Church World Service, Trees for Life and the Educational Concerns for Hunger Organization are the three nongovernmental organizations that have endorsed the aphorism "Natural nutrition for tropics" to encourage or increase the consumption of various plants as nutrient source including *Moringa oleifera* (Fahey 2005). Extraction of bioactive components of all parts of the moringa showed different medicinal properties.

In many parts of the world fresh and dried leaves of *M. oleifera* both are used for fortification in meals to increase its nutritional value. For instance moringa is used as a food fortificant in many soups (Babayeju et al. 2014), herbal biscuits (Alam et al. 2014), cake (Kolawole et al. 2013), yogurt (Hekmat et al. 2015), weaning foods (Arise et al. 2014), amala (Karim et al. 2015) and bread (Chinma et al. 2014).

This review firstly summarizes the information about bioactive compounds and their nutritional composition of *M. oleifera*. Secondly, the use of moringa in medical sciences and then fortification of moringa in meals are presented.

14.2 Nutritional Composition of Moringa with Bioactive Compounds

Moringa oleifera has many nutritional benefits due to bioactive compounds present in different parts of this plant. Moreover, moringa also contains many micro and macro nutrients that is useful in combating malnutrition worldwide. In this section we reviewed the nutritional composition and the amount of nutrients found in every part of moringa plant.

14.2.1 Leaves

The dried leaves of Moringa have crude protein content of 30.3% (Table 14.1) and has 19 amino acids (Table 14.2). Alanine was present in a highest value of 3.033% among amino acids present in leaves. The least content was cysteine with 0.01%. Some studies have reported varying crude protein contents that has a range between 16, 22.42, 23.27, 27.4 and 40% (Sarwatt et al. 2004; Nouala et al. 2006; Oduro et al. 2008; Sanchez-Machado et al. 2009).

Calcium has a value of 3.65% and potassium is 1.5%, and in macro-minerals phosphorus had the value of 0.30% (Table 14.3). Eight ounces of milk provides 300–400 mg of calcium while moringa leaves can provide 1000 mg of calcium and

Table 14.1 Chemical composition of dried leaves of Moringa (*M. oleifera Lam*)

Serial number	Nutritive value in %	Dry leaf
1.	Moisture	9.533
2.	Fat	6.50
3.	Total polyphenols	2.02
4.	Crude protein	30.29
5.	Condensed tannins (mg/g)	3.12

Adapted from Moyo et al. (2011)

Table 14.2 Amino acids composition of dried Moringa leaves (*M. oleifera Lam.*)

Serial number	Amino acid	Quantity	Serial number	Amino acid	Quantity
1.	Arginine	1.78	**11.**	Methionine	0.297
2.	Glycine	1.533	**12.**	Alanine	3.033
3.	Aspartic acid	1.43	**13.**	Isoleucine	1.177
4.	Proline	1.203	**14.**	Histidine	0.716
5.	Serine	1.087	**15.**	Phenylalanine	1.64
6.	Threonine	1.357	**16.**	Tryptophan	0.486
7.	Valine	1.413	**17.**	Leucine	1.96
8.	Tyrosine	2.650	**18.**	Cysteine	0.01
9.	Glutamic acid	2.53	**19.**	Lysine	1.637
10.	HO-Proline	0.093			

Adapted from Moyo et al. (2011)

moringa powder gives more than 4000 mg calcium. Beef contains 2 mg of iron while moringa powder contains 28 mg of iron. According to a report moringa gives more iron than spinach (Fuglie 2005). Zinc is the important mineral for DNA and RNA synthesis and for the proper growth of sperm cells hence, it is necessary to take it in a proper and required amount. It was found that 25.5–31.03 mg of zinc/kg is present in *M. oleifera* leaves that is the per day requirement in diet (Moyo et al. 2011).

Among micro-minerals, iron has a value of 490 mg/kg and selenium is 3.63 mg/kg. The value of copper is 8.25 mg/kg. It was found that dried Moringa leaves contain 17 fatty acids. It was also found that α-linolenic acid had a value of 44.57%, heneicosanoic had a value of 14.41%, g-linolenic had a value of 0.20%, palmiteic had a value of 0.17% and capric acid had a value of 0.07%. Vitamin E was present in amount of 77 mg/100 g, Beta-carotene had a value of 18.5 mg/100 g (Moyo et al. 2011). According to a recent study M. oleifera leaves are important source of vitamin A in children (Lopez-Teros et al. 2017).

Table 14.3 Fatty acids composition of dried Moringa leaves (*M. oleifera Lam.*)

Serial number	Fatty acids	Quantity	Serial number	Fatty acids	Quantity
1.	Ether extract	6.50	**14.**	Vaccenic (C18:1c7)	0.36
2.	Stearic acid (C18:0)	2.13	**15.**	α-Linolenic (C18:3c9,12,15(n-3)	44.57
3.	Heneicosanoic (C21:0)	14.41	**16.**	Lignoceric (24:0)	2.91
4.	Palmitoleic (C16:1c9)	0.17	**17.**	Arachidic (C20:0)	1.61
5.	Palmitic (C16:0)	11.79	**18.**	Lauric (C12:0)	0.58
6.	Capric (C10:0)	0.07	**19.**	Total saturated fatty acids (SFA)	43.31
7.	Margaric (C17:0)	3.19	**20.**	Total poly unsaturated fatty acids (PUFA)	52.21
8.	Myritic (C14:0)	3.66	**21.**	PUFA: SFA (PUFA:SFA)	1.21
9.	Oleic (C18:1c9)	3.96	**22.**	Total mono unsaturated fatty acids (MUFA) 4.48 1.984	4.48
10.	g-Linolenic (C18:3c6,9,12 (n-6)	0.20	**23.**	PUFA: MUFA (PUFA:MUFA)	14.80
11.	Linoleic (C18;2c9,12(n-6)	7.44	**24.**	Total Omega-6 fatty acids (n-6)	7.64
12.	Tricosanoic (C23:0)	0.66	**25.**	n-6/n-3 0.17 0.016	0.17
13.	Behenic (C22:0)	1.24	**26.**	Total Omega-3 fatty acids (n-3)	44.57

Adapted from Moyo et al. (2011)

14.2.2 Flower

Flowers of *M. oleifera* contains 18.92% protein, 2.91% lipids, 32.45% of dietary fibre and 36.04% of nonstructural carbohydrates. Total concentration of amino acids in dry matter of flower of moringa is 74.5 mg/g (Sanchez-Machado et al. 2010). Flowers of moringa has flavonoids (3.5 mg/g) and anthocyanin (52.80 mg/g) due to which it has potential therapeutic uses, kaempferol and quercetin are flavonoids i-e compounds having phenolic hydroxyl groups that has antioxidant action (Vats and Gupta 2017). Flowers contain 31% amino acids. The hydroethanolic extract of *M. oleifera* flower significantly stifled the secretion and expression of tumor necrosis factor-alpha (TNF-α), NO, prostaglandin E_2 (PGE_2), interleukin-(IL-) 6, IL-1β, inducible NO synthase (iNOS), cyclooxygenase-2 (COX-2) and nuclear factor-kappa B (NF-κB). However, depending on concentration (100 μg/mL and 200 μg/mL) it significantly enhanced the production of IL-10 and IκB-α (inhibitor of κB) (Tan et al. 2015). Dried and fresh flowers of *M. oleifera* are used for the preparation of tea. It has Kandamino acids and Ca and it also shows hypecholesterolemic properties. When fried the taste of flowers are like a mushroom. Flowers of M. oleifera also act as anti-arthritic agents and can be used for the cure of cold and urinary issues (Masih et al. 2019).

14.2.3 Pods

Immature pods contain 19.34% of protein, 1.28% of lipids, 46.78% of dietary fiber and 24.98% of nonstructural carbohydrates (Sanchez-Machado et al. 2010) they are fibrous thus can help in the treatment of digestive issues and in the treatment of colon cancer (Oduro et al. 2008). Total content of essential amino acids in immature pods is 30% (Sanchez-Machado et al. 2010). A study showed that taking 1 cup of fresh, sliced pods i.e. 100 g has 157% of vitamin C which is the daily requirement of an adult. It is also used to enhance the shelf life and for the storage of food. Pods have anti-helminthic properties and thus are used for treating spleen and liver infections. It is also used for the treatment of articular pains (pain in the joints) (Masih et al. 2019).

14.2.4 Seeds

M. oleifera seeds has a crude protein content of 28.02%, 33.78% of crude lipid, 28.77% of carbohydrates, 94.74 mg/100 g of ascorbic acid, 2.84 mg/100 g of calcium and 129.03 mg/100 g of sodium (Igwilo et al. 2017). According to a recent research on rats it was found that seed extracts of *M. oleifera* showed anti-fibric effects on liver fibrosis (Hamza,2010). In rats, against CCl_4 induced liver fibrosis, it shows remarkable protective effects. This was confirmed by the biochemical analysis of hydroxyproline which is a marker of deposition of collagen in liver.

High content of cystine and methionine is reported in seeds of *M. oleifera* which is close to that for eggs and milk. Proteins of *M. oleifera* seeds is highly digestible i-e 93% because it has urease activity and free from inhibitors of trypsin. Powder of M. oleifera seeds is used to clarify the turbid water as it is a natural flocculent. Further, Moringa seed kernels has 40% of oil and has a high quality of fatty acid composition (oleic acid >70%) so it captured the interest of scientists. After refining it has a significant resistance to oxidative degradation (Masih et al. 2019).

14.2.5 Roots

When the seedling of *M. oleifera* is 60 cm long a sauce can be prepared from the its roots which is similar to that of horseradish sauce due to this *M. oleifera* is also known as the horseradish tree. Root of *M. oleifera* is properly grind up and salt and vinegar is added in it after removing the bark as harmful substances are present in it. Therefore, consumption of roots is very low. (Daniell et al. 2011). The sauce is stored in refrigerator and the gum obtained from the bark is used for the seasoning of food (Mulugeta and Fekadu 2014).

14.3 Medical Advantages and Applications of Moringa oleifera

M. oleifera is considered as the universal remedy. It can be used for the treatment of more than 300 diseases. Phytochemicals present in moringa makes it a good medicinal agent. In this section we reviewed the role of moringa in different conditions. In this section we reviewed the use of moringa to treat certain diseases like diabetes, cancer and liver diseases.

14.3.1 Anti-diabetic Properties of Moringa

Moringa shows its role in the cure of Type I and Type II diabetes. In Type I diabetes production of insulin in body is stopped or inhibited, a hormone that maintain blood glucose levels in body. While in Type II insulin resistance occurred due to the dysfunction of beta cells which produce insulin it does not able to sense glucose levels hence blood glucose levels rise. It was observed through many studies that *M. oleifera* can be an anti-diabetic agent. Study in rats showed that aqueous extract of *M. oleifera* is used to treat the streptozotocin induced (Type I diabetes) and insulin resistant (Type II diabetes) (Divi et al. 2012). According to anther study researchers gave moringa seed powder to STZ induced diabetes and found that their fasting glucose level decreases in blood (Al-Malki and EI Rabey 2015). It was found by a research that in serum antioxidant enzymes increased when rats were given 500 mg/kg of moringa seed powder as a result the ROS level decreases in beta cells caused by STZ induction (Mbikay 2012)· A research is done on humans in which 46 individuals with Type II diabetes were selected, it was noticed that their fasting glucose level was reduced by 28% and postprandial glucose level were reduced by 26% by giving them 8 g of *M. oleifera* leaf powder for 40 days, when they were compared with untreated individuals (Kumari 2010). It can be figured out that flavonoids like quercetin and phenolics have scavenging effect on ROS in mitochondria that protects beta cells and as a result kept hyperglycemia under control (Al-Malki and EI Rabey 2015).

A study on 35 diabetic individuals showed that giving 4.6 g tablets of *M. oleifera* leaf extracts for 50 days to them has increased their high density lipoprotein levels and decreased their total cholesterol (Nambiar et al. 2010).

14.3.2 Anti-cancer Properties of Moringa

Studies has shown that cancer cell growth hampers with *M. oleifera*, it can be act as anti-neo-proliferative agent. Solvent and soluble extracts of leaves are proved as effective anticancer agent. Furthermore, reactive oxygen species production by

M. oleifera is specified and it targets only cancerous cells, due to this it is perfect anticancer agent. According to a study the extracts enhanced the expression of glutathione S-transferase that in turn inhibit the express of antioxidants (Tiloke et al. 2013). Moringa leaf extracts shows both antioxidants and anticancerous effects by inducing reactive oxygen species. The compounds in leaves due to which it has anticancer properties are niazimicin, glucosinolates, benzyl isothiocyanate (Hermawan et al. 2012). Benzyl isothiocyanate is found to be associated with cancer. Research showed that benzyl isothiocyanate cause intracellular reactive oxygen species, which then cause cell death. Hence, this could be a reason that makes moringa a good anticancer agent (Gopalakrishan et al. 2016). *In vitro* anticancer activity with dichloromethane and methanolic *Moringa oleifera* leaf extract against human colorectal adenocarcinoma, hepatocellular carcinoma, breast adenocarcinoma showed nontoxic effect on human fibroblast (Nakamura et al. 2002). Another study showed the effect of methanolic and hydromethanolic moringa leaf extracts on model of melanoma of mouse. The researchers found that tumor growth is delayed when mouse were given 500 mg/kg for 15 days of this extract and hence lifespan of mouse increased (Miyoshi et al. 2004). Due to its bioactive compounds in these extracts it has anticancer property like hexadecenoic acidethyl ester (Al-Asmari et al. 2015).

14.3.3 Anti-inflammatory and Immunomodulatory Activity of Moringa

According to a study when 600 mg/kg dose of methanolic extract of *M. oleifera* is given it reduce 82.28% of paw edema (Saleem et al. 2020). Another study on guinea pigs were showed that when they were treated with *M. oleifera* seed extracts of butanol it interrupted their acetyl-choline induced bronchospasms and airway inflammation by modifying Th1/Th2 cytokines (Mahajan et al. 2009). There are many bioactive compounds of *Moringa oleifera* due to which it has anti-inflammatory property like quercetin which involves in inhibiting the activation of NF-kB that is the important step to unfettered inflammatory process (Das et al. 2012). Besides this, many other bioactive compounds of moringa are responsible for its inti-inflammatory process like phenolics and flavonoids.

Quercetin and leaf extract of *M. oleifera* reduced IL-6 and TNF-α release and involves in the regulation of the expression of C-reactive protein, IFN-γ and iNOS in rats (Das et al. 2012). Same result was found from isothiocyanate in *M. oleifera* leaves which has noticeable decrease in the production of pro-inflammatory mediators by RAW macrophages specifically IL-1 β, iNOS, TNF-α and NO (Waterman et al. 2014).

In terms of immunomodulatory effects of *Moringa oleifera*, it is observed in study on rats that ethanolic *M. oleifera* stimulated cellular and humoral immunity and reduced cyclophosphamide-induced immunosuppression (Gupta et al. 2010).

14.3.4 Antioxidant Activity of Moringa

Whole tree of moringa has nutritional importance and used for many medicinal purposes but the antioxidant activity is particularly high in seeds, leaf and pods as compared to other parts. Especially in leaves the high content of phenols and flavonoids reduce the oxidation damage to major biomolecules by the induction of deoxyribose degradation and inhibiting the lipid peroxidation and action of nitric oxide thus preventing the generation of free radicals (Sasikala et al. 2010).

A study on 60 postmenopausal women demonstrated that their serum levels of malondialdehyde were decreased that is generated due to lipid peroxidation and levels of superoxide dismutase, glutathione peroxidase and ascorbic acid which are the indicators of antioxidant property were increased by the supplementation of *Moringa oleifera* leaf extract for 3 months (Kushwaha et al. 2012).

14.3.5 Hepatoprotective Activity

In most of the studies it was found that *M. oleifera* has significant hepatoprotective properties. A study was conducted on rats to determine hepatoprotective activity of *n*-hexane, ethyl acetate(EtOAc), dichloromethane, aqueous fractions and *n*-butanol fractions of methanol extract of leaves of *Moringa oleifera* against carbon tetrachloride(CCl_4)-induced liver injury. The *n*-hexane followed by dichloromethane and aqueous fractions were significantly regulate levels of lipid peroxide by-products and serum enzyme activities in the liver. Dichloromethane, *n*-hexane and aqueous fractions were more effective against CCl_4-induced hepatotoxicity (Atta et al. 2018). Moreover, in albino rats chloroform and methanol extracted from leaves of *M. oleifera* protects liver damage due to CCl4. Along with leaves of *M. oleifera*, root and flowers of this plant also acquire strong hepatoprotective activity. According to a recent study on the effect seed extracts of *M. oleifera* on liver fibrosis showed that it has the potential to abate liver fibrosis. In this study liver fibrosis id induced by CCl4 and concurrent administration of *M. oleifera* seed extract is done. Hence as a result elevation of globulin level and serum aminotransferase activities is controlled by *M. oleifera* seed extract that is induced by CCl4. (Hamza 2010).

14.3.6 Wound Healing Potential of M. oleifera

Activation of endothelial cells, macrophages and fibroblasts involves in the wound healing action when a cell responses to injury. When proliferation of fibroblast is occurred then the restoration of function and structure is started in the site of wound. *In vitro* study was conducted to explore and identify the bioactive compounds that are responsible for the wound healing action and potential of *M. oleifera* in wound healing. The study was included proliferation, cell viability and wound scratch test assays. For the identification of bioactive compounds in *M. oleifera* for wound healing analysis done for it was Liquid chromatography-mass spectrometry/mass spectrometry and High performance liquid chromatography (LC-MS/MS). The results of this study showed that cell viability, proliferation and migration of human dermal fibroblasts had been increased when it was treated with aqueous fraction of *M. oleifera*. This study in the crude methanolic extract with HPLC and LC-MS/MS revealed that in addition with quercetin and kaempferol compounds *M. oleifera* also contain a major bioactive compound that is Vicenin-2 which was confirmed by using Ultra-Violet (UV)spectroscopic methods and HPLC with standard Vicenin-2. This *in vitro* study proposed that wound healing is enhanced by using bioactive fraction of *M. oleifera* that contains Vicenin-2 compound (Muhammad et al. 2013).

14.3.7 Hypotensive Effects

Many bioactive compounds are present in leaves of *M. oleifera* which are involved to stabilize the blood pressure that are mustard oil glycosides, nitrile and thiocarbamate glycosides. A study on rats showed that compounds that are, niazimicin, niazinin A, niazinin B and niazinin A + B have potential to lower the blood pressure that is possibly mediated through antagonist effect of calcium (Dubey et al. 2013).

14.4 Fortification of Food with M. oleifera

It is reported in many studies that *M. oleifera* is used for fortification in many food to enhance their nutritional value, for example in amala, cereal gruel and bread. In this section we reviewed the use of moringa as food fortificant to increase the nutrient content of certain foods.

14.4.1 Amala

Due to the low nutritional level of amala it is fortified by many fortificants such as soybean flour (Jimoh and Olatidoye 2009), *M. oleifera* leaf powder (Karim et al. 2015) and distillers spent grain (Awoyale et al. 2010). It is reported that amala used to prepared yam flour is fortified with *M. oleifera* leaf powder at different concentrations of 2.5%, 5%, 7.5% and 10%. It is observed that approximately 48% of protein content of amala is increased by 10% addition of leaf powder. Similarly following the addition of *M. oleifera* leaf powder calcium, magnesium, iron, potassium and sodium contents were also increased.

14.4.2 Cereal Gruel

Cereal gruel is a used in breakfast by adults and consumed as complementary food for infants. When powder of leaves of *M. oleifera* or powder of its flower is added to cereals, vitamin A content of cereal gruel in increased by 15 folds (Olorode et al. 2013). Other nutrients like protein, iron, calcium and phosphorous also increased at significant level (Abioye and Aka 2015; Olorode et al. 2013). It is reported by researchers while working with white maize that when 15% *M. oleifera* leaf powder is added, protein content is increased to 94% (Olorode et al. 2013).

14.4.3 Bread

The addition of *M. oleifera* seed, leaf and flower powder to bread increased its nutritional value (Chinma et al. 2014; Ogunsina et al. 2010; Sengev et al. 2013). When 5% of *Moringa oleifera* leaf powder is put into wheat flour bread its protein content is increased to 54% and crude fiber content is increased to approximately 56% (Sengev et al. 2013). According to another report fortification of *M. oleifera* leaf powder in bread increased its crude fiber content to 88% and protein content to 17% (Chinma et al. 2014). By the fortification of 15% *M. oleifera* seed flour into wheat flour bread increased its protein content to 67% (Ogunsina et al. 2010).

14.4.4 Biscuits

One of the use of *M. oleifera* leaves and seeds is the fortification of cookies and wheat biscuits (Ogunsina et al. 2010; Alam et al. 2014; Dachana et al. 2010; Manaois et al. 2013). By 20% addition of *M. oleifera* leaves, the protein content is increased compared to the unfortified control. When the cookies were fortified with 10%, protein content increased by 45%, and when it was fortified with 20% of *M. oleifera* seed flour, protein content was increased by 90% (Ogunsina et al. 2010). Wheat cookies that are fortified with *M. oleifera* seed flour increased protein content as compared to *M. oleifera* leaf flour; of approximately 1% increase according Alam et al. (2014), and of 22% according to Dachana et al. (2010), which is still lower than leaves flour fortification giving a 45% increase. In reported values the variations is due to the flour prepared from seeds and on the other hand from leaves of *M. oleifera*, chemical composition of wheat flour and the ingredients used for the preparation of biscuits like eggs. In rice cookies 5% fresh *M. oleifera* leaf flour increased protein content at higher extent as compared to the 5% dried *M. oleifera* leaf flour that are approximately 26% and 14% respectively (Manaois et al. 2013).

14.5 Conclusion

M. oleifera has proteins, minerals, vitamins, fatty acids resources that is necessary for proper and normal functioning of body and is used to combat with malnutrition. Moringa fulfill many of the nutrient requirements of human body. *M. oleifera* is famous by the name of miracle tree because of significant medicinal properties that are due to bioactive compounds present in different parts moringa. Due to medicinal properties moringa is used as curative and therapeutic food. Moringa treat many serious diseases like cancer and diabetes due to which mortality and morbidity rate decreases. Moringa has anti-inflammatory, antioxidant, immunomodulatory and hepatoprotective activity. Due to higher nutrient contents moringa is used for the fortification of food for various purposes. Moringa is fortified in different food items to increase the nutritional value. Hence moringa is used to improve nutrition and health of people in Sub-Saharan countries.

References

Abioye V, Aka M (2015) Proximate composition and sensory properties of Moringa fortified maize-ogi. Nutr Food Sci. https://doi.org/10.4172/2155-9600.S12-001

Alam M, Alam M, Hakim M, Huq AO, Moktadir SG (2014) Development of fiber enriched herbal biscuits: a preliminary study on sensory evaluation and chemical composition. Int J Nutr Food Sci 3:246–250

Al-Asmari AK, Albalawi SM, Athar MT, Khan AQ, Al Shahrani H, Islam M (2015) Moringa oleifera as an anti-cancer agent against breast and colorectal cancer cell lines. PLoS One 10(8):e0135814. https://doi.org/10.1371/journal.pone.0135814

Al-Malki AL, El Rabey HA (2015) The antidiabetic effect of low doses of *Moringa oleifera* Lam. seeds on streptozotocin induced diabetes and diabetic nephropathy in male rats. Biomed Res Int 2015:1–13. https://doi.org/10.1155/2015/381040

Arise A, Arise R, Sanusi M, Esan O, Oyeyinka S (2014) Effect of *Moringa oleifera* flower fortification on the nutritional quality and sensory properties of weaning food. Croat. J Food Sci Technol 6:65–71

Aslam M, Anwar F, Nadeem R, Rashid U, Kazi TG, Nadeem M (2005) Mineral composition of *Moringa oleifera* leaves and pods from different regions of Punjab, Pakistan. Asian J Plant Sci 4:417–421. https://doi.org/10.3923/ajps.2005.417.421

Atta AH, Nasr SM, Almaweri AH, Sedky D, Mohammad AM, Desouky HM, Shalaby MA (2018) Phytochemical, antioxidant and hepatoprotective effects of different fractions of *Moringa oleifera* leaves methanol extract against liver injury in animal model. Asian Pac J Trop Med 11(7):423–429. https://doi.org/10.4103/1995-7645.237186

Awoyale W, Maziya-Dixon B, Sanni L, Shittu TA (2010) Nutritional and sensory properties of amala supplemented with distiller's spent grain (DSG). J Food Agric Environ 8:66–70

Babayeju A, Gbadebo C, Obalowu M, Otunola G, Nmom I, Kayode R, Toye A, Ojo F (2014) Comparison of Organoleptic properties of egusi and efo riro soup blends produced with moringa and spinach leaves. Food Sci Qual Manag 28:15–18

Chinma C, Abu J, Akoma S (2014) Effect of germinated tigernut and moringa flour blends on the quality of wheat-based bread. Food Process Preserv 38:721–727. https://doi.org/10.1111/jfpp.12023

Dachana K, Rajiv J, Indrani D, Prakash J (2010) Effect of dried moringa (Moringa oleifera Lam.) leaves on rheological, microstructural, nutritional, textural and organoleptic characteristics of cookies. J Food Qual 33:660–677. https://doi.org/10.1111/j.1745-4557.2010.00346.x

Daniell SK, Sixl W, Gudrit S, Fuchs W (2011) Fitomedicina on the use of Moringa oleifera as a medicinal plant in India and the Philippines. Egészségtudomány 55:57–63

Das N, Sikder K, Ghosh S, Fromenty B, Dey S (2012) *Moringa oleifera* Lam. leaf extract prevents early liver injury and restores antioxidant status in mice fed with high-fat diet. Indian J Exp Biol 50(1):404–412

Divi SM, Bellamkonda R, Dasireddy SK (2012) Evaluation of antidiabetic and antihyperlipidemic potential of aqueous extract of *Moringa oleifera* in fructose fed insulin resistant and STZ induced diabetic wistar rats: a comparative study. Asian J Pharm Clin Res 5:67–72

Dubey DK, Dora J, Kumar A, Gulsan RK (2013) A multipurpose tree-Moringa oleifera. Int J Pharm Chem Sci 2:415–423

Fahey JW (2005) Moringa oleifera: a review of the medical evidence for its nutritional, therapeutic and prophylactic properties. Trees life J 1:1–33

Fuglie LJ (2005) The Moringa Tree: a local solution to malnutrition Church World Service in Senegal, 2005

Gopalakrishan L, Doriya K, Kumar DS (2016) Moringa oleifera: a review on nutritive importane and its medicinal application. Food Sci Wellness 5:49–56. https://doi.org/10.1016/j.fshw.2016.04.001

Gupta A, Gautam MK, Singh RK, Kumar MV, Rao CV, Goel RK (2010) Immunomodulatory effect of Moringa oleifera Lam. extract on cyclophosphamide induced toxicity in mice. Indian J Exp Biol 48(1):1157–1160

Hamza AA (2010) Ameliorative effects of *Moringa oleifera* Lam seed extract on liver fibrosis in rats. Food Chem Toxicol 48:345–355. https://doi.org/10.1016/j.fct.2009.10.022

Hekmat S, Morgan K, Soltani M, Gough R (2015) Sensory evaluation of locally-grown fruit purees and inulin fibre on probiotic yogurt in mwanza, Tanzania and the microbial analysis of probiotic yogurt fortified with Moringa oleifera. J Health Popul Nutr 33:60–67

Hermawan A, Nur KA, Sarmoko, Dewi D, Putri P, Meiyanto E (2012) Ethanolic extract of *Moringa oleifera* increased cytotoxic effect of doxorubicin on HeLa cancer cells. J Net Rem 12:108–114

Igwilo IO, Okonkwo JC, Ugochukwu GC, Ezekwesili CN, Nwenyi V (2017) Comparative studies on the nutrient composition and antinutritional factors in different parts of Moringa oleifera plant found in Awka, Nigeria. The Bioscientist 5:1–12

Jimoh K, Olatidoye O (2009) Evaluation of physicochemical and rheological characteristics of soybean fortified yam flour. J Appl Biosci 13:703–706

Karim O, Kayode R, Oyeyinka S, Oyeyinka A (2015) Physico chemical properties of stiff dough 'amala' prepared from plantain (Musa Paradisca) flour and Moringa (Moringa oleifera) leaf powder. Food Health Dis 4:48–58

Kolawole F, Balogun M, Opaleke D, Amali H (2013) An evaluation of nutritional and sensory qualities of wheat-moringa cake. Agrosearch 13:87–94

Kumari DJ (2010) Hypoglycaemic effect of Moringa oleifera and Aza-dirachta indica in type 2 diabetes mellitus. Bioscan 5(20):211–214

Kushwaha S, Chawla P, Khurana DS (2012) Effect of supplementation of drumstick (*Moringa oleifera*) and amaranth (*Amaranthus tricolor*) leaves powder on lipid profile in postmenopausal women. Int J Sci Res Publ 2(11):162–168

Lopez-Teros V, Ford JL, Green MH, Tang G, Grusak MA, Quihui-Cota L, Muzhingi T, Paz-Cassini M, Astiazaran-Garcia H (2017) Use of a "Super-child" approach to Assess the vitamin A equivalence of Moringa oleifera leaves, develop a compartmental model for vitamin A kinetics, and estimate vitamin A total body stores in young Mexican Children. J Nutr 147:2356–2363

Mahajan SG, Banerjee A, Chauhan BF, Padh H, Nivsarkar M, Mehta AA (2009) Inhibitory effect of n-butanol fraction of *Moringa oleifera* Lam. seeds on ovalbumin-induced airway inflammation in a Guinea pig model of asthma. Int J Toxicol 28(6):519–527. https://doi.org/10.1177/1091581809345165

Manaois RV, Morales AV, Abilgos-Ramos RG (2013) Acceptability, shelf life and nutritional quality of moringa-supplemented rice crackers. Philipp J Crop Sci 38:1–8

Masih LP, Singh S, Elamathi S, Anandhi P, Abraham T (2019) Moringa: a multipurpose potential crop – a review. Proc Indian Nat Sci Acad 85(3):589–601. https://doi.org/10.16943/ptinsa/2019/49579

Mbikay M (2012) Therapeutic potential of Moringa oleifera leaves in chronic hyperglycemia and dyslipidemia: a review. Front Pharmacol 3:1–12. https://doi.org/10.3389/fphar.2012.00024

Miyoshi N, Uchida K, Osawa T, Nakamura Y (2004) A link between benzyl isocyanate-induced cell cycle arrest and apoptosis: involvement of mitogen-activated protein kinases in the Bcl-2 phosphorylation. Cancer Res 64:2134–2142. https://doi.org/10.1158/0008-5472

Moyo B, Masika PJ, Hugo A, Muchenje V (2011) Nutritional characterization of Moringa (Moringa oleifera Lam.) leaves. Afr J Biotechnol 10(60):12925–12933. https://doi.org/10.5897/AJB10.1599

Muhammad AA, Pauzi NAS, Arulselvan P, Abas F, Fakurazi S (2013) In vitro wound healing potential and identification of bioactive compounds from *Moringa oleifera* Lam. BioMed Res Int. 2013, Article ID 974580, 10 pages. https://doi.org/10.1155/2013/974580

Mulugeta G, Fekadu A (2014) Industrial and agricultural potentials of Moringa. J Nat Sci Res 4:14

Nakamura Y, Kawakami M, Yoshihiro A, Miyoshi N, Ohighasi H, Kawai K (2002) Involvement of mitochondrial death pathway in chemo preventive benzyl isocyanate-induced apoptosis. J Biol Chem 277:8492–8499. https://doi.org/10.1074/jbc.M109760200

Nambiar VS, Guin P, Parnami S, Daniel M (2010) Impact of antioxidants from drumstick leaves on the lipid profile of hyperlipidemics. J Herb Med Toxicol 4(1):165–172

Nouala FS, Akinbamijo OO, Adewum A, Hoffman E, Muetzel S, Becker K (2006) The influence of Moringa oleifera leaves as substitute to conventional concentrate on the in vitro gas production and digestibility of groundnut hay. Livest Res Rural Dev 18:9

Oduro I, Ellis WO, Owusu D (2008) Nutritional potential of two leafy vegetables: Moringa oleifera and Ipomoea batatas leaves. Sci Res Essays 3(2):057–060

Ogunsina B, Radha C, Indrani D (2010) Quality characteristics of bread and cookies enriched with debittered Moringa oleifera seed flour. Int J Food Sci Nutr 62:185–194. https://doi.org/10.3109/09637486.2010.526928

Oliveira JTA, Silveira SB, Vasconcelos KM, Cavada BS, Moreira RA (1999) Compositional and nutritional attributes of seeds from the multiple purpose tree *Moringa oleifera* Lamarck. J Sci Food Agric 79(6):815–820. https://doi.org/10.1002/(SICI)1097-0010(19990501)79

Olorode O, Idowu M, Ilori O (2013) Effect of benoil (Moringa oleifera) leaf powder on the quality characteristics of 'Ogi'. Am J Food Nutr 3:83–89. https://doi.org/10.5251/ajfn.2013.3.2.83.89

Saleem A, Saleem M, Akthar MF (2020) Antioxidant, anti-inflammatory and antiarthritic potential of *Moringa oleifera* Lam: an ethnomedicinal plant of Moringaceae family. South Afr J Bot 128:246–256. https://doi.org/10.1016/j.sajb.2019.11.023

Sanchez-Machado DI, Nunez-Gastelum JA, Reyes-Moreno C, RamirezWong B, Lopenz-Cervantes J (2009) Nutritional quality of edible parts of Moringa oleifera. Food Anal Method. https://doi.org/10.1007/s1261009-9106-Z

Sanchez-Machado DI, Nunez-Gastelum JA, Reyes-Moreno C, Ramirez-Wong B, Lopez-Cervantes J (2010) Nutritional quality of edible parts of Moringa oleifera. Food Anal Methods 3:175–180

Sarwatt SV, Milang'ha MS, Lekule FP, Madalla N (2004) Moringa oleifera and cottonseed cake as supplements for smallholder dairy cows fed Napier grass. Lives Res Rural Dev 16:13–20

Sasikala V, Rooban BN, SGS P, Sahasranamam V, Abraham A (2010) *Moringa oleifera* prevents selenite-induced cataractogenesis in rat pups. J Ocul Pharmacol Ther 26(5):441–447. https://doi.org/10.1089/jop.2010.0049

Sengev A, Abu J, Gernah D (2013) Effect of Moringa oleifers leaf powder supplementation on some quality characteristics of wheat bread. Food Nutr Sci 4:270–275. https://doi.org/10.4236/fns.2013.43036

Tan WS, Arulselvan P, Karthivashan G, Fakurazi S (2015) Moringa oleifera flower extract suppresses the activation of inflammatory mediators in lipopolysaccharide-stimulated RAW 264.7

Macrophages via NF-κB Pathway. Mediators Inflamm. Article ID 720171, 11 pages. https://doi.org/10.1155/2015/720171
Thurber MD, Fahey JW (2010) Adoption of *Moringa oleifera* to combat under nutrition viewed through the lens of diffusion of innovations theory. Ecol Food Sci Nutr 48:1–13. https://doi.org/10.1080/03670240902794598
Tiloke C, Phulukdaree A, Chuturgoon AA (2013) The Antiproliferative effect of *Moringa oleifera* crude aqueous leaf extract on cancerous human alveolar epithelial cells, BMC complement. Altern Med 13:226–233
Vats S, Gupta T (2017) Evaluation of bioactive compounds and antioxidant potential of hydro-ethanolic extract of *Moringa oleifera* Lam. from Rajasthan, India. Physiol Mol Biol Plants 23(1):239–248
Waterman C, Cheng DM, Rojas-Silva P, Poulev A, Dreifus J, Lila MA (2014) Stable, water extractable isothiocyanates from Moringa oleifera leaves attenuate inflammation in vitro. Phyto Chem 103:114–122. https://doi.org/10.1016/j.phytochem.2014.03.028

Index

E. Lichtfouse (ed.), *Sustainable Agriculture Reviews 52*, Sustainable Agriculture Reviews 52, https://doi.org/10.1007/978-3-030-73245-5

T

U

V

The manufacturer's authorised representative in the EU is Springer Nature Customer Service Centre GmbH, Europaplatz 3, 69115 Heidelberg, Germany. If you have any concerns regarding our products, please contact ProductSafety@springernature.com

Printed and bound by CPI Group (UK) Ltd, Croydon, CR0 4YY
30/09/2026
02225593-0013